Discrete Mathematics and Theoretical Computer Science

Springer
London
Berlin
Heidelberg
New York
Barcelona
Hong Kong
Milan
Paris
Singapore
Tokyo

Gheorghe Ştefănescu

Network Algebra

Springer

Gheorghe Ştefănescu
Faculty of Mathematics, University of Bucharest, Str. Academiei 14,
RO-70109 Bucharest, Romainia

ISBN 1-85233-195-X Springer-Verlag Berlin Heidelberg New York

British Library Cataloguing in Publication Data
Ştefănescu, Gheorghe
Network algebra. - (Discrete mathematics and theoretical
Computer science)
1. Algebraic logic 2. Computer science - Mathematics
I. Title
004'.0151'13
ISBN 185233195X

Library of Congress Cataloging-in-Publication Data
A catalog record for this book is available from the Library of Congress

Typesetting: Camera ready by author
Printed and bound at the Athenæum Press Ltd., Gateshead, Tyne & Wear
34/3830-543210 Printed on acid-free paper SPIN 10739465

In heaven,
to the memory of

Calvin C. Elgot, Stephen C. Kleene

and my mother Maria

and on earth, to

Cristina and

Andrei

Preface

This book is devoted to a general, algebraic study of networks and their behaviour. The term network is used in a broad sense here as consisting of a collection of interconnecting cells.

Models Two radically different interpretations of this enlarged notion of network are particularly relevant. First, virtual networks are obtained using the Cantorian interpretation in which at most one cell is active at a given time. With this interpretation, Network Algebra covers the classical models of control, including finite automata or flowchart schemes. In a second Cartesian interpretation, each cell is always active. This implies models for reactive and concurrent systems such as Petri nets or data-flow networks may be covered as well.

These two interpretations may be mixed. A sketch of the resulting much more complicated Mixed Network Algebra models is included in the last part of the book.

Algebraic structures The results are presented in the unified framework of the calculus of flownomials. This is an abstract calculus very similar to the classical calculus of polynomials.

The kernel structure is the *BNA* ("Basic Network Algebra"). After their introduction in the context of control flowcharts setting [Şte86b], the BNA axioms were rediscovered in various fields ranging from circuit theory to action calculi, from data-flow networks to knot theory (traced monoidal categories), from process graphs to functional programming.

In general, the involved algebraic structures are BNAs enriched with branching constants and additional specific axioms. The branching constants are specified by xy with $x \in \{a, b, c, d\}$ and $y \in \{\alpha, \beta, \gamma, \delta\}$; the axioms are divided into three groups: weak, strong, and enzymatic axioms.

As already said, an important feature of the book is the uniform presentation of many apparently unrelated models which have appeared in the literature. A few such structures are described below.

(1) *BNA* (case $a\alpha$-weak, $a\alpha$-strong, $a\alpha$-enzymatic): any transformation which preserves the graph-isomorphism relation is valid (no duplication or removal of wires or cells is permitted);
(2) *xy-symocats with feedback* (case xy-weak, $a\alpha$-strong, $a\alpha$-enzymatic): in these

cases duplication or removal is permitted for wires, but not for network cells;
(3) *Elgot theory* (case $b\delta$-weak, $a\delta$-strong, $a\delta$-enzymatic): in this setting, networks may be completely unfolded to get "regular" trees; Elgot theory axioms capture the transformation rules that are valid for these trees;
(4) *Kleene theory* (case $d\delta$-weak, $d\delta$-strong, $d\delta$-enzymatic): in this setting, the input-output sequences modeling network behaviour define "regular" languages; Kleene theory captures all the identities which are valid for regular languages (flownomial and regular expressions are equivalent in this context);
(5) *Park theory* (case $b\delta$-weak, $b\delta$-strong, $b\delta$-enzymatic): this is a setting in-between Elgot and Kleene theories; it deals with input-output behaviour, but in the deterministic case;
(6) *Căzănescu–Ungureanu theory* (case $b\delta$-weak, $a\delta$-strong, $a\alpha$-enzymatic): this structure is a BNA over a (co)algebraic theory; Floyd–Hoare logic for program correctness may be developed in this setting;
(7) *Conway theory* (case $d\delta$-weak, $d\delta$-strong, $a\alpha$-enzymatic): this structure is a BNA over a matrix theory; this setting suffices for proving Kleene theorems;
(8) *Milner theory* (case $d\delta$-weak, $a\delta$-strong, $a\delta$-enzymatic): this setting may be seen as a lifting with weak constants of the Elgot theory to the nondeterministic case; this setting is used to build-up process algebra.

This ability of the calculus of flownomials to cover many particular algebras may be a good motivation for the reader to follow the, maybe too abstract, Part II of the book.

Results The main results presented in the book include: axiomatizations for isomorphic networks without or with branching constants; axiomatizations for several classes of relations; axiomatizations for regular trees or regular languages; the Universality Theorems; the Structural Theorems; correctness of Floyd–Hoare logic; the duality between flowcharts and circuits; correctness of ACP with respect to bisimulation semantics; axiomatization for input–output behaviour of deterministic data-flow networks; a Kleene-like theorem for Petri net languages; a construction of the free distributive category; the space-time duality principle.

How to use the book The book is mainly self-contained. While no prior knowledge of specific fields is required, a mature education in algebra is helpful for reading Part II. Moreover, a priori knowledge of the models studied in Part III may provide a good insight into the role of the presented results; furthermore, it may awaken interest in applying such algebraic methods to describe other relevant results of those particular fields.

The material contained in the book may be used as a basis for computer science courses on several levels. The author has used the material to teach both undergraduate and graduate courses at the University of Bucharest, GKLI (the Graduate School for Logic in Computer Science) Munich and Kyushu University.

Each of Chapters 7–11 may be used as a four-week introduction to the algebraic theory of the corresponding models. The first two models (flowcharts and automata) may be presented at the undergraduate level, while the other three models (process algebras, data-flow networks, Petri nets) fit better at the graduate level.

Part II may be used for a one-semester graduate course, if certain more complicated theorems, such as the Structural Theorems for cases $b\delta$ and $d\delta$, are taken for granted. Part II may be extended to a one-year course, including all proofs, as well as additional material from Chapters 7 and 8.

Certain topics from Part IV may be used as teaching material for postgraduate courses.

Most of the sections are accompanied sets of exercises and problems. They range from straightforward to very difficult, sometimes open, problems; an approximation of their difficulty is indicated using a numeric scale from 1 to 100, whose meaning will be explained in the Introduction.

For researchers, several open problems and proposed research subjects are included in the book. In particular, many interesting open problems are raised by the mixed network algebra model sketched in Part IV.

More on the structure of the book may be found in the Introduction.

Acknowledgment I would like to express my warmest thanks to the people who have helped me to write this book. A few key persons are J.A. Bergstra, M. Broy, V.E. Căzănescu, and Y. Kawahara. Either directly or indirectly, they had a crucial role in the present achievement. (A more complete list may be found in the Introduction.)

The book was written in several stages. The kernel of the book is Part II; it has been written when the author was at the Institute of Mathematics of the Romanian Academy; an English version of Part II was completed during the author's 1994 research stage at the Technical University of Munich. An intermediary version was completed in 1997 when the author was Visiting Professor at Kyushu University and GKLI (the Graduate School for Logic in Computer Science), Munich. The final version was completed in 1999 in Bucharest and Munich.

Finally, many thanks go to C. Calude and J.A. Goguen for the facilities to publish the book in the Discrete Mathematics and Theoretical Computer Science series, and to Mrs Rebecca Mowat from Springer-Verlag, London for help in its production.

Bucharest, February, 2000 GŞ

Table of contents

Part I

An introduction to Network Algebra

Brief overview of the key results

There is one thing stronger than all the armies in the world, and that is an idea whose time has come.
Victor Hugo

In this introduction we briefly present the *calculus of flownomials*, a unified algebraic theory for various types of networks. This approach is the result of an effort to integrate various algebraic approaches used in computer science.

We may agree: a variety of models is desirable. Then, one may use the most appropriate one for the task. But the models should not be completely unrelated. If so, there will be no transfer of information, results, techniques, etc. with the undesired result of duplication of effort and lost of time.

As a source of inspiration, one may look at the marvelous world of polynomials. There is a lot of freedom to accommodate various models, but still within a clear and unique (or coherent) algebraic framework.

Flownomials, to be presented here, were designed with the explicit intention to inherit some of the qualities of polynomials. The motivation for their introduction comes from particular concrete models of computation, but the result is an abstract mathematical model that may be used whenever the underlying rules are valid. (A striking example is the coincidence of the kernel BNA structure of the calculus of flownomials [Şte86b, Şte90] with the later introduced traced monoidal categories occurring in the study of quantum groups [JSV96].)

Flownomials may be used as a mechanism for translating facts from one field to another, or, furthermore, may transform themselves into their own field from where translation to particular models is straightforward.

Regular expressions

A basic calculus for sequential computation is provided by Kleene's *calculus of regular expressions.* It has a simple syntax and a simple semantics. The regular expressions over a set of atomic symbols X, denoted as $\mathsf{RegExp}(X)$, are given by the following grammar

$$E ::= E + E \mid E \cdot E \mid E^* \mid 0 \mid 1 \mid x(\in X)$$

Syntax convention. Here, the standard Backus–Naur shape for presenting the syntax is used. This means an expression is either 0,1, or x (in X), or star of an expression, or sum or product of such expressions. ◊

In the standard semantics regular expressions are interpreted as languages, i.e., sets of strings. The operations are interpreted as "union," "catenation," and "iterated catenation," respectively, while the constants are interpreted as "empty set" and "empty string," respectively.

This calculus is a basic calculus for sequential computation and has a broad range of applications in the design and analysis of circuits and programming languages, study of automata and grammars, analysis of flowchart schemes, semantics of programming languages, etc.

However, the calculus has a strong, but less visible, restriction. Namely, there is one more operation used by the calculus of regular expressions which is often hidden by the above simple syntax and semantics. In order to handle complex objects like automata, grammars, circuits, etc. one has to add one more operation to the above syntax, namely a *matrix building operator*: if $a_{ij}, i \in [m], j \in [n]$ ($[k]$ denotes the set $\{1, 2, \ldots, k\}$) are expressions, then the matrix (a_{ij}) is an (extended) expression. Such an expression is then used to specify the behaviour of a complex object with m inputs and n outputs.

By changing the point of view from "elements" to such "complex objects," one may see that the above construction is based on the following restriction:

Matrix theory rules: reducing m-to-n to 1-to-1 networks

- Let $f : m \to n$ denote the behaviour of a network F with m input ports and n output ports.
- For $i \in [m], j \in [n]$, let f_{ij} denote the behaviour of the network obtained by restricting F to its i-th input port and j-th output port only.
- By the matrix theory rules, f is uniquely determined by these f_{ij}.

This is a very strong restriction indeed and it makes the calculus unsuited for many computation models. In particular, as is well-known, in such a framework one may rather freely copy or discard network cells, a questionable property in certain cases.

Iteration theories

An algebraic setting in between regular expressions and the calculus of flownomials is provided by an algebraic theory modeling flowchart schemes or data-flow networks. In such a case a weaker hypothesis may be used. In the form to be presented below the hypothesis fits with the semantic models of flowchart schemes. The case of data-flow models is dual, i.e., a many-output component is specified as a tuple of single-output components.

> *Coalgebraic theory rules: reducing m-to-n to 1-to-n networks*
>
> • Let $f : m \to n$ denote the behaviour of a network F with m input ports and n output ports.
> • For $i \in [m]$, let f_i denote the behaviour of the network obtained from F by considering only its i-th input port (all the output ports are preserved).
> • By colgebraic theory rules, f is uniquely determined by these f_i.

The resulting algebraic calculus is more general than the calculus of regular expressions and it is useful to study certain semantic models. However, it is still restrictive and no simple syntactic model similar to Kleene's model of regular expressions exists.

Flownomials

One of the main objectives of this book is to present the *calculus of flownomials.* It may be seen as an extension of the calculus of regular expressions to the case of many-input/many-output atoms.

> *Symocat rules: atomic m-to-n networks*
>
> Using symocats (symmetric strict monoidal categories), no reductions of a network behaviour $f : m \to n$ to certain parts is a priori possible. The network is atomic; its behaviour cannot be decomposed into simpler parts.

The calculus of flownomials is an abstract calculus for networks (= labeled directed hyper-graphs) and their behaviours. It starts with two families of doubly-ranked elements:

- a family of variables $X = \{X(a,b)\}_{a,b \in M}$; these variables denote "black boxes," i.e., atomic elements with two types of connecting ports: input ports (described by a) and output ports (described by b).
- a connection structure $T = \{T(a,b)\}_{a,b \in M}$; the elements in T model known processes between the input/output interfaces; e.g., its elements may model the flow of messages that are transmitted via the connecting channels between the input/output pins; this is the "known" (or "interpreted") part of the model; the flownomial operations are supposed to be already defined on this part.

$M = (M, \star, \epsilon)$ is a monoid modeling network interfaces; if M is a free monoid, then an interface may be considered as a string of atomic ports/pins. We also use the functions $i : \ldots \to M$ and $o : \ldots \to M$ that give the corresponding interfaces for inputs and outputs, respectively; these may be used for network cells in X, connecting arrows in T, or general networks over X and T.

The syntax of the calculus of flownomials may now be given. $\mathsf{FlowExp}[X,T]$ is specified by:

$$E ::= E \star E \mid E \cdot E \mid E \uparrow^c \mid \mathsf{I}_a \mid {}^a\mathsf{X}^b \mid \wedge_k^a \mid \vee_a^k \mid x(\in X) \mid f(\in T)$$

where $a, b, c \in M$ and $k \in \mathbb{N}$. The syntax is explained below. Before, notice the simple way regular expressions fit in this setting.

From flownomial to regular expressions. Indeed, the regular expression setting is a particular instance of this one, e.g., $M = (\mathbb{N}, +, 0)$ and $0 = \wedge_0^1 \cdot \vee_1^0$, $1 := \mathsf{I}_1 (= \wedge_1^1 = \vee_1^1)$, $f + g := \wedge_2^1 \cdot (f \star g) \cdot \vee_1^2$, $f \cdot g := f \cdot g$, $f^* := [\vee_1^2 \cdot (1 + f) \cdot \wedge_2^1] \uparrow^1$. ◊

The collection of *flownomial expressions* is obtained starting with the elements in X and T, certain constants (to be described below), and applying three operations:

juxtaposition "$\star$", *composition* "$\cdot$", and *feedback* "$\uparrow$"

Juxtaposition $f \star g$ is always defined and $i(f \star g) = i(f) \star i(g)$ and $o(f \star g) = o(f) \star o(g)$. The composite $f \cdot g$ is defined only in the case $o(f) = i(g)$ and, in such a case, $i(f \cdot g) = i(f)$ and $o(f \cdot g) = o(g)$. The feedback $f \uparrow^c$ is defined only in the case $i(f) = a \star c$ and $o(f) = b \star c$, for some $a, b \in M$ and, in such a case, $i(f \uparrow^c) = a$ and $o(f \uparrow^c) = b$. (Technically, a condition $a \star c = a' \star c \Rightarrow a = a'$ is necessary for the feedback case; it is valid when M is a free monoid.)

The *acyclic case* refers to the case without feedback.

The support theory for connections T should "contain" certain finite binary relations. (They are finite since our syntactic representation of networks is finite; semantically, they may be instantiate as infinite relations, e.g., if the data associated with a sort form an infinite set.) The main classes used below are: bijections, functions, partial functions and relations. At the abstract level, this means that we should have some constants generating these relations. The collection of constants used for this purpose is:

identity I_a, *(block) transposition* ${}^a\mathsf{X}^b$, *(block) ramification* $\wedge_k^a$, and *(block) identification* $\vee_a^k$, where $a, b \in M$ and $k \in \mathbb{N}$

The types of constants are as follows: $i(\mathsf{I}_a) = o(\mathsf{I}_a) = a$; $i({}^a\mathsf{X}^b) = a \star b$, $o({}^a\mathsf{X}^b) = b \star a$; $i(\wedge_k^a) = a$, $o(\wedge_k^a) = a \star \ldots \star a$($k$ times); $i(\vee_a^k) = a \star \ldots \star a$($k$ times), $o(\vee_a^k) = a$.

Each of the above classes of relations (and 13 other classes) may be generated using certain sub-collections of these constants only. To specify them one has to *restrict* the using of the branching constants $\wedge_m$ and $\vee^n$ as follows:

Restrictions on branching degrees
• Restriction $x = a$ means the constants of the type $\wedge_m^a$ always have $m = 1$; restriction $x = b$ means $m \leq 1$; $x = c$, means $m \geq 1$; and $x = d$ means no restriction (arbitrary m).
• Restriction y is defined in a similar manner, using n of the constants $\vee_a^n$ and the corresponding Greek letters.

The notation xy, where $x \in \{a,b,c,d\}$ and $y \in \{\alpha,\beta,\gamma,\delta\}$, refers to the class of relations generated with branching constants satisfying restriction xy. For example, restriction $a\alpha$ refers to bijections, $a\delta$ to functions, $b\delta$ to partial functions, and $d\delta$ to relations.

For flownomials we may single out certain abstract algebraic structures to play the role the ring structure is playing for polynomials. They are extensions of a basic algebraic structure, called *BNA (Basic Network Algebra), or aα-flow*, which uses I and X in its definition only (no branching constants). The critical axioms used to define these extensions are:

(1) *commutation of the branching constants to the other elements*

$$\begin{array}{rcl} f \cdot \wedge_k^b & = & \wedge_k^a \cdot (kf) \\ \vee_a^k \cdot f & = & (kf) \cdot \vee_b^k \end{array}$$

where $f : a \to b$ and $kf = f \star \ldots \star f$, k-times.

This axiom scheme is used in two ways:

(1a) In the *weak* variant it is used only for certain ground terms f's.

Abstract relations. A *ground term*, or an *abstract relation*, is a term written with the flownomial operations and certain constants in $\mathsf{I}, \mathsf{X}, \wedge_k, \vee^k$. A *ground xy-term*, or an *abstract xy-relation*, is a term which may be written using only constants fulfilling restriction xy. ◇

(1b) In the *strong* variant it applies to arbitrary f's.

(2) *enzymatic axiom for the looping operation*

$$f \cdot (\mathsf{I}_b \star z) = (\mathsf{I}_a \star z) \cdot g \quad \text{implies} \quad f \uparrow^c = g \uparrow^d$$

where $z : c \to d$ is an abstract relation and $f : a \star c \to b \star c$, $g : a \star d \to b \star d$ are arbitrary elements.

♡ The above commutations 1a, 1b allow us to use branching constants, with their intended role, within acyclic processes. The present enzymatic axiom allows them to be used in a cyclic environment, too. ◇

Roughly speaking:

- an *xy-symocat* (shorthand for *xy-symmetric strict monoidal category*) is the acyclic structure defined using the weak commutation for all xy-terms;
- in a *strong xy-symocat*, the strong commutation axioms hold for all xy-terms;
- finally, an *xy-flow* is: (1) a BNA, which is (2) a strong xy-symocat, and (3) obeys the enzymatic axiom whenever z is an abstract xy-relation.

A large number of algebraic structures may be obtained by choosing the branching constants and certain weak, strong or enzymatic axioms for them. This freedom gives flexibility to the calculus which is well-suited to model various kind of networks and equivalences.

Classification using weak/strong/enzymatic scheme

We may classify the main algebraic structures used in the book as

$$\text{BNA} + x_1y_1\text{-weak} + x_2y_2\text{-strong} + x_3y_3\text{-enzymatic}$$

The first restriction x_1y_1 shows what branching constants may be used, the second x_2y_2 which branching constants strongly commute with arbitrary arrows, and the third x_3y_3 for which branching constants the enzymatic rule may be applied. (Usually $x_1y_1 \geq x_2y_2 \geq x_3y_3$, where "$\geq$" means more constants, hence weaker restriction.)

♡ In September 1994, Jan Bergstra suggested using the generic "network algebra" name for this sort of algebraic structure. ◇

Polynomials may be seen as classes of equivalent polynomial expressions. In a similar manner, *flownomials* are defined as classes of equivalent flownomial expressions under an appropriate relation of equivalence; however, we will have various classes of flownomials, i.e., xy-flownomials, for various restrictions xy. Moreover, in both calculi there are two equivalent ways to introduce such equivalences: (1) by using the rules of the appropriate algebra, or (2) by using normal form representations.

For the first way, the standard equivalence $\sim_{xy}$ on flownomial expressions may be introduced as the congruence generated by the commutation of xy-constants with variables in X in the class of congruence relations fulfilling the enzymatic rule.

♡ This means, if two terms as in the premise of the enzymatic axiom are equivalent, then the corresponding terms in the conclusion of the enzymatic axiom are equivalent, too. ◇

The second way to define such an equivalence is to use *normal form flownomial expressions.* Such an expression over X and T is of the following type:

$$((\mathsf{I}_a \star x_1 \star \ldots \star x_k) \cdot f) \uparrow^{i(x_1)\star\ldots\star i(x_k)}$$

with $f : a\star o(x_1)\star\ldots\star o(x_k) \to b\star i(x_1)\star\ldots\star i(x_k)$ in T and $x_1, \ldots, x_k$ in X. Then, two normal form flownomial expressions $F = ((\mathsf{I}_a\star x_1\star\ldots\star x_m)\cdot f) \uparrow^{i(x_1)\star\ldots\star i(x_m)}$ and $G = ((\mathsf{I}_a \star y_1 \star \ldots \star y_n) \cdot g) \uparrow^{i(y_1)\star\ldots\star i(y_n)}$ are *similar* via a relation $r : m \to n$ iff

(i) $(i,j) \in r \Rightarrow x_i = y_j$;
(ii) $f \cdot (\mathsf{I}_b \star i(r)) = (\mathsf{I}_a \star o(r)) \cdot g$.

Here $i(r)$ represents the "block" extension of r to inputs (e.g., for a free monoid M, if r relates two variables x_i and y_j, then $i(r)$ relates the 1st input of x_i to the 1st input of y_j, the 2nd to the 2nd, and so on); similarly $o(r)$ denotes the block extension of r to outputs.

Simulation has some good properties: it is reflexive, transitive, and compatible with the flownomial operations; unfortunately, it is not symmetric. The generated congruence, denoted $\overset{xy}{\Longleftrightarrow}$, is just the equivalence relation generated by simulation via xy-relations.

It turns out that $\overset{xy}{\Longleftrightarrow}$ coincides with $\sim_{xy}$ in the basic cases studied in this book: $a\alpha, a\delta, b\delta, d\delta$ (sometimes the proofs require certain additional conditions). Generally speaking, $\sim_{xy}$ is coarser.

A basic model is $[T, X]_{xy}$, i.e., the algebra of normal form flownomial expressions modulo $\overset{xy}{\Longleftrightarrow}$ equivalence. Another one is $\mathsf{FlowExp}[X, T]_{xy}$ consisting of classes of $\sim_{xy}$-equivalent normal form flownomials over X and T. The above fact shows that these models are isomorphic in the basic cases $a\alpha, a\delta, b\delta, d\delta$ studied in the book.

Names. Many algebraic structures are used within the book. They are described by short and quite similar notations, which sometime may not be easily distinguished. So, it may be useful to have alternative names for certain basic cases. The following alternative names are proposed for some of them, with a brief description of the basic features of these models.

First a few graph-isomorphism models.

BNA Case ($a\alpha$-weak, $a\alpha$-strong, $a\alpha$-enzymatic)
This is the basic algebraic structure we are working with. In this setting, one may represent cyclic networks. The cells in X may have multiple input or output pins. The signature allows us to use only bijective relations. More sophisticated connections may be used, but only within the connection theory T. Any transformation which preserves the graph-isomorphism relation is valid. This is a linear-like setting: no duplication or removal of network cells is permitted.

xy-symocats with feedback Case (xy-weak, $a\alpha$-strong, $a\alpha$-enzymatic)
Only 9 cases are allowed for xy, namely those with $x = b$, or $y = \beta$, or $xy \in \{a\alpha, d\delta\}$. In these cases branching constants are added to the syntactic part of the calculus. They come with appropriate algebraic rules, which were designed in such a way: (1) to be strong enough to allow usual relations with their (angelic) operations to be mapped in these structures, and (2) to preserve the graph-isomorphism setting. Hence, duplication or removing is permitted for connections/wires, but not for network cells.

Together with BNA, the next 3 cases enter into the category of *pure* (or natural) structures, i.e., they are particular xy-flows. In these cases, all three conditions (weak, strong, enzymatic) are applied for the same class of relations, with a slight exception for the weak case, where actually the closure of the restriction to feedback has to be used.

Elgot theory Case ($b\delta$-weak, $a\delta$-strong, $a\delta$-enzymatic)
In this case, by the $a\delta$-strong rules one can safely restrict to single-input networks, variables, etc. As a byproduct, such networks may be completely unfolded to get "regular" trees. Elgot theory axioms capture the transfor-

mation rules that are valid for these trees. Rather different networks may produce the same unfolding tree, but the enzymatic rule (together with the other ones) suffices to identify them.

Kleene theory Case ($d\delta$-weak, $d\delta$-strong, $d\delta$-enzymatic)
Kleene theory does the same as Elgot theory, but for input–output computation sequences (traces) for nondeterministic networks. In this case, due to the $d\delta$-strong rules, one can safely restrict to ordinary single-input/single-output networks, cells, etc. For these networks input–output sequences may be defined, producing the associated regular languages. Kleene theory captures all the identities which are valid in this model or regular languages. Actually, flownomials and regular expressions are equivalent in this context. Again, the enzymatic rule proves to be of crucial importance in the proof of axioms' completeness.

Park theory Case ($b\delta$-weak, $b\delta$-strong, $b\delta$-enzymatic)
This is a setting in-between Elgot and Kleene theories. It deals with input–output behaviour, but in the deterministic case. Hence matrices cannot be used here. While this case has not (yet!) the widely recognized value of the two cases above, it has its own beauty. For instance, the Structural Theorem modeling deterministic minimization is an interesting and nontrivial extension of the one for Elgot theories, but still the proof is clean, with a rather natural additional "identification-free" hypothesis.

Three more cases deserve some special attention, and maybe a special name as we just propose. In these structures, there is a mismatch between the restrictions used for the weak, strong, and enzymatic rules.

Căzănescu–Ungureanu theory Case ($b\delta$-weak, $a\delta$-strong, $a\alpha$-enzymatic)
Căzănescu–Ungureanu theory is a BNA over a coalgebraic theory. Modulo some translations between the presentations, Căzănescu–Ungureanu structure was introduced in [CU82] under the name of "algebraic theory with iterate." Floyd–Hoare logic for program correctness may be developed in this setting.

Conway theory Case ($d\delta$-weak, $d\delta$-strong, $a\alpha$-enzymatic)
This case is similar to the above one, so a Conway theory is a BNA over a matrix theory. This setting suffices for proving Kleene theorems. (In [BÉ93b] the name Conway matrix theory is used for this structure, while the above Căzănescu–Ungureanu structure is called Conway theory. We prefer to use the present more discriminating names for them.)

Milner theory Case ($d\delta$-weak, $\dot{a}\delta$-strong, $a\delta$-enzymatic)
This may be seen either as a lifting (with weak constants) of the Elgot theory to the nondeterministic case, or as a relaxation of Kleene theory, preserving the strong and enzymatic rules just for functions. This setting is used to build-up process algebra.

Basic results

The main results on the calculus of flownomials presented in the book are:

(1) Expressiveness results: Various classes of networks are represented by flownomial expressions over appropriate connecting theories. E.g.: (a) usual nondeterministic networks (with atoms in X) are represented by flownomials over (X and) $\mathbb{R}$el; deterministic over $\mathbb{F}$n, partial and deterministic over $\mathbb{P}$fn, etc. (b) networks having connecting channels that pass values in a set D are represented by flownomials over substructures in AddRel(D)/MultRel(D); (c) higher order flownomials are represented by flownomials over other flownomials.

(2) Axiomatization results: Correct (i.e., sound and complete) axiomatizations for the classes of equivalent networks with respect to various natural equivalence relations are presented, e.g.,

- BNA axioms are correct for graph isomorphism — case ($a\alpha$-weak, $a\alpha$-strong, $a\alpha$-enzymatic);
- xy-symocat with feedback for graph isomorphism with various constants — case (xy-weak, $a\alpha$-strong, $a\alpha$-enzymatic);
- Elgot theory for deterministic input behaviour, or regular trees — case ($b\delta$-weak, $a\delta$-strong, $a\delta$-enzymatic);
- Milner theory for nondeterministic input behaviour, or nondeterministic networks modulo bisimulation — case ($d\delta$-weak, $a\delta$-strong, $a\delta$-enzymatic);
- Park theory for deterministic IO behaviour — case ($b\delta$-weak, $b\delta$-strong, $b\delta$-enzymatic);
- Kleene theory for nondeterministic IO behaviour, or regular languages — case ($d\delta$-weak, $d\delta$-strong, $d\delta$-enzymatic).

(3) Uniformity: All the axiomatization results above are proved in an abstract setting, namely in the case the connecting wires are represented by morphisms in an appropriate abstract theory. Hence we have uniform proofs that may be used, for example, to the case of connections made by simple wiring relations, or by message passing channels, or by other flownomials. On the other hand, the proofs are ordered within a natural framework: from the simplest case of BNAs, to the most complicated case of Kleene theories.

(4) Universality: In such an abstract setting the correctness problem consists of the preservation of the algebraic structure when one passes from connections to classes of equivalent networks. This problem is solved by theorems of the following type (in cases $b\delta$ and $d\delta$ the proofs are done under some mild additional conditions for the connecting theory T): If T is an xy-flow, then $[X,T]_{xy}$ is xy-flow; moreover, $[X,T]_{xy}$ forms the xy-flow freely generated by adding X to T.

If one replaces "flownomials" by "polynomials" and "xy-flow" by "ring," then one gets the classical universal property for polynomials. In particular, this result shows that one may use the standard algebraic refinement methods.

(5) Special networks: The calculus of flownomials is general enough to cover various types of networks, including flowchart schemes, automata, data-flow networks, Petri nets. (A process algebra version, re-cast in the timed-NA framework, may, perhaps, be included as well; in the current approach, process algebra has no "identities.")

(6) The approach is variable-free with respect to interface references. As a byproduct a high degree of modularity is achieved. One may freely shifts pieces of networks from one part to another.

Mixed calculi

Appropriate instances of the NA models may be found for control (flowcharts), space (data-flows) and time (processes). A very promising line for current research is to mix such models in order to have a unique calculus for all these features of the computing systems. This may lead towards an algebraic calculus for (concurrent, object-oriented) software systems.

Structure of the book

Contents. The kernel of the book consists of two parts: Part II (Chapters 2–6), which contains results on general, abstract networks, and Part III (Chapters 7–11) dedicated to particular networks. Except for these, there is an introduction (Part I), a final part (Part IV), and an Appendix containing certain deferred technical proofs.

Except for this Introduction, Part I contains Chapter 1. This is a rather unorthodox technical introduction to the calculus of flownomials, in the sense that it gives a short tour of the basic features of the calculus. The usual reader is expected to give a light reading to the chapter (in the sense that many facts are to be taken for granted) and to return later, when the relevant material from Parts II and III is described in more detail. The active reader is invited to try to fill in his own proofs; this is particularly useful to get a clear feeling of the relative power of the weak, strong, and enzymatic axioms; this kind of reader is advised to start with a proof of the relationship between the 9 axiomatic systems presented in Section 1.7, provided he has some background on regular algebras and iteration theories.

The kernel of the book is Part II, containing Chapters 2–6. In Chapter 2 the BNA structure is introduced, and the soundness and correctness of the BNA axioms for (abstract) networks modulo graph isomorphism is proved.

Chapter 3 adds branching constants to the BNA signature, together with the corresponding weak axioms. The main axiomatization results of Chapter 2 are extended to networks with branching constants.

In Chapter 4, the strong and enzymatic axioms for branching constants are introduced. They allow us to duplicate or remove network cells. Some technical results are included as well.

Elgot theories, as axiomatizations for input behaviour (regular trees), are studied in Chapter 5. A basic technical result is the Structural Theorem for the equivalence relation generated by simulation via functions.

Extensions to the input–output behaviour of deterministic networks (Park theories) and of nondeterministic networks (Kleene theories) are presented in Chapter 6. Similar, but more complicated, Structural Theorems are provided also for these cases. Part II ends with a table containing the basic axioms used in the calculus of flownomials.

Algebraic introductions to particular networks are included in Part II. These chapters have been written with the aim to be as independent as possible. The material was selected according to the author's preferences and to its fitness to the Network Algebra approach. The reader is invited to handle more results on these particular models within the Network Algebra setting.

Chapter 7 is dedicated to flowchart schemes. The main result here is the soundness and completeness of the Floyd–Hoare logic, when it is interpreted in Căzănescu–Ungureanu theories. A few interesting duality results between flowcharts and circuits are presented. Section 7.7 requires a good understanding of Part II.

In Chapter 8 we describe certain fundamental results on the algebraic theory of finite automata. The main result is an axiomatization for regular languages. Along the way, algebraic versions for Kleene theorems, the minimization, and the power-set construction for obtaining deterministic automata out of nondeterministic ones are described. Except for Section 8.11, this chapter is fairly independent.

A brief presentation of process algebra, in its ACP version, is described in Chapter 9. The main result here is the soundness and completeness theorem for ACP with respect to bisimulation semantics; it is proved in full detail. The verification of the ABP (Alternating Bit Protocol) is included as a sample for process algebra application.

Chapter 10 is dedicated to data-flow networks. Both synchronous and asynchronous models are presented. For asynchronous networks, axiomatizations for networks with various branching constants are presented, both for deterministic and nondeterministic networks. In the deterministic case, an axiomatization for the input–output behaviour is presented, as well. This chapter is more eclectic than the other chapters in Part III; one reason may be that, while this model is rather popular, there are no widely accepted texts presenting the model.

In Chapter 11, a brief tour of Petri nets is included. The main result is the coincidence of Petri net languages and the languages represented by concurrent regular expressions.

The bulk of Part IV is Chapter 12 which briefly describes the Mixed Network Algebra models. They are obtained mixing network algebra models for control, space, and time. We present: axiomatizations for mixed relations; a construction with "plans" for free distributive categories; an application of the mixed setting to program parallelization; and a space-time duality thesis. Many open questions related to this model exist; some of them are included in the text.

The book ends with an Appendix containing certain deferred technical proofs.

The bibliography includes most of the relevant, directly connected papers, but no serious attempt towards a complete list has been made; in certain cases, special attention has been directed to trace the history of evolving ideas.

Notation. The main chapters are denoted by numbers from 1 to 12, while the appendix is denoted by V. The references to Theorems, Lemmas, etc. use a unique counter within a chapter; their full labeling is using as prefix the chapter number. The proofs end with □.

There are some parentheses, whose meaning is as follows: must be read ♠ ... ♣; may be read ♡ ... ◇; optional ⦶ ... ⦶.

Most of the sections are accompanied by exercises and problems. An approximation of their difficulty is indicated using a numeric scale from 1 to 100, whose meaning is the following: 1 min. (1–5); 10 min. (5–10); 1 hour (10–15); 1 day (15–20); 1 week (20–30); 1 month (30–50); 1 year (50–100); up to the reader (- -); open (**).

Site. I will keep a web-page with additional information on the book at the following address `http://funinf.math.unibuc.ro/~ghstef/na`

Acknowledgments

The hereby presented calculus of flownomials is the result of the attempt I began in the winter of '85/'86 to integrate Elgot's algebraic theories of flowchart schemes and Kleene's calculus for finite automata. I gratefully acknowledge here the deep influence of Elgot's papers and of Conway's book on regular Kleene algebra on the design of the present calculus.

Many persons have contributed in an explicit or implicit way to this achievement.

First of all, V.E. Căzănescu had a significant contribution both to the development of this calculus and to its dissemination from the very beginning via the lectures he has given at the University of Bucharest. In a later stage, J.A. Bergstra and M. Broy have contributed to the development of the multiplicative, data-flow based interpretation of flownomials. Recently, R. Grosu and D. Lucanu have helped in developing the mixed model.

For keeping me informed with the relevant literature of the field (a vital problem before '89) I am grateful especially to D.B. Benson, S.L. Bloom, B. Courcelle, Z. Esik, J.A. Goguen, J.W. Klop, R. Milner, J.C. Shepherdson, and M. Wirsing.

The material has been used to teach both undergraduate and graduate courses at the University of Bucharest, GKLI (the Graduate School for Logic in Computer Science) Munich and Kyushu University. I am grateful to the students for their valuable feedback.

Parts of the book were presented at various meetings, including: CWI and University of Amsterdam, 1990, 1992, 1994; University of Augsburg, 1997; LaBRI - University of Bordeaux I, 1992; CAAP, Nice, France, 1986; EECS (East European Category Seminar) Predela, Bulgaria, 1988; FCT conferences: Szged, Hungary, 1993 and Dresden, Germany, 1995; IFIP WG 2.2 meeting, Udine, Italy, 1999; Japan Advanced Institute for Science and Technology, 1997; International Congress of "Logic, Philosophy and Methodology of Science," Moscow, USSR, 1987; Technical University Munich, 1992, 1994, 1997–8; RelMiCS meetings: Daghstul, Germany, 1994; Parati, Brazil, 1995; Hammamet, Tunisia, 1997; and Warsaw, Poland, 1998; Joint Conference on Discrete Mathematics and Applied Mathematics, Seta, Japan, 1997; SLACS'97 - Symbolic Logics and Computer Science, Fukuoka, Japan, 1997; Turku Center for Computer Science, Turku, Finland, 1999; Utrecht University, 1994. Among others, the following have helped me in this respect: R.J. Bach, J.A. Bergstra, M. Broy, B. Courcelle, Z. Esik, A. Haeberer, G. Hotz, A. Jaoua, Y. Kawahara, B. Möller, E. Orlowska, and G. Schmidt.

The book was written in several stages. The kernel of the book is Part II; it has been written when the author was at the Institute of Mathematics of the Romanian Academy; an English version of Part II was completed during the author's 1994 DAAD research stage at the Technical University of Munich. An intermediary version was completed in 1997 when the author was Visiting Professor at Kyushu University and GKLI (the Graduate School for Logic in Computer Science), Munich. The final version was completed in 1999 in Bucharest and Munich.

The project of writing this book would perhaps have failed without the fundamental help made by the following troika: IMAR, TUM, and DOI-K. (1) The fifteen years 1980–1995 I spent at the Mathematical Institute of the Romanian Academy (IMAR) were of great importance for this project; the institute provided a unique environment, with its top level members and a complete freedom in choosing the research subject. (2) Thanks to the help of Prof. M. Broy, between 1994–1999 I had the possibility to visit several times the Technical University Munich (TUM); among other, many drafts of the book were completed at TUM. (3) Finally, due to Prof. Y. Kawahara and the Japanese government, I was able to spend the fall of 1997 as a Visiting Professor at the Department of Informatics of Kyushu University (DOI-K); this provided me with a Toshiba laptop and a financial independence which made possible the completion of the project.

C. Dima, B. Möller, S. Rudeanu, and B. Warinschi made many useful, especially stylistic, suggestions on different versions of the manuscript.

Finally, many thanks go to C.S. Calude and J.A. Goguen for the facilities to publish the book in the Discrete Mathematics and Theoretical Computer Science series, and to Mrs Rebecca Mowat from Springer-Verlag, London for help in its production.

1. Network Algebra and its applications

> You can not teach a man anything; you can only help him find it within himself.
>
> Galileo

In this chapter we give a brief introduction to the calculus of flownomials. This calculus is very similar to the classical calculus of polynomials. It provides an algebraic model for networks abstractly seen as labeled directed hyper-graphs. More precisely, networks may be identified with flownomial expressions modulo Basic Network Algebra (BNA) axioms.

To this end it is shown that: (1) usual, concrete networks may by represented by flownomial expressions using appropriate relations for connecting network cells; and (2) BNA rules are sound and complete with respect to the graph-isomorphism equivalence on networks.

Further, a few extensions of the BNA setting are briefly presented. One is the full Network Algebra (NA) setting; this setting contains branching constants, too.

The next one takes into account the possibility of using different looping operations (e.g., iteration or repetition) provided the graph-isomorphism setting is replaced by a stronger one. The advantages of the use of feedback over the use of iteration or repetition are broadly outlined here. In particular, equivalent presentations for BNAs over the stronger algebraic or matrix theories are presented providing an enlightening of the algebraic properties of looping operations.

Finally, certain natural further extensions of BNA, called xy-flows, to cope with the axiomatization of various kinds of network behaviours are considered.

This chapter is a technical introduction to the basic kernel of network algebra. The proof of many of the results stated either implicitly or explicitly in the present chapter will be given in the next chapters in extended, abstract settings. The reader is advised to give a first light reading to what follows and to return to this chapter later for a deeper understanding.

1.1 Algebra of finite relations

The algebra of polynomials is based on a ring algebra on numbers. In a similar way the algebra of flownomials is based on a network algebra (NA) on relations. In this section the basic NA structure on finite relations is presented.

Working with relations. Let S be a sort set. The elements in the free monoid $(S^\star, \star, \epsilon)$ are denoted by $a = a_1 \star \ldots \star a_{|a|}$, where $|a|$ is the length of a and $a_i \in S$, for all $i \in [|a|]$ (we let $[n]$ denote the set $\{1, 2, \ldots, n\}$). $\mathbb{R}\mathrm{el}_S$ is defined by

$$\mathbb{R}\mathrm{el}_S(a, b) = \{f \subseteq [|a|] \times [|b|] : (i, j) \in f \Rightarrow a_i = b_j\}, \quad \text{for } a, b \in S^\star$$

An element in $\mathbb{R}\mathrm{el}_S(a, b)$ is called a *finite S-sorted relation*, also denoted $f : a \to b$ or $a \xrightarrow{f} b$.

Motivation for some decision design. These relations are binary relations between finite sets. Each set is considered as an *interface* consisting of left-to-right ordered *pins* or *channels*; it is represented by an initial segment of natural numbers, say, $[k] = \{1, 2, \ldots, k\}$ ($k = 0$ is allowed). The first set represents the *input interface* and the second one the *output interface.* In addition, a sort is attached to each input or output interface pin.

The role of the sorts should be clear at the level of interfaces: they provide us with the possibility to have different types for data attached to different channels of the input/output interfaces.

The condition that relations should be sorted requires some motivation. It says that, if two pins are connected, then the corresponding sorts are equal.

As a consequence of this condition, finite sorted relations may be naturally mapped in more complex (eventually infinite) relations obtained by the instantiation of the sorts with appropriate data sets. They are then thought as relations which preserve data along the connected input/output pins. No such natural mapping is possible for a relation connecting different sorts. Indeed, a sort may be instantiated with an arbitrary set. Then, it is not clear what natural relation may be between the data associated to the source and the target points of an edge connecting different sorts. ◇

A network algebra structure may be defined on finite, sorted relations as follows. The NA operations of *juxtaposition* $\star$, *composition* $\cdot$ and *feedback* $\uparrow$ are defined by:

- $f \star g = f \cup \{(|a| + i, |b| + j) : (i, j) \in g\} : a \star c \to b \star d$, if $a \xrightarrow{f} b, c \xrightarrow{g} d$;
- $f \cdot g = \{(i, k) : \exists j \in [|b|].(i, j) \in f \text{ and } (j, k) \in g\} : a \to c$, if $a \xrightarrow{f} b \xrightarrow{g} c$;
- $f \uparrow^s = \{(i, j) : i \in [|a|],\ j \in [|b|] \text{ and either } [(i, j) \in f] \text{ or }$ $[(i, |b| + 1) \in f \text{ and } (|a| + 1, j) \in f]\} : a \to b$, if $a \star s \xrightarrow{f} b \star s$ and $s \in S$. This operation is extended to $\uparrow^a, a \in S^\star$ by $f \uparrow^\epsilon = f,\ f \uparrow^{a \star b} = (f \uparrow^b) \uparrow^a$.

The NA constants are I (*identity*), X (*transposition*), $\vee$ and $\top$ (*identification constants* of branching degree 2 and 0, respectively), and $\wedge$ and $\perp$ (*ramification*

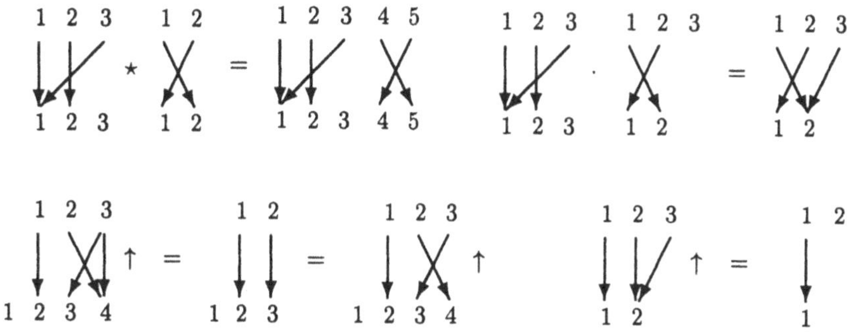

Fig. 1.1. Operations on relations

constants of branching degree 2 and 0, respectively). They act on blocks and have the following meaning in $\mathbb{R}$el:

- $\mathsf{I}_a = \{(i,i) : i \in [|a|]\} : a \to a$
- ${}^a\mathsf{X}^b = \{(i, |a| + i) : i \in [|a|]\} \cup \{(|a| + i, i) : i \in [|b|]\} : a \star b \to b \star a$
- $\vee_a = \{(i,i) : i \in [|a|]\} \cup \{(|a| + i, i) : i \in [|a|]\} : a \star a \to a$
- $\top_a = \emptyset : \epsilon \to a$
- $\wedge^a = \{(i,i) : i \in [|a|]\} \cup \{(i, |a| + i) : i \in [|a|]\} : a \to a \star a$
- $\perp^a = \emptyset : a \to \epsilon$

Fig. 1.1 illustrates how the operations act on relations.

The meaning of the operations. The NA operations are illustrated in Fig. 1.1; pictures for the constants are given in the examples below. Juxtaposition is disjoint union, but with the resulting relation represented in the standard way; hence the indices of the interfaces in g are shifted appropriately. Composition is the usual relational composition. (Unary) feedback $f \uparrow^s$ connects the extreme right output to the extreme right input and thereafter *both pins are hidden*; then an input–output pair is in $f \uparrow^s$ if they are either directly connected in f, or are connected via feedback. Identities are simple connecting wires between input and output. Transposition ${}^a\mathsf{X}^b$ switches the pins in a with those in b. Finally, branching constants are used; they allow us to identify, create, copy, or discard pins from an interface. ◇

Examples. (1) To keep the intuition close to the formalism, a useful graphical representation of relations is used throughout the book. Our convention is that edges are directed and they go from top to bottom. As an example the relation $\{(1,1),(1,3),(3,2),(4,1),(4,5)\} \in \mathbb{R}\mathrm{el}_{\{s,t,u,v\}}(s \star t \star v \star s \star v,\ s \star v \star s \star u \star s)$ is represented in Fig. 1.2. This is a sorted relation, i.e., the sorts associated to the input and output pins of each pair are equal. Indeed, for $(3,2)$ both the input and the output pins have sort v, while for the other pairs the common sort is s.

Fig. 1.2. A many-sorted relation $f = \{(1,1),(1,3),(3,2),(4,1),(4,5)\} : s\star t\star v\star s\star v \to s\star v\star s\star u\star s$; a NA representation is $f = (\Lambda^s_2\star\Lambda^t_0\star\Lambda^v_1\star\Lambda^s_2\star\Lambda^v_0)\cdot(\mathsf{I}_s\star({}^s\mathsf{X}^v\star\mathsf{I}_s)\cdot(\mathsf{I}_v\star{}^s\mathsf{X}^s)\cdot({}^v\mathsf{X}^s\star\mathsf{I}_s)\star\mathsf{I}_s)\cdot(\mathsf{V}^2_s\star\mathsf{V}^1_v\star\mathsf{V}^1_s\star\mathsf{V}^0_u\star\mathsf{V}^1_s)$; the bracket representation is $f = ([13]_s[]_t 2_v[15]_s[]_v; 4_u)$

(2) The symbols I_a, ${}^a\mathsf{X}^b$, T_a, V_a, Λ^a, $\perp^a$ have been chosen to suggest this graphical representation. In the case of a unique sort, the free monoid $S^\star$ may be identified with the monoid of natural numbers. In such a case the constants may be graphically represented as follows:

$\mathsf{I}_1 =$ , ${}^1\mathsf{X}^1 =$ , $\mathsf{V}_1 =$ , $\mathsf{T}_1 =$ , $\Lambda^1 =$ , and $\perp^1 =$ .

If in such a constant a number n (or m) greater than 1 appears, then the corresponding edge is replaced by a compact group of n (resp. m) parallel arrows. For example,

${}^1\mathsf{X}^2 =$ 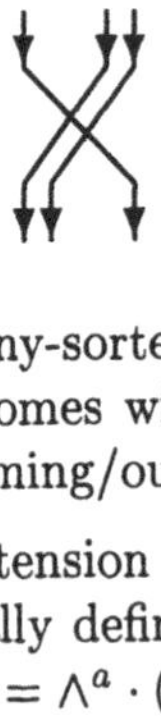 $\mathsf{V}_2 =$ 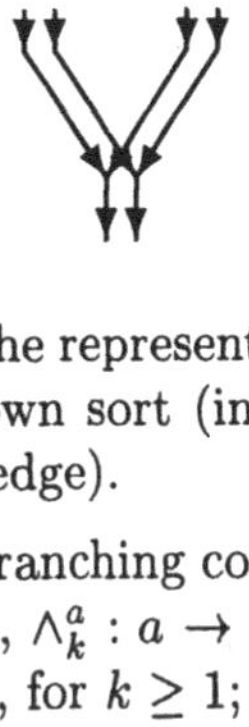 $\perp^3 =$ 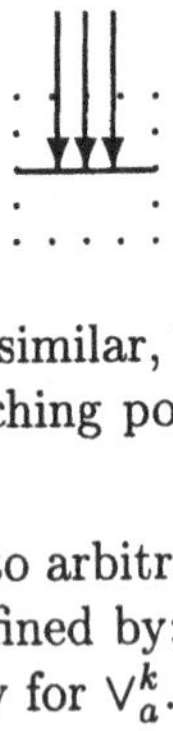 .

In the many-sorted case the representation is similar, but each edge of the drawn relation comes with its own sort (in a branching point the sorts are equal for each incoming/outgoing edge).

(3) An extension of the branching constants to arbitrary branching degrees may be naturally defined, e.g., $\Lambda^a_k : a \to ka$ is defined by: $\Lambda^a_0 = \perp^a$; $\Lambda^a_1 = \mathsf{I}_a$; $\Lambda^a_2 = \Lambda^a$; $\Lambda^a_{k+1} = \Lambda^a \cdot (\Lambda^a_k \star \mathsf{I}_a)$, for $k \geq 1$; similarly for V^k_a.

Notice that the following identity is valid: $f = (\Lambda^s_2\star\Lambda^t_0\star\Lambda^v_1\star\Lambda^s_2\star\Lambda^v_0)\cdot[(\mathsf{I}_s\star({}^s\mathsf{X}^v\star\mathsf{I}_s)\cdot(\mathsf{I}_v\star{}^s\mathsf{X}^s)\cdot({}^v\mathsf{X}^s\star\mathsf{I}_s)\star\mathsf{I}_s)]\cdot(\mathsf{V}^2_s\star\mathsf{V}^1_v\star\mathsf{V}^1_s\star\mathsf{V}^0_u\star\mathsf{V}^1_s)$. As we shall see in Chapter 3 such a decomposition "ramifications–bijection–identifications" (called *quasi-nf representation*), is generally valid showing that the NA operators are complete for representing all finite relations.

(4) A, sometimes shorter, *bracket representation* $f = ([13]_s[]_t 2_v[15]_s[]_v; 4_u)$, will be used as well. This representation is based on an input-to-output parsing of a relation. It is obtained in the following way. The outputs connected to an input are collected in square brackets (in increasing order and separated by ","), the

input sort is put as an index, and then these square brackets are listed in round brackets from left to right for all inputs; we get $([1,3]_s[]_t[2]_v[1,5]_s[]_v)$. Next, the outputs which are not related to any inputs are listed separately, after a ";" separator, in increasing order, together with their associated sorts; we get $([1,3]_s[]_t[2]_v[1,5]_s[]_v;4_u)$. Finally, a pair of matching square brackets is omitted when it surrounds a unique element and, to shorten the representation, the "," separator is omitted in square brackets if there are no more than 9 outputs and no confusion may arise. This way the given representation $f = ([13]_s[]_t2_v[15]_s[]_v;4_u)$ is obtained.

♡ Notice that the quasi-nf representation of relations is autodual, while the bracket representation is not. (The duality here corresponds to reversing the sense of arrows; it will be formally defined in Chapter 3.) ◇

(5) Not all natural subsets of relations in $\mathbb{R}\mathrm{el}_S$ are closed to all NA operations. More precisely, $\mathbb{B}\mathrm{i}_S$ (bijections), $\mathbb{I}\mathrm{n}_S$ (injections), $\mathbb{P}\mathrm{Sur}_S$ (partially defined surjections), and $\mathbb{P}\mathrm{fn}_S$ (partially defined functions) are closed with respect to all NA operations, while $\mathbb{S}\mathrm{ur}_S$ (surjections) and $\mathbb{F}\mathrm{n}_S$ (functions) are closed with respect to juxtaposition and composition, only. The last (lower-right) example in Fig. 1.1 shows that $\mathbb{S}\mathrm{ur}_S$ and $\mathbb{F}\mathrm{n}_S$ are not closed with respect to feedback.

Exercises and Problems (sec. 1.1)

1. Find a NA term representation for the relation $\{(1,4),\ (2,2),\ (3,1),\ (3,3),\ (4,3)\} : b\star b\star a\star a \to a\star b\star a\star b.$ (04)
2. Type-check the following NA expressions:
   (1) ${}^b\mathsf{X}^{b\star a\star a}\cdot(\mathsf{I}_b\star{}^a\mathsf{X}^a\star\mathsf{I}_b)\cdot{}^{b\star a}\mathsf{X}^{a\star b}\cdot(\mathsf{I}_a\star{}^b\mathsf{X}^{b\star a})$
   (2) ${}^b\mathsf{X}^{b\star a\star a}\cdot({}^b\mathsf{X}^a\star{}^a\mathsf{X}^a\uparrow^a\star\mathsf{I}_b)$
   (3) ${}^b\mathsf{X}^{b\star a\star a}\cdot(\mathsf{I}_b\star{}^a\mathsf{X}^a\star\mathsf{I}_b)\cdot({}^b\mathsf{X}^a\star{}^b\mathsf{X}^a)$
   (4) $({}^b\mathsf{X}^b\star{}^a\mathsf{X}^a)\cdot(\mathsf{I}_{b\star b}\star{}^a\mathsf{X}^a\uparrow^a\star\mathsf{I}_a)$
   (5) $(((\mathsf{I}_b\star\wedge^b)\cdot({}^b\mathsf{X}^b\star\mathsf{I}_b)\cdot(\mathsf{I}_b\star\vee_b))\uparrow^b\star\mathsf{I}_{b\star a\star a})$
   For those which are correct compute the resulting relation. (07)
3. Check the validity of the decomposition $(\wedge_2^s\star\wedge_0^t\star\wedge_1^v\star\wedge_2^s\star\wedge_0^v)\cdot(\mathsf{I}_s\star({}^s\mathsf{X}^v\star\mathsf{I}_s)\cdot(\mathsf{I}_v\star{}^s\mathsf{X}^s)\cdot({}^v\mathsf{X}^s\star\mathsf{I}_s)\star\mathsf{I}_s)\cdot(\vee_s^2\star\vee_v^1\star\vee_s^1\star\vee_u^0\star\vee_s^1)$ for $f = \{(1,1),(1,3),(3,2),(4,1),(4,5)\} : s\star t\star v\star s\star v\to s\star v\star s\star u\star s.$ (07)
4. Refine s by $v\star t$ in the above example; that is, replace each occurence of a sort s by $v\star t$ and each arrow of sort s by a group of two parallel arrows (one for v and one for t). Find the bracket representation of the resulting relation. How may this refinement be defined using the bracket representation of relations? (15)
5. Define the converse relation $f^\vee = \{(j,i) : (i,j)\in f\}\in\mathbb{R}\mathrm{el}_S(b,a)$, for $f\in\mathbb{R}\mathrm{el}_S(a,b)$. Find a procedure to pass from the bracket representation of f to the one of $f^\vee$. (12)
6. Get definitions for the NA operations and constants based on the bracket representation of relations. Use these definitions to pass from the represen-

tation of f in Exercise 1 to its bracket representation $([13]_s[]_t2_v[15]_s[]_v;4_u)$. (15)

7. Write down efficient algorithms for operating with relations. (- -)

1.2 Basic Network Algebra, BNA

In this section both Basic Network Algebra (BNA for short) and the associated $a\alpha$-flow structure are introduced.

BNA. The full NA signature, already used in the previous section to put an algebraic structure on relations, consists of three operations (juxtaposition, composition, and feedback) and four types of constant (identities, transpositions, identifications, and ramifications). Leaving out branching constants we get the *BNA signature*.

Definition 1.1. A *Basic Network Algebra (BNA)* specification consists of the BNA signature $\star, \cdot, \uparrow, \mathsf{I}, \mathsf{X}$ and the *BNA rules*, presented in Table 1.1.

Table 1.1. BNA

I.	Axioms for symocats	II.	Axioms for feedback
B1	$f \star (g \star h) = (f \star g) \star h$	R1	$f \cdot (g \uparrow^c) \cdot h = ((f \star \mathsf{I}_c) \cdot g \cdot (h \star \mathsf{I}_c)) \uparrow^c$
B2	$\mathsf{I}_\epsilon \star f = f = f \star \mathsf{I}_\epsilon$	R2	$f \star g \uparrow^c = (f \star g) \uparrow^c$
B3	$f \cdot (g \cdot h) = (f \cdot g) \cdot h$	R3	$(f \cdot (\mathsf{I}_b \star g)) \uparrow^c = ((\mathsf{I}_a \star g) \cdot f) \uparrow^d$
B4	$\mathsf{I}_a \cdot f = f = f \cdot \mathsf{I}_b$		for $f : a \star c \to b \star d, \quad g : d \to c$
B5	$(f \star f') \cdot (g \star g') = f \cdot g \star f' \cdot g'$	R4	$f \uparrow^\epsilon = f$
B6	$\mathsf{I}_a \star \mathsf{I}_b = \mathsf{I}_{a \star b}$	R5	$(f \uparrow^b) \uparrow^a = f \uparrow^{a \star b}$
B7	${}^a\mathsf{X}^b \cdot {}^b\mathsf{X}^a = \mathsf{I}_{a \star b}$	R6	$\mathsf{I}_a \uparrow^a = \mathsf{I}_\epsilon$
B8	${}^a\mathsf{X}^{b \star c} = ({}^a\mathsf{X}^b \star \mathsf{I}_c) \cdot (\mathsf{I}_b \star {}^a\mathsf{X}^c)$	R7	${}^a\mathsf{X}^a \uparrow^a = \mathsf{I}_a$
B9	$(f \star g) \cdot {}^c\mathsf{X}^d = {}^a\mathsf{X}^b \cdot (g \star f)$		
	for $f : a \to c, \quad g : b \to d$		

Comments. (1) Notice that BNA is just a convenient set of axioms to start with. It is not the simplest set. E.g., R3 may be restricted to the case g's are transpositions and an equivalent system of axioms is obtained. Moreover, like in the case of Boolean algebra, there are many different equivalent ways to present this structure, in particular using different sets of operators. Some of them will be described in some detail in the following within the book.

(2) The BNA rules may be graphically illustrated as in Fig. 1.3.

(3) Axioms B1–2 show that juxtaposition is a monoidal operation, having I_ϵ as a neutral element. B3–4 state a similar property for composition, hence a BNA is a category. B5–6 describe the interaction between juxtaposition and composition. A structure that fulfils B1–6 is known as a *strict monoidal category*, cf. [Lan71].

The next three axioms B7–9 are on transpositions. The resulting algebraic structure is known as a symmetric strict monoidal category, cf. [Lan71], shortly named here as *symocat*; it was also used in [Hot65] and subsequent papers, e.g., under the name of *x-category*. Of these axioms, B7 shows that the transposition of a with b, followed by the one of b with a, leads to the identity. B8 shows that the transposition of a over $b \star c$ may be done in two steps: first transpose a over b, then transpose it over c. Finally, B9 explain the effect of composing transpositions with arbitrary cells: a transposition may be used to "commute" two cells, while keeping fixed their input/output interfaces in the network.

(3) The key axioms for the BNA structure are the axioms for feedback R1–7. R1 describes a commutation of feedback with composition. R2 describes a similar property for juxtaposition. R3 shows the invariance of networks with respect to the shift of blocks on feedback. R4–5 describe how to deal with repeated feedbacks. Finally, R6–7 show how feedback acts on constants.

(4) A special mention on the simplicity of axioms R4–5 deserves to be made here. When it is to be described with different looping operations as iteration or repetition, this property leads to very complicated formulae. It is a crucial merit of the feedback operation to drastically simplfy it; see the last section of this chapter for the details which illustrate this point.

(5) A very important property of the BNA axioms is that they lay in a sort of *linear setting*. We mean the following. Each term in an axiom contains a unique occurrence of a variable and all variables that occur in one side of an equation occur in the other side of the equation as well. This lack of implicit duplication or creation/deletion of variables is very important to keep the starting network algebra setting as simple and general as possible. Later on one may explicitely add duplications, deletions, etc.

(6) Notice that $\mathbb{R}\mathrm{el}_S$ is a BNA when it is structured with BNA operations and constants such as in the previous section.

(7) Except for the axioms for transpositions, these BNA axioms are those originally introduced in [Şte86b]. The structure defined by the BNA axioms has been rediscovered in various settings. One specific paper is [JSV96]; there and in certain subsequent papers [Abr96], [Jef97] the following names for the feedback axioms are used: tightening for R1, superposing for R2, sliding for R3, vanishing for R4–R5, yanking for R7.

(8) Looping operations similar to our feedback may be found in various places, e.g., the "linear mechanism" in [Con71] or the feedback of [Bai76], but they are not presented in an axiomatic setting using symocats. If stronger structures than symocats are used, then the possibility of getting an algebra for flowgraphs themselves is lost.

Abstract networks. It is well-known that natural numbers may be used to build up polynomials, but they are not enough to study their properties and, therefore, they need to be extended, e.g., to integer, rational, real, or complex numbers.

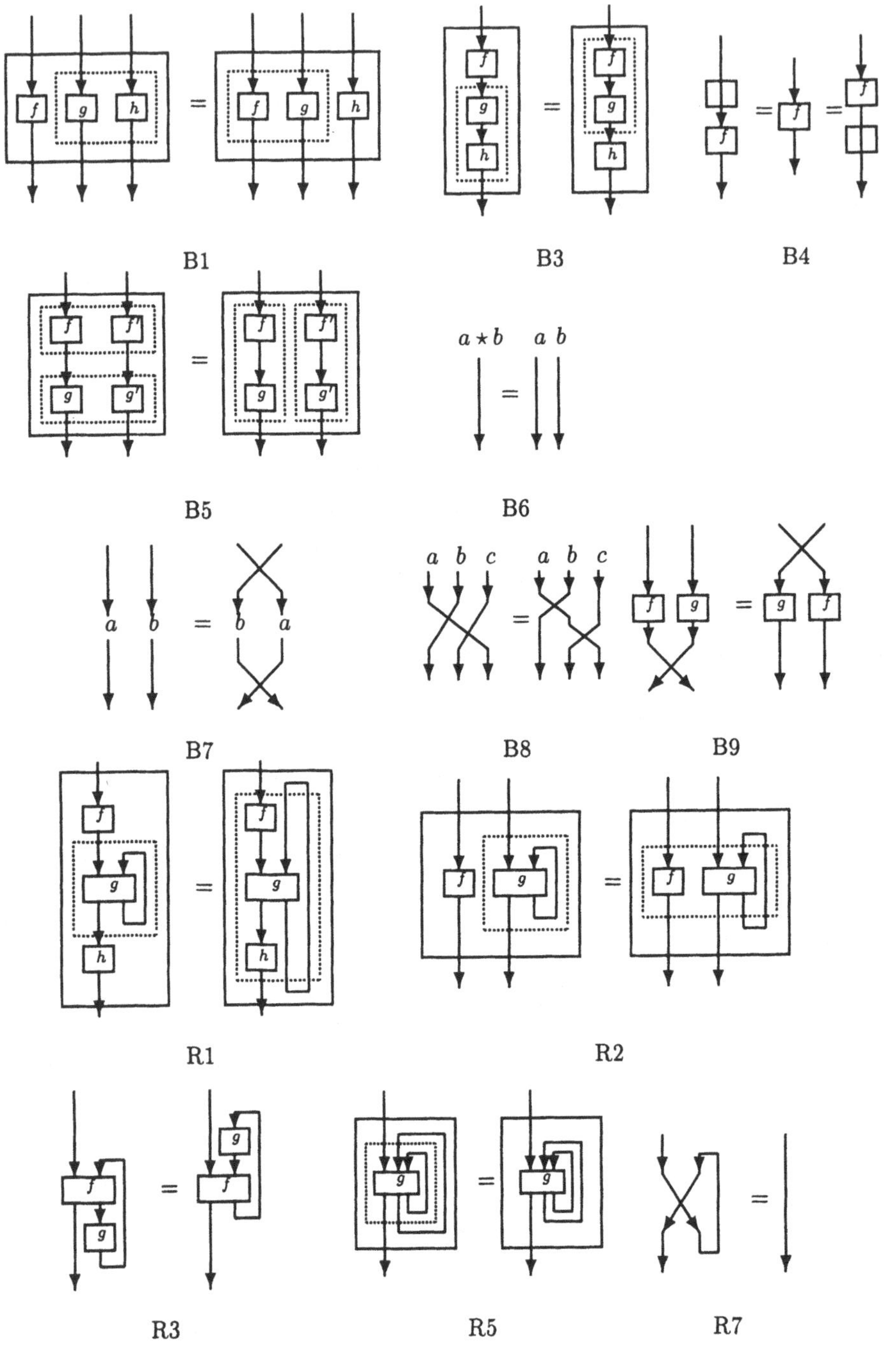

Fig. 1.3. Graphical representation of BNA rules

Similarly, finite sorted relations in $\mathbb{R}\mathrm{el}_S$ may be used to build up networks (or flownomials; we will see soon how to do this), but they are not enough to study their properties. (Networks are considered here from an intuitive point of view. They will be formally introduced later.) Generaly speaking, networks will consist of cells connected with arrows from *an arbitrary* BNA; such general networks are called *abstract networks.*

We illustrate the need for abstract networks with two cases.

A first situation is related to *network semantics.* One may specify the behaviour of the network cells using a particular semantic domain. These semantic domains occur as BNAs, but more complex than $\mathbb{R}\mathrm{el}_S$. It may happen quite often that one has to give the semantics in 2 steps. When only certain cells of a network are interpreted and the remaning ones are left still uninterpreted one gets more *abstract networks.* In these networks the arrows/elements in the given semantic domain are used to connect the network's still uninterpreted cells.

A similar situation is at the syntactic lavel, where one has to consider *network refinement.* In such a situation one gets higher order networks, where certain network cells are connected via other networks. Fortunately, as we shall see quite often, networks themselves form a BNA; hence such higher order networks are just abstract networks over particular BNAs.

An example: additive relations over a state space. The network semantics is exemplified with the standard relational semantics of the networks, provided the flowchart scheme interpretation is used.

Let D be the set of memory states of a computing device. If variables in X are interpreted as state-transforming relations, then the interpretation of a network/flownomial expression $E : m \to n$ with cells in X, connections in $\mathbb{R}\mathrm{el}$, m input pins, and n output pins is a relation $E_I \subseteq ([m] \times D) \times ([n] \times D)$ denoting the relational semantics of the corresponding flowchart program. Namely, $((j,d),(j',d')) \in E_I$ iff "if one runs the program obtained from the expression E and the interpretation I starting from the j-th input of the program and using the initial memory state d, then the program finishes on the j'-exit of the program within the memory state d', eventually."

This semantics may be easily described in a *compositional* way, provided a BNA model is initially built up of these state-transforming relations. This is an additive, relational semantic model $\mathsf{AddRel}(D)$, defined by the sets $\mathsf{AddRel}(D)(m,n) = \{r\colon r \subseteq ([m] \times D) \times ([n] \times D)\}, m,n \in \mathbb{N}$. It will be described in the full NA setting in Section 1.6. The BNA operations are simple generalizations of the corresponding definitions in $\mathbb{R}\mathrm{el}$, with one exception: feedback. Feedback's definition is more tricky.

Additive feedback on relations. For $r \in \mathsf{AddRel}(D)(m+1,n+1)$, the feedback $r\uparrow \in \mathsf{AddRel}(D)(m,n)$ is defined by relations $(r\uparrow)_{i,j} = r_{i,j} \cup (r_{i,n+1} \cdot r^*_{m+1,n+1} \cdot r_{m+1,j})$, for $i \in [m]$, $j \in [n]$. A few explanations are necessary here: a relation $r \in \mathsf{AddRel}(D)(m,n)$ is reduced to a family of relations $r_{i,j} \subseteq D \times D$, for $i \in [m]$, $j \in [n]$, where $r_{i,j} =$

$\{(d, d') : ((i, d), (j, d')) \in r\}$; next, "$*$" denotes the reflexive–transitive closure, i.e., for $s \subseteq D \times D$, $s^* = 1_D \cup s \cup s \cdot s \cup s \cdot s \cdot s \cup \ldots$, where $1_D = \{(d, d) : d \in D\}$). ◇

$\mathbb{R}$el is naturally mapped in AddRel(D). Seen from the AddRel(D) perspective, $\mathbb{R}$el represents the straight-line programs where only GOTO statements are used. No memory changes accompany the redirection of control flow. $\mathbb{R}$el may also be identified with AddRel($\bullet$), additive relations over a unique state "$\bullet$".

BNAs/$a\alpha$-flows. We start by slightly clarifying the overlap between two names used for this basic algebraic structure, that is, BNA vs. $a\alpha$-flow (to be formally defined below).

First, and most important, both "BNA" and "$a\alpha$-flow" names are used for *the same* algebraic structure. Next, the name "BNA" is better suited when one refers to the given set of axioms or when one wants to emphasize this particular graph-isomorphism setting. On the other hand, the name "$a\alpha$-flow" is preferred when the abstract algebraic structure specified by the BNA axioms is considered or when this structure is to be considered as a particular xy-flow.

When the interface monoid is left unspecified it is considered to be the additive monoid $\mathbb{N}$. An algebraic structure $(T, \star, \cdot, \uparrow, \mathsf{I}_m, {}^m\mathsf{X}^n)$ obeying BNA rules is also called *$a\alpha$-flow (over $\mathbb{N}$)*. A *morphism of $a\alpha$-flows over* $\mathbb{N}$, say $H : T \to T'$, is then defined by a family of applications $H_{m,n} : T(m, n) \to T'(m, n)$, for $m, n \in \mathbb{N}$, which preserve the operations and the constants.

In the general case, an arbitrary monoid $(M, \star, \epsilon)$ may be used to specify the input/output interfaces. Moreover, the morphisms are allowed to change these interfaces. These lead to the following basic definitions.

Definition 1.2. (BNA/$a\alpha$-flow) A *BNA or $a\alpha$-flow over M* is an abstract structure $(T, \star, \cdot, \uparrow, \mathsf{I}_a, {}^a\mathsf{X}^b)$ given by (below $a, b, c \in M$):

- a family of *arrows* $T = \{T(a, b)\}_{a,b \in M}$;
- constants: *identities* $\mathsf{I}_a \in T(a, a)$ and *transpositions* ${}^a\mathsf{X}^b \in T(a \star b,\ b \star a)$;
- operations: *juxtaposition* $\star : T(a, b) \times T(c, d) \to T(a \star c,\ b \star d)$, *composition* $\cdot : T(a, b) \times T(b, c) \to T(a, c)$, and *feedback* $\uparrow^c : T(a \star c,\ b \star c) \to T(a, b)$

which fulfil the axioms in Table 1.1.

A *morphism of $a\alpha$-flows* $H : T \to T'$, where T is an M-$a\alpha$-flow and T' is an M'-$a\alpha$-flow, is given by a morphism of monoids $h : M \to M'$ and a family of applications $H_{m,n} : T(m, n) \to T'(h(m), h(n))$, for $m, n \in M$ preserving the operations, i.e.,

$$H(f \star g) = H(f) \star H(g); \quad H(f \cdot g) = H(f) \cdot H(g); \quad H(f \uparrow^p) = H(f) \uparrow^{h(p)};$$
$$H(\mathsf{I}_a) = \mathsf{I}_{h(a)}, \quad H({}^m\mathsf{X}^n) = {}^{h(m)}\mathsf{X}^{h(n)}$$

Taking BNAs as objects, together with the above defined morphisms, one gets a category. This category will be subsequently specialized and studied for particular BNAs. The reader who is familiar with well-known categorical properties, will find that many network algebra results may be described, perhaps more elegantly, in categorical terms. However, to keep the approach simple and to have a close relation with the simple and deep calculus of polynomials, we have decided to follow the current, somehow more classical, presentation.

An important property of BNA, which is also relevant from the categorical point of view, is

> $\mathbb{B}\mathrm{i}$ *in BNAs*
>
> If T is an M-$a\alpha$-flow and $h : S^\star \to M$ is a morphism of monoids, then there exists a unique morphism of $a\alpha$-flows $H : \mathbb{B}\mathrm{i}_S \to T$ which acts as h on objects; see Theorem 2.7.

This shows that usual sorted bijections are naturally mapped in BNAs.

Exercises and Problems (sec. 1.2)

1. Show that B7+B9 may be replaced by B9′: $f \star g = {}^a\mathsf{X}^b \cdot (g \star f) \cdot {}^d\mathsf{X}^c$, for $a \xrightarrow{f} c$, $b \xrightarrow{g} d$. Show that ${}^a\mathsf{X}^\epsilon = \mathsf{I}_a$. (07)
2. Prove that BNA axioms are valid in $\mathbb{R}\mathrm{el}_S$ (defined in the previous section). (07)
3. Show that both (1) $\mathsf{AddRel}(D)$ and (2) $\mathbb{P}\mathrm{fn}(D)$ (i.e., the classes of partially defined functions in $\mathsf{AddRel}(D)$) are BNAs. (09)
4. Let U, V, W be finite vector spaces; consider a linear function $f : V \otimes U \to W \otimes U$; take as feedback the (generalized) trace $f \uparrow^U: V \to W$, where $f \uparrow^U$ is the linear function given by $(f \uparrow^U)_{i,j} = \sum_k f_{i\otimes k, j\otimes k}$. Show that linear functions (with this feedback and the natural symocat structure) have a BNA structure. (12)
5. Are the axioms independent? (**)

1.3 Flownomial expressions

In this section flownomial expressions are introduced, together with the associated BNA calculus. Examples of flownomial representations for certain networks are given. The meaning of the BNA calculus is briefly indicated and some examples of working with flownomials are presented.

Flownomial expressions. We have defined operations and constants on relations in $\mathbb{R}\mathrm{el}_S$. In order to construct the calculus of flownomials we still need a family X of doubly-ranked sets $X = \{X(a,b)\}_{a,b\in S^\star}$. An element $x \in X(a,b)$ is considered as a *variable* network/cell with $|a|$ input pins of sorts $a_1, \ldots, a_{|a|}$ and $|b|$ output pins of sorts $b_1, \ldots, b_{|b|}$.

Definition 1.3. (flownomial expressions over T) Let T be a BNA over a monoid $(S^\star, \star, \epsilon)$, for instance $\mathbb{R}\mathrm{el}_S$. For $a, b \in S^\star$, the sets $\mathsf{FlowExp}[X,T](a,b)$ of *flownomial expressions of type $a \to b$ over X and T* are defined as follows:

(i) a variable x in $X(a,b)$ or a symbol denoting a connection f in $T(a,b)$ is a flownomial expression of type $a \to b$;

(ii) compound expressions: if $E^1 : a \to b$, $E^2 : c \to d$, $E^3 : b \to c$ and $E : a \star s \to b \star s$, $(a, b, c \in S^\star, s \in S)$ are flownomial expressions of the indicated type, then

$$(E^1 \star E^2) : a \star c \to b \star d$$
$$(E^1 \cdot E^3) : a \to c$$
$$(E \uparrow^s) : a \to b$$

are flownomial expressions of the indicated type;

(iii) all the flownomial expressions are obtained by using rules (i) and (ii).

In what follows we shall omit many parentheses in flownomial expressions by declaring that: feedback has the strongest binding power, then composition, then summation; summation and composition are supposed to be associative. Thus for example,

$$((x \star x) \star ((x \star y) \cdot ((f \uparrow^s) \uparrow^t))) \quad \text{is written} \quad x \star x \star (x \star y) f \uparrow^{t \star s}$$

This also illustrates that $\uparrow^a$ means $\uparrow^{a_{|a|}} \ldots \uparrow^{a_1}$ and that $\cdot$ may be omitted.

♡ Recall that $\mathbb{R}\mathrm{el}$ is a BNA, hence flownomials may be defined over an arbitrary subset $T \subseteq \mathbb{R}\mathrm{el}$ which contains the BNA constants and is closed with respect to the BNA operations. E.g., the calculus may be done over $\mathbb{B}\mathrm{i}$, $\mathbb{I}\mathrm{n}$, $\mathbb{P}\mathrm{Sur}$, or $\mathbb{P}\mathrm{fn}$. ◇

BNA calculus. The calculus based on BNA/$a\alpha$-flow axioms is formally defined as follows.

First, a flownomial expression involving only elements in T is identified with the corresponding element computed in T. Then, consider the congruence relation $\sim_{a\alpha}$ generated on $\mathsf{FlowExp}[X,T]$ by this rule together with the BNA rules. Finally, an *$a\alpha$-flownomial over X and T* is a $\sim_{a\alpha}$-congruence class of flownomial expressions.

♡ There will be several extensions of BNA, notably the xy-flow structures. Consequently, in certain specific contexts flownomial expressions will be considered modulo certain $\sim_{xy}$ equivalences that are coarser than $\sim_{a\alpha}$. By default, a flownomial will mean an $a\alpha$-flownomial. ◇

Graphical interpretation. Every flownomial expression f in $\mathsf{FlowExp}[X,T]$, for $T \subseteq \mathbb{R}\mathrm{el}$ may be graphically interpreted as a graph $\mathrm{GR}(f)$ as is shown in Fig. 1.4. The pictures in Fig. 1.4 represent some particular, but typical, examples using one-sorted interfaces. The reader should have no problems to deal with networks having many-sorted interfaces; he only has to be careful to connect channels of the same type.

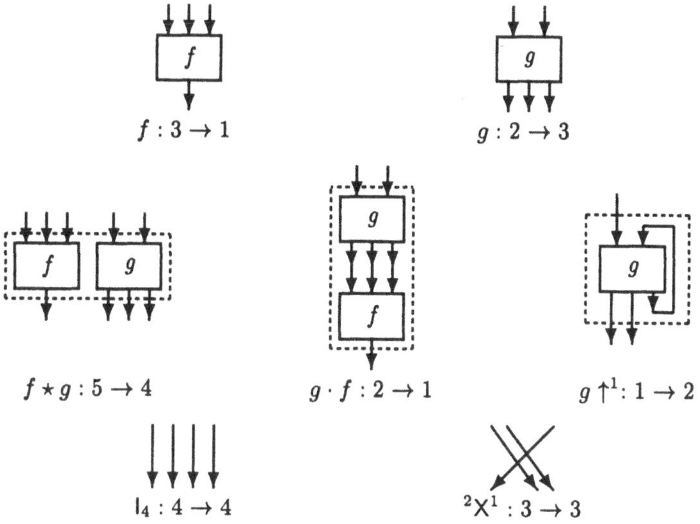

Fig. 1.4. Operations and constants of BNA

Networks and their representation. Flownomial expressions may be used to represent networks. All networks are represented by such expressions. The efficiency of this representation constitutes an important question which requires more investigation. A few examples are given here and in the exercises to follow showing the advantages and the limits of the current representation.

On network representation. As the example below suggests, for certain regular networks it is possible to give short and elegant representations. However, as one probably expects, for irregular networks the formulas tend to become longer and longer. An aim here may be to find short and efficient representations with a uniform complexity for all networks. To have an idea on what we mean here, the following analogy with the representation of numbers/polynomials is useful. Our current relation/network representation is like a Roman representation of natural numbers; the open problem is to find an Arabic-like representation. ◇

Example 1.4. In this example we show that for certain regular networks it is possible to give short and elegant representations. The following abbreviations are used: iterated sequential composition is $\mathrm{K}_{i=m}^{n} f_i = f_m \cdot \ldots \cdot f_n$ (m and n are arbitrary here, so the indices may either increase or decrease; in the formula below the index for the last K decreases from $k-1$ to 1); juxtaposition to the nth is $nf = f \star \ldots \star f$ (n times). Then denote

$$r_{k,l} = ((\mathrm{K}_{i=1}^{k-1}(\mathsf{I}_{k-i} \star if \star \mathsf{I}_{l-i}) \cdot \mathrm{K}_{i=0}^{l-k}(\mathsf{I}_i \star kf \star \mathsf{I}_{l-k-i}) \cdot \mathrm{K}_{i=k-1}^{1}(\mathsf{I}_{l-i} \star if \star \mathsf{I}_{k-i})) \cdot {}^{l}\mathsf{X}^{k}) \uparrow^{l}$$

where $k < l$ and $f : 2 \to 2$. The instance $r_{3,4}$ is illustrated in Fig. 1.5.

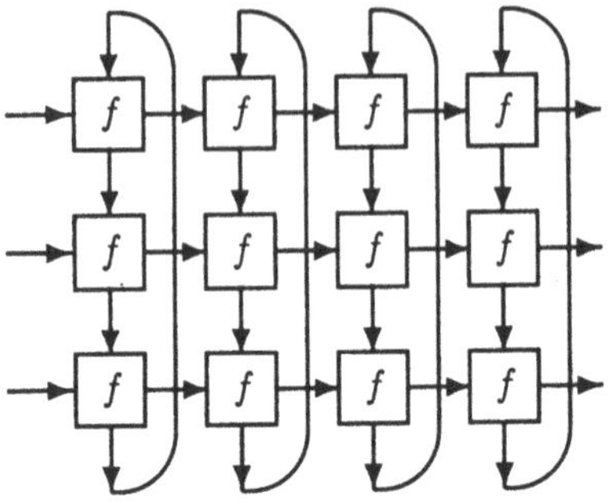

Fig. 1.5. A regular network and its NA representation (the picture corresponds to $r_{3,4}$): $r_{k,l} = ((\mathrm{K}_{i=1}^{k-1}(\mathsf{l}_{k-i} \star if \star \mathsf{l}_{l-i}) \cdot \mathrm{K}_{i=0}^{l-k}(\mathsf{l}_i \star kf \star \mathsf{l}_{l-k-i}) \cdot \mathrm{K}_{i=k-1}^{1}(\mathsf{l}_{l-i} \star if \star \mathsf{l}_{k-i})) \cdot {}^l\mathsf{X}^k) \uparrow^l$

The meaning of BNA calculus. BNA is a calculus for networks. That is, the usual networks may be identified with flownomials modulo BNA axioms. This claim is based on two facts.

(1) *Fitness of operations (expressiveness)*: The (usual, concrete) nondeterministic networks built up with atoms in X coincide with the graphical representations of flownomial expressions in $\mathsf{FlowExp}[X, \mathbb{R}\mathrm{el}]$. Other particular classes of networks may be represented by flownomial expressions using subsets of $\mathbb{R}\mathrm{el}$, e.g., $\mathbb{P}\mathrm{fn}$ for partially defined, deterministic networks, $\mathbb{F}\mathrm{n}$ for deterministic networks, $\mathbb{B}\mathrm{i}$ for networks with bijective connections (no branching connections), etc.

(2) *Fitness of BNA rules*: Let $T \subseteq \mathbb{R}\mathrm{el}$ be a substructure of $\mathbb{R}\mathrm{el}$ which is closed with respect to BNA operations, for example $T = \mathbb{P}\mathrm{fn}$, or $T = \mathbb{R}\mathrm{el}$. Then, the following results are valid: (i) *(soundness)* Two flownomial expressions over T which are equivalent via BNA rules have isomorphic graphical representations; and (ii) *(completeness)* Two flownomial expressions over T which have isomorphic graphical representations may be proved equivalent using BNA rules.

The proofs of these results will be given in the next chapters in a more general, abstract setting.

♡ Using pictures one may easily check the soundness part. But for the completeness part, as well as for the expressiveness result, we need the whole machinery of the normal form representations. (A flownomial expression is in a *normal form* if it is written as $((\mathsf{l}_a \star x_1 \star \ldots \star x_k) \cdot f) \uparrow^r$ where $r = a_1 \star \ldots \star a_k$, $x_1 \in X(a_1, b_1), \ldots, x_k \in X(a_k, b_k)$ and $f \in T(a \star b_1 \star \ldots \star b_k,\ b \star a_1 \star \ldots \star a_k)$.) ◇

Examples. Actually a unique, but important, example, is given here. It shows that the well-known "fixed-point equation," which is used to unfold loops, actually may be split into two parts: an application of a simple acyclic rule $f \cdot \wedge^a = \wedge^{b \star a} \cdot (f \star f)$, followed by a (graph-isomorphic) BNA transformation.

Example 1.5. We prove here that certain expressions representing the networks in Fig. 1.6 (b) and (c) are BNA equivalent:

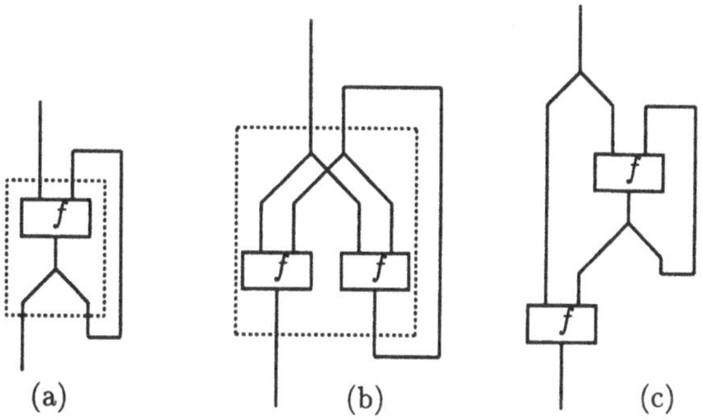

Fig. 1.6. Graph isomorphism + $f \cdot \wedge^a = \wedge^{b\star a} \cdot (f \star f) \Rightarrow$ fixed-point equation

$$
\begin{aligned}
& [\wedge_2^{b\star a} \cdot (f \star f)] \uparrow^a \\
&= [(\wedge_2^b \star \wedge_2^a) \cdot (\mathsf{I}_b \star {}^b\mathsf{X}^a \star \mathsf{I}_a) \cdot (\mathsf{I}_{b\star a} \star f) \cdot (f \star \mathsf{I}_a)] \uparrow^a && \text{B5 + identity in } \mathbb{R}\mathrm{el} \\
&= \wedge_2^b \cdot (\mathsf{I}_{b\star b} \star \wedge_2^a) \cdot (\mathsf{I}_b \star {}^b\mathsf{X}^a \star \mathsf{I}_a) \cdot (\mathsf{I}_{b\star a} \star f)] \uparrow^a \cdot f && \text{R1} \\
&= \wedge_2^b \cdot [(\mathsf{I}_b \star {}^b\mathsf{X}^a \star \mathsf{I}_a) \cdot (\mathsf{I}_{b\star a} \star f) \cdot (\mathsf{I}_{b\star a} \star \wedge_2^a)] \uparrow^{a\star a} \cdot f && \text{R3} \\
&= \wedge_2^b \cdot [(\mathsf{I}_b \star {}^b\mathsf{X}^a \star \mathsf{I}_a) \cdot (\mathsf{I}_{b\star a} \star f) \cdot (\mathsf{I}_{b\star a} \star \wedge_2^a)] \uparrow^a \uparrow^a \cdot f && \text{R5} \\
&= \wedge_2^b \cdot [(\mathsf{I}_b \star {}^b\mathsf{X}^a) \cdot (\mathsf{I}_{b\star a} \star f \cdot \wedge_2^a) \uparrow^a] \uparrow^a \cdot f && \text{R1} \\
&= \wedge_2^b \cdot [(\mathsf{I}_b \star {}^b\mathsf{X}^a) \cdot (\mathsf{I}_b \star \mathsf{I}_a \star (f \cdot \wedge_2^a) \uparrow^a \ )] \uparrow^a \cdot f && \text{R2} \\
&= \wedge_2^b \cdot (\mathsf{I}_b \star [(f \cdot \wedge_2^a) \uparrow^a \star \mathsf{I}_a] \cdot {}^a\mathsf{X}^a) \uparrow^a \cdot f && \text{B9} \\
&= \wedge_2^b \cdot [\mathsf{I}_b \star (f \cdot \wedge_2^a) \uparrow^a] \cdot (\mathsf{I}_b \star ({}^a\mathsf{X}^a) \uparrow^a) \cdot f && \text{R1,R2} \\
&= \wedge_2^b \cdot (\mathsf{I}_b \star (f \cdot \wedge_2^a) \uparrow^a) \cdot f && \text{R7}
\end{aligned}
$$

Exercises and Problems (sec. 1.3)

1. Show that $r_{3,4}$ in Example 1.4 represents the network in Fig. 1.5, indeed. (05)

2. Find the graphical representation for the following expressions (kf means $f\star \overset{k}{\ldots} \star f$; f^k means $f\cdot \overset{k}{\ldots} \cdot f$; $\varphi_{m,n}$ is the bijection which naturally rearranges m groups of n elements in n groups of m elements (see Notation 3.7 for a precise definition):
   - $E1 = [(nf) \cdot {}^1\mathsf{X}^{n-1}] \uparrow^n$; (ring buffer; $f : 1 \to 1$)
   - $E2 = \varphi_{2,4} \cdot 4C_1 \cdot [2(\mathsf{I}_1 \star {}^1\mathsf{X}^1 \star \mathsf{I}_1)] \cdot \varphi_{4,2} \cdot 4C_2 \cdot [2\varphi_{2,2}] \cdot 4C_3$; (commutator Banyan; $C_i : 2 \to 2$ maps an input datum to the 1st output port if its i-th bit is 0, otherwise to the 2nd port; in the case of collision only one datum is forwarded)
   - $E3 = [\mathsf{dn}\star 2\mathsf{up}\star \mathsf{dn}] \cdot [2\varphi_{2,2} \cdot (2\mathsf{dn}\star 2\mathsf{up})]^2 \cdot [\varphi_{2,4} \cdot 4\mathsf{dn}]^3$; (commutator Batcher; $\mathsf{up} : 2 \to 2$ is thought of as a sorting "up" cell, while dn do a "down" sorting)
   - $E4 = [\varphi_{2,n}(nf \cdot \varphi_{n,2} \star C_n)(\mathsf{I}_n \star {}^n\mathsf{X}^n)] \uparrow^{2n}$; (star topology; $f : 2 \to 2$ and $C_n : n \to n$). (10)

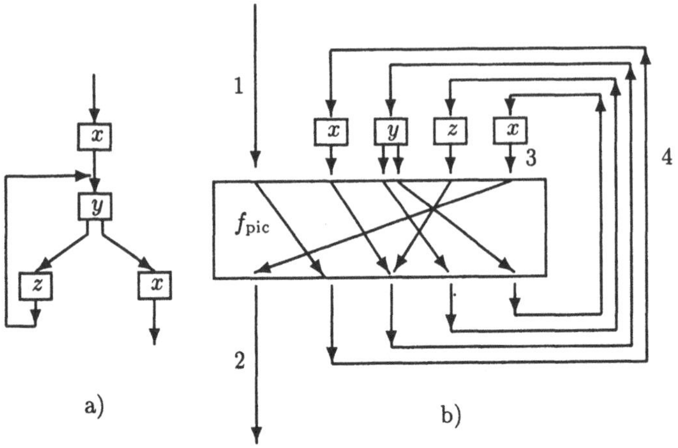

Fig. 1.7. The normal form representation of networks

3. Sorting networks: Let
$$N = (1365247)(2C \star \mathsf{I}_1 \star C)(1635427)(2C \star \mathsf{I}_1 \star C)(1324657)(3C \star \mathsf{I}_1)$$
with $2 \xrightarrow{C} 2$ such that $C(a,b)$ is (a,b) if $a \leq b$, othervise (b,a); (1365247) is the bracket representation of the bijection that maps 1 in 1, 2 in 3, 3 to 6, etc. Show that N can sort "bitonic" sequences, i.e., if $a_1 \geq a_2 \geq a_3 \geq a_4 \leq a_5 \leq a_6 \leq a_7$, then $N(a_1, a_2, \ldots, a_7) = (a_{\tau(1)}, a_{\tau(2)}, \ldots, a_{\tau(7)})$ for a bijection τ with $a_{\tau(1)} \leq a_{\tau(2)} \cdots \leq a_{\tau(7)}$. (12)
4. Take network pictures from various places (circuit designs, Petri nets, flowchart schemes, etc.) and give flownomial representations for them. (- -)

1.4 Concrete vs. abstract networks

In this section the relationship between flownomial expressions and networks with (concrete) connections in $\mathbb{R}\text{el}_S$ is established. It is shown that such a concrete network may be represented by a normal form flownomial expression, unique up to an isomorphism (simulation via a bijection). This property is taken as the definition of *abstract networks*, in the case when the connecting arrows are from an arbitrary BNA.

The normal form representation. Every network (labelled directed hypergraph) with atoms in X, for example the one in Fig. 1.7(a), may be (re)drawn in a standard way as in Fig. 1.7(b). The arrows of the rectangle f_{pic} in Fig. 1.7(b) may be specified using a relation

$$f \in \mathbb{R}\text{el}(m \star o(x_1) \star \ldots \star o(x_k),\ n \star i(x_1) \star \ldots \star i(x_k))$$

where m gives the number and the sorts of the inputs into the network, n similarly for the outputs of the network, $x_1 \star x_2 \star \ldots \star x_k$ is a chosen string of all the atoms in the network, and $o(x)$ (resp. $i(x)$), for an x in X, gives the number and the sorts of the inputs into (resp. the outputs from) the atom x.

Hence networks, drawn in such normal form, are in a bijective correspondence with the pairs

$$(x_1 \star \ldots \star x_k, f), \text{ where } x_1 \star \ldots \star x_k \text{ is a word in } X^\star,\ x_1, \ldots, x_k \in X$$
$$\text{and } f \in \mathbb{R}\mathrm{el}(m \star o(x_1) \star \ldots \star o(x_k),\ n \star i(x_1) \star \ldots \star i(x_k))$$

An expression such as

$$((\mathsf{I}_m \star x_1 \star \ldots \star x_k) \cdot f) \uparrow^{i(x_1)\star \ldots \star i(x_k)}$$

with x_i and f as above is called a *normal form expression* and the corresponding pair

$$(x_1 \star \ldots \star x_k,\ f)$$

is called a *nf-pair*. The set of m-to-n nf-pairs is denoted by $[X, T](m, n)$.

Example of normal forms. For the example in Fig. 1.7, let s denotes the (unique) sort of the channels. The network in Fig. 1.7(a) may be represented, e.g., as $x[(1_s 1_s)y(z \star x)(2_s 1_s)] \uparrow^s$. For its normal form in Fig. 1.7(b) the relation $f_{\mathrm{pic}} : s \star s \star s \star s \star s \star s \to s \star s \star s \star s \star s$ is $(2_s 3_s 4_s 5_s 3_s 1_s)$; hence the associated normal form expression is $[(\mathsf{I}_s \star x \star y \star z \star x) \cdot (2_s 3_s 4_s 5_s 3_s 1_s)] \uparrow^{s \star s \star s \star s}: s \to s$. ◇

The following result is now obvious.

Fact 1.6. *Concrete networks built up with atoms in X coincide with the networks represented by flownomial expressions over appropriate classes of relations, e.g.,*

nondeterministic networks	↔	*relations* ($\mathbb{R}\mathrm{el}$)
deterministic, partial networks	↔	*partial functions* ($\mathbb{P}\mathrm{fn}$)
deterministic networks	↔	*functions* ($\mathbb{F}\mathrm{n}$)
networks with bijective connections	↔	*bijections* ($\mathbb{B}\mathrm{i}$)

□

Relating isomorphic normal form representations. We look for a mathematical formulation of the property that two nf-pairs represent the same network. That is, the equivalence relation on nf-pairs we are going to capture here is *graph-isomorphism.*

For a given network and a given linear order of its atoms there is *one and only one* nf-pair representing that network. So the problem is to relate different nf-pairs obtaining by varying the linear order used to display the network atoms in the nf-pairs.

This equivalence is captured by simulation via bijections. Before the formal definition we fix a notation. If $u : [k] \to [k]$, then $i(u)$ denotes its extension with respect to the inputs of $x_1, \ldots, x_k$; in the tuple notation this means $(u(1)_{i(x_1)} u(2)_{i(x_2)} \ldots u(k)_{i(x_k)})$. Similarly for output extension $o(u)$.

Definition 1.7. Two nf-pairs $F = (x_1 \star \ldots \star x_k, f)$ and $G = (y_1 \star \ldots \star y_k, g)$ in $[X,T](m,n)$ are *similar via a bijections* $u : [k] \to [k]$ (denoted $F \to_u G$) if

(i) $j = u(i)$ implies $x_i = y_j$ and
(ii) $f(\mathsf{l}_n \star i(u)) = (\mathsf{l}_m \star o(u))g$.

♡ The first condition says that the bijection relates cells with the same label; the second condition shows that, making abstraction of the permutation u, the connections are the same in F and G. Instead of bijections, one may equivalently use chains of simulations via simple transpositions. ◇

Simulation via bijections is denoted $\to_{a\alpha}$ or $\xrightarrow{a\alpha}$. It is an equivalence relation. Now the following proposition is obvious.

Proposition 1.8. *Two nf-pairs F and F' represent the same network if and only if they are similar via a bijection, i.e., iff $F \xrightarrow{a\alpha} F'$.* □

Abstract networks. The above properties of concrete networks over an $\mathbb{R}\mathrm{el}_S$ may be lifted, by definition, to an abstract setting, where connections between network cells are specified by elements in an arbitrary BNA. We get the following notion of "abstract network."

Let $[X,T]$ denote the set of nf-pairs (normal form pairs) over X and a BNA T. A notion of simulation via bijections, denoted $\xrightarrow{a\alpha}$, may be introduced on $[X,T]$ as above. Then, $[X,T]_{a\alpha}$ denotes $[X,T]/_{\xrightarrow{a\alpha}}$. An *abstract network* or *aα-flownomial* is an element in $[X,T]_{a\alpha}$, where T is an arbitrary BNA.

On simulation definition. The introduction of the notion of simulation via bijections makes sense because we already said that bijections may be mapped in arbitrary BNAs. In addition, we will see in Chapter 2 that the BNA structure on T may be naturally lifted to abstract networks, hence $[X,T]_{a\alpha}$ will be a BNA, too. Actually, $[X,T]_{a\alpha}$ is the free BNA obtained by adding X to T, a property similar to the universal property of polynomials. ◇

EXERCISES AND PROBLEMS (SEC. 1.4)

1. Use BNA rules to show that the flownomial expressions $x[(1_s1_s)y(z \star x)(2_s1_s)] \uparrow^s$ and $[(x \star z)(1_s1_s)y(2_s1_s)] \uparrow^s \cdot x$ associated to the picture in Fig. 1.7(a), are equivalent to $[(\mathsf{l}_s \star x \star y \star z \star x)\cdot(2_s3_s4_s5_s3_s1_s)] \uparrow^{s\star s\star s\star s\star s}: s \to s$, the expression associated to the normal form picture in Fig. 1.7(b). (12)

1.5 Network algebra, NA

In this section the signature of the full Network Algebra setting is presented. The specific NA rules for the branching constants are not stated here; see Chapter 3.

In Fig. 1.8, the meaning of (block extensions of) the additional constants for branching connections $\wedge$, $\perp$, $\vee$ and $\top$ is illustrated by means of a graphical representation.

$\wedge^3 : 3 \to 6$ $\perp^3 : 3 \to 0$ $\vee_2 : 4 \to 2$ $\top_2 : 0 \to 2$

Fig. 1.8. Additional constants for branching connections

Not too much on these constants will be given here. Their (angelic) interpretation in $\mathbb{R}$el leads to identities written as ground terms and which are valid when interpreted in $\mathbb{R}$el. An axiomatization consists of the following groups of laws: (1) *associativity* (ramifications followed by ramifications may be collected in a unique step; the same for identifications), (2) *commutativity* (the order of edges in a branching point does not matter, hence these edges may be commuted leaving the term unchanged), (3) *commutation* (identifications followed by ramifications may be commuted to have first ramifications, then identifications), and finally (4) *idempotency* (a k ramification is annihilated by a forthcoming k identification). This is further extended to the cyclic case.

Axioms for branching constants and the bracket representation. It may be interesting to see how these axioms are related to the bracket representation of relations. By associativity, all ramifications in an input are collected in a unique bracket [...]; by commutativity we may order them to be increasing in such a bracket; by commutation [...] and (...) are commuted, as in $(1_a 1_a)[1_a 2_a 3_a] = ([1_a 2_a 3_a][1_a 2_a 3_a])$ (this is the key rule to allow for the composite of two bracket representations to be reduced to such a bracket representation); finally idempotency shows that $[1_a 2_a](1_a 1_a) = (1_a)$, reducing multiple edges between the same input–output pair.

Notice that the bracket representation is asymmetric, while the axioms are symmetric with respect to ramification or identification constants. In other words, in the bracket representation the axioms look somehow different from the identification point of view. ◇

One more thing deserves to be mentioned here. Besides the usual angelic relation operators, certain different, *demonic* operators may be defined on $\mathbb{R}$el. For them, different axioms for the branching constants are to be used. (See Appendix C.) Hence, in this delicate point related to branching constants where we already had critical axioms to be selected for weak, strong, or enzymatic properties, the situation is even worse: one has to consider one more option, i.e., angelic vs. demonic behaviour for branching constants!

But this may be seen as a further argument for the present NA approach. We mean, if the meaning of the branching constants is so complicated, then it is much better to have a calculus where their role is clearly separated, and the branching constants are not used in an implicit way. (Looping operators like iteration or repetition, to be presented in Section 7, implicitely use such branching constants.) NA has such a discriminating property.

EXERCISES AND PROBLEMS (SEC. 1.5)

1. Describe the meaning of the (associativity, commutativity,...) axioms using the bracket representation, but the identification's point of view. (08)

1.6 Control, space, time: 3 faces of NA models

Control. Here we briefly describe the standard network algebra operators in their additive form needed for modeling the behavioural aspects of an agent. The resulting axiomatic system is denoted by NA. Sometimes we use the equivalent acronym AddNA to emphasize the additive meaning of the operators. Its relational semantics $\mathsf{AddRel}(D)$ is described, as well.

For this additive interpretation the network algebra operators are denoted by

$$\oplus \quad \cdot \quad \uparrow_{\oplus} \quad \mathsf{I} \quad \mathsf{X} \quad \bullet\!\!<_k \quad {}_k\!>\!\!\bullet$$

To describe this state-transforming relational semantics we start with a set D of value-vectors which represent the *memory states* of a computing device. If variables in X are interpreted as state-transforming relations, then a flownomial $E : m \to n$ over X is interpreted as a relation $E_I \subseteq mD \times nD$; mD denotes the disjoint union of m copies of D; its elements are denoted either as (j, d) or as $j.d$. The meaning is $((j, d), (j', d')) \in E_I$ iff "if one runs the program obtained from the expression E and the interpretation I starting with input j of the program and initial memory state d, then the program finishes on the exit j' of the program in memory state d', eventually".

This observation leads to the construction of a model $\mathsf{AddRel}(D)$ defined as follows (in a many sorted case). Suppose S is a set of atomic sorts and $D = \{D_s\}_{s \in S}$ is an S-sorted set of data. The types of the program interfaces in this additive case are modeled by elements in the additive monoid $S^{\oplus} = (S^{\oplus}, \oplus, 0)$ freely generated by S. Hence a general sort for an additive interface is represented by a word $a = a_1 \oplus \ldots \oplus a_{|a|}$, where a_i is the i-th letter of a. To each sort a one may naturally associate a semantic domain for data of type a as follows. If $a \in S^{\oplus}$ is as above then we denote by aD the disjoint sum $D_{a_1} \oplus \ldots \oplus D_{a_{|a|}}$. A generic element x in this latter set will be denoted as $i.x_i$.

The disjoint sum is associative up to isomorphisms.

♠ The notation we just indicated for n-ary disjoint sums does not reflect naturally this property, namely a renaming is necesary for this. Look, e.g., to $A \oplus (B \oplus C) = A \oplus B \oplus C = (A \oplus B) \oplus C$; the elements in C are referred to as 2.2.c on left, 3.c in middle, and 2.c on right. Perhaps a $\langle\ldots\rangle$ notation to mimic the product notation $(\ldots)$ is better. ♣

On the other hand the operation is not considered implicitely commutative. That is, for the first property the natural isomorphism is not explicitly stated, while for the latter one it is done via additive transpositions χ.

$\mathsf{AddRel}(D)$ denotes the "additive" relational structure defined by

$$\mathsf{AddRel}(D)(a,b) = \{f:\ f \subseteq aD \times bD\}$$

for $a, b \in S^{\oplus}$.

The network algebra operators have the following meaning in $\mathsf{AddRel}(D)$:

- *Summation:* for $f: a \to b$ and $g: c \to d$ the sum $f \oplus g: a \oplus c \to b \oplus d$ is $f \oplus g = \{(x,y):\ \text{either } x = 1.x_1, y = 1.y_1, \text{ and } (x_1,y_1) \in f \text{ or } x = 2.x_2, y = 2.y_2, \text{ and } (x_2,y_2) \in g\}$.
- *Composition:* for $f: a \to b$ and $g: b \to c$ the composite $f \cdot g: a \to c$ is $f \cdot g = \{(x,z):\ \text{there exists } y \text{ such that } (x,y) \in f \text{ and } (y,z) \in g\}$.
- *Feedback:* In order to define the feedback first we notice that a relation $f: a \to b$ is given by a family of relations $\{f_{i,j}\}_{i \in [|a|],\ j \in [|b|]}$, where

  $$f_{i,j} = \{(d,d'):\ (i.d, j.d') \in f\} \subseteq D_{a_i} \times D_{b_j}$$

  As usual, r^* denotes the reflexive–transitive closure of a relation r.
  For $f: a \oplus s \to b \oplus s$, $a, b \in S^{\oplus}$, $s \in S$ the feedback $f \uparrow^s_{\oplus}: a \to b$ is defined by

  $$(f \uparrow^s_{\oplus})_{i,j} = f_{i,j}\ \cup\ f_{i,|b\oplus s|} \cdot f^*_{|a\oplus s|,|b\oplus s|} \cdot f_{|a\oplus s|,j}$$

  for $i \in [|\,a\,|]$, $j \in [|\,b\,|]$.
  The definition is extended to arbitrary sorts by $f \uparrow^0_{\oplus} = f$, $\quad f \uparrow^{a\oplus s}_{\oplus} = (f \uparrow^s_{\oplus}) \uparrow^a_{\oplus}$.
- *Identity:* $\mathsf{I}_a: a \to a$ is defined by
  $\mathsf{I}_a = \{(x,x):\ x \in D_a\}$
- *Additive Transposition:* ${}^b_a\chi: a \oplus b \to b \oplus a$ is defined by
  ${}^b_a\chi = \{(x,y):\ \text{either } x = 1.x_1, y = 2.y_2 \text{ and } x_1 = y_2 \text{ or } x = 2.x_2, y = 1.y_1 \text{ and } x_2 = y_1\}$
- *Additive Ramification:* ${}_a{\bullet\!\!<}_k: a \to ka$ defined by
  ${}_a{\bullet\!\!<}_k = \{(x, i.x):\ i \in [k],\ x \in D_a\}$
- *Additive Identification:* ${}_k{>\!\!\bullet}_a: ka \to a$ defined by
  ${}_k{>\!\!\bullet}_a = \{(x,y):\ \text{if } x \text{ is } i.x_i, \text{ then } y = x_i\}$

We present below an example to show how this model may be used to give the semantics of sequential, flowchart programs.

Example 1.9. We describe here a flowchart program for computing the binomial coefficients $\binom{k}{n}$:

Binom program. Take four variables n, k, y, z, interpreted as natural numbers. Let $S = \{s_1, s_2, s_3, s_4\}$ be a sort set and suppose the associated data sets are

$$D_{s_1} = D_n \times D_k;\ D_{s_2} = D_n \times D_k \times D_y \times D_z;\ D_{s_3} = D_k \times D_y \times D_z;\ D_{s_4} = D_z$$

all D_n, D_k, D_y, D_z being in this case $\mathbb{N}$. Then, there are 5 cells

$$i_1 : s_1 \to s_2;\ t_1 : s_2 \to s_2 \oplus s_2;\ i_2 : s_2 \to s_3;\ t_2 : s_3 \to s_3 \oplus s_3;\ e : s_3 \to s_4$$

The relational semantics for these cells is

$$\begin{array}{l} (n,k) \xrightarrow{i_1} (n,k,n,1); \\ (n,k,y,z) \xrightarrow{t_1} \text{if } y \leq n-k \text{ then } 1.(n,k,y,z) \text{ else } 2.(n,k,y-1,z*y); \\ (n,k,y,z) \xrightarrow{i_2} (k,1,z); \\ (k,y,z) \xrightarrow{t_2} \text{if } y > k \text{ then } 1.(k,y,z) \text{ else } 2.(k,y+1,z/y); \\ (k,y,z) \xrightarrow{e} (z) \end{array}$$

Take the normal form flownomial expression

$$E = [(\mathsf{l}_{s_1} \oplus i_1 \oplus t_1 \oplus i_2 \oplus t_2 \oplus e)(2_{s_1}3_{s_2}4_{s_2}3_{s_2}5_{s_3}6_{s_3}5_{s_3}1_{s_4})] \uparrow^{s_1 \oplus s_2 \oplus s_2 \oplus s_3 \oplus s_3}$$

Then $|E|(n,k) = \binom{k}{n}$, when $k \leq n$, where $|E|$ is the function in $\mathsf{AddRel}(D)$ associated to E induced by the given interpretation of the cells.
An equivalent representation is: $F = i_1 \cdot [(1_{s_2}1_{s_2})t_1] \uparrow^{s_2} \cdot i_2 \cdot [(1_{s_3}1_{s_3})t_2] \uparrow^{s_3} \cdot e.$ ◊

As one can see from this example, the network cells can change the state space, hence the state space may be different in different points of the program. The state space has a certain multiplicative structure, but this is irrelevant here! What is important is the fact that *the statements are atomic.* They are not decomposed in smaller actions which may act, say, in parallel on pieces of the state space. Hence we have a pure additive model over monadic states. (The "atomicity" condition is relaxed in Chapter 12 where mixed network algebras are presented.)

Space. We present a multiplicative version MultNA of the standard network algebra operators. It may be used to model the pure architectural (or distributed, parallel) features of a system.

The symbols

$$\otimes \quad \cdot \quad \uparrow_\otimes \quad \mathsf{I} \quad \mathsf{X} \quad \mathsf{R}_k \quad \mathsf{V}^k$$

will be used for NA operations and constants in the case the multiplicative interpretation is intended. The semantic, relational model $\mathsf{MultRel}(D)$ is a model for parallel data transformers. Such a transformer $f : m \to n$ acts on an m-tuple of input data and produces an n-tuple of output data. In this model one may naturally define the parallel and sequential composition operators and certain constants: identity, transposition, ramification constants (copy and sink), and identification constants (source and equality test). Next the feedback may be defined using the set of all possible invariants for the feedback channel(s). (Under additional hypotheses iterative definitions for the multiplicative feedback may be introduced, too; see Chapters 7 and 10.)

In the following we will use an extension of this setting to the many sorted case. Let S be a set of atomic sorts and $D = \{D_s\}_{s\in S}$ be an S-sorted set of data. For arbitrary sorts denoting multiplicative interfaces we use the multiplicative monoid $S^{\otimes} = (S^{\otimes}, \otimes, 1)$ freely generated by S. Hence such a sort is specified by a word $a = a_1 \otimes \ldots \otimes a_{|a|}$, where a_i is the i-th letter of a. For $a \in S^{\otimes}$ let us denote by D^a the product $D_{a_1} \times \ldots \times D_{a_{|a|}}$. Finally the "multiplicative" structure $\mathsf{MultRel}(D)$ is defined by

$$\mathsf{MultRel}(D)(a,b) = \{f\colon\ f \subseteq D^a \times D^b\}$$

for $a, b \in S^{\otimes}$.

In $\mathsf{MultRel}(D)$ the network algebra operations and constants are interpreted as follows:

- *Product:* for $f : a \to b$ and $g : c \to d$ the product $f \otimes g : a \otimes c \to b \otimes d$ is $f \otimes g = \{((x_1,x_2),(y_1,y_2))\colon\ (x_1,y_1) \in f \text{ and } (x_2,y_2) \in g\}$.
- *Composition:* for $f : a \to b$ and $g : b \to c$ the composite $f \cdot g : a \to c$ is $f \cdot g = \{(x,z)\colon\ \text{there exists } y \text{ such that } (x,y) \in f \text{ and } (y,z) \in g\}$.
- *Feedback:* For $f : a \otimes c \to b \otimes c$ we define $f \uparrow^c_{\otimes}\colon a \to b$ as the maximal fixed point, i.e., $f \uparrow^c_{\otimes} = \{(x,y)\colon \exists z \in D^c.\ ((x,z),(y,z)) \in f\}$.
- *Identity:* $\mathsf{I}_a : a \to a$ is defined by $\mathsf{I}_a = \{(x,x)\colon\ x \in D^a\}$.
- *Multiplicative Transposition:* ${}^a\mathsf{X}^b : a \otimes b \to b \otimes a$ is defined by ${}^a\mathsf{X}^b = \{((x,y),(y,x))\colon\ x \in D^a \text{ and } y \in D^b\}$.
- *Multiplicative Ramification:* $\wedge^a_k : a \to a^k$ is defined by $\wedge^a_k = \{(x,(x,\ldots,x))\colon\ x \in D^a\}$.
- *Multiplicative Identification:* $\vee^k_a : a^k \to a$ is defined by $\vee^k_a = \{((x,\ldots,x),x)\colon\ x \in D^a\}$.

An example is given below to show how this model may be used to study circuits.

Example 1.10. In the following we specify a logical net for binary multiplication. It is an example of an acyclic, combinatorial circuit.

Multiplication circuit. The circuit C below uses the standard boolean cells $\mathsf{and}, \mathsf{or}, \mathsf{xor} : 2 \to 1$ (xor is "exclusive or"). It also uses a unique sort s whose writing is omitted. C is

$$(\wedge^2_3 \otimes \wedge^2_2)\{[\mathsf{and} \otimes \mathsf{or} \otimes (\mathsf{I}_2 \otimes \mathsf{and} \cdot \wedge^1_2)({}^2\mathsf{X}^1 \otimes \mathsf{I}_1)](\mathsf{I}_1 \otimes \mathsf{and})\mathsf{or} \otimes (\mathsf{I}_1 \otimes \mathsf{xor})\mathsf{xor}) \otimes \mathsf{xor}\}$$

It has 4 inputs (a_1, a_0, b_1, b_0) and 3 outputs (c_2, c_1, c_0); the aim of the circuit is to compute binary multiplication, i.e., $\overline{a_1a_0} \times \overline{b_1b_0} = \overline{c_2c_1c_0}$. Indeed, if one considers the set $D_s = \{0,1\}$ associated to s and the usual interpretation for the cells $\mathsf{and}, \mathsf{or}, \mathsf{xor}$ in

$\mathsf{MultRel}(D)(2,1)$, then the interpretation $|C| \in \mathsf{MultRel}(D)(4,3)$ is a function satisfying $\overline{a_1a_0} \times \overline{b_1b_0} = \overline{c_2c_1c_0}$. ◇

Time. Here we describe a timed network algebra model. TimeNA is just a different notational version of the multiplicative network algebra MultNA; it is used to describe the pure temporal aspects of a system.

The timed network algebra operators are induced by the operation $\frown$ which acts both on interfaces and on morphisms. Appropriate sequential composition, feedback, and branching constants may be introduced. Actually, the full timed version of the network algebra $\mathsf{Mult_T NA}$ is orthogonal to the spatial, parallel one and contains the temporal operators

$$\frown,\ \cdot,\ \uparrow^{\frown},\ \mathsf{I},\ \mathsf{X},\ \propto_k,\ {}_k\!\!>\!\!\circ$$

In this model, the interface is interpreted as a model for *job request*. For an atomic sort a, D_a is thought of as representing data expanded in time; an $w \in D_a$ specifies the time slots at which a request of doing action a is present. The arrows $f : a \to b$ are interpreted as *job transformers*: the agent whose behaviour is represented by f transforms an input job request of type a to an output job request of type b.

The meaning of the temporal operators is the following:

- *Piping composition*: The operator $\frown$ on temporal interfaces is simply the product. On morphisms, $\frown$ describes the "parallel temporal" operator: if $f : a \to b$ and $f' : a' \to b'$, then $f \frown f' : a \frown a' \to b \frown b'$ means apply f to the part of the job request of type a and f' to the following part of type a'. The resulting output job streams of type b and b', respectively, are collected.
- *Timed sequential composition*: This operation models the causal propagation of actions; it is a kind of unidirectional interaction. If $f : a \to b$ and $f' : b \to c$, then $f \cdot f' : a \to c$ consist of the application of f to the input job request of type a and then of f' to the resulting job stream of type b produced by f.
- *Timed feedback*: As in the case of multiplicative feedback, this operation is more difficult to define. Combined with timed sequential composition, it models a possibly circular interaction between two processes. As in the case of the above model, one may consider here an infinite spatial extension of a process (a "wave") and to define the timed feedback similar to the definition of the usual multiplicative feedback in the synchronous case.
- *Timed identity*: $\mathsf{I}_a : a \to a$ simply passes the job-stream request.
- *Timed transposition*: ${}^a_b\mathsf{X} : a \frown b \to b \frown a$ simply switches the parts of the job-stream request of type a and b, respectively. It has nothing paradoxical in it. Indeed, the time is not interchanged, just our recording of the time information. (The time information related to the presence or absence of a request of type a at various time slots is provided by the corresponding value $w \in D_a$.)

- *Timed ramification:* ${}_a{\propto}_k : a \to a^{(k)}$ is a retiming constant which allow a job stream request to be repeated a number of times. Stuttering of job streams, but also removal of some of them, may be expressed using this constant.
- *Timed identification:* ${}_k{>\!\!\circ}_a : a^{(k)} \to a$ gives the dual retiming process. As in the multiplicative space-based case, it may be somehow problematic. It requires us to identify repeated job requests of type a, and it is undefined if they are not equal.

This model is less standard than the above two models for behaviour or space. It is described in more detail in Chapter 12. Carefully developed, it may be used to recast process algebra in a relational framework: processes are to be considered as "job transformers."

EXERCISES AND PROBLEMS (SEC. 1.6)

1. Show that $|E|$ in Example 1.9 indeed computes the binomial coefficients. (10)
2. Take other simple flowchart programs and check their correctness using the AddRel model. (- -)
3. Show the circuit in Example 1.10 correctly computes binary multiplication. (10)
4. Describe, in network algebra terms, circuits for binary multiplication of numbers of arbitrary length. (12)

1.7 Feedback, iteration, and repetition

This section provides a comparison of three algebraic looping operations: Kleene's repetition (star), Elgot's iteration (dagger), and the already familiar feedback (up arrow). (1) First, we briefly present algebraic and matrix theories. They give a different acyclic basis for iteration and repetition, respectively. (2) Next, we give a translation between iteration and feedback within the algebraic theory setting; moreover, equivalent axiomatic systems for BNAs over algebraic theories are presented. (3) Then, we add a translation between iteration and repetition within the matrix theory setting; moreover, equivalent axiomatic systems for BNAs over matrix theories are presented. (4) Finally, some advantages of the use of feedback over the use of iteration or repetition are emphasized.

Looping operations. To start with, (right-hand-side) feedback $\uparrow^a$ maps arrows from $T(a \star b, a \star c)$ to $T(b, c)$; its intended meaning is to connect the leftmost output of sort a to the leftmost input of sort a and (important!) after such a connection *both connected input or output pins are hidden.*

Iteration $\dagger$ maps arrows from $T(a, a \star b)$ to $T(a, b)$; its intended meaning is to connect the leftmost output of sort a to the input and, very important again, after such a connection *the connected output pins are hidden, while the input ones are kept visible.*

Repetition $*$ maps arrows from $T(a, a)$ to $T(a, a)$; its intended meaning is to connect the output to the input and, very important again, after such a connection *both the input and the output pins are kept visible.*

Two simple consequences follow from this presentation. First, iteration uses an implicit identification constant to provide further access to the input of an arrow; actually, $f^\dagger = \uparrow^a (\vee_a f)$. Repetition, even worse, implicitly requires both ramification and identification constants for keeping its input/output pins visible; see below for its definition in terms of feedback.

Second, it's only feedback which agrees with the usual relational composition. Indeed, in $f \cdot g$ there is an implicit connection between the output of f and the input of g. In agreement with the feedback's definition all these connected interfaces are hidden; for instance the output of f is not preserved in the output of the composite. (Polycategories of Lambek consider such a possibility.) As a side-effect, composition is easily expressed in terms of feedback, e.g., as $f \cdot g = [(f \star g)(21)] \uparrow$.

Versions of these looping operations, with algebraic counterparts, have been encountered in many places in the literature. A few more examples are:

$f^* g = \wedge^1 \cdot (\mathsf{I}_1 \star \uparrow^1 (\vee_1 \cdot f \cdot \wedge^1)) \cdot \vee_1 \cdot g$ — binary star in [Kle56];
$f^* = \wedge^1 \cdot (\mathsf{I}_1 \star (\vee_1 \cdot f \cdot \wedge^1) \uparrow^1) \cdot \vee_1, \quad f : 1 \to 1$ — unary star in [CEW58];
$\mu f = (f \cdot \wedge^m) \uparrow^m, \ f : n \star m \to m$ — "feedback" in [Bro93] (actually, a dual iteration);
reflexion — (a version of our standard feedback) in [Mil94].

Successive looping. Feedback turned out to express more naturally certain properties. Take, for instance a composed system that may be decomposed in pieces. Then one may apply the looping operation once, or successively on pieces. A fundamental property says the same result is obtained, provided the method is properly applied. It may be expressed in the following ways using repetition, iteration, and feedback, respectively. (The formulae to follow, describing this property, are parts of the axiomatic systems presented in Tables 1.2–1.4.)

For repetition the formula, known as the "matrix formula," is

$$\text{R3.2} \quad \begin{bmatrix} f & g \\ h & i \end{bmatrix}^* = \begin{bmatrix} f^* + f^* \cdot g \cdot (h \cdot f^* \cdot g + i)^* \cdot h \cdot f^* & f^* \cdot g \cdot (h \cdot f^* \cdot g + i)^* \\ (h \cdot f^* \cdot g + i)^* \cdot h \cdot f^* & (h \cdot f^* \cdot g + i)^* \end{bmatrix}$$

♡ The right-hand-side formula may be seen as given all paths in an automaton. Suppose we have an automaton with two states 1 and 2 and transitions represented by $1 = f \cdot 1 + g \cdot 2$; $2 = h \cdot 1 + i \cdot 2$. All paths from 2 to 2 are of the following shape. Repeat the following two typical paths from 2 to 2: either i (go directly from 2 to 2 via i), or $h \cdot f^* \cdot g$ (indirectly, go with h in 1, repeat f there, and return in 2 via g). The result is specified by the lower-right term of the matrix above. The other terms are justified in a similar way. ◊

For iteration the formula, known as the "pairing axiom," is

I3.2 $\langle f, g\rangle^\dagger = \langle f^\dagger \cdot \langle (g \cdot \langle f^\dagger, \mathsf{I}_{b\star c}\rangle)^\dagger, \mathsf{I}_c\rangle, (g \cdot \langle f^\dagger, \mathsf{I}_{b\star c}\rangle)^\dagger\rangle$

♡ It corresponds to the following method for solving systems of equations. Suppose one has the system (1) $x = f(x, y, z)$; (2) $y = g(x, y, z)$. The right-hand-side formula specifies the result obtained by the following procedure: first, solve equation (1) leading to, say, $x = h(y, z)$; replace this in (2) to get $y = g(h(y, z), y, z)$; solve this to get the final solution for y, say, $y = i(z)$; replace that in $x = h(y, z)$ to get the final solution for x. ◊

Finally, for repetition the formula is

F1.2 $\uparrow^b\uparrow^a f = \uparrow^{a\star b} f$

♡ This is by far the simplest way to describe this property. ◊

(Co)algebraic and matrix theories. We assume the reader has some familiarity with the use of symocats (symmetric strict monoidal categories), (co)algebraic theories, and matrix theories. They are presented in detail in Chapters 2 and 4.

Recall that a symocat T is defined by the acyclic BNA axioms B1–B9.

A symocat T is a *coalgebraic theory* if some constants $\top_a \in T(\epsilon, a)$ and $\vee_a \in T(a \star a, a)$ are given that obey certain axioms which are not described here; see Chapter 4. The key property of coalgebraic theories is that a tupling operation $\langle\ ,\ \rangle : T(a, c) \times T(b, c) \to T(a \star b, c)$ may be introduced as

$$\langle f, g\rangle = (f \star g) \cdot \vee_c$$

Then, there is a unique way to write an arrow $h \in T(a \star b, c)$ as $\langle f, g\rangle$, with f, g as above.

A coalgebraic theory T is a *matrix theory* if some further constants $\bot^a \in T(a, \epsilon)$ and $\wedge^a \in T(a, a \star a)$ are given and certain additional axioms are valid; see again Chapter 4. In a matrix theory T a target-tupling operation $[\ ,\ ] : T(a, b) \times T(a, c) \to T(a, b \star c)$ may be introduced as

$$[f, g] = \wedge^a \cdot (f \star g)$$

The key property of matrix theories T is that one may define a matrix-building operation which maps a quadruple of morphisms $f : a \to c, g : a \to d, h : b \to c$ and $i : b \to d$ into $\begin{bmatrix} f & g \\ h & i \end{bmatrix} \in T(a \star b, c \star d)$ defined to be either

$$\langle [f, g], [h, i]\rangle \quad \text{or} \quad [\langle f, h\rangle, \langle g, i\rangle]$$

Then, there is a unique way to write an arrow j in $T(a\star b, c\star d)$ as $j = \begin{bmatrix} f & g \\ h & i \end{bmatrix}$ with f, g, h and i as above.

♡ In a matrix theory T we may also define a sum operation $+ : T(a,b) \times T(a,b) \to T(a,b)$ and some constants $0^a_b \in T(a,b)$ as follows

$$f + g = \wedge^a \cdot (f \star g) \cdot \vee_b \qquad 0^a_b = \bot^a \cdot \top_b$$

It follows that $(T(a,a), +, \cdot, 0^a_a, \mathsf{I}_a)$ is a semiring. Actually, a matrix theory over the monoid of natural numbers may also be presented as the theory of matrices over a semiring, see [Elg76a]. ♢

Axioms for looping operations. (1) To fit easier with iteration, we are using here the left-hand-side feedback. For a symocat T and a (left-hand-side) feedback operation $\uparrow^a$: $T(a \star b, a \star c) \to T(b, c)$ we consider two sets of axioms: F1 (consisting of 7 axioms F1.1–7) and F2 (consisting of 5 axioms F2.1–5); actually, for F1, a coalgebraic theory structure on T is required.

(2) Next, suppose T is a coalgebraic theory. Let $\dagger : T(a, a \star b) \to T(a, b)$ be an iteration operation (the result is written: $f^\dagger$). Four systems of axioms I1–I4 involving iteration are given below.

(3) Finally, suppose that T is a matrix theory. Let $* : T(a,a) \to T(a,a)$ be a repetition operation (the result is written: f^*). Three systems of axioms R1–R3 involving repetition are given below.

(4) It is by now well-known that a BNA is a symocat in which a feedback operation is given fulfilling the axioms R1–7 in Table 1.1. A *BNA over a coalgebraic theory* (resp. *over a matrix theory*) is a coalgebraic theory (resp. a matrix theory) in which a feedback operation is given fulfilling the BNA axioms for feedback.

(5) BNAs are particular xy-flows, namely aa-flows. General xy-flows, to be used starting with Chapter 4, are defined using an enzymatic rule. Sometimes this property may be described using the other looping operations. Here are the corresponding definitions.

A morphism $y : a \to b$ is called $\uparrow$*-enzymatic* if for every $f : a \star c \to a \star d$ and $g : b \star c \to b \star d$ the equality $f \cdot (y \star \mathsf{I}_d) = (y \star \mathsf{I}_c) \cdot g$ implies $\uparrow^a f = \uparrow^b g$. It is called $\dagger$*-enzymatic* if for every $f : a \to a \star c$ and $g : b \to b \star c$ the equality $f \cdot (y \star \mathsf{I}_c) = y \cdot g$ implies $f^\dagger = y \cdot g^\dagger$. Finally, it is $*$*-enzymatic* if for every $f : a \to a$ and $g : b \to b$ the equality $f \cdot y = y \cdot g$ implies $f^* \cdot y = y \cdot g^*$.

Results: equivalent systems of axioms. The following axiomatic systems F1–F2, I1–I4 and R1–R3 are considered.

Results:

(1) The axioms for feedback in F1 are equivalent to the BNA axioms R1–7, when T is an arbitrary symocat. If T actually is a coalgebraic theory, then the axiomatic systems for feedback F1–F2 are equivalent, too.

(2) Natural translations between feedback and iteration may be defined; see the next subsection. The axiomatic systems F1–F2 and I1–I4 are equivalent,

Table 1.2. Axiomatic systems for feedback for BNAs over coalgebraic theories

F1.1	$\uparrow^a ({}^a\mathsf{X}^a) = \mathsf{I}_a$	F2.1	$\uparrow^a ({}^a\mathsf{X}^a \cdot (\mathsf{I}_a \star f)) = f$
F1.2	$\uparrow^b \uparrow^a f = \uparrow^{a \star b} f$	F2.2	$\uparrow^b \uparrow^a f = \uparrow^{a \star b} f$
F1.3	$\uparrow^{a\star b} g = \uparrow^{b \star a} f$ where	F2.3	$\uparrow^{a\star b} g = \uparrow^{b \star a} f$ where
	$g = ({}^a\mathsf{X}^b \star \mathsf{I}_c) \cdot f \cdot ({}^b\mathsf{X}^a \star \mathsf{I}_d)$		$g = ({}^a\mathsf{X}^b \star \mathsf{I}_c) \cdot f \cdot ({}^b\mathsf{X}^a \star \mathsf{I}_d)$
F1.4	$(\uparrow^a f) \cdot g = \uparrow^a (f \cdot (\mathsf{I}_a \star g))$	F2.4	$\uparrow^a (f \star \mathsf{T}_d) = \uparrow^a f \star \mathsf{T}_d$
F1.5	$g \cdot (\uparrow^a f) = \uparrow^a ((\mathsf{I}_a \star g) \cdot f)$	F2.5	$\uparrow^a \langle f, g\rangle = g \cdot \langle \uparrow^a \langle f, \mathsf{I}_a \star \mathsf{T}_c\rangle, \mathsf{I}_c\rangle$
F1.6	$\uparrow^a f \star g = \uparrow^a (f \star g)$		for $f : a \to a \star c, g : b \to a \star c$
F1.7	$\uparrow^a \mathsf{I}_a = \mathsf{I}_\epsilon$		

Table 1.3. Axiomatic systems for Căzănescu–Ungureanu theories, presented in terms of iteration

I1.1	$(f \cdot (\mathsf{V}_a \star \mathsf{I}_b))^\dagger = f^{\dagger\dagger}$	I2.1	$f \cdot \langle f^\dagger, \mathsf{I}_b\rangle = f^\dagger$
I1.2	$(f \cdot (g \star \mathsf{I}_c))^\dagger = f \cdot \langle (g \cdot f)^\dagger, \mathsf{I}_c\rangle$	I2.2	$(f \cdot (\mathsf{V}_a \star \mathsf{I}_b))^\dagger = f^{\dagger\dagger}$
	for $f : a \to b \star c, g : b \to a$	I2.3	$g \cdot (f \cdot (g \star \mathsf{I}_c))^\dagger = (g \cdot f)^\dagger$
I1.3	$(f \cdot (\mathsf{I}_a \star g))^\dagger = f^\dagger \cdot g$		for $f : a \to b \star c, g : b \to a$
		I2.4	$(f \cdot (\mathsf{I}_a \star g))^\dagger = f^\dagger \cdot g$
I3.1	$(\mathsf{T}_a \star \mathsf{I}_a)^\dagger = \mathsf{I}_a$	I4.1	$(\mathsf{T}_a \star f)^\dagger = f$
I3.2	$\langle f, g\rangle^\dagger = \langle f^\dagger \cdot \langle h, \mathsf{I}_c\rangle, h\rangle$	I4.2	$\langle f, g\rangle^\dagger = \langle f^\dagger \cdot \langle h, \mathsf{I}_c\rangle, h\rangle$
	where $h = (g \cdot \langle f^\dagger, \mathsf{I}_{b\star c}\rangle)^\dagger$		where $h = (g \cdot \langle f^\dagger, \mathsf{I}_{b\star c}\rangle)^\dagger$
I3.3	$({}^a\mathsf{X}^b \cdot f \cdot ({}^b\mathsf{X}^a \star \mathsf{I}_c))^\dagger = {}^a\mathsf{X}^b \cdot f^\dagger$	I4.3	$({}^a\mathsf{X}^b \cdot f \cdot ({}^b\mathsf{X}^a \star \mathsf{I}_c))^\dagger = {}^a\mathsf{X}^b \cdot f^\dagger$
I3.4	$(f \cdot (\mathsf{I}_a \star g))^\dagger = f^\dagger \cdot g$	I4.4	$(f \star \mathsf{T}_c)^\dagger = f^\dagger \star \mathsf{T}_c$

provided T is a coalgebraic theory and those translations between feedback and iteration are used; moreover, a morphism is $\uparrow$-enzymatic iff it is $\dagger$-enzymatic, for the corresponding iteration operator.

Table 1.4. Axiomatic systems for Conway theories, presented in terms of repetition

R1.1	$(f+g)^* = (f^* \cdot g)^* \cdot f^*$	R2.1	$f^* = \mathsf{I}_a + f \cdot f^*$
R1.2	$(f \cdot g)^* = \mathsf{I}_a + f \cdot (g \cdot f)^* \cdot g$	R2.2	$(f+g)^* = (f^* \cdot g)^* \cdot f^*$
		R2.3	$(f \cdot g)^* \cdot f = f \cdot (g \cdot f)^*$

R3.1	$(0_a^a)^* = \mathsf{I}_a$
R3.2	$\begin{bmatrix} f & g \\ h & i \end{bmatrix}^* = \begin{bmatrix} f^* + f^* \cdot g \cdot w \cdot h \cdot f^* & f^* \cdot g \cdot w \\ w \cdot h \cdot f^* & w \end{bmatrix}$
	where $w = (h \cdot f^* \cdot g + i)^*$
R3.3	$({}^a\mathsf{X}^b \cdot f \cdot {}^b\mathsf{X}^a)^* = {}^a\mathsf{X}^b \cdot f^* \cdot {}^b\mathsf{X}^a$

(3) Natural translations between iteration and repetition may be defined, as well; see next subsection. The axiomatic systems F1–F2, I1–I4, and R1–R3 are equivalent, provided T is a matrix theory and those translations between feedback, iteration, and repetition are used; moreover, a morphism is $\uparrow$-enzymatic iff it is $\dagger$-enzymatic iff it is $*$-enzymatic, for the corresponding iteration and repetition operators.

Hint for proofs. The proof of these equivalences may be found in Chapters 7 and 8. My advice is that the reader should try to produce his own proofs. That way he may have a deeper feeling on the role and the power of the involved axioms.

The proofs in Chapters 7 and 8 have the following structure

$$\begin{array}{ccccccc} & & & & \mathrm{F1} & & \mathrm{F2}\\ & & & & | & & | \\ \mathrm{I2} & — & \mathrm{I1} & — & \mathrm{I3} & — & \mathrm{I4} \\ | & & | & & | & & \\ \mathrm{R2} & & \mathrm{R1} & & \mathrm{R3} & & \end{array}$$

. The lines show the directly proved equivalences. Except for the equivalences between the second and the third line (counted from top to bottom), which require a matrix theory setting, the other proofs use the coalgebraic theory rules only. The equivalence of F1 with the standard BNA axioms R1–7, in the general symocat setting, may be proved directly, but in the book it follows from the equivalence between BNAs and LR-flows over $\mathbb{B}\mathrm{i}$ proved in Appendix A.

We describe below the translations and the results in more detail.

Translations between looping operations. Let T be a coalgebraic theory, $\mathbf{Fd}(T)$ the set of all left-hand-side feedback operations defined on T, and $\mathbf{It}(T)$ the set of all iteration operations defined on T.

♡ Notice that a feedback, iteration, or repetition operation actually is a family of such operations indexed by the elements of the interface monoid. ◇

We define two translations $\alpha : \mathbf{Fd}(T) \to \mathbf{It}(T)$ and $\beta : \mathbf{It}(T) \to \mathbf{Fd}(T)$ as follows:

- $\alpha(\uparrow)$ is the iteration operation that maps $f \in T(a, a \star b)$ to $\uparrow^a \langle f, \mathsf{l}_a \star \top_b \rangle$;

- $\beta(\dagger)$ is the feedback operation that maps $f = \langle f_1, f_2 \rangle \in T(a \star b, a \star c)$ to $f_2 \langle f_1^\dagger, \mathsf{l}_c \rangle$, where $f_1 : a \to a \star c$ and $f_2 : b \to a \star c$.

Let:
- $\mathbf{Fd}_i(T)$ be the subset of feedback operations in $\mathbf{Fd}(T)$ that obey F2.5;
- $\mathbf{Fd}_r(T)$ be the subset of feedback operations in $\mathbf{Fd}(T)$ that obey F1.4–6;
- $\mathbf{It}_r(T)$ be the subset of iteration operations in $\mathbf{It}(T)$ that obey I3.4.

Finally, let us consider the following restrictions of this translations: : $\mathbf{Fd}_i(T) \xrightarrow{\alpha_i} \mathbf{It}(T) \xrightarrow{\beta_i} \mathbf{Fd}_i(T)$; $\mathbf{Fd}_r(T) \xrightarrow{\alpha_r} \mathbf{It}_r(T) \xrightarrow{\beta_r} \mathbf{Fd}_r(T)$.

Theorem 1.11. *Suppose T is a coalgebraic theory. Then:*
a) The restrictions $\alpha_i, \beta_i, \alpha_r$ and β_r are (totally defined) bijective functions. Moreover, α_i is the inverse of β_i and α_r of β_r.
b) For $k \in [4]$, $\dagger$ satisfies I4.k *iff $\beta(\dagger)$ satisfies* F2.k.
c) For $k \in [4]$, $\dagger$ satisfies I3.k *iff $\beta(\dagger)$ satisfies* F1.k.
d) y is $\dagger$-functorial iff y is $\beta(\dagger)$-functorial.

Let T be a matrix theory and $\mathbf{Rp}(T)$ be the set of all the repetition operations defined on T. We use the following translations $\sigma : \mathbf{It}(T) \to \mathbf{Rp}(T)$ and $\tau : \mathbf{Rp}(T) \to \mathbf{It}(T)$ defined by

- $\sigma(\dagger)$ is the repetition operation that maps $f \in T(a, a)$ to $[f, \mathsf{l}_a]^\dagger$;

- $\tau(*)$ is the iteration operation that maps $f = [f_1, f_2] \in T(a, a \star b)$ to $f_1^* \cdot f_2$, where $f_1 : a \to a$ and $f_2 : a \to b$.

Finally, let us consider the restrictions $\mathbf{It}_r(T) \xrightarrow{\sigma_r} \mathbf{Rp}(T) \xrightarrow{\tau_r} \mathbf{It}_r(T)$ induced by σ and τ.

Theorem 1.12. *Suppose T is a matrix theory. Then:*
a) The restrictions σ_r and τ_r are (totally defined) bijective functions. Moreover, σ_r is the inverse of τ_r.
b) For $k \in [3]$, $*$ satisfies R3.k *iff $\tau(*)$ satisfies* I3.k.
c) For $k \in [3]$, $*$ satisfies R2.k *iff $\tau(*)$ satisfies* I2.k.
d) For $k \in [2]$, $*$ satisfies R1.k *iff $\tau(*)$ satisfies* I1.k.
e) y is $*$-functorial iff y is $\tau(*)$-functorial.

Some conclusions. A few more advantages of the use of feedback over the use of iteration or repetition deserve to be singled out here.

First, the proper acyclic context for the use of feedback is a symocat, for iteration a coalgebraic theory, and for repetition a matrix theory. Hence the feedback may be used in a more general context than iteration or repetition.

Second, in the context of matrix theories there is a bijection between repetition operations and those iteration operations *obeying* axiom I3.4. Hence *iteration is more expressive than repetition* as it displays some properties of looping operations that are hidden by repetition. Similarly, in the context of coalgebraic theories there is a bijection between iteration operations and those feedback operations *obeying* axiom F2.5. Hence *feedback is more expressive than iteration* as it displays some properties of looping operations that are hidden by iteration.

Third, as we already noted, some properties are easier to express in terms of feedback, e.g., the property expressed by the "matrix formula" R3.2 or by the "pairing axiom" I3.2 is expressed in terms of feedback as F1.2.

Finally, the comparison of the axiom systems that we have displayed in this section enlightens the role of some misterious axioms as R1.1–2 used in the setting of regular algebras (resp. of the axioms I3.1–4 used in the setting of iteration theories). Namely, they provide a combination of complete axioms for acyclic behaviour given by the matrix theory rules (resp. coalgebraic theory rules) with complete axioms for cyclic flowgraphs modulo graph-isomorphism only. Clearly, something is missing here: we need a device that allows us to cope with the behaviour of cyclic processes. In our study, such a device is provided by the enzymatic rule.

Exercises and Problems (sec. 1.7)

1. Use network algebra terms to describe labeled, weighted directed graphs. (*Hint*: These are graphs with labeled edges and nodes; each edge has a label f of type $1 \xrightarrow{f} 1$ and a weight $w(f)$; each node has a label X of type $m \xrightarrow{X} n$, for appropriate $m, n \geq 1$, and it is to be interpreted as $\vee_1^m \cdot \wedge_n^1$.) (09)

2. The *shortest paths* problem requires we find the paths with the smallest weight connecting two nodes. Show that the shortest path problem may be algebraically solved within a Conway theory which satisfies the additional axioms:

1. $\wedge_2(f \star g)\vee^2 = f$, provided $w(f) \leq w(g)$;
2. $(\vee^m \cdot \wedge_n(\mathsf{I}_{n-1} \star f)) \uparrow^1 = (\vee^m \cdot \wedge_n(\mathsf{I}_{n-1} \star \wedge_0^1 \cdot \vee_1^0)) \uparrow^1$, provided $w(f) \geq 0$. (15)

3. Apply the method suggested by the above comments to find the shortest paths from x_4 to x_1 in the network
$$[x_4(a_5 \star a_7 x_5(a_6 \star a_8))] \uparrow^1 [(\mathsf{I}_1 \star {}^1\mathsf{X}^1)(x_2(a_1 \star a_3) \star \mathsf{I}_1)(\mathsf{I}_1 \star x_3(a_2 \star a_4))] \uparrow^1 \cdot x_1$$
where $x_4 = x_5 = \wedge_2^1$, $x_2 = x_3 = \vee_1^2 \cdot \wedge_2^1$, $x_1 = \vee_1^2 \cdot \wedge_0^1$ and the weights of $a_1, \ldots, a_9$ are 1,4,2,1,8,2,2,4, respectively. (11)

1.8 Network behaviours as xy-flows

A brief account on network behaviour algebras in presented here. The key axioms of Elgot theories (as a model for the input behaviour), and Park theories (which do the same for deterministic input–output behaviour) are illustrated.

A brief account on network behaviour algebras is presented here. Not so much is included because this part may be studied in other frameworks as well, e.g., using axiomatized iteration or repetition. It is not typical to the NA approach. Actually, there are rather deep results in this setting. We only want to show that an easy extension of the NA graph-isomorphism setting is enough to cope with such problems. What we need is to add appropriate strong and enzymatic rules.

The basic algebraic structures used in network algebra approach to model network behaviours are xy-flows. They will be described in full detail in the forthcoming chapters. Here is one example to illustrate the enzymatic rule. Together with the weak and strong commuting rules, it gives the troika which discriminate between various algebras modeling network behaviour.

xy-flow axioms, recalled. Besides the BNA axioms, the following critical axioms for the branching constants of type xy are used:

– commutation of the branching constants to the other elements: $f \cdot \wedge_k^b = \wedge_k^a \cdot (kf)$, $\vee_a^k \cdot f = (kf) \cdot \vee_b^k$, where $f : a \to b$; this is used in two ways: the weak variant, where f's are certain abstract relations and the strong variant, where f's are arbitrary.

– enzymatic axiom for the looping operation: $f \cdot (\mathsf{I}_b \star z) = (\mathsf{I}_a \star z) \cdot g$ implies $f \uparrow^c = g \uparrow^d$, where $z : c \to d$ is an abstract relation and $f : a \star c \to b \star c$, $g : a \star d \to b \star d$ are arbitrary elements. ◇

Let $x : s \to 3s$, $y : s \to s$, $z : 2s \to s$; suppose $z = (z_1 \star z_2)\vee_s^2$, with $z_1, z_2 : s \to s$. Consider the following network

$$F = [\vee_s^2 x(\mathsf{I}_s \star y \star x)] \uparrow^s \cdot [(\mathsf{I}_{3s} \star y \star x \star y)(1_s 3_s 2_s 3_s 2_s 5_s 4_s 4_s)] \uparrow^{2s} \cdot [((\mathsf{I}_s \star \vee_s^2 y)z) \uparrow^s \star \mathsf{I}_s]$$

Its unfolding tree, describing the input behaviour, may be represented by the term

$\underline{x(z_1(yz_2)(yz_2)..,y,x((yz_2)(yz_2)..,y,}\ \underline{x(z_1(yz_2)(yz_2)..,y,x((yz_2)(yz_2)..,y),}\bullet\bullet))))$

where $\bullet\bullet$ means that the underlined term is repeated forever.

In an Elgot theory, F may be reduced to

$$G = [\mathsf{V}_s^2 x(\mathsf{I}_{2s} \star x)] \uparrow^s \cdot(1_s 3_s 2_s 3_s) \cdot ([(\mathsf{I}_s \star \mathsf{V}_s^2 y)z] \uparrow^s \star y)$$

Elgot theory uses strong commutation and enzymatic rules for functions. The cells $x \star y$ in the middle term of F are not accessible. They may be eliminated with the enzymatic rule for injections as follows:

Deleting nonaccessible cells. Notice that:
$(\mathsf{I}_{4s} \star \underline{\mathsf{V}_{2s}^0})(\mathsf{I}_{3s} \star y \star x \star y)(1_s 3_s 2_s 3_s 2_s 5_s 4_s 4_s)$
$= (\textit{strong rule on cells})(\mathsf{I}_{3s} \star y)(\mathsf{I}_{7s} \star \mathsf{V}_{4s}^0)(1_s 3_s 2_s 3_s 2_s 5_s 4_s 4_s)$
$= (\mathsf{I}_{3s} \star y)(1_s 3_s 2_s 3_s)(\mathsf{I}_{3s} \star \underline{\mathsf{V}_{2s}^0})$
By the enzymatic rule for the underlined injective enzyme V_{2s}^0 it follows that
$[(\mathsf{I}_{3s} \star y \star x \star y)(1_s 3_s 2_s 3_s 2_s 5_s 4_s 4_s)] \uparrow^{2s} = (\mathsf{I}_{3s} \star y)(1_s 3_s 2_s 3_s) \uparrow^{0s}$
Actually, the last term is equal to one without feedback, i.e., $(\mathsf{I}_{3s} \star y)(1_s 3_s 2_s 3_s)$. ◇

Identifying cells. Now, back to F, moving through the output and identifying with the strong commuting rule both occurences of y one gets G. ◇

No more signifiant transformations of G are possible in an Elgot theory, but we will return soon to this setting. One may easily see that the same unfolding tree is obtained for G.

In a Park theory, F may be further reduced to

$$H = (\mathsf{V}_s^2 x) \uparrow^s \cdot(\wedge_0^s \star y)$$

The novelty of this setting is the strong commutation and enzymatic rules for the converse of injections. These allow us to delete noncoaccessible cells. This may be applied to the last term in G as follows:

Deleting noncoaccessible cells. Notice that
$(\mathsf{I}_s \star \mathsf{V}_s^2 y)z\underline{\wedge_0^s}$
$= (\textit{strong rule on cells})(\mathsf{I}_s \star \mathsf{V}_s^2 y)\wedge_0^{2s}$
$= (\textit{strong rule on cells})\wedge_0^{3s}$
$= (\mathsf{I}_{2s} \star \underline{\wedge_0^s})\wedge_0^{2s}$
By the enzymatic rule for the underlined enzyme $\underline{\wedge_0^s}$ (converse of injection) one gets
$[(\mathsf{I}_s \star \mathsf{V}_s^2 y)z] \uparrow^s = \wedge_0^{2s} \uparrow^{0s} (= \wedge_0^{2s})$ ◇

Hence G is equivalent to $I = [\mathsf{V}_s^2 x(\mathsf{I}_{2s} \star x)] \uparrow^s \cdot([]_s 1_s []_s 1_s)y$.

By a graph-isomorphism transformation (actually to get the normal form for $[\mathsf{V}_s^2 x(\mathsf{I}_{2s} \star x)] \uparrow^s \cdot([]_s 1_s []_s 1_s))$, I is equal to $[(\mathsf{I}_1 \star x \star x)(2_s []_s 1_s 3_s []_s 1_s 2_s)] \uparrow^{2s}$.

We may return to the Elgot theory setting (or stay in the current extended Park setting) to apply the enzymatic rule for surjections in order to identify both occurrences of x:

Identifying cells. One may see that
$(\mathsf{I}_1 \star x \star x)(2_s[]_s 1_s 3_s []_s 1_s 2_s)(\mathsf{I}_s \star \underline{\mathsf{V}^2_s})$
$= (\mathsf{I}_1 \star x \star x)(\mathsf{I}_1 \star \mathsf{V}^2_{3s})(2_s[]_s 1_s 2_s)$
$= (\textit{strong rule on cells})(\mathsf{I}_1 \star \underline{\mathsf{V}^2_s})(\mathsf{I}_1 \star x)(2_s[]_s 1_s 2_s)$
By the enzymatic rule for the underlined surjective enzyme $\underline{\mathsf{V}^2_s}$ it follows that
$[(\mathsf{I}_1 \star x \star x)(2_s[]_s 1_s 3_s []_s 1_s 2_s)] \uparrow^{2s} = [(\mathsf{I}_1 \star x)(2_s[]_s 1_s 2_s)] \uparrow^s$ ◊

This shows $I = [(\mathsf{I}_1 \star x)(2_s[]_s 1_s 2_s)] \uparrow^s \cdot y$. A graph-isomorphism transformation shows that this final term is equal to $(\mathsf{V}^2_s x) \uparrow^s \cdot (\wedge^s_0 \star y)$; hence we are done.

Notice that in this Park theory, after the unfolding, one more transformation is done on the resulting trees: the subtrees without outputs are identified with the empty tree. Hence the deterministic IO behaviour of F is given by the tree represented by the term

$$\underline{x(\bot, y,}\ \underline{x(\bot, y,} \bullet\bullet))$$

where, as before, •• means that the underlined term is repeated forever.

♡ Some transformations are more easily done using certain more general results. For instance, it is known that the strong rules extend from cells to networks, hence non-(co)accessible subnetworks may be eliminated at once in the corresponding steps; however, our aim here was to use the basic enzymatic rule and the strong rule on cells, only. ◊

BNA, iteration theories, regular algebra. In order to get an algebraic theory of computation one needs an axiomatic looping operation. This may be repetition, iteration, or feedback. The proper acyclic context for repetition is a matrix theory, for iteration a (co)algebraic theory and for feedback a symocat.

A *BNA* is a structure which satisfies all the identities (written in terms of juxtaposition, composition, feedback, and constants $\mathsf{I}_a, {}^a\mathsf{X}^b$) which are valid for networks. This model is more related with the algorithms themselves than with their behaviours. Regular algebras and iteration theories, to be briefly introduced below, occur as particular BNAs.

An *iteration theory*, cf. Bloom, Elgot and Wright [BEW80a], is a structure which satisfies all the identities (written in terms of tupling, composition, iteration, and certain constants $\mathsf{I}_a, 0_a, \pi^a_i$) which are valid in the theories of regular trees. An axiomatization for iteration theories was found by Ésik [Ési80]. This algebra is intended as a model for the input behaviour of deterministic computation. See Chapters 5 and 7 for more on this.

A *regular algebra*, cf., e.g., Conway [Con71], is a structure which satisfies all the identities (written in terms of union, composition, repetition, and constants 0,1) which are valid in the algebras of regular events. This algebra is intended as a model for the input–output behaviour of nondeterministic computation. An axiomatization for regular algebra is given in Chapter 8; see also Chapter 6.

1.9 Mixed Network Algebra, MixNA

A proposal for a unified algebraic calculus for (concurrent, object-oriented) software systems is briefly sketched here; more may be found in Chapter 12.

A mixed network algebra may be developed combining the 3 faces of NA interpretations described in a previous section (1.6): control, space, and time. In this more complicated setting a software component has two multiplicative interfaces, one for spaces and one for time, and one additive for control.

Vertical composition along the component states is like in imperative programs. Horizontal composition along the process interaction interfaces (job streams) describes the components interaction as in an OO system.

Vertical feedback models the repetition of components invocation in time generating classes of similar memory instances – these are usual "states" (one arrives at the same point of a program, but at a different time). Horizontal feedback models the repetition of components invocation in space generating classes of similar job requirements – these are the usual OO programming "classes" (one arrives at the same point of a program, but at a different space, hence a new object of the class is considered).

For spatial parallelism, a true parallel data-flow like model comes along with the horizontal juxtaposition operation. For temporal parallelism (piped processes), a true parallel job-flow like model comes along with the vertical juxtaposition operation.

Multiplicative operators distribute over the additive ones, while space and time are dual features. These may be used to keep the algebraic laws at a reasonable complexity size.

The above is more like a dream. The model seems to be very interesting, but a lot of work has to be done to get enough results for a calculus for concurrent object-oriented systems similar to the calculus provided by the (unmixed) network algebra for more modest settings.

More information on the proposed model and a few partial results are inserted in the final chapter of the book. The results are rather positive, keeping the above view/project still as a valid option for a unified algebraic calculus for software systems.

Comments, problems, bibliographic remarks

Comments. A few general comments on certain particular features of the present algebraic framework may be useful.

Polynomials vs. flownomials. The calculus of polynomials is a calculus well suited to model the *quantitative* feature of the world. It is based on quantity, size, number. The basic operation is the addition of numbers. This and the derived operations (subtraction, multiplication, etc.) show how the world quantitatively changes.

The target of the calculus of flownomials is to model the *relational* features of the world. It is based on structure, skeleton, relation. It seems the basic operation is the connection of relations. This and the derived operation (for instance, juxtaposition, composition and feedback) show how the world structurally changes.

Notes. Most of the material used in the first half of this book was presented in draft form in [Şte91], [Şte94].

The representation of finite relations as enriched symocats used in the first section is taken from [CŞ91, CŞ94]; it will be described in detail in Chapter 3.

Basic Network Algebra was introduced in [Şte86b, Şte90]; it was originally called "biflow," and introduced without axioms for bijections; axioms for bijections have been described in [CŞ88a, CŞ88b] leading to the current presentation. The using of the "Basic Network Algebra" name for this algebra was suggested by Jan Bergstra in 1994. Trace monoidal categories are introduced in [JSV96]; their usefulness for computer science problems, especially related to distributed computation, were emphasized in [Abr96].

The normal form flownomial expressions are introduced in [Şte86b, Şte90]. In a general form, flownomial expressions are presented in [CŞ90a]. Their presentation here follows the one in Chapter B, Sec. 1–2 of [Şte91]. Abstract arrows for connecting flow-graphs are used in [CU82]; they were used in all of our subsequent papers on flowchart schemes and flownomials, including [Şte87a], [Şte87b], [CŞ88a], [CŞ90a], [CŞ92].

The Network Algebra framework was presented in [CŞ90a]; some technical details were deferred to [CŞ92].

The control or space based network algebra models are classical, but may be presented in a slightly different format here; the time-based model seems to be new.

The section comparing looping operations is based on [CŞ95]. The first axiomatic looping operation was "Kleene's star" introduced in [Kle56] and used as a key operation of regular algebras, see [Con71]. Another axiomatic looping operation is "Elgot dagger" introduced in [Elg75] as the key operation of iterative algebraic theories.

Iterative and iteration theories are presented in detail in [BÉ93b]. Căzănescu–Ungureanu theories were introduced in [CU82] under the name of "algebraic theories with iterate"; in Bloom and Ésik's book [BÉ93b], depending on the form of the looping operations, they are called either "Conway theories" or "Conway feedback theories." Conway theories are implicitly present in [Con71]: they are obtained from the classical axioms C1–C14 in [Con71], dropping C14; in [BÉ93b] they are called "Conway matrix theories."

Matrix and algebraic theories are classical structures; see, e.g., [Law63], [Elg76a], [TWW79]. In its full generality, the enzymatic rule for describing network or flowchart behaviour was introduced in [AM80], under the "functoriality axiom" name.

Mixed network algebra models are described in more detail in Chapter 12; their study had been started in [Şte97a], [Şte97b], [Şte98b].

Example 1.4 is from [BMŞ97]. Example 1.5 describing the splitting of fixed-point equations in graph-isomorphism transformations and cell duplications is based on [Şte87a]. A useful book for information on computer networks is [Tan96]; the examples of Exercise 1.3.2 are from that book. Sorting networks, such as the ones used in Exercise 1.3.3, may be found in [Knu73]. Matrices are often used for solving shortest paths problems, see, e.g., [Tom72]; the result in Exercise 1.7. 2 seems to be new.

Part II

Relations, flownomials, and abstract networks

2. Networks modulo graph isomorphism

> A law is valuable not because it is law, but because there is right in it.
>
> Henry Ward Beecher

This chapter introduces the kernel of the network algebra, i.e., an algebraic theory of networks modulo graph isomorphism.

We start with a technical result related to algebraic presentations for bijections. It is proved that bijections are naturally mapped in arbitrary BNAs.

Then the basic model $[X,T]_{a\alpha}$ is introduced: it consists of isomorphic normal form flownomial expressions (or shortly, nf-pairs) with cells in X and connections in T. Appropriate BNA operations on nf-pairs are introduced.

A concrete network may be identified with a class of isomorphic nf-pairs, provided $T \subseteq \mathbb{R}\mathrm{el}_S$. Such a property is taken as *the definition of networks* in the general abstract setting. Hence elements in $[X,T]_{a\alpha}$ represent networks. As we will see, this implies that *flownomial expressions over X and T modulo BNA axioms represent networks.*

Two fundamental results are proved showing that BNA axioms are correct for networks modulo graph isomorphism. (1) **Soundness theorem:** if T is a BNA, then $[X,T]_{a\alpha}$ is a BNA. Therefore if the axioms are sound on the connecting relations, then they are sound on isomorphic networks; and (2) **Completeness theorem:** if two flownomial expressions have isomorphic normal forms, then their equality is deductible from the BNA axioms.

The proof of completeness is rather short and I urge the reader to follow it. It is similar to the deduction of the rules for defining ring operations on normal form polynomial expressions from the ring axioms.

The same similarity with polynomials is present in the soundness part. The basic fact is the preservation of the BNA structure by passing from connecting relations to flownomials. While the proof is not complicated, it is fairly long; the interested reader may try to give a direct proof for this result. Our proof is somewhat simpler, but it uses a weaker LR-flow presentation for BNAs. It is presented in Appendix B; Appendix A shows the equivalence of the BNA and LR-flow axioms. (In a LR-flow general composition and transpositions are left out and only particular compositions with bijections are used.)

2.1 Symocats

In this section *symocats* are introduced. This is the restriction of the BNA structure to the acyclic setting. Then we briefly describe two symocat structures on relations $\mathsf{AddRel}(D)$ and $\mathsf{MultRel}(D)$.

"Symocat" is a shorthand name for symmetric strict monoidal category of MacLane, cf. [Lan71]; they also occurred in [Hot65] and subsequent papers under the name of x-categories.

A bijection is a relation, hence it may be represented as a 0-1 matrix. While this observation has important applications, it also has some limitations. The matrix theory rules are too strong. So the aim is to find *weaker* algebraic structures in which bijections, or more generally relations, may be mapped. Symocats do this job for bijections; enriched with branching constants they do the job for other relations, too.

Why not matrices?. An appropriate algebraic structure for modeling matrices is the semiring structure where one may freely copy or discard elements using distributivity or zero-laws: $x(y+z) = xy + xz$, $0x = 0$, etc. However, in many important cases one cannot freely copy or discard network cells. As we have seen in the previous chapter and we will see quite often in the coming chapters, in such cases a graph-isomorphism equivalence on networks is more appropriate to provide a general and useful algebraic framework. The acyclic structure associated with the graph-isomorphism framework turned out to coincide with Mac Lane's symocats. ◇

Definition 2.1. (1) A *category* is a structure $(B, \cdot, \mathsf{I}_a)$ given by a class of *objects* $\mathcal{O}(B)$, sets of *arrows* $B(a,b)$ with a,b objects, *identities* $\mathsf{I}_a \in B(a,a)$, and *composition* $\cdot : B(a,b) \times B(b,c) \to B(a,c)$ satisfying axioms B3–B4 in Table 2.1.

(2) A category $(B, \cdot, \mathsf{I}_a)$ is *strict monoidal* if: the objects of B form a monoid $(\mathcal{O}(B), \star, \epsilon)$; a monoidal operation on arrows is given (also denoted by "$\star$"): $\star : B(a,b) \times B(c,d) \to B(a \star c, b \star d)$; and axioms B1–B2 and B5–B6 in Table 2.1 hold.

(3) A *symocat* (symmetric strict monoidal category) $(B, \star, \cdot, \mathsf{I}, \mathsf{X})$ is a strict monoidal category enriched with *(block) transposition* constants ${}^a\mathsf{X}^b \in B(a \star b, b \star a)$ for a,b objects in B and such that the remaining axioms B7–B9 in Table 2.1 hold.

(4) A *morphism between symocats* $H : B \to B'$ is a functor whose restriction to objects is a monoid morphism and which commutes with the monoidal operation on morphisms and constants ${}^a\mathsf{X}^b$.

♡ To be explicit, a symocat morphism $H : B \to B'$ is given by a morphism of monoids $h : \mathcal{O}b(B) \to \mathcal{O}b(B')$ and a family of applications $H : B(a,b) \to B'(h(a), h(b))$ obeying the commutation rules: $H(f \cdot g) = H(f) \cdot H(g)$; $H(f \star g) = H(f) \star H(g)$; $H(\mathsf{I}_a) = \mathsf{I}_{h(a)}$; $H({}^a\mathsf{X}^b) = {}^{h(a)}\mathsf{X}^{h(b)}$. ◇

Table 2.1. Axioms for symocats

B1	$f \star (g \star h) = (f \star g) \star h$	B6	$\mathsf{I}_a \star \mathsf{I}_b = \mathsf{I}_{a \star b}$
B2	$\mathsf{I}_\epsilon \star f = f = f \star \mathsf{I}_\epsilon$	B7	${}^a\mathsf{X}^b \cdot {}^b\mathsf{X}^a = \mathsf{I}_{a \star b}$
B3	$f \cdot (g \cdot h) = (f \cdot g) \cdot h$	B8	${}^a\mathsf{X}^{b \star c} = ({}^a\mathsf{X}^b \star \mathsf{I}_c) \cdot (\mathsf{I}_b \star {}^a\mathsf{X}^c)$
B4	$\mathsf{I}_a \cdot f = f = f \cdot \mathsf{I}_b$	B9	$(f \star g) \cdot {}^c\mathsf{X}^d = {}^a\mathsf{X}^b \cdot (g \star f)$
B5	$(f \star f') \cdot (g \star g') = f \cdot g \star f' \cdot g'$		for $f : a \to c, g : b \to d$

(5) Finally, the ground terms in a symocat (i.e., terms without variables) are called ***abstract bijections.***

The axioms for symocats are presented in Table 2.1; they were also presented in Table 1.1 as part of the BNA axioms.

EXERCISES AND PROBLEMS (SEC. 2.1)

1. A semi-symocat is the structure axiomatized by B1–4, B6–8, and weaker (non-commutative) forms for B5 and B9, i.e., in B5 f' or g should be an identity and in B9 f or g should be an identity. Show that a semi-symocat is a symocat iff $(\mathsf{I}_c \star f)(g \star \mathsf{I}_d) = g \star f$ for $f : a \to b,\ g : c \to d$. (15)

2.2 Bijections in symocats

The main result of this section shows that the sorted finite bijections form a free symocat, hence they are naturally mapped in arbitrary symocats.

Symmetries play a fundamental role in many branches of mathematics. There are well-known results on algebraic presentation of their group; see, e.g., [CM72]. However, these results are concentrated on bijections of a *fixed* length, say on a set with n elements. In the context of Network Algebra we need an algebraic presentation for *all* bijections at once, not for separate groups of bijections. The totality of bijections is what interests us.

Why block transpositions?. One design point that requires some explanation is the decision to use block transpositions instead of simple transpositions; in the one-sorted case the question is whether to use ${}^m\mathsf{X}^n$ for arbitrary numbers m, n, rather than ${}^1\mathsf{X}^1$ only. Simple transpositions suffice to generate all bijections; see Exercise 3 for a presentation of bijections by means of these constants only. Our choice is determined by the general graph-isomorphism equivalence. In this case network cells cannot be reduced to single-input/single-output cells. Hence block transpositions are necessary to permute them, see axiom B9. With simple transpositions such a property is more difficult to describe. ◇

Some technical observations. As Mifsud observed in [Mif96], B9 implies: ${}^{a}\mathsf{X}^{\epsilon} = {}^{a}\mathsf{X}^{\epsilon\star\epsilon} = ({}^{a}\mathsf{X}^{\epsilon}\star\mathsf{I}_{\epsilon})(\mathsf{I}_{\epsilon}\star{}^{a}\mathsf{X}^{\epsilon}) = {}^{a}\mathsf{X}^{\epsilon}.{}^{a}\mathsf{X}^{\epsilon}$; hence $\mathsf{I}_a = {}^{a}\mathsf{X}^{\epsilon}.{}^{\epsilon}\mathsf{X}^{a} = {}^{a}\mathsf{X}^{\epsilon}.{}^{a}\mathsf{X}^{\epsilon}.{}^{\epsilon}\mathsf{X}^{a} = {}^{a}\mathsf{X}^{\epsilon}$. Consequently, the following identity holds in a symocat:

$$\text{B10} \quad {}^{a}\mathsf{X}^{\epsilon} = \mathsf{I}_a$$

Using B7, it follows that the following identities hold in a symocat:

$$\text{B8}' \quad {}^{a\star b}\mathsf{X}^{c} = (\mathsf{I}_a \star {}^{b}\mathsf{X}^{c}) \cdot ({}^{a}\mathsf{X}^{c} \star \mathsf{I}_b)$$
$$\text{B10}' \quad {}^{\epsilon}\mathsf{X}^{a} = \mathsf{I}_a$$

Lemma 2.2. *The following identities also hold in a symocat:*

(1) ${}^{a}\mathsf{X}^{b\star c\star d} \cdot (\mathsf{I}_b \star {}^{c}\mathsf{X}^{d} \star \mathsf{I}_a) = (\mathsf{I}_{a\star b} \star {}^{c}\mathsf{X}^{d}) \cdot ({}^{a}\mathsf{X}^{b\star d} \star \mathsf{I}_c) \cdot (\mathsf{I}_{b\star d} \star {}^{a}\mathsf{X}^{c})$

(2) $({}^{a\star b}\mathsf{X}^{c} \star \mathsf{I}_d) \cdot (\mathsf{I}_{c\star a} \star {}^{b}\mathsf{X}^{d}) = (\mathsf{I}_{a\star b} \star {}^{c}\mathsf{X}^{d}) \cdot (\mathsf{I}_a \star {}^{b}\mathsf{X}^{d} \star \mathsf{I}_c) \cdot {}^{a\star d\star b}\mathsf{X}^{c}$

PROOF: (1) Using B9 and then B8, we get

$$\begin{aligned} {}^{a}\mathsf{X}^{b\star c\star d} \cdot ((\mathsf{I}_b \star {}^{c}\mathsf{X}^{d}) \star \mathsf{I}_a) &= (\mathsf{I}_a \star (\mathsf{I}_b \star {}^{c}\mathsf{X}^{d})) \cdot {}^{a}\mathsf{X}^{b\star d\star c} \\ &= (\mathsf{I}_{a\star b} \star {}^{c}\mathsf{X}^{d})\ ({}^{a}\mathsf{X}^{b\star d} \star \mathsf{I}_c)\ (\mathsf{I}_{b\star d} \star {}^{a}\mathsf{X}^{c}) \end{aligned}$$

(2) Using in turn B7, B8′, and B9, we get

$$\begin{aligned} ({}^{a\star b}\mathsf{X}^{c} \star \mathsf{I}_d)\ (\mathsf{I}_{c\star a} \star {}^{b}\mathsf{X}^{d}) &= (\mathsf{I}_{a\star b} \star {}^{c}\mathsf{X}^{d})\ (\mathsf{I}_{a\star b} \star {}^{d}\mathsf{X}^{c})\ ({}^{a\star b}\mathsf{X}^{c} \star \mathsf{I}_d)\ (\mathsf{I}_{c\star a} \star {}^{b}\mathsf{X}^{d}) \\ &= (\mathsf{I}_{a\star b} \star {}^{c}\mathsf{X}^{d}) \cdot {}^{a\star b\star d}\mathsf{X}^{c} \cdot (\mathsf{I}_c \star (\mathsf{I}_a \star {}^{b}\mathsf{X}^{d})) \\ &= (\mathsf{I}_{a\star b} \star {}^{c}\mathsf{X}^{d})\ (\mathsf{I}_a \star {}^{b}\mathsf{X}^{d} \star \mathsf{I}_c) \cdot {}^{a\star d\star b}\mathsf{X}^{c} \end{aligned}$$
□

Mapping bijections in symocats. As specified in the previous chapter, the elements in the free monoid $S^\star$ are denoted by $a = a_1 \star \ldots \star a_{|a|}$, where $|a|$ is the length of the word a and $a_i \in S$, for all $i \in [|a|]$. $\mathbb{B}\mathrm{i}_S$ denotes the family of finite S-sorted bijections in $\mathbb{R}\mathrm{el}_S$, i.e.,

$$\mathbb{B}\mathrm{i}_S(a,b) = \{f \subseteq [|a|] \times [|b|] : f \text{ bijective function and } (i,j) \in f \Rightarrow a_i = b_j\}$$

for $a, b \in S^\star$. The symocat operations on $\mathbb{B}\mathrm{i}_S$, inherited from the standard BNA structure on $\mathbb{R}\mathrm{el}_S$, were also defined in Chapter 1.

Theorem 2.3. *(mapping bijections in symocats)*
For every symocat $(B, \star, \cdot, \mathsf{I}_a, {}^{a}\mathsf{X}^{b})$ and for every morphism of monoids $h : S^\star \to \mathcal{O}b(B)$, there exists a unique morphism of symocats $H : \mathbb{B}\mathrm{i}_S \to B$ whose restriction to objects is h.

PROOF: (a) First, we provide an appropriate definition for H. On objects H coincides with h. It remains to find an appropriate definition of H on morphisms.

We prove by induction on n that there is a function H which maps a $g \in \mathbb{B}\mathrm{i}_S(c,b)$ with $|c| \leq n$ in $H(g) \in B(h(c),h(b))$ such that the following properties are satisfied:

1) $H(\mathsf{I}_a) = \mathsf{I}_{h(a)}$ if $|a| \leq n$
2) $H({}^a\mathsf{X}^b) = {}^{h(a)}\mathsf{X}^{h(b)}$ if $|a \star b| \leq n$
3) $H(f \cdot g) = H(f) \cdot H(g)$ if $f \in \mathbb{B}\mathrm{i}_S(a,b)$, $g \in \mathbb{B}\mathrm{i}_S(b,c)$ and $|a| \leq n$
4) $H(f \star g) = H(f) \star H(g)$ if $f \in \mathbb{B}\mathrm{i}_S(a,b)$, $g \in \mathbb{B}\mathrm{i}_S(c,d)$ and $|a \star c| \leq n$

For $n = 1$, define $H(\mathsf{I}_\epsilon) = \mathsf{I}_\epsilon$ and $H(\mathsf{I}_s) = \mathsf{I}_{h(s)}$, for $s \in S$. It is obvious that the above properties 1–4 hold.

Suppose that we have already defined H for all $g \in \mathbb{B}\mathrm{i}_S(c,b)$ with $|c| < n$ such that properties 1–4 are valid. Now, we define H for bijections of length n as follows:

♠ *Definition:* Let $f \in \mathbb{B}\mathrm{i}_S(a,b)$ with $|a| = n$. For an $i \in [|a|]$ take the decomposition $a = a' \star a_i \star a''$ with $|a'| = i-1$ and the decomposition $b = b' \star b_{f(i)} \star b''$ with $|b'| = f(i) - 1$. Then there exists a unique $f_i \in \mathbb{B}\mathrm{i}_S(a' \star a'', b' \star b'')$ such that

$$f = (\mathsf{I}_{a'} \star {}^{a_i}\mathsf{X}^{a''})\ (f_i \star \mathsf{I}_{a_i})\ (\mathsf{I}_{b'} \star {}^{b''}\mathsf{X}^{a_i})$$

(f_i is the bijection obtained from f dropping the i-th input, the $f(i)$-th output and the edge that connect them.) Now it is natural to define

$$H(f) = E_i$$

where $E_i = (\mathsf{I}_{h(a')} \star {}^{h(a_i)}\mathsf{X}^{h(a'')})\ (H(f_i) \star \mathsf{I}_{h(a_i)})\ (\mathsf{I}_{h(b')} \star {}^{h(b'')}\mathsf{X}^{h(a_i)})$. ♣

We have to show that this definition is correct, i.e., the morphism E_i does not depend on i. To prove this, we consider two different indices $i, j \in [|a|]$, say with $i < j$. Let $a = a' \star a_i \star a'' \star a_j \star a'''$ with $|a'| = i - 1$ and $|a' \star a_i \star a''| = j - 1$.

There are two cases: (i) $f(i) < f(j)$ and (ii) $f(i) > f(j)$. Since the difference between them is small, we prove the correctness of the definition only for the second case.

Suppose therefore that $f(i) > f(j)$. The output interface b may be decomposed as $b = b' \star a_j \star b'' \star a_i \star b'''$ with $|b'| = f(j) - 1$ and $|b' \star a_j \star b''| = f(i) - 1$. Then there exists a unique $g \in \mathbb{B}\mathrm{i}_S(a' \star a'' \star a''',\ b' \star b'' \star b''')$ such that both identities below hold:

$$f_i = (\mathsf{I}_{a' \star a''} \star {}^{a_j}\mathsf{X}^{a'''})\ (g \star \mathsf{I}_{a_j})\ (\mathsf{I}_{b'} \star {}^{b'' \star b'''}\mathsf{X}^{a_j})$$

$$f_j = (\mathsf{I}_{a'} \star {}^{a_i}\mathsf{X}^{a'' \star a'''})\ (g \star \mathsf{I}_{a_i})\ (\mathsf{I}_{b' \star b''} \star {}^{b'''}\mathsf{X}^{a_i})$$

(i.e., g is the bijection obtained from f dropping the inputs i, j, the outputs $f(i), f(j)$ and the edges that connect them). Using the inductive hypothesis we get

$$H(f_i) = (\mathsf{I}_{h(a' \star a'')} \star {}^{h(a_j)}\mathsf{X}^{h(a''')})\ (H(g) \star \mathsf{I}_{h(a_j)})\ (\mathsf{I}_{h(b')} \star {}^{h(b'' \star b''')}\mathsf{X}^{h(a_j)})$$

and similarly for $H(f_j)$. Then E_i may be rewritten as

$$
\begin{aligned}
(1)\ \ E_i \;=\;& (\mathsf{I}_{h(a')} \star {}^{h(a_i)}\mathsf{X}^{h(a''\star a_j\star a''')})\,(\mathsf{I}_{h(a'\star a'')} \star {}^{h(a_j)}\mathsf{X}^{h(a''')} \star \mathsf{I}_{h(a_i)}) \\
&\cdot (H(g) \star \mathsf{I}_{h(a_j\star a_i)})\,(\mathsf{I}_{h(b')} \star {}^{h(b''\star b''')}\mathsf{X}^{h(a_j)} \star \mathsf{I}_{h(a_i)})\,(\mathsf{I}_{h(b'\star a_j\star b'')} \star {}^{h(b''')}\mathsf{X}^{h(a_i)})
\end{aligned}
$$

and E_j as

$$
\begin{aligned}
E_j \;=\;& (\mathsf{I}_{h(a'\star a_i\star a'')} \star {}^{h(a_j)}\mathsf{X}^{h(a''')})\,(\mathsf{I}_{h(a')} \star {}^{h(a_i)}\mathsf{X}^{h(a''\star a''')} \star \mathsf{I}_{h(a_j)}) \\
&\cdot (H(g) \star \mathsf{I}_{h(a_i\star a_j)})\,(\mathsf{I}_{h(b'\star b'')} \star {}^{h(b''')}\mathsf{X}^{h(a_i)} \star \mathsf{I}_{h(a_j)})\,(\mathsf{I}_{h(b')} \star {}^{h(b''\star a_i\star b''')}\mathsf{X}^{h(a_j)})
\end{aligned}
$$

By axiom B7, $\mathsf{I}_{h(a_i\star a_j)} = {}^{h(a_i)}\mathsf{X}^{h(a_j)} \cdot \mathsf{I}_{h(a_j\star a_i)} \cdot {}^{h(a_j)}\mathsf{X}^{h(a_i)}$, hence (use B6)

$$
\begin{aligned}
(2)\ \ E_j \;=\;& (\mathsf{I}_{h(a'\star a_i\star a'')} \star {}^{h(a_j)}\mathsf{X}^{h(a''')})\,(\mathsf{I}_{h(a')} \star {}^{h(a_i)}\mathsf{X}^{h(a''\star a''')} \star \mathsf{I}_{h(a_j)}) \\
&\cdot (\mathsf{I}_{h(a'\star a''\star a''')} \star {}^{h(a_i)}\mathsf{X}^{h(a_j)})\,(H(g) \star \mathsf{I}_{h(a_j\star a_i)})\,(\mathsf{I}_{h(b'\star b''\star b''')} \star {}^{h(a_j)}\mathsf{X}^{h(a_i)}) \\
&\cdot (\mathsf{I}_{h(b'\star b'')} \star {}^{h(b''')}\mathsf{X}^{h(a_i)} \star \mathsf{I}_{h(a_j)})\,(\mathsf{I}_{h(b')} \star {}^{h(b''\star a_i\star b''')}\mathsf{X}^{h(a_j)}).
\end{aligned}
$$

The product of the first two factors in (1) is equal to the product of the first three factors in (2) and the product of the last two factors in (1) is equal to the product of the last three factors in (2). (This follows from Lemma 2.2 and axiom B6.) Hence $E_i = E_j$.

To conclude, we may define $H(f)$ by

$$
\begin{aligned}
H(f) = &(\mathsf{I}_{h(a_1\star\ldots\star a_{i-1})} \star {}^{h(a_i)}\mathsf{X}^{h(a_{i+1}\star\ldots\star a_n)})\,(H(f_i) \star \mathsf{I}_{h(a_i)}) \\
&\cdot (\mathsf{I}_{h(b_1\star\ldots\star b_{f(i)-1})} \star {}^{h(b_{f(i)+1}\star\ldots\star b_n)}\mathsf{X}^{h(a_i)})
\end{aligned}
$$

for an arbitrary i and the definition is correct.

(b) In the remaining part of the proof we check that H fulfils conditions 1–4. Condition 1 obviously holds. For 2, in the nontrivial case $a \neq \epsilon$, say $a = c \star s$ with $s \in S$, we see that

$$
\begin{aligned}
H({}^{a}\mathsf{X}^{b}) &= (\mathsf{I}_{h(c)} \star {}^{h(s)}\mathsf{X}^{h(b)})\,(H({}^{c}\mathsf{X}^{b}) \star \mathsf{I}_{h(s)})\,(\mathsf{I}_{h(b\star c)} \star \mathsf{I}_{h(s)}) \\
&= (\mathsf{I}_{h(c)} \star {}^{h(s)}\mathsf{X}^{h(b)})\,({}^{h(c)}\mathsf{X}^{h(b)} \star \mathsf{I}_{h(s)}) \\
&= {}^{h(c)\star h(s)}\mathsf{X}^{h(b)} \\
&= {}^{h(a)}\mathsf{X}^{h(b)}
\end{aligned}
$$

For 3, assume $f \in \mathbb{B}\mathrm{i}_S(a,b)$, $g \in \mathbb{B}\mathrm{i}_S(b,c)$ with $|a| = n$. Take an $i \in [|a|]$ and write $a = a' \star a_i \star a''$ with $|a'| = i-1$, $b = b' \star a_i \star b''$ with $|b'| = f(i)-1$, and $c = c' \star a_i \star c''$ with $|c'| = g(f(i)) - 1$. If g_i denoted the bijections associated with g in the same way the bijections f_i were associated with f, then

$$
f \cdot g = (\mathsf{I}_{a'} \star {}^{a_i}\mathsf{X}^{a''})\,(f_i \cdot g_{f(i)} \star \mathsf{I}_{a_i})\,(\mathsf{I}_{c'} \star {}^{c''}\mathsf{X}^{a_i}).
$$

Hence,

$$
\begin{aligned}
H(f)\cdot H(g) &= (\mathsf{I}_{h(a')} \star {}^{h(a_i)}\mathsf{X}^{h(a'')})\,(H(f_i) \star \mathsf{I}_{h(a_i)})\,(\mathsf{I}_{h(b')} \star {}^{h(b'')}\mathsf{X}^{h(a_i)}) \\
&\quad \cdot (\mathsf{I}_{h(b')} \star {}^{h(a_i)}\mathsf{X}^{h(b'')})\,(H(g_{f(i)}) \star \mathsf{I}_{h(a_i)})\,(\mathsf{I}_{h(c')} \star {}^{h(c'')}\mathsf{X}^{h(a_i)}) \\
&= (\mathsf{I}_{h(a')} \star {}^{h(a_i)}\mathsf{X}^{h(a'')})\,(H(f_i \cdot g_{f(i)}) \star \mathsf{I}_{h(a_i)})\,(\mathsf{I}_{h(c')} \star {}^{h(c'')}\mathsf{X}^{h(a_i)}) \\
&= H(f\cdot g)
\end{aligned}
$$

Property 4 is easy to verify; its proof is omitted.

The proof is almost finished. We have defined a symocat morphism $H : \mathbb{B}\mathrm{i}_S \to B$. From its definition it is obvious that it is the unique symocat morphism whose restriction to objects coincides with h. □

This theorem will be used often in the following. It provides a mechanism to map finite bijections in arbitrary symocats relative to interface decompositions. These decompositions may vary from case to case according to the action of the monoid morphism h on objects. If the interfaces are kept fixed, then the mapping is unique. We state this particular case more formally.

Let $\mathbf{SyMoCat}_{S^*}$ be the category of symocats with a fixed monoid of objects $S^\star$ and using symocat morphisms which preserve the objects. Then, by Theorem 2.3 above, it follows that

Corollary 2.4. *$\mathbb{B}\mathrm{i}_S$ is an initial object in* $\mathbf{SyMoCat}_{S^*}$. □

Working with bijections. From the results of the previous subsection, we get a few important heuristics.

Computation rule for abstract bijections.

Computation rule for abstract bijections
If two expressions built up with "$\cdot$", "$\star$", I_a, ${}^a\mathsf{X}^b$ $(a, b \in S^\star)$ specify the same S-sorted bijection when they are interpreted in $\mathbb{B}\mathrm{i}_S$, then they specify the same element when they are interpreted in an arbitrary symocat.

An example. To illustrate this point we consider an arbitrary expression $g : a \star b \star c \to a' \star b' \star c'$. Then, both morphisms $E = (\mathsf{I}_a \star {}^x\mathsf{X}^{b\star y\star c})\ (\mathsf{I}_{a\star b} \star {}^y\mathsf{X}^c \star \mathsf{I}_x)\ (g \star \mathsf{I}_{y\star x})\ (\mathsf{I}_{a'} \star {}^{b'\star c'}\mathsf{X}^y \star \mathsf{I}_x)\ (\mathsf{I}_{a'\star y\star b'} \star {}^{c'}\mathsf{X}^x)$ and $E' = (\mathsf{I}_{a\star x\star b} \star {}^y\mathsf{X}^c)\ (\mathsf{I}_a \star {}^x\mathsf{X}^{b\star c} \star \mathsf{I}_y)\ (\mathsf{I}_{a\star b\star c} \star {}^x\mathsf{X}^y) \cdot (g \star \mathsf{I}_{y\star x})\ (\mathsf{I}_{a'\star b'\star c'} \star {}^y\mathsf{X}^x)\ (\mathsf{I}_{a'\star b'} \star {}^{c'}\mathsf{X}^x \star \mathsf{I}_y)\ (\mathsf{I}_{a'} \star {}^{b'\star x\star c'}\mathsf{X}^y)$ represent the same bijection when they are interpreted in $\mathbb{B}\mathrm{i}_{\{a,b,c,a',b',c',x,y\}}(a \star x \star b \star y \star c,\ a' \star y \star b' \star x \star c')$ (see Figure 2.1). Hence E and E' represent the same morphism when they are interpreted in an arbitrary symocat. (A formal proof of this statement was given in Lemma 2.2.) ◇

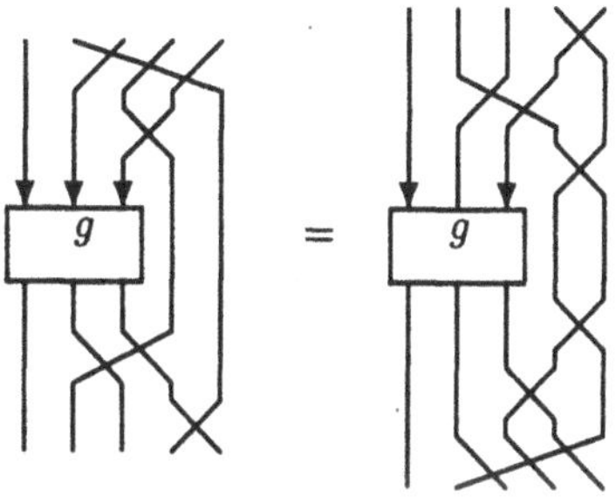

Fig. 2.1. Example of bijections

Notating abstract bijections. Another consequence of Theorem 2.3 is the possibility to use a short *bracket notation* for abstract bijections. It is the specialization of the general bracket representation of relations to this particular case of bijections. In brief, take a fresh name x_i for each object a_i in the list $a_1, \ldots, a_n$ and apply the theorem for the mapping $x_i \to a_i, i \in [n]$ and bijection φ.

> *Notation for abstract bijections*
>
> If $\varphi : [n] \to [n]$ is a (usual) bijection and $a_1, \ldots, a_n$ are n, not necessarily different, objects in a symocat B, then
>
> $$(\varphi(1)_{a_1}\varphi(2)_{a_2}\ldots\varphi(n)_{a_n}) \in B(a_1 \star \ldots \star a_n,\ a_{\varphi^{-1}(1)} \star \ldots \star a_{\varphi^{-1}(n)})$$
>
> denotes the *abstract bijection* in B obtained in the following way:
> (1) Take n distinct symbols $x_1, \ldots, x_n$ and let
>
> $$\overline{\varphi} \in \mathbb{B}\mathrm{i}_{\{x_1,\ldots,x_n\}}(x_1 \star \ldots \star x_n,\ x_{\varphi^{-1}(1)} \star \ldots \star x_{\varphi^{-1}(n)})$$
>
> be the multi-sorted bijection which coincides with φ.
> (2) Apply Theorem 2.3 to get the symocat morphism H which extends the mapping $x_i \to a_i, i \in [n]$.
> (3) Define
>
> $$(\varphi(1)_{a_1}\varphi(2)_{a_2}\ldots\varphi(n)_{a_n}) = H(\overline{\varphi}).$$

With this notation the morphisms E and E' above may be simply written as

$$E = E' = (1_a 5_x 2_b 4_y 3_c)\ (g \star \mathsf{I}_{y\star x})\ (1_{a'} 3_{b'} 5_{c'} 2_y 4_x).$$

♡ A quick reading of a bijection out of the notation like $(1_a 5_x 2_b 4_y 3_c)$ is as follows. The numbers (counted from left to right) represent where the arrows *go*: 1st input to 1st output, 2nd input to 5th output, and so on. The sorts may be collected from the outputs; indeed, the bijection is sorted, hence 5_x also says that the input going to the 5-th output has sort x. But the sorts are more easily obtained from the inputs: the input sorts are simply the ones attached to the numbers from left to right. In our case it is the sequence $a \star x \star b \star y \star c$. ♢

Proposition 2.5. *(commutation: bijections – morphisms)*
If φ is a usual bijection $\varphi : [n] \to [n]$ and $f_i \in B(a_i, b_i), i \in [n]$ are morphisms in a symocat B, then

$$(f_1 \star \ldots \star f_n) \cdot (\varphi(1)_{b_1} \ldots \varphi(n)_{b_n}) = (\varphi(1)_{a_1} \ldots \varphi(n)_{a_n}) \cdot (f_{\varphi^{-1}(1)} \star \ldots \star f_{\varphi^{-1}(n)})$$

PROOF: Every bijection ϕ is a composite of transpositions of the type $\mathsf{I}_r \star {}^1\mathsf{X}^1 \star \mathsf{I}_{n-r-2}$. Hence, it is enough to prove the proposition for such bijections. In such a particular case the identity may be written as $(f_1 \star \ldots \star f_r \star f_{r+1} \star f_{r+2} \star f_{r+3} \star \ldots \star f_n) \cdot (\mathsf{I}_{b_1\star\ldots\star b_r} \star {}^{b_{r+1}}\mathsf{X}^{b_{r+2}} \star \mathsf{I}_{b_{r+3}\star\ldots\star b_n}) = (\mathsf{I}_{a_1\star\ldots\star a_r} \star {}^{a_{r+1}}\mathsf{X}^{a_{r+2}} \star \mathsf{I}_{a_{r+3}\star\ldots\star a_n}) \cdot (f_1 \star \ldots \star f_r \star f_{r+2} \star f_{r+1} \star f_{r+3} \star \ldots \star f_n)$ and obviously holds in B due to axiom B9. □

We finish this section with some computation rules for the bracket representation of abstract bijections.

Remark 2.6. Computation rule for bracket representation of abstract bijections:

(1) $\mathsf{I}_a = (1_a)$

(2) $(\sigma(1)_{a_1} \dots \sigma(m)_{a_m}) \star (\tau(1)_{b_1} \dots \tau(n)_{b_n})$
$= (\sigma(1)_{a_1} \dots \sigma(m)_{a_m} (\tau(1) + m)_{b_1} \dots (\tau(n) + m)_{b_n})$

(3) If $a_i = b_{\sigma(i)}, \forall i \in [n]$, then
$(\sigma(1)_{a_1} \dots \sigma(n)_{a_n}) \cdot (\tau(1)_{b_1} \dots \tau(n)_{b_n}) = (\tau(\sigma(1))_{a_1} \dots \tau(\sigma(n))_{a_n})$

(4) $(\sigma(1)_{a_1} \dots \sigma(i)_{a'_i \star a''_i} \dots \sigma(n)_{a_n}) = (\tau(1)_{a_1} \dots \tau(i)_{a'_i} \tau(i+1)_{a''_i} \dots \tau(n+1)_{a_n})$,
where for $j \in [n+1]$

$$\tau(j) = \begin{cases} \sigma(j) & \text{when } 1 \le j \le i & \text{and } \sigma(j) \le \sigma(i) \\ \sigma(j)+1 & \text{when } 1 \le j \le i & \text{and } \sigma(j) > \sigma(i) \\ \sigma(j-1) & \text{when } i < j \le n+1 & \text{and } \sigma(j-1) < \sigma(i) \\ \sigma(j-1)+1 & \text{when } i < j \le n+1 & \text{and } \sigma(j-1) \ge \sigma(i) \end{cases}$$

(5) $(\sigma(1)_{a_1} \dots \sigma(i)_0 \dots \sigma(n)_{a_n}) = (\tau(1)_{a_1} \dots \tau(i-1)_{a_{i-1}} \tau(i)_{a_{i+1}} \dots \tau(n-1)_{a_n})$,
where for $j \in [n-1]$

$$\tau(j) = \begin{cases} \sigma(j) & \text{when } 1 \le j < i & \text{and } \sigma(j) < \sigma(i) \\ \sigma(j)-1 & \text{when } 1 \le j < i & \text{and } \sigma(j) > \sigma(i) \\ \sigma(j+1) & \text{when } i \le j \le n-1 & \text{and } \sigma(j+1) < \sigma(i) \\ \sigma(j+1)-1 & \text{when } i \le j \le n-1 & \text{and } \sigma(j+1) > \sigma(i) \end{cases}$$

A few example of refinements are: $(2_a 1_{b\star c}) = (3_a 1_b 2_c)$; $(3_{a\star b} 1_c 2_\epsilon 4_d) = (2_a 3_b 1_c 4_d)$; etc.

EXERCISES AND PROBLEMS (SEC. 2.2)

1. Prove the validity of the computing rules given in Remark 2.6. (07)
2. Let S_n be the group of symmetries on a set with n elements. If p_i denotes the transposition of the i-th and $i+1$-th elements, show that the following two axioms give a presentation for S_n: $p_i p_i = 1$ and $p_i p_{i+1} p_i p_{i+1} p_i p_{i+1} = 1$. (15)
3. A new presentation for bijections is given by the strict monoidal category axioms B1–6 and L1–2 (for atomic transpositions ${}^1\mathsf{X}^1$):
L1. ${}^1\mathsf{X}^1 \cdot {}^1\mathsf{X}^1 = \mathsf{I}_2$;
L2. $({}^1\mathsf{X}^1 \star \mathsf{I}_1)(\mathsf{I}_1 \star {}^1\mathsf{X}^1)({}^1\mathsf{X}^1 \star \mathsf{I}_1) = (\mathsf{I}_1 \star {}^1\mathsf{X}^1)({}^1\mathsf{X}^1 \star \mathsf{I}_1)(\mathsf{I}_1 \star {}^1\mathsf{X}^1)$.
Prove this. Show that the rewriting system obtained by the orientation of the equations from left to right is confluent and terminating. (10)

2.3 Bijections in BNAs

The aim of this section is to point out the important fact that (many-sorted) bijections provide a model for the initial BNA structure.

Not so much on BNAs is included here; they were presented in detail in the previous chapter; the reader has to (re)read the first 3 sections of Chapter 1, if necessary. In this section we present the result on mapping bijections in BNAs.

Mapping bijections in BNA models. This result follows by a lifting of the previous result on mapping bijections in symocats to the cyclic context. This implies that bijections may be mapped in BNA models.

Theorem 2.7. *(mapping bijections in BNA models)*
For every BNA $(B, \star, \cdot, \uparrow, \mathsf{I}_a, {}^a\mathsf{X}^b)$ *and for every morphism of monoids* $h : S^* \to \mathcal{O}b(B)$ *there exists a unique morphism of BNA structures* $H : \mathbb{B}\mathrm{i}_S \to B$ *whose restriction to the objects coincides with* h.

PROOF: (sketch) This is part of a more general result (Theorem 3.15) which provides mapping for various classes of finite relations in enriched BNAs. We sketch here the basic lines of the proof that apply to the present case of bijections. The result heavily uses the acyclic mapping H of bijections in symocats given in Theorem 2.3. This mapping H is (uniquely) extended to the cyclic context as follows. In the case a feedback is applied to a bijection $f : a \star s \to a \star s$ with $s \in S$ there are 2 situations: (1) either $f(|a| + 1) = |a| + 1$ or (2) not. In the first case $f = g \star \mathsf{I}_s$ and $f \uparrow^s = g$. With the BNA axioms for juxtaposition and its action on identities one gets $H(f) \uparrow^{h(s)} = H(g) = H(f \uparrow^s)$. In the second case, using some permutations on the input and the output a-parts, one may suppose that $f = g \star {}^s\mathsf{X}^s$; the deduction is similar, but now axiom R7 on the effect of feedback on transpositions is used. □

EXERCISES AND PROBLEMS (SEC. 2.3)

1. Fill in the details for the proof of Theorem 2.7. (08)
2. Prove that R3 follows from its weaker form where g is X, while the other BNA axioms are present. (10)

2.4 Semantic models: I. BNA structure

Additive (Cantorian) relations: I. symocat structure. The interpretation described in this paragraph provides the fundamentals for the standard state-based semantics for sequential programs. It is obtained using the disjoint-union operation as the interpretation of the symocat juxtaposition operation. The essential feature of this model is that each time a unique pin in a network interface is used. This is the mathematical counterpart of the well-known fact that in a (sequential) program each time the control is located to a unique component.

We recall a few definitions from Chapter 1, tailored on the symocat level.

ⓓ For this additive interpretation, the symocat operators are denoted by

$$\oplus \quad \cdot \quad \mathsf{I} \quad \mathsf{X}$$

To describe its relational semantics we start with a set D of value-vectors which represent the *memory states* of a computing device. If an interpretation of the variables in X is given, then we get an interpretation of a flownomial $E : m \to n$ over X as a relation $E_I \subseteq mD \times nD$ (mD denotes the disjoint union of m copies of D; its elements are denoted either as (j, d) or as $j.D$) with the meaning that $((j, d), (j', d')) \in E_I$ iff "if one runs the program obtained from the expression E and the interpretation I starting with input j of the program and initial memory state d, then the program finishes on the exit j' of the program in memory state d', eventually".

Formally, the $\mathsf{AddRel}(D)$ model is defined as follows (in a many sorted case). Suppose that S is a set of atomic sorts and $D = \{D_s\}_{s\in S}$ is an S-sorted set of data. The types of the program interfaces in this additive case are modeled by elements in the additive monoid $S^\oplus = (S^\oplus, \oplus, 0)$ freely generated by S. Hence a general sort for an additive interface is represented by a word $a = a_1 \oplus \ldots \oplus a_{|a|}$, where a_i is the i-th letter of a. To each sort a one may naturally associate a semantic domain for data of type a as follows. If $a \in S^\oplus$ is as above then we denote by aD the disjoint sum $D_{a_1} \sqcup \ldots \sqcup D_{a_{|a|}}$. A generic element in this later set will be denoted as $i.x$. The disjoint sum is considered associative, up to isomorphisms, but not commutative. Now $\mathsf{AddRel}(D)$ denotes the "additive" relational structure defined by

$$\mathsf{AddRel}(D)(a, b) = \{f : f \subseteq aD \times bD\}$$

for $a, b \in S^\oplus$.

The symocat operators have the following meaning in $\mathsf{AddRel}(D)$:

Summation: for $f : a \to b$ and $g : c \to d$ the sum $f \oplus g : a \oplus c \to b \oplus d$ is
$f \oplus g = \{(x, y) :\ (x = 1.x_1, y = 1.y_1,$ and $(x_1, y_1) \in f)$ or $(x = 2.x_2, y = 2.y_2,$ and $(x_2, y_2) \in g)\}$

Composition: for $f : a \to b$ and $g : b \to c$ the composite $f \cdot g : a \to c$ is
$f \cdot g = \{(x, z) :$ there exists y such that $(x, y) \in f$ and $(y, z) \in g\}$

Identity: $\mathsf{I}_a : a \to a$ is defined by
$\mathsf{I}_a = \{(x, x) : x \in D_a\}$

Additive Transposition: ${}^b_a\mathsf{X} : a \oplus b \to b \oplus a$ is defined by
$${}^b_a\mathsf{X} = \{(x, y) :\ (x = 1.x_1, y = 2.y_2 \text{ and } x_1 = y_2) \text{ or } (x = 2.x_2, y = 1.y_1 \text{ and } x_2 = y_1)\}$$ ⓓ

Fact 2.8. *$\mathsf{AddRel}_S(D)$ is a symocat.* □

Mapping finite sorted relations in $\mathsf{AddRel}_S(D)$*:* Notice that $\mathbb{R}\mathrm{el}_S$ is naturally mapped in an arbitrary $\mathsf{AddRel}_S(D)$ by the application

$$f \longrightarrow \{((i, x), (j, x)) : i \in [|a|], j \in [|b|], (i, j) \in f, \text{ and } x \in D_{a_i}\}$$

for $a, b \in S^\oplus$.

♡ This means a usual relation in $\mathbb{R}\mathrm{el}_S$ is interpreted as a state transforming relation which actually does not change the state. The relation f is "sorted", hence this application is correctly defined. It also preserves the symocat operations. ◇

In the deterministic case, the standard model for the interpretation of flownomials is $\mathbb{P}\mathrm{fn}_S(D)$. This is the substructure of $\mathsf{AddRel}_S(D)$ consisting of partially defined functions in $\mathsf{AddRel}_S(D)$.

We present below an example to show how this model may be used to give the semantics of sequential, flowchart programs. (It also uses feedback; see Chapter 1 or the end of this section.)

Example 2.9. A flowchart program for computing the binomial coefficients $\binom{k}{n}$ is described here:

⦶ Take four variables n, k, y, z, interpreted as natural numbers.
Let $S = \{s_1, s_2, s_3, s_4\}$ be a sort set and suppose the associated data sets are

$$D_{s_1} = D_n \times D_k;\ D_{s_2} = D_n \times D_k \times D_y \times D_z;\ D_{s_3} = D_k \times D_y \times D_z;\ D_{s_4} = D_z$$

all D_n, D_k, D_y, D_z being in this case $\mathbb{N}$. Then, there are 5 cells

$$i_1 : s_1 \to s_2;\ t_1 : s_2 \to s_2 \oplus s_2;\ i_2 : s_2 \to s_3;\ t_2 : s_3 \to s_3 \oplus s_3;\ e : s_3 \to s_4$$

The relational semantics for these cells is

$$\begin{array}{l}
(n,k) \xrightarrow{i_1} (n,k,n,1);\\
(n,k,y,z) \xrightarrow{t_1} \mathsf{if}\ y \leq n-k\ \mathsf{then}\ 1.(n,k,y,z)\ \mathsf{else}\ 2.(n,k,y-1,z*y);\\
(n,k,y,z) \xrightarrow{i_2} (k,1,z);\\
(k,y,z) \xrightarrow{t_2} \mathsf{if}\ y > k\ \mathsf{then}\ 1.(k,y,z)\ \mathsf{else}\ 2.(k,y+1,z/y);\\
(k,y,z) \xrightarrow{e} (z)
\end{array}$$

Take the normal form flownomial expression

$$E = [(\mathsf{I}_{s_1} \oplus i_1 \oplus t_1 \oplus i_2 \oplus t_2 \oplus e)(2_{s_1} 3_{s_2} 4_{s_2} 3_{s_2} 5_{s_3} 6_{s_3} 5_{s_3} 1_{s_4})] \uparrow^{s_1 \oplus s_2 \oplus s_2 \oplus s_3 \oplus s_3}$$

Then $|E|(n,k) = \binom{k}{n}$, when $k \leq n$, where $|E|$ is the function in $\mathsf{AddRel}(D)$ associated with E according to the given interpretation of the cells. ⦶

As one can see from this example, the network cells may change the state space, hence the state space may be different at different points of the program. The state space has a certain multiplicative structure, but this is irrelevant here! What is important is the fact that the statements are atomic. They are not decomposed into smaller actions which act, say, in parallel on pieces of the state space. Hence we have a pure additive model over monadic states.

Multiplicative relational models I: symocat structure. Now we describe a model $\mathsf{MultRel}(D)$ of parallel data transformers. Such a transformer $f : m \to n$ acts on an m-tuple of input data and produces an n-tuple of output data. This model gives the standard semantics for the interpretation of Cartesian networks. In this model, the symocat operations are defined using the multiplicative interpretation of the monoidal operation. As in the Cantorian case, we describe the model in the many sorted case.

This setting may be used to represent logical nets or combinatorial circuits.

♡ It may be extended to model arbitrary circuits or data-flow networks as follows. Each network interface pin is considered to have a proper temporal extension. At a semantic level each sort $s \in S$ is interpreted as consisting of streams of data. The multiplicative symocat interpretation is put on top of these constructions to model the interaction between the network cells. ◊

Again, we recall a few definitions on MultRel(D) from Chapter 1, tailored on the symocat level.

⏀ The symbols

$$\otimes \quad \cdot \quad \mathsf{I} \quad \mathsf{X}$$

will be used for symocat operations and constants in the case the multiplicative interpretation is intended. The semantic, relational model MultRel(D) is a model for parallel data transformers. Such a transformer $f : m \to n$ acts on an m-tuple of input data and produces an n-tuple of output data. In this model one may naturally define the parallel and sequential composition operators and certain constants: identity, transposition. Later on ramification constants (copy and sink), identification constants (source and equality test), and feedback will be defined, too.

Now we describe an extension of this setting to the many sorted case. Let S be a set of atomic sorts and $D = \{D_s\}_{s \in S}$ be an S-sorted set of data. For arbitrary sorts denoting multiplicative interfaces we use the multiplicative monoid $S^\otimes = (S^\otimes, \otimes, 1)$ freely generated by S. Hence such a sort is specified by a word $a = a_1 \otimes \ldots \otimes a_{|a|}$, where a_i is the i-th letter of a. For $a \in S^\otimes$ let us denote by D^a the product $D_{a_1} \times \ldots \times D_{a_{|a|}}$. The "multiplicative" relational structure MultRel(D) is defined by

$$\mathsf{MultRel}(D)(a,b) = \{f : f \subseteq D^a \times D^b\}$$

for $a, b \in S^\otimes$.

In MultRel(D) the network algebra operations and constants are interpreted as follows:

Product: for $f : a \to b$ and $g : c \to d$ the product $f \otimes g : a \otimes c \to b \otimes d$ is
$f \otimes g = \{((x_1, x_2), (y_1, y_2)) : (x_1, y_1) \in f \text{ and } (x_2, y_2) \in g\}$

Composition: for $f : a \to b$ and $g : b \to c$ the composite $f \cdot g : a \to c$ is
$f \cdot g = \{(x, z) : \text{ there exists } y \text{ such that } (x, y) \in f \text{ and } (y, z) \in g\}$

Identity: $\mathsf{I}_a : a \to a$ is defined by
$\mathsf{I}_a = \{(x, x) : x \in D^a\}$

Multiplicative Transposition: ${}^a\mathsf{X}^b : a \otimes b \to b \otimes a$ is defined by
${}^a\mathsf{X}^b = \{((x, y), (y, x)) : x \in D^a \text{ and } y \in D^b\}$ ⏀

Fact 2.10. $\mathsf{MultRel}_S(D)$ *is a symocat.* □

Mapping finite sorted relations in $\mathsf{MultRel}_S(D)$*:* How may finite, sorted relations in $\mathbb{R}\mathrm{el}_S$ be mapped in $\mathsf{MultRel}_S(D)$? The mapping is simple,

$$f \longrightarrow \{((x_1, \ldots, x_{|a|}), (y_1, \ldots, y_{|b|})) : (i, j) \in f \Rightarrow x_i = y_j\}$$

for $a, b \in S^\otimes$. The sorted relation f "forces" the components of the interpretation to have equal data on connected pins. The condition that f preserves the sorts is necessary to have a correct definition.

An example is given below to show how this model may be used to study circuits.

Example 2.11. The following formula is used to specify a logical net (acyclic, combinatorial circuit) for binary multiplication.

ⓘ The circuit C below uses the standard boolean cells $\mathsf{and}, \mathsf{or}, \mathsf{xor} : 2 \to 1$ (xor is "exclusive or"). It also uses a unique sort s whose writing is omitted. C is

$$(\wedge_3^2 \otimes \wedge_2^2)\{[\mathsf{and} \otimes \mathsf{or} \otimes (\mathsf{l}_2 \otimes \mathsf{and} \cdot \wedge_2^1)({}^2\mathsf{X}^1 \otimes \mathsf{l}_1)](\mathsf{l}_1 \otimes \mathsf{and})\mathsf{or} \otimes (\mathsf{l}_1 \otimes \mathsf{xor})\mathsf{xor}) \otimes \mathsf{xor}\}$$

It has 4 inputs (a_1, a_0, b_1, b_0) and 3 outputs (c_2, c_1, c_0); the aim of the circuit is to compute binary multiplication, i.e., $\overline{a_1a_0} \times \overline{b_1b_0} = \overline{c_2c_1c_0}$. Indeed, if one consider the set $D_s = \{0, 1\}$ associated to s and the usual interpretation for the cells and, or, xor in $\mathsf{MultRel}(D)(2,1)$, then the interpretation $|C| \in \mathsf{MultRel}(D)(4,3)$ is a function satisfying $\overline{a_1a_0} \times \overline{b_1b_0} = \overline{c_2c_1c_0}$. ⓘ

Additive relations: II. BNA structure. The feedback operator has the following meaning in $\mathsf{AddRel}(D)$:

- *Feedback:* In order to define the feedback first we notice that a relation $f : a \to b$ is given by a family of relations $\{f_{i,j}\}_{i\in[|a|],\ j\in[|b|]}$, where

$$f_{i,j} = \{(d, d') : (i.d, j.d') \in f\} \subseteq D_{a_i} \times D_{b_j}$$

  As usual, r^* denotes the reflexive-transitive closure of a relation r.
  For $f : a \oplus s \to b \oplus s,\ a, b \in S^\oplus,\ s \in S$ the feedback $f \uparrow_\oplus^s : a \to b$ is defined by

$$(f \uparrow_\oplus^s)_{i,j} = f_{i,j}\ \cup\ f_{i,|b\oplus s|} \cdot f^*_{|a\oplus s|,|b\oplus s|} \cdot f_{|a\oplus s|,j}$$

  for $i \in [|\,a\,|],\ j \in [|\,b\,|]$.
  The definition is extended to arbitrary sorts by $f \uparrow_\oplus^0 = f,\quad f \uparrow_\oplus^{a\oplus s} = (f \uparrow_\oplus^s) \uparrow_\oplus^a$.

Multiplicative relations: II. BNA structure. A general feedback operator may be defined in $\mathsf{MultRel}(D)$ as follows:

- *Feedback:* For $f : a \otimes c \to b \otimes c$ we define $f \uparrow_\otimes^c : a \to b$ as the maximal fixed point, i.e., $f \uparrow_\otimes^c = \{(x, y) : \exists z \in D^c.\ ((x, z), (y, z)) \in f\}$.

It is also possible to define iterative multiplicative feedback operators. For such a definition additional structure on data is required. E.g., one may use streams and minimal fixed-point definition of the feedback as in Chapter 10. Another possibility is to consider subsets/classes of a given universe; then an iterative multiplicative definition dual to the one used in the additive case may be given; see Section 7.7.

EXERCISES AND PROBLEMS (SEC. 2.4)

1. Prove the stated Facts 2.8 and 2.10, i.e., $\mathsf{AddRel}_S(D)$ and $\mathsf{MultRel}_S(D)$ are indeed BNAs. (08)

2. Show that the mapping of $\mathbb{R}\mathrm{el}_S$ in $\mathsf{AddRel}_S(D)$ is an embedding (injective). (05)
3. Find examples showing that the mapping of $\mathbb{R}\mathrm{el}_S$ in $\mathsf{MultRel}_S(D)$ is not an embedding. (08)
4. For a matrix A define the exponential e^A by $e^A = \mathsf{I} + A + A^2/2! + A^3/3! \ldots$. Show that $det(e^A) = e^{Tr(A)}$, where det is the determinant, and Tr is the classical trace of a matrix, i.e., the sum of its diagonal elements. May this be lifted to the case of generalized trace of Exercise 1.2.4? (14)

2.5 Other presentations of BNAs

In this section we briefly describe certain alternatives in defining the BNA structure. The emphasis is on the possibility to use different sets of operators, rather than to find equivalent, perhaps simpler, sets of axioms within a given syntactic framework. In more detail, the equivalent LR-flow presentations is described; this uses left- and right-compositions with bijections, rather than general BNA composition.

LR-flows. In this subsection we present a simpler axiomatic system for BNAs, called LR-flow. The main idea is to use the relation $f \cdot g = [(f \star g) \cdot {}^n\mathsf{X}^p] \uparrow^n$, which is valid in a BNA. It shows an intimate relationship exists between feedback and composition. Based on this connection composition may be reduced to weaker forms, e.g., left and right compositions with bijections. With such a reduction composition is drastically simplified. BNA appears as a sort of "bimodule over $\mathbb{B}$i."

♡ This relation is quite natural, since in many semantic models feedback is thought of as a repeated composition. The structure resulting by replacing composition by juxtaposition and feedback is simpler. However, there is a price to pay. First, finite bijections and general BNA arrows lay in different worlds. Their composition depends on a decomposition of the appropriate interface, hence often one has to play with the refinement rules related to these decompositions. The second complication is that two instances of a composition $f \cdot g$ occur, depending on whether f or g is bijection, the other one being a general arrow. As a consequence one BNA axiom sometimes may have several instances in the new setting. ♢

The basic idea is the following. If one has an arrow $f : m \to n$ in a symocat T and a usual bijection $\phi : k \to k$, is it possible to compose them? (Note that k is a number, while m, n are objects in T.) The answer is yes, but relative to a decomposition of the interface of f in k pieces. To fix the matter, suppose we want to make a composition $f \cdot \phi$. If a decomposition of n is give as $n_1 \star \ldots \star n_k$, then the bijection acts on the output interface of f and "permutes" the corresponding pieces.

Definition 2.12. A LR-*flow over* $\mathbb{B}$i (left-right flow over $\mathbb{B}$i) is similar to a BNA, except that composition is replaced with: (1) RHS-composition with bijections in $\mathbb{B}i$: for $f \in T(m,n)$, $n = n_1 \star \ldots \star n_k$, and $\phi \in \mathbb{B}i(k,k)$,

$$f \rhd_{n_1,\ldots,n_k} \phi \in T(m,\ n_{\phi^{-1}(1)} \star \ldots \star n_{\phi^{-1}(k)})$$

and (2) LHS-composition with bijections in $\mathbb{B}i$: for $f \in T(m,n)$, $m = m_1 \star \ldots \star m_k$, and $\phi \in \mathbb{B}i(k,k)$

$$\phi \lhd_{m_1,\ldots,m_k} f \in T(m_{\phi(1)} \star \ldots \star m_{\phi(k)},\ n)$$

The LR-flow axioms are mainly instances of the BNA axioms, i.e.,
B1, B2, B6, B9 are left unchanged;
B3, B4, B5, R1 have 3, 2, 2, 4 instances, respectively;
R3 is simplified (g is X);
B7, B8, R6, R7 are not necessary;
and finally add two new axioms:
R-refine: $f \rhd_{n_1,\ldots,n_i^1\star\ldots\star n_i^r,\ldots,n_k} \phi = f \rhd_{n_1,\ldots,n_i^1,\ldots,n_i^r,\ldots,n_k} (\phi(1)_1 \ldots \phi(i)_r \ldots \phi(k)_1)$
L-refine: similar

The axioms are fully described in Table V.1 of Appendix A.

The main reason for introducing this new structure is that it provides an equivalent presentation for BNA where composition is simplified. Namely,

Theorem 2.13. *LR-flows and BNAs are equivalent.*

The proof is given in Appendix A. It is based on the following translation:

- If $(\{T(m,n)\}_{m,n\in M}, \star, \cdot, \uparrow, \mathsf{I}_m, {}^m\mathsf{X}^n)$ is a BNA, then $(\{T(m,n)\}_{m,n\in M}, \star, \lhd, \rhd, \uparrow, \mathsf{I}_m)$ is a LR-flow, where

  $f \rhd_{n_1,\ldots,n_k} \phi = f \cdot (\phi(1)_{n_1} \ldots \phi(k)_{n_k})$ and
  $\psi \lhd_{m_1,\ldots,m_k} f = (\psi(1)_{m_{\psi(1)}} \ldots \psi(k)_{m_{\psi(k)}}) \cdot f$;
- If $(\{T(m,n)\}_{m,n\in M}, \star, \lhd, \rhd, \uparrow, \mathsf{I}_m)$ is a LR-flow, then $(\{T(m,n)\}_{m,n\in M}, \star, \cdot, \uparrow, \mathsf{I}_m, {}^m\mathsf{X}^n)$ is a BNA, with $\cdot$ and X defined by:

  $f \cdot g = [(f \star g) \rhd_{n,p} {}^1\mathsf{X}^1] \uparrow^n$ for $m \xrightarrow{f} n \xrightarrow{g} p$
  ${}^p\mathsf{X}^q = \mathsf{I}_{p\star q} \rhd_{p,q} {}^1\mathsf{X}^1$

Based on this result $f \rhd_{n_1,\ldots,n_k} \phi$ may be identified with $f \cdot (\phi(1)_{n_1} \ldots \phi(k)_{n_k})$; in the first term ϕ is a concrete bijection, while in the second term it is seen as an abstract bijection included in T.

Other presentations for BNA. We just described an equivalent system of axioms for BNA which was based on a weaker form for composition, but the feedback was let unchanged. It is also possible to use a more liberal feedback which connect an arbitrary output port to an arbitrary input port. In such a setting, composition and transpositions may be completely eliminated. See the section's exercises and the included references for more information.

EXERCISES AND PROBLEMS (SEC. 2.5)

1. For $f : a \star b \star c \to d \star b$ define a upper-extended feedback $f \uparrow^{a(b)c} : a \star c \to d$ which feedback the indicated b part of the output ports to the emphasized b part of the input ports. Find an equivalent presentation for BNA using this feedback, juxtaposition, RHS-composition with bijections, and identities. (20)

2. For $f : a \star b \star c \to a' \star b \star c'$ consider an extended feedback $f \uparrow^{a(b)c}_{a'(b)c'} : a \star c \to a' \star c'$. Show that the following axioms define an equivalent presentation for BNA using this feedback, juxtaposition, and identities (for simplicity, the monoidal operation "$\star$" on interfaces is omitted and the neutral element is denoted by 0):

   1. $f \star (g \star h) = (f \star g) \star h$
   2. $f \star \mathsf{I}_0 = f = \mathsf{I}_0 \star f$
   3. $\mathsf{I}_a \star \mathsf{I}_b = \mathsf{I}_{ab}$
   4. $f \uparrow^{a(0)b}_{c(0)d} = f$
   5. $f \uparrow^{a(uv)b}_{c(uv)d} = f \uparrow^{a(u)vb}_{c(u)vd} \uparrow^{a(v)b}_{c(v)d}$
   6. $f \uparrow^{a(u)b}_{c(u)d} \star g = (f \star g) \uparrow^{a(u)ba'}_{c(u)db'}$
   7. $f \star g \uparrow^{a(u)b}_{c(u)d} = (f \star g) \uparrow^{a'a(u)b}_{b'c(u)d}$
   8. $f \uparrow^{a(u)bvc}_{a'(u)b'vc'} \uparrow^{ab(v)c}_{a'b'(v)c'} = f \uparrow^{aub(v)c}_{a'b'(v)c'} \uparrow^{a(u)bc}_{a'(v)b'c'}$
   9. $f \uparrow^{a(u)bvc}_{a'vb'(u)c'} \uparrow^{ab(v)c}_{a'(v)b'c'} = f \uparrow^{aub(v)c}_{a'(v)b'uc'} \uparrow^{a(u)bc}_{a'b'(u)c'}$
   10. $(\mathsf{I}_a \star f) \uparrow^{a(a)0}_{0(a)b} = f = (f \star \mathsf{I}_b) \uparrow^{a(b)0}_{0(b)b}$
   11. $\mathsf{I}_a \uparrow^{0(a)0}_{0(a)0} = \mathsf{I}_0$
   12. $(f \star g) \uparrow^{a(b)0}_{0(b)c} = (g \star f) \uparrow^{0(b)a}_{c(b)0}$

   Hint: To get a BNA define $f \cdot g = (f \star g) \uparrow^{a(b)0}_{0(b)c}$ and ${}^{a}\mathsf{X}^{b} = (\mathsf{I}_{aba}) \uparrow^{ab(a)0}_{0(a)ba}$; see [CŞ90b] for more on this exercise. (20)

2.6 Network representation; model $[X, T]_{a\alpha}$

The basic model for networks $[X, T]_{a\alpha}$ is described here. The first step is to construct the model $[X, T]$ of network representations (or nf-pairs). Then simulation via bijections is introduced to capture the isomorphism relation on representations. Next BNA operations are extended from representations to networks, thought of as classes of isomorphic nf-pairs. This way the basic $[X, T]_{a\alpha}$ model is finally obtained. The section ends with examples of BNA computations using normal form flownomial expressions.

Network representation: model $[X, T]$. We have seen that each network F may be represented by a normal form flownomial expression

$$((\mathsf{I}_m \star x_1 \star \ldots \star x_k) \cdot f) \uparrow^{i(x_1 \star \ldots \star x_k)}$$

where $x_1, \ldots, x_k$ are the cells of F and f represents its connecting relation. For convenience, we use a shorthand notation for these normal form flownomial expressions: an *nf-pair* represents a network by a pair $(x_1 \star \ldots \star x_k, f)$ containing its basic ingredients. All the other information may be inferred from this pair.

Definition 2.14. ($[X,T]$) Let X be a doubly ranked set and T a BNA. $[X,T]$ is defined by the sets

$$[X,T](m,n) = \{(x_1 \star \ldots \star x_k, f) : \text{ with } x_1,\ldots,x_k \text{ variables in } X \text{ and } \\ f \in T(m \star o(x_1) \star \ldots \star o(x_k),\ n \star i(x_1) \star \ldots \star i(x_k))\}$$

When $k = 0$, $x_1 \star \ldots \star x_k$ is interpreted as the empty word $\epsilon \in X^\star$. An element in $[X,T]$ is called *network representation*, or *nf-pair*, for short.

Connections and simple cell variables are seen as elements in $[X,T]$ via the mappings

$$E_X(x) = (x, {}^m\mathsf{X}^n), \text{ for } x \in X(m,n) \text{ and} \\ E_T = (\epsilon, f), \text{ for } f \in T(m,n)$$

The BNA structure on nf-pairs.

In this subsection we show how BNA operations may be defined on normal form flownomial expressions. The basic $[X,T]$ model of network representations is obtained.

In the definitions below we use the convention that $\underline{x}, i(\underline{x}), \underline{x}', \ldots$ denote $x_1 \star \ldots \star x_k$, $i(x_1) \star \ldots \star i(x_k)$, $x'_1 \star \ldots \star x'_{k'}, \ldots$, respectively.

The BNA operations are defined in $[X,T]$ as follows:

- JUXTAPOSITION: for $(\underline{x}, f) \in [X,T](m,n)$ and $(\underline{x}', f') \in [X,T](p,q)$

$$(\underline{x}, f) \star (\underline{x}', f') = (\underline{x} \star \underline{x}',\ (\mathsf{I}_m \star {}^p\mathsf{X}^{o(\underline{x})} \star \mathsf{I}_{o(\underline{x}')})(f \star f')(\mathsf{I}_n \star {}^{i(\underline{x})}\mathsf{X}^q \star \mathsf{I}_{i(\underline{x}')}))$$

  This operation is illustrated in Fig. 2.2.

- (RIGHT-HAND-SIDE) FEEDBACK IN $[X,T]$: for $(\underline{x}, f) : m \star p \to n \star p$

$$(\underline{x}, f)\uparrow^p = (\underline{x},\ [(\mathsf{I}_m \star {}^{o(\underline{x})}\mathsf{X}^p) f (\mathsf{I}_n \star {}^p\mathsf{X}^{i(\underline{x})})]\uparrow^p)$$

  This operation is illustrated in Fig. 2.3.

- COMPOSITION IN $[X,T]$: for $(\underline{x}, f) : m \to n$ and $(\underline{x}', f') : n \to p$

$$(\underline{x}, f) \cdot (\underline{x}', f') = (\underline{x} \star \underline{x}',\ (f \star \mathsf{I}_{o(\underline{x}')}) \cdot (\mathsf{I}_n \star {}^{i(\underline{x})}\mathsf{X}^{o(\underline{x}')}) \cdot (f' \star \mathsf{I}_{i(\underline{x})}) \cdot (\mathsf{I}_p \star {}^{i(\underline{x}')}\mathsf{X}^{i(\underline{x})}))$$

  This operation is illustrated in Fig. 2.4.

BNA operations from LR-flow operations. We show how the above operations may be obtained from a simpler LR-flow structure on $[X,T]$.

As we have pointed out the LR-flow structure may be seen as a simpler presentation for BNAs, where left and right compositions with bijections are used instead of general BNA composition.

♠ This structure is defined in full detail in Appendix A, where its equivalence with the standard BNA presentation is proved. ♣

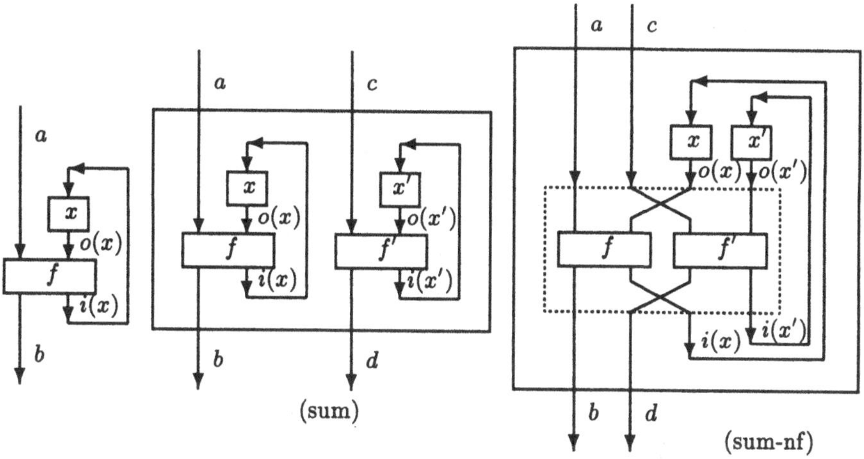

Fig. 2.2. Operations on normal forms representations: Juxtaposition
In the left picture, the schematic graphical representation for normal form flownomial expressions is presented. The result of juxtaposition applied to two normal form flownomial expressions is presented in the middle picture. The resulting normal form picture for juxtaposition is presented in the final picture of the figure.
Similar pictures may be drawn for the other LR-flow operations. See also Fig. 2.4, 2.3 for the remaining BNA operations.

If T is a BNA, then it is naturally a LR-flow. First we use this to define LR-flow operations on $[X,T]$. Then we use the passing from LR-flows to BNAs to obtain the definitions for the BNA operations on $[X,T]$.

In the definition of a BNA or LR-flow, one may use a Left Feedback "$\uparrow$ _" instead of the standard (Right) Feedback "_ $\uparrow$". The connection between these two feedbacks is given by the following relations

$$\uparrow^p f = [{}^1\mathsf{X}^1 \triangleleft_{p,m} f \triangleright_{p,n} {}^1\mathsf{X}^1] \uparrow^p \quad \text{for } f : p\star m \to p\star n$$
$$f \uparrow^p = \uparrow^p [{}^1\mathsf{X}^1 \triangleleft_{m,p} f \triangleright_{n,p} {}^1\mathsf{X}^1] \quad \text{for } f : m\star p \to n\star p.$$

- JUXTAPOSITION: for $(\underline{x}, f) \in [X,T](m,n)$ and $(\underline{x}', f') \in [X,T](p,q)$

$$(\underline{x}, f) \star (\underline{x}', f') = (\underline{x} \star \underline{x}',\ (1324) \triangleleft_{m,o(\underline{x}),p,o(\underline{x}')} (f \star f') \triangleright_{n,i(\underline{x}),q,i(\underline{x}')} (1324))$$

- RIGHT COMPOSITION WITH ARROWS FROM $\mathbb{B}$i: for $(\underline{x}, f) \in [X,T](m,n)$, $P = (n_1,\dots,n_k)$ a partition of n, and $\phi \in \mathbb{B}\mathrm{i}(k,k)$

$$(\underline{x}, f) \triangleright_P \phi = (\underline{x},\ f \triangleright_{P,i(\underline{x})} (\phi \star \mathsf{I}_1))$$

- LEFT COMPOSITION WITH ARROWS FROM $\mathbb{B}$i: for $(\underline{x}, f) \in [X,T](n,p)$, $P = (n_1,\dots,n_k)$ a partition of n, and $\phi \in \mathbb{B}\mathrm{i}(k,k)$

$$\phi \triangleleft_P (\underline{x}, f) = (\underline{x},\ (\phi \star \mathsf{I}_1) \triangleleft_{P,o(\underline{x})} f)$$

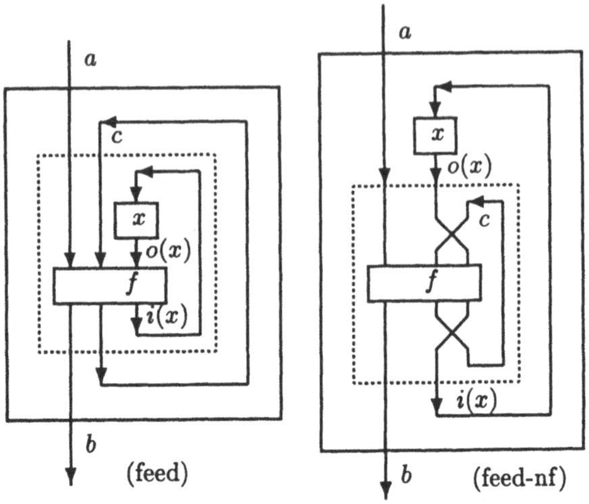

Fig. 2.3. Operations on normal forms representations: Feedback
The result of feedback applied to a normal form flownomial expression is presented in the left part of the figure and the picture with the normal form for feedback is presented in the right part of the figure.

– LEFT-HAND-SIDE FEEDBACK: for $(\underline{x}, f) \in [X, T](p \star m, p \star n)$

$$\uparrow^p (\underline{x}, f) = (\underline{x}, \uparrow^p f)$$

The other BNA operations in $[X, T]$ may be obtained as follows:

• RHS feedback on $[X, T]$ (use the relationship between LHS- and RHS-feedbacks):

If $(\underline{x}, f) : m \star p \to n \star p$ with $\underline{x} : r \to s$, then

$$\begin{aligned}
(\underline{x}, f) \uparrow^p &= \uparrow^p [(21) \lhd_{m,p} (\underline{x}, f) \rhd_{n,p} (21)] \\
&= (\underline{x}, \uparrow^p [(213) \lhd_{m,p,s} f \rhd_{n,p,r} (213)]) \\
&= (\underline{x}, [(21) \lhd_{p,m\star s} [(213) \lhd_{m,p,s} f \rhd_{n,p,r} (213)] \rhd_{p,n\star r} (21)] \uparrow^p) \\
&= (\underline{x}, [(132) \lhd_{m,p,s} f \rhd_{n,p,r} (132)] \uparrow^p)
\end{aligned}$$

Consequently, the following identity holds: for $(\underline{x}, f) : m \star p \to n \star p$

$$(\underline{x}, f) \uparrow^p = (\underline{x}, [(\mathsf{I}_1 \star {}^1\mathsf{X}^1) \lhd_{m,p,o(\underline{x})} f \rhd_{n,p,i(\underline{x})} (\mathsf{I}_1 \star {}^1\mathsf{X}^1)] \uparrow^p)$$

which coincides, modulo slightly different notations, with the definition already given in the previous subsection.

• General composition on $[X, T]$: The definition for composition in $[X, T]$ is obtained using the representation of composition in terms of juxtaposition and feedback and the just proved rule for "$\_ \uparrow^n$" as follows.

If $(\underline{x}, f) : m \to n$ with $\underline{x} : r \to s$ and $(\underline{x}', f') : n \to p$ with $\underline{x}' : r' \to s'$, then

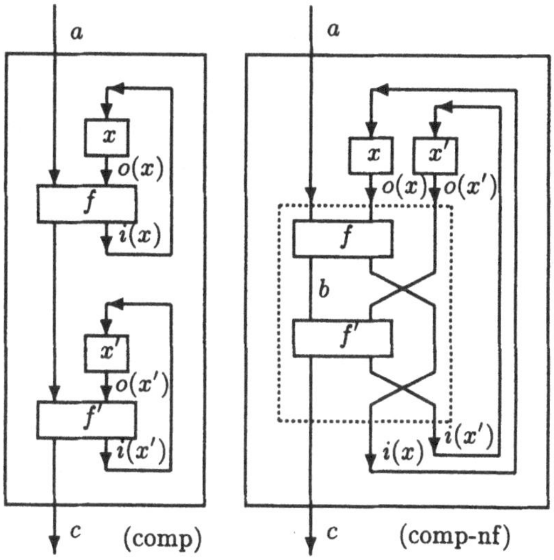

Fig. 2.4. Operations on normal forms representations: Composition
The result of composition applied to two normal form flownomial expressions is presented in the left picture in the figure and the picture with the normal form for composition is presented in the right part of the figure.

$$
\begin{aligned}
(\underline{x}, f)\cdot(\underline{x}', f') &= [((\underline{x}, f)\star(\underline{x}', f'))\triangleright_{n,p}(21)]\uparrow^n\\
&= (\underline{x}\star\underline{x}', g)
\end{aligned}
$$

where

$$
\begin{aligned}
g &= [(1243)\triangleleft_{m,s,n,s'}(f\star f')\triangleright_{n,r,p,r'}(4213)]\uparrow^n\\
&= [(f\star \mathsf{I}_{s'\star n})\cdot(\mathsf{I}_{n\star r}\star {}^{s'}\mathsf{X}^n\cdot f')\cdot(4_n 2_r 1_p 3_{r'})]\uparrow^n\\
&= (f\star\mathsf{I}_{s'})\cdot[(\mathsf{I}_{n\star r}\star{}^{s'}\mathsf{X}^n)\cdot{}^n\mathsf{X}^{r\star n\star s'}\cdot(\mathsf{I}_r\star f'\star\mathsf{I}_n)\cdot{}^{r\star p\star r'}\mathsf{X}^n\cdot(4_n 2_r 1_p 3_{r'})]\uparrow^n\\
&= (f\star\mathsf{I}_{s'})\cdot[(4_n 1_r 3_{s'} 2_n)\cdot(\mathsf{I}_r\star f'\star\mathsf{I}_n)\cdot(2_r 1_p 3_{r'} 4_n)]\uparrow^n\\
&= (f\star\mathsf{I}_{s'})\cdot[(4_n 1_r 3_{s'} 2_n)\cdot((\mathsf{I}_r\star f')\cdot(2_r 1_p 3_{r'})\star\mathsf{I}_n)]\uparrow^n\\
&= (f\star\mathsf{I}_{s'})\cdot(4_n 1_r 3_{s'} 2_n)\uparrow^n\cdot(\mathsf{I}_r\star f')\cdot(2_r 1_p 3_{r'})\\
&= (f\star\mathsf{I}_{s'})\cdot(2_n 1_r 3_{s'})\cdot{}^r\mathsf{X}^{n\star s'}\cdot(f'\star\mathsf{I}_r)\cdot{}^{p\star r'}\mathsf{X}^r\cdot(2_r 1_p 3_{r'})\\
&= (f\star\mathsf{I}_{s'})\cdot(\mathsf{I}_n\star{}^r\mathsf{X}^{s'})\cdot(f'\star\mathsf{I}_r)\cdot(\mathsf{I}_p\star{}^{r'}\mathsf{X}^r)
\end{aligned}
$$

We just found the following relation which complete the deduction of the BNA operations on nf-pairs from the LR-flow operations: for $(\underline{x}, f) : m \to n$ and $(\underline{x}', f') : n \to p$

$$
(\underline{x}, f)\cdot(\underline{x}', f') = (\underline{x}\star\underline{x}',\ \ (f\star\mathsf{I}_{o(\underline{x}')})\cdot(\mathsf{I}_n\star{}^{i(\underline{x})}\mathsf{X}^{o(\underline{x}')})\cdot(f'\star\mathsf{I}_{i(\underline{x})})\cdot(\mathsf{I}_p\star{}^{i(\underline{x}')}\mathsf{X}^{i(\underline{x})}))
$$

Isomorphic nf-pairs. The equivalence relation on nf-pairs we are using in this chapter is *(graph-)isomorphism*. It is captured by simulation via bijections. Before the formal definition, we fix a notation: If $u : [k] \to [k]$, then $i(u)$ denotes its extension with respect to the inputs of $x_1, \ldots, x_k$; in the bracket notation, this means $(u(1)_{i(x_1)} u(2)_{i(x_2)} \ldots u(k)_{i(x_k)})$. Similarly for output extension $o(u)$.

Definition 2.15. We say that two nf-pairs $F = (x_1 \star \ldots \star x_k, f)$ and $G = (y_1 \star \ldots \star y_k, g)$ in $[X,T](m,n)$ are ***similar via a bijection*** $u : [k] \to [k]$, and write $F \to_u G$, if

$$\text{(i) } u(j) = j' \text{ implies } x_j = y_{j'} \text{ and}$$
$$\text{(ii) } f(\mathsf{I}_n \star i(u)) = (\mathsf{I}_m \star o(u))$$

The resulting relation $\{(F,G):$ there exists a bijection u such that $F \to_u G\}$ is denoted by $\stackrel{a\alpha}{\longrightarrow}$ or $\stackrel{\mathbb{B}\mathrm{i}}{\longrightarrow}$.

Proposition 2.16. *Simulation via bijections* $\stackrel{a\alpha}{\longrightarrow}$ *is an equivalence relation and it is compatible with the LR-flow/BNA operations in the following sense:*

$$\begin{array}{llllllll}
a) & F' \star F & \to_{u \star \mathsf{I}_k} & F'' \star F; & F \star F' & \to_{\mathsf{I}_k \star u} & F \star F'' \\
b) & F' \triangleright_P \phi & \to_u & F'' \triangleright_P \phi; & \phi \triangleleft_P F' & \to_u & \phi \triangleleft_P F'' \\
 & F' \cdot F & \to_{u \star \mathsf{I}_k} & F'' \cdot F; & F \cdot F' & \to_{\mathsf{I}_k \star u} & F \cdot F'' \\
c) & \uparrow^p F' & \to_u & \uparrow^p F''; & F' \uparrow^p & \to_u & F'' \uparrow^p
\end{array}$$

whenever $F' \to_u F''$, $F = (x_1 \star \ldots \star x_k, f)$, *and the corresponding operations make sense.*

The simple proof is omitted. This show that BNA/LR-flow operations are well defined on the quotient structure $[X,T]/_{\stackrel{a\alpha}{\longrightarrow}}$.

Definition 2.17. $[X,T]_{a\alpha}$ denotes the quotient structure $[X,T]/_{\stackrel{a\alpha}{\longrightarrow}}$. An element in $[X,T]_{a\alpha}$ is called ***abstract network*** over X and T.

This quotient structure will be a veritable BNA/LR-flow, but we have to pass the task of checking the validity of BNA/LR-flow axioms; try to give your own proof – see Exercise 1.

Exercises and Problems (sec. 2.6)

1. Prove that the BNA axioms, except for B5, B9, and R3 hold in $[X,T]$, whenever T is a BNA. Show that the remaining ones hold in $[X,T]_{a\alpha}$. (12)
2. Prove Proposition 2.16. (10)
3. Give examples of abstract networks over $\mathsf{AddRel}(D)$ and $\mathsf{MultRel}(D)$. (- -)

2.7 Working with flownomials

Representing relations. Our current representation of relations is based on an input-to-output parsing. Each input may be associated with zero, one, or more outputs, say $i_1, \ldots, i_k$. We list them as $[i_1 \ldots i_k]_a$, where a is the attached sort. (If there is only one element in the list, the square parentheses may be omitted.) These partial informations E_i for each input i are collected in a tuple-notation $(E_1 E_2 \ldots)$. For example, the relation

$$\{(1,1),(1,3),(2,2),(2,4),(3,3)\} : a \star b \star a \star d \to a \star b \star a \star b$$

is represented as

$$([13]_a [24]_b 3_a []_d)$$

The representation is short and as far as the input–output relation is concerned it works well. However, it has the drawback that if a pin of the output interface is not related to any input pin, its sort information can not be inferred from the notation. For this reason, the notation is slightly extended by inserting this information. The final representation looks as follows: $(E_1 E_2 \ldots E_n; E)$, where E is the list of the unreachable output pins together with their attached sorts. To have an example again, notice that

$$\{(1,1),(1,5),(2,3),(2,4),(3,5)\} : a \star b \star a \star d \to a \star c \star b \star b \star a \star b$$

is represented as

$$([15]_a [34]_b 5_a []_d ; 2_c 6_b)$$

Working with flownomials via nf-pairs. The proof of the equality in Example 1.5 was made using BNA rules in an arbitrary way. Actually the equality of $a\alpha$-flownomials may be algorithmically checked using a simple decision procedure based on their evaluation in $[X,T]_{a\alpha}$. The algorithm consists of two steps:

- evaluate each expression in $[X,T]$,
- check whether the resulting normal forms expressions are isomorphic (similar via bijections).

Examples. Below we show how this method may be applied to check the equality in Example 1.5. In this example, the connections are made with elements in $\mathbb{P}\mathrm{fn}^{-1}$ (converses of partial functions).

Example 2.18. The nf-pair associated with a flownomial expression is obtained from the nf-pairs of its subexpressions using the BNA operations on nf-pairs.

For $x : b \star a \to a$ and the expression $\wedge_2^b \cdot (\mathsf{I}_b \star (x \cdot \wedge_2^a) \uparrow^a) \cdot x$ the procedure gives:

- $x = (x,\ (2_b 3_a 1_a)) : b \star a \to a$
- $x \cdot \wedge_2^a = (x,\ (3_b 4_a [12]_a)) : b \star a \to 2a$

– $(x \cdot \wedge_2^a) \uparrow^a = (x,\ (2_b[13]_a)) : b \to a$
– $\mathsf{I}_b \star (x \cdot \wedge_2^a) \uparrow^a = (x,\ (1_b 3_b[24]_a)) : 2b \to b \star a$
– $\wedge_2^b \cdot (\mathsf{I}_b \star (x \cdot \wedge_2^a) \uparrow^a) = (x,\ ([13]_b[24]_a)) : b \to b \star a$
– $[\wedge_2^b \cdot (\mathsf{I}_b \star (x \cdot \wedge_2^a) \uparrow^a)] \cdot x = (x \star x,\ ([24]_b[35]_a 1_a)) : b \to a$

For $[\wedge_2^{b\star a} \cdot (x \star x)] \uparrow^a$ one gets

– $x = (x,\ (2_b 3_a 1_a)) : b \star a \to a$
– $x \star x = (x \star x,\ (3_b 4_a 5_b 6_a 1_a 2_a)) : 2(b \star a) \to 2a$
– $\wedge_2^{b\star a} \cdot (x \star x) = (x \star x,\ ([35]_b[46]_a 1_a 2_a)) : b \star a \to 2a$
– $(\wedge_2^{b\star a} \cdot (x \star x)) \uparrow^a = (x \star x,\ ([24]_b 1_a [35]_a)) : b \to a$

Notice that $(\mathsf{I}_b \star {}^a\mathsf{X}^a) \cdot ([24]_b[35]_a 1_a) = ([24]_b 1_a [35]_a) \cdot (\mathsf{I}_a \star {}^{b\star a}\mathsf{X}^{b\star a})$. Indeed, $(1_b 3_a 2_a) \cdot ([24]_b[35]_a 1_a) = ([24]_b 1_a [35]_a) = ([24]_b 1_a [35]_a) \cdot (1_a 4_b 5_a 2_b 3_a)$. Hence the resulting normal forms are similar and the starting expressions are equivalent.

Exercises and Problems (sec. 2.7)

Find normal forms for the following flownomial expressions.

1. $x[(1_s 1_s) y (z \star x)(2_s 1_s)] \uparrow^s$ and $[(x \star z)(1_s 1_s) y (2_s 1_s)] \uparrow^s \cdot x$, where $s \xrightarrow{x,z} s$, $s \xrightarrow{y} 2s$; are they isomorphic? (10)
2. $(\wedge^1 \star \wedge^1 \star \mathsf{I}_1 \star \perp^1) \cdot (\mathsf{I}_1 \star \vee_1 \cdot (\top_1 \star y) \cdot \vee_1 \star x \cdot [(\wedge^2 \cdot (\mathsf{I}_2 \star \vee_1)) \uparrow^1])$, where $2 \xrightarrow{x} 1$, $1 \xrightarrow{y} 1$. (10)
3. $[\vee_s^2 x (\mathsf{I}_s \star y \star x)] \uparrow^s \cdot [(\mathsf{I}_{3s} \star y \star x \star y)(1_s 3_s 2_s 3_s 2_s 5_s 4_s 4_s)] \uparrow^{2s} \cdot [(\mathsf{I}_s \star \vee_s^2 y) z] \uparrow^s$, where $s \xrightarrow{x} 3s$, $s \xrightarrow{y} s$, $2s \xrightarrow{z} s$. (10)

2.8 BNA soundness

In this section we give an abstract proof of the soundness theorem. It is shown that the BNA structure is preserved when one passes from connections to abstract networks. The proof is based on the equivalent LR-flow presentation of BNA described in full detail in Appendices A and B.

Abstract networks and BNA soundness problem. It is clear that one may use pictures in order to show that the BNA rules are sound for concrete networks, i.e., for networks with connections in a $\mathbb{R}\mathrm{el}_S$. We presented such pictures in Chapter 1. But there is also a more formal way to prove this result. This procedure contains two steps:

(i) Prove that BNA axioms hold on involved connecting relations (e.g., one may use a substructure in a $\mathbb{R}\mathrm{el}_S$ containing constants "I_a" and "${}^a\mathsf{X}^b$" and closed with respect to the BNA operations "$\star$", "$\cdot$" and "$\uparrow$"; such structures inherit from $\mathbb{R}\mathrm{el}_S$ the property of being a BNA),

(ii) Show that if the BNA axioms hold in T, then they hold on networks over X and T.

This frame for the proof of the soundness theorem for concrete networks may be applied as well to abstract networks. Before stating the problem more precisely, we have to recall a few notations and definitions.

Abstract networks. (1) $\xrightarrow{a\alpha}$ denotes simulation via bijections; (2) $[X,T]$ denotes the algebra of nf-pairs (normal form pairs) over X and T; (3) $[X,T]_{a\alpha}$ denotes $[X,T]/\underset{\rightarrow}{a\alpha}$. (4) An *abstract network* is an element in $[X,T]_{a\alpha}$, where T is an arbitrary BNA. ◇

The difficult part (ii) of the soundness problem takes the following form:

Is $[X,T]_{a\alpha}$ *a BNA, whenever* T *is a BNA?*

The answer is positive. Our (shorter) proof uses the LR-flow equivalent presentation for BNA and it is given in detail in Appendix B.

Theorem 2.19. $[X,T]_{a\alpha}$ *is a BNA, whenever* T *is a BNA.*

⦶ Proof sketch. This is a fundamental result and its proof is presented in full detail in Appendix B, but using the simpler equivalent LR-flow axioms. For a direct proof one may proceed as follows. It may be seen that all BNA axioms, except for B5, B9, and R3, hold in $[X,T]$. For B5, B9, and R3 isomorphic nf-pairs are obtained as evaluations in $[X,T]$ for the right- and left-hand-side, respectively. We illustrate the proof with two verifications:

• For B1 suppose three pairs in $[X,T]$ are given, namely $(\underline{x},f): m \to n$ with $\underline{x}: r \to s$, $(\underline{x}',f'): m' \to n'$ with $\underline{x}': r' \to s'$, and $(\underline{x}'',f''): m'' \to n''$ with $\underline{x}'': r'' \to s''$. Then

$$[(\underline{x},f)\star(\underline{x}',f')]\star(\underline{x}'',f'') = (\underline{x}\star\underline{x}'\star\underline{x}'',\ g)$$

where

$$\begin{aligned} g &= (1_{m\star m'}3_{m''}2_{s\star s'}4_{s''})[(1_m3_{m'}2_s4_{s'})(f\star f') \\ &\quad \cdot(1_n3_r2_{n'}4_{r'})\star f''](1_{n\star n'}3_{r\star r'}2_{n''}4_{r''}) \\ &= (1_m2_{m'}5_{m''}3_s4_{s'}6_{s''})[(1_m3_{m'}2_s4_{s'}5_{m''}6_{s''})(f\star f'\star f'') \\ &\quad \cdot(1_n3_r2_{n'}4_{r'}5_{n''}6_{r''})](1_n2_{n'}4_r5_{r'}3_{n''}6_{s''}) \\ &= (1_m3_{m'}5_{m''}2_s4_{s'}6_{s''})(f\star f'\star f'')(1_n4_r2_{n'}5_{r'}3_{n''}6_{r''}) \end{aligned}$$

In a similar way one gets $(\underline{x},f)\star[(\underline{x}',f')\star(\underline{x}'',f'')] = (\underline{x}\star\underline{x}'\star\underline{x}'',\ g)$; hence axiom B1 holds in $[X,T]$.

• B9, actually the equivalent form B9$'$, holds in $[X,T]_{a\alpha}$. If $(\underline{x},f): m \to n$ with $\underline{x} = x_1\star\ldots\star x_k : r \to s$ and $(\underline{x}',f'): m' \to n'$ with $\underline{x}' = x_1'\star\ldots\star x_{k'}' : r' \to s'$, then the right-hand-side term of B9$'$ is

$$(2_m1_{m'})[(\underline{x}',f')\star(\underline{x},f)](2_{n'}1_n) = (\underline{x}'\star\underline{x},\ g)$$

where $g = (\underline{x}'\star\underline{x},\ (3_m1_{m'}2_{s'}4_s)(f'\star f)(2_{n'}3_{r'}1_n4_r))$.

The last nf-pair is similar via a block transposition to the nf-pair obtained from the evaluation of the left-hand-side term in B9$'$, namely

$$(\underline{x}, f) \star (\underline{x}', f') = (\underline{x} \star \underline{x}',\ h)$$

where $h = (\underline{x} \star \underline{x}',\ (1_m 3_{m'} 2_s 4_{s'})(f \star f')(1_n 3_r 2_{n'} 4_{r'}))$.

More precisely, we show that $(\underline{x}' \star \underline{x},\ g) \rightarrow_{(2_{k'} 1_k)} (\underline{x} \star \underline{x}',\ h)$

Condition (i) in the definition of simulation clearly holds. The validity of the second condition (ii) follows by: $(1_m 2_{m'} 4_s 3_{s'}) g (1_n 2_{n'} 4_{r'} 3_r) = (3_m 1_{m'} 4_s 2_{s'})(f' \star f)(2_{n'} 4_{r'} 1_n 3_r) = (1_m 3_{m'} 2_s 4_{s'})(f \star f')(1_n 3_r 2_{n'} 4_{r'}) = h$. ⦶

As a corollary of this result one may iterate the construction to obtain higher-order flownomials, i.e., flownomials using as connections other flownomials.

BNA Soundness. Based on the above theorem we get the main result of this section, i.e.,

Theorem 2.20. *(Soundness Theorem) If two flownomial expressions are equivalent via BNA axioms, then they represent the same network.* □

EXERCISES AND PROBLEMS (SEC. 2.8)

1. Complete the direct proof that $[X, T]_{a\alpha}$ is a BNA, whenever T is a BNA. (12)
2. Prove Theorem 2.19 using the equivalent BNA presentations of Section 2.5. (18)

2.9 BNA completeness

This section contains the BNA completeness issues, i.e., it is proved that two flownomial expressions are equivalent via BNA axioms, whenever they represent the same network.

A concrete network may be represented by an nf-pair unique up to simulation via bijections. This follows by definition for abstract networks. BNA axioms are complete for

- every expression may be brought to a normal form using rules that are deductible from BNA axioms;
- simulation via bijections, which relates isomorphic normal forms, is deductible from BNA axioms.

These proofs are given in the abstract setting, i.e., when the connecting theory is an arbitrary BNA.

Bringing expressions to normal form.

In this subsection we prove that every expression may be brought to a normal form by using the rules NF1–NF5 in Table 2.2.

These rules are nothing else than BNA operations on $[X, T]$, but written with normal form flownomial expressions instead of nf-pairs.

Table 2.2. Rules for bringing an expression f to a normal form $\mathsf{nf}(f)$

NF1	$\mathsf{nf}(f) = (\mathsf{I}_m \cdot f) \uparrow^{\epsilon}$	for $f \in T(m,n)$
NF2	$\mathsf{nf}(x) = [(\mathsf{I}_m \star x) \cdot {}^m\mathsf{X}^n] \uparrow^m$	for $x \in X(m,n)$
	In the following rules NF3–NF5, $\underline{x}$, $o(\underline{x})$, $\underline{x}'$, etc. will denote $x_1 \star \ldots \star x_k$, $o(x_1) \star \ldots \star o(x_k)$, $x'_1 \star \ldots \star x'_{k'}$ etc., respectively.	
NF3	$\mathsf{nf}([(\mathsf{I}_m \star \underline{x}) \cdot f] \uparrow^{i(\underline{x})} \star [(\mathsf{I}_p \star \underline{x}') \cdot f'] \uparrow^{i(\underline{x}')}) = [(\mathsf{I}_{m\star p} \star \underline{x} \star \underline{x}') \cdot g] \uparrow^{i(\underline{x})\star i(\underline{x}')}$	
	where $g = (\mathsf{I}_m \star {}^p\mathsf{X}^{o(\underline{x})} \star \mathsf{I}_{o(\underline{x}')}) \cdot (f \star f') \cdot (\mathsf{I}_n \star {}^{i(\underline{x})}\mathsf{X}^q \star \mathsf{I}_{i(\underline{x}')})$ (in T)	
NF4	$\mathsf{nf}([(\mathsf{I}_m \star \underline{x}) \cdot f] \uparrow^{i(\underline{x})} \cdot [(\mathsf{I}_n \star \underline{x}') \cdot f'] \uparrow^{i(\underline{x}')}) = [(\mathsf{I}_m \star \underline{x} \star \underline{x}') \cdot g] \uparrow^{i(\underline{x})\star i(\underline{x}')}$	
	where $g = (f \star \mathsf{I}_{o(\underline{x}')}) \cdot (\mathsf{I}_n \star {}^{i(\underline{x})}\mathsf{X}^{o(\underline{x}')}) \cdot (f' \star \mathsf{I}_{i(\underline{x})}) \cdot (\mathsf{I}_p \star {}^{i(\underline{x}')}\mathsf{X}^{i(\underline{x})})$	
NF5	$\mathsf{nf}([(\mathsf{I}_{m\star p} \star \underline{x}) \cdot f] \uparrow^{i(\underline{x})} \uparrow^p) = [(\mathsf{I}_m \star \underline{x}) \cdot g] \uparrow^{i(\underline{x})}$	
	where $g = [(\mathsf{I}_m \star {}^{o(\underline{x})}\mathsf{X}^p) \cdot f \cdot (\mathsf{I}_n \star {}^p\mathsf{X}^{i(\underline{x})})] \uparrow^p$	

Proposition 2.21. *Every expression may be brought to a normal form with rules NF1–NF5 in Table 2.2.*

PROOF: It is clear that every expression may be brought to a normal form using NF1–NF5 applied from left to right. □

Notation 2.22. $\equiv_{nf}$ denotes the congruence relation generated on flownomial expressions in $\mathsf{FlowExp}[X, T]$ by the rules NF1–NF5.

These rules may be applied to a normal form expression starting from the definition of the mapping of X and T in $[X, T]$ to get the following result:

$$[(\mathsf{I}_m \star E_X(x_1) \star \ldots \star E_X(x_k)) \cdot E_T(f)] \uparrow^{i(x_1)\star\ldots\star i(x_k)} = (x_1 \star \ldots \star x_k,\ f)$$

Simulation via bijections. The rule of simulation via transpositions SIM_{Tr} is given in Table 2.3. This can be easily extended to arbitrary bijections. The corresponding rule for bijections is denoted $\mathrm{SIM}_{\mathbb{B}\mathrm{i}}$.

Table 2.3. Simulation via transpositions

SIM$_{Tr}$: $[(\mathsf{I}_m \star x_1 \star x_2 \star x_3 \star x_4) \cdot f] \uparrow^{i(x_1 \star x_2 \star x_3 \star x_4)}$
$= [(\mathsf{I}_m \star x_1 \star x_3 \star x_2 \star x_4) \cdot f'] \uparrow^{i(x_1 \star x_3 \star x_2 \star x_4)}$
where x_1, x_2, x_3, x_4 are from $X^\star$, f is from T, and
$f' = (\mathsf{I}_m \star \mathsf{I}_{o(x_1)} \star {}^{o(x_3)}\mathsf{X}^{o(x_2)} \star \mathsf{I}_{o(x_4)}) \cdot f \cdot (\mathsf{I}_n \star \mathsf{I}_{i(x_1)} \star {}^{i(x_2)}\mathsf{X}^{i(x_3)} \star \mathsf{I}_{i(x_4)})$

Deductions for normalization and simulation rules. First we show that certain identities suggested by these rules hold in a BNA.

Lemma 2.23. *The following identities hold in a BNA:*
Nf1 $f = (\mathsf{I}_m \cdot f) \uparrow^\epsilon$ *for* $f : m \to n$
Nf2 $f = ((\mathsf{I}_m \star f) \cdot {}^m\mathsf{X}^n) \uparrow^m$ *for* $f : m \to n$
Nf3 *for* $g : r \to s$, $f : m \star s \to n \star r$, $g' : r' \to s'$ *and* $f' : p \star s' \to q \star r'$

$$[(\mathsf{I}_m \star g) \cdot f] \uparrow^r \star [(\mathsf{I}_p \star g') \cdot f'] \uparrow^{r'}$$
$$= [(\mathsf{I}_{m\star p} \star g \star g') \cdot (\mathsf{I}_m \star {}^p\mathsf{X}^s \star \mathsf{I}_{s'}) \cdot (f \star f') \cdot (\mathsf{I}_n \star {}^r\mathsf{X}^q \star \mathsf{I}_{r'})] \uparrow^{r \star r'}$$

Nf4 *for* $g : r \to s$, $f : m \star s \to n \star r$, $g' : r' \to s'$ *and* $f' : n \star s' \to p \star r'$

$$[(\mathsf{I}_m \star g) \cdot f] \uparrow^r \cdot [(\mathsf{I}_p \star g') \cdot f'] \uparrow^{r'}$$
$$= [(\mathsf{I}_m \star g \star g') \cdot (f \star \mathsf{I}_{s'}) \cdot (\mathsf{I}_n \star {}^r\mathsf{X}^{s'}) \cdot (f' \star \mathsf{I}_r) \cdot (\mathsf{I}_p \star {}^{r'}\mathsf{X}^r)] \uparrow^{r \star r'}$$

Nf5 *for* $g : r \to s$ *and* $f : m \star p \star s \to n \star p \star r$,

$$([(\mathsf{I}_{m\star p} \star g) \cdot f] \uparrow^r) \uparrow^p = [(\mathsf{I}_m \star g) \cdot (\ (\mathsf{I}_m \star {}^s\mathsf{X}^p) \cdot f \cdot (\mathsf{I}_n \star {}^p\mathsf{X}^r)\ ) \uparrow^p] \uparrow^r$$

Sim$_{Tr}$ *for* $f : m \star s_1 \star s_2 \star s_3 \star s_4 \to n \star r_1 \star r_2 \star r_3 \star r_4$ *and* $g_j : r_j \to s_j$ $\forall j \in [4]$

$$[(\mathsf{I}_m \star g_1 \star g_2 \star g_3 \star g_4) \cdot f] \uparrow^{r_1 \star r_2 \star r_3 \star r_4} = [(\mathsf{I}_m \star g_1 \star g_3 \star g_2 \star g_4)$$
$$\cdot (\mathsf{I}_{m \star s_1} \star {}^{s_3}\mathsf{X}^{s_2} \star \mathsf{I}_{s_4}) \cdot f \cdot (\mathsf{I}_{m \star r_1} \star {}^{r_2}\mathsf{X}^{r_3} \star \mathsf{I}_{r_4})] \uparrow^{r_1 \star r_3 \star r_2 \star r_4}$$

PROOF: In the proof, we use the following identity

R2′ $f \uparrow^p \star g = [(\mathsf{I}_m \star {}^{m'}\mathsf{X}^p) \cdot (f \star g) \cdot (\mathsf{I}_n \star {}^p\mathsf{X}^{n'})] \uparrow^p$ for $f : m \star p \to n \star p$, $g : m' \to n'$.

which follows from the BNA axioms. Identity Nf1 follows from R4. For Nf2 note that

$$
\begin{aligned}
[(\mathsf{I}_m \star f) \cdot {}^m\mathsf{X}^n] \uparrow^m &= [{}^m\mathsf{X}^m \cdot (f \star \mathsf{I}_m)] \uparrow^m && \text{by B9} \\
&= ({}^m\mathsf{X}^m) \uparrow^m \cdot f && \text{by R1} \\
&= \mathsf{I}_m \cdot f && \text{by R7} \\
&= f
\end{aligned}
$$

Nf3 may be proved as follows:

$$[(\mathsf{I}_m \star g)f]\uparrow^r \star[(\mathsf{I}_p \star g')f']\uparrow^{r'}$$
$$= [(\mathsf{I}_m \star {}^p\mathsf{X}^r)\ [(\mathsf{I}_m \star g)f \star (\mathsf{I}_p \star g')f']\uparrow^{r'} \cdot(\mathsf{I}_n \star {}^r\mathsf{X}^q)]\uparrow^r \qquad \text{R2}', \text{ then R2}$$
$$= [(\mathsf{I}_m \star {}^p\mathsf{X}^r \star \mathsf{I}_{r'})\ ((\mathsf{I}_m \star g)f \star (\mathsf{I}_p \star g')f')\ (\mathsf{I}_n \star {}^r\mathsf{X}^q \star \mathsf{I}_{r'})]\uparrow^{r'}\uparrow^r \qquad \text{R1}$$
$$= [(\mathsf{I}_m \star {}^p\mathsf{X}^r \cdot (g \star \mathsf{I}_p) \star g')\ (f \star f')\ (\mathsf{I}_n \star {}^r\mathsf{X}^q \star \mathsf{I}_{r'})]\uparrow^{r\star r'} \qquad \text{B6}$$
$$= [(\mathsf{I}_m \star \mathsf{I}_p \star g \star g')\ (\mathsf{I}_m \star {}^p\mathsf{X}^s \star \mathsf{I}_{s'})\ (f \star f')\ (\mathsf{I}_n \star {}^r\mathsf{X}^q \star \mathsf{I}_{r'})]\uparrow^{r\star r'} \qquad \text{B9,R1}$$

Nf4 may be proved as follows:

$$[(\mathsf{I}_m \star g)\ f]\uparrow^r \cdot[(\mathsf{I}_n \star g')\ f']\uparrow^{r'}$$
$$= [([(\mathsf{I}_m \star g)\ f]\uparrow^r \star \mathsf{I}_{r'}) \cdot (\mathsf{I}_n \star g')\ f']\uparrow^{r'} \qquad \text{by R1}$$
$$= [[(\mathsf{I}_m \star {}^{r'}\mathsf{X}^r)\ ((\mathsf{I}_m \star g)\ f \star \mathsf{I}_{r'})\ (\mathsf{I}_n \star {}^r\mathsf{X}^{r'})]\uparrow^r \cdot(\mathsf{I}_n \star g')\ f']\uparrow^r\uparrow^{r'} \qquad \text{by R2}'$$
$$= [(\mathsf{I}_m \star {}^{r'}\mathsf{X}^r)\ ((\mathsf{I}_m \star g)\ f \star \mathsf{I}_{r'})\ (\mathsf{I}_n \star {}^r\mathsf{X}^{r'})\ ((\mathsf{I}_n \star g')\ f' \star \mathsf{I}_r)]\uparrow^r\uparrow^{r'} \qquad \text{by R1}$$
$$= [(\mathsf{I}_m \star {}^{r'}\mathsf{X}^r)\ ((\mathsf{I}_m \star g)\ f \star \mathsf{I}_{r'})\ (\mathsf{I}_n \star {}^r\mathsf{X}^{r'}\ (g' \star \mathsf{I}_r))\ (f' \star \mathsf{I}_r)]\uparrow^{r'\star r} \qquad \text{by B6}$$
$$= [(\mathsf{I}_m \star {}^{r'}\mathsf{X}^r)\ ((\mathsf{I}_m \star g)\ f \star \mathsf{I}_{r'})\ (\mathsf{I}_n \star (\mathsf{I}_r \star g')\ {}^r\mathsf{X}^{s'})\ (f' \star \mathsf{I}_r)]\uparrow^{r'\star r} \qquad \text{by B9}$$
$$= [(\mathsf{I}_m \star {}^{r'}\mathsf{X}^r)\ (\mathsf{I}_m \star g \star g')\ (f \star \mathsf{I}_{r'})\ (\mathsf{I}_n \star {}^r\mathsf{X}^{s'})\ (f' \star \mathsf{I}_r)]\uparrow^{r'\star r} \qquad \text{by B6}$$
$$= [(\mathsf{I}_m \star g \star g')\ (f \star \mathsf{I}_{s'})\ (\mathsf{I}_n \star {}^r\mathsf{X}^{s'})\ (f' \star \mathsf{I}_r)\ (\mathsf{I}_p \star {}^{r'}\mathsf{X}^r)]\uparrow^{r\star r'} \qquad \text{by R3}$$

Before Nf5 we prove Sim_{Tr}:

$$[(\mathsf{I}_m \star g_1 \star g_3 \star g_2 \star g_4)\ (\mathsf{I}_{m\star s_1} \star {}^{s_3}\mathsf{X}^{s_2} \star \mathsf{I}_{s_4}) \cdot f \cdot (\mathsf{I}_{n\star r_1} \star {}^{r_2}\mathsf{X}^{r_3} \star \mathsf{I}_{r_4})]\uparrow^{r_1\star r_3\star r_2\star r_4}$$
$$= [(\mathsf{I}_{m\star r_1} \star {}^{r_3}\mathsf{X}^{r_2} \star \mathsf{I}_{r_4})\ (\mathsf{I}_m \star g_1 \star g_2 \star g_3 \star g_4) \cdot f \cdot (\mathsf{I}_{n\star r_1} \star {}^{r_2}\mathsf{X}^{r_3} \star \mathsf{I}_{r_4})]\uparrow^{r_1\star r_3\star r_2\star r_4}$$
$$= [(\mathsf{I}_m \star g_1 \star g_2 \star g_3 \star g_4) \cdot f \cdot (\mathsf{I}_{n\star r_1} \star {}^{r_2}\mathsf{X}^{r_3} \star \mathsf{I}_{r_4}) \cdot (\mathsf{I}_{n\star r_1} \star {}^{r_3}\mathsf{X}^{r_2} \star \mathsf{I}_{r_4})]\uparrow^{r_1\star r_2\star r_3\star r_4}$$
$$= [(\mathsf{I}_m \star g_1 \star g_2 \star g_3 \star g_4) \cdot f]\uparrow^{r_1\star r_2\star r_3\star r_4}$$

Finally, Nf5 is a particular instance of Sim_{Tr}, namely when $g_1 = g_4 = \mathsf{I}_\epsilon$ and $g_2 = \mathsf{I}_p$. □

From this lemma we get the following:

Proposition 2.24. *NF1–NF5 and* $SIM_{\mathbb{B}\mathrm{i}}$ *are deductible from BNA axioms.* □

Notice that the normal form associated with a flownomial expression is unique. The argument is simple. By a well-known universal algebra result, the quotient algebra $\mathsf{FlowExp}[X,T]/_{\equiv_{nf}}$ is a free model for expressions modulo NF1–NF5. But the algebra of nf-pairs $[X,T]$ is a model for NF1–NF5, too. Two different normal form expressions are interpreted as different elements in $[X,T]$, hence they must also be different in the free model $\mathsf{FlowExp}[X,T]/_{\equiv_{nf}}$.

Completeness Theorem. We collect the results of this section to prove the following completeness theorem:

Theorem 2.25. *(completeness of the BNA rules)*
If two expressions represent the same abstract network, then they are equivalent via BNA axioms.

PROOF: An expression E can be brought to the normal form $\mathsf{nf}(E)$ using NF1–NF5. By the above results (Lemma 2.24) these rules are deductible from BNA axioms. With the Soundness Theorem 2.20 this implies that E and the resulting normal form $\mathsf{nf}(E)$ represent the same element in $[X,T]_{a\alpha}$.

Now, if E and E' represent the same abstract network S, then the corresponding normal form expressions $\mathsf{nf}(E)$ and $\mathsf{nf}(E')$ are equivalent. (They represent S.) The definition of abstract network guarantees the existance of a simulation via a bijection connecting $\mathsf{nf}(E)$ and $\mathsf{nf}(E')$. Lemma 2.24 shows that simulation rule is deductible from BNA axioms. Therefore E and E' are connected via BNA axioms and the completeness theorem is proved. □

EXERCISES AND PROBLEMS (SEC. 2.9)

1. Why is the left-hand side of NF5 not a normal form? (04)
2. Prove that the following identity holds in $[X,T]$: for E_X, E_T such as in Definition 2.14:
$[(\mathsf{l}_m \star E_X(x_1) \star \ldots \star E_X(x_k)) \cdot E_T(f)] \uparrow^{i(x_1)\star\ldots\star i(x_k)} = (x_1 \star \ldots \star x_k,\ f)$. (10)
3. Prove the completeness theorem using the equivalent BNA presentations described in Section 2.5 and Exercise 2.5.2. (18)

2.10 Networks as $a\alpha$-flownomials

In this section we present a universal theorem for (abstract) networks.

Abstract networks may be seen either as (1) classes of isomorphic representations, i.e., elements in $[X,T]_{a\alpha}$ or as (2) classes of flownomial expressions modulo BNA axioms, i.e., elements in $\mathsf{FlowExp}[X,T]/_{\text{BNA}}$. This double view on networks is justified by the BNA Soundness and Completness Theorems and leads to the following result. Before this, notice that we also use the shorter acronym $a\alpha$-flownomials for flownomial expressions modulo BNA axioms; the corresponding equivalence is denoted by $\sim_{a\alpha}$ or $_{\text{BNA}}$.

Theorem 2.26. *(meaning of flownomials modulo BNA axioms)*
The algebras $\mathsf{FlowExp}[X,T]/_{\text{BNA}}$ *and* $[X,T]_{a\alpha}$ *are isomorphic. Hence flownomials over X and T modulo BNA axioms may be identified with the abstract networks built up with atoms in X and connections in T.*

PROOF: By Theorem 2.20 $[X,T]_{a\alpha}$ is a BNA. Hence the evaluation morphism

$$\mathcal{E} : \mathsf{FlowExp}[X,T] \to [X,T]_{a\alpha}$$

induces a morphism at the level of $a\alpha$-flownomials

$$\mathcal{E}^{a\alpha} : \mathsf{FlowExp}[X,T]/_{\mathrm{BNA}} \to [X,T]_{a\alpha}$$

Obviously, $\mathcal{E}^{a\alpha}$ is a surjective morphism. On the other hand, Theorem 2.25 asserts that two expressions that have the same evaluation in $[X,T]_{a\alpha}$ are BNA equivalent. Consequently, $\mathcal{E}^{a\alpha}$ is also an injective morphism, hence it is an isomorphism. □

The construction of the quotient algebra $\mathsf{FlowExp}[X,T]/_{\mathrm{BNA}}$ is the standard construction of the free algebra satisfying BNA axioms. The following universal property of $a\alpha$-flownomials is now obvious.

Theorem 2.27. *(universal property of $a\alpha$-flownomials / abstract networks)*
Let M be a monoid and T a BNA over M. Then:

(i) $[X,T]_{a\alpha}$ is a BNA over M, and

(ii) There exists an embedding (I_X, I_T) of (X,T) in $[X,T]_{a\alpha}$, where I_X is a function and I_T is a BNA morphism such that for every BNA Q, function ϕ_X, and BNA morphism ϕ_T, there exists a unique BNA morphism $\phi^\sharp : [X,T]_{a\alpha} \to Q$ such that $I_T \cdot \phi^\sharp = \phi_T$ and $I_X \cdot \phi^\sharp = \phi_X$. □

Actually, the above theorem shows that the BNA algebra $[X,T]_{a\alpha}$ is the direct categorical sum of the BNA algebra T and the BNA algebra freely generated by X. On the way we have got a construction for the free BNA generated by a set of variables. Indeed, if $T = \mathbb{B}\mathrm{i}$, the initial BNA, then the above theorem produces the following result:

Corollary 2.28. $[X,\mathbb{B}\mathrm{i}]_{a\alpha}$ *is the free BNA generated by X.*

We finish this chapter with a simple corollary of the fact that all deterministic programs are represented by flownomial expressions over $\mathbb{P}\mathrm{fn}$.

Theorem 2.29. *(computability completeness)*
Let F be a set of basic statements in $\mathbb{P}\mathrm{fn}(D)$, where D is the set of memory states. The set of functions computed by the programs built up with statements from F coincides with the set of functions represented by flownomial expressions over F and $\mathbb{P}\mathrm{fn}$. □

Hence flownomial expressions have full computability power, if one starts with an appropriate set of basic functions. A simple well-known case, presented in [DW81], is when $D = \mathbb{N}^{\mathbb{N}}$ and F consists of functions $\mathsf{plus\text{-}1}$, $\mathsf{minus\text{-}1}$, and $\mathsf{test\text{-}to\text{-}0}$ for each components. (The memory state is atomic, hence copies of these functions are to be provided in the basis F for each memory location. Within a mixed additive-multiplicative setting, briefly presented in Chapter 12, functions $\mathsf{plus\text{-}1}$, $\mathsf{minus\text{-}1}$, and $\mathsf{test\text{-}to\text{-}0}$ for individual memory cells suffice.)

Comments, problems, bibliographic remarks

Beyond BNA? At first sight, it seems that the BNA axioms are rather general. So the question "are there useful settings weaker than BNA?" may appear superfluous. However, there are at least two contexts where one may find that the BNA framework is too narrow.

Quantum groups and traced monoidal categories. The first example is provided by quantum groups developed by Joyal, Street, and Verity [JSV96]. In this setting, the acyclic structure has to be weakened. More precisely, one is interested in spatially oriented crossings, hence a kind of "knot theory" is used. The crossing constant X has now two versions: X^+, say, in the case when the upper-left-to-bottom-right arrow is over the other and X^- in the opposite case. The symocat general axiom $\mathsf{X}\mathsf{X} = \mathsf{I}$ has to be dropped. It is valid for opposite crossings, but not for two occurrences of the same crossing. (E.g., $\mathsf{X}^+\mathsf{X}^- = \mathsf{I}$ is valid, but $\mathsf{X}^+\mathsf{X}^+ = \mathsf{I}$ is not valid.) As a byproduct, some other BNA axioms have also to be changed.

Non-commutative BNAs. The second example is generated by an attempt to have a unique normal form for networks, that is, to drop commuting rule(s) applied to "heavy" elements. This setting was developed by Căzănescu. Roughly speaking, the left-to-right order of the heavy elements should be preserved. Such a setting may be useful to model the placement of the chips in a circuit.

In such a "non-commutative" setting, a distinction is made between cells (the heavy elements) and wires (the light elements): cell commutation is forbidden, but commutation is permitted for wires. In technical terms this means that axioms B6 and B10 should be weakened: in B6 one of the morphisms f' or g should be an identity; in B10 f or g should be an identity; the axiom R3 should be used in its weaker form when only bijections are allowed to be shifted along the feedback; the other BNA axioms are let unchanged. The resulting algebraic structure is called *flow*. The BNA structure (or *biflow*, or $a\alpha$-flow, as it was also called in the literature) is a "commutative" flow, i.e., a flow which fulfils $(\mathsf{I} \star f)(g \star \mathsf{I}) = g \star f$.

Notes. Symocats were independently introduced in [Hot65] (called x-categories) and [Lan71] (with their current "symmetric strict monoidal category" name). A first application to the algebraic study of flowcharts is given in [ES82].

Classical presentations for bijections as symmetric groups may be found in [CM72]. Theorem 2.3 on mapping bijections in symocats is from [CŞ88a]; its extension to BNAs (Theorem 2.7) is included in [CŞ88b, CŞ89].

The present formal model of networks/flowcharts as nf-pairs $[X, T]$ is from [Şte86b]; it is based on the model in [CU82], which, subsequently, is based on Elgot's model of "normal descriptions," cf. [Elg75].

This chapter closely follows Chapter B, Sections 3–6 of [Şte91]. Its main results, the soundness and completeness theorems, are from [Şte86b]; the included proof using the equivalent LR-flow presentation of the BNA structure is from that paper, too. This result is based on previous work dealing with the axiomatization of flowchart schemes, including [CU82, BÉ85]. See also [Bar87, CŞ88a, CŞ89].

With different sets of operators, various algebras for flow graphs appear in [Mil79, Par87, CŞ90b].

Classical computation models are presented in [DW81]. For the categorical background, we recommend [Lan71], [PP79], or [BW90b]. For classical algebra, a good textbook is [LB67]. The trace operator and its relationship with matrix exponential operator is presented [Arn78]. A presentation of quantum groups, but prior to the introduction of trace monoidal categories, is [Kas95]. For knots and related problems, see [Kau95].

Exercise 2.2.3 is from [Laf92]. For Exercise 2.2.4 see [Arn78]; for Exercise 2.5.2 see [CŞ90b].

3. Algebraic models for branching constants

Si vous dites aux grandes personnes: ≪J'ai vu une belle maison en briques roses, avec des géraniums avec fenêtres et des colombes sur le toit...≫ elles ne parviennent pas à s'imaginer cette maison. Il faut leur dire: ≪J'ai vu une maison de cent mille francs.≫ Alors elles s'écrient: ≪Comme c'est joli!≫

Antoine de Saint-Exupery

We start here the study of the branching structure of networks. The subject is very complex and will be developed in many other forthcoming chapters.

In this chapter we introduce branching constants (ramifications and identifications) to the BNA signature, hence the full NA syntax for the calculus of flownomials is obtained.

The properties of the branching constants are quite different in different models and it is somehow difficult to get a uniform presentation. In this book we are mainly using branching constants with *angelic* relation operators. In this, say, standard view on relations, an input–output pair is connected in a composed relation if there is a path connecting them. See Appendix C for some different *demonic* approaches.

The main result of the first part consists of axiomatizations for 16 classes of finite relations. These classes may be obtained by intersections of the following basic classes: *total, surjective, univocal,* and *injective* relations. They may also be characterized by certain natural restrictions on the ramification degree of the branching constants, called *xy-restrictions* ($x \in \{a, b, c, d\}$, $y \in \{\alpha, \beta, \gamma, \delta\}$).

The algebras which appear in this way are symocats enriched with branching constants and specific equations, briefly named *xy-symocats.*

In the second part we extend the axiomatization for networks modulo isomorphism from the BNA to the NA setting. That is, branching constants are present at the syntactic level of the flownomial expressions. The key result is provided by certain axiomatizations for relations as BNAs enriched with branching constants.

The resulting algebraic structures are called *xy-symocats with feedback* and are axiomatized by: (1) the BNA axioms, (2) the *xy*-symocat axioms, and (3) certain new axioms which show how the feedback acts on these branching constants.

3.1 xy-symocats (xy-weak rules)

We briefly present here the results on the axiomatization of relations in the acyclic setting. Actually, this is done for 16 natural classes of relations, which are introduced via both syntactic restrictions or semantic properties. The corresponding algebraic structures are symocats enriched with branching constants and specific equations, briefly called xy-symocats.

Finite sorted relations. We recall the definition of $\mathbb{R}\mathrm{el}_S$. The set of S-sorted relations from a to b is

$$\mathbb{R}\mathrm{el}_S(a,b) = \{f : f \subseteq [|a|] \times [|b|] \text{ and } [(i,j) \in f \Rightarrow a_i = b_j]\}$$

Notational conventions. There is a small notational problem with the symocat monoidal operation, called juxtaposition and denoted by $\star$. At the semantic level, we have two basic different interpretations for symocats: one additive where $\star$ is sum, denoted $\oplus$, and the other multiplicative where $\star$ is product, denoted $\otimes$. Most of the results of this chapter hold for both interpretations. In the presentation there are some notational problems, e.g., how to denote and name repeated occurrences of $\star$. We decided to base the proofs on the additive interpretation, still preserving the $\star$ symbol and the general symocat setting as much as possible. E.g., ka denotes $a\star \overset{k}{\ldots} \star a$; $\sum_{i\in[n]} f_i$ denotes $f_1 \star \ldots \star f_n$; etc. $\diamond$

Acyclic NA structure on $\mathbb{R}\mathrm{el}_S$. The acyclic NA signature consists of

$$\star, \cdot, \mathsf{I}, \mathsf{X}, \wedge_k, \vee^k$$

The basic operations of juxtaposition and composition and constants I, X are well known to us by now. For the new part, the general branching constants are defined as follows (they already appeared in Chapter 1):

- IDENTIFICATION:

$$\vee_a^k = \{(i,i) : i \in [|a|]\} \cup \ldots \cup \{((k-1)|a| + i, i) : i \in [|a|]\} \in \mathbb{R}\mathrm{el}_S(ka, a)$$

- RAMIFICATION:

$$\wedge_k^a = \{(i,i) : i \in [|a|]\} \cup \ldots \cup \{(i, (k-1)|a| + i) : i \in [|a|]\} \in \mathbb{R}\mathrm{el}_S(a, ka)$$

♠ In the case $k = 0$ the identification constant $\vee_a^k$ is also called SOURCE and denoted by $\top_a = \emptyset \in \mathbb{R}\mathrm{el}_S(\epsilon, a)$, where ϵ is the empty word. The ramification constant $\wedge_k^a$ for $k = 0$ is called SINK and denoted by $\perp^a = \emptyset \in \mathbb{R}\mathrm{el}_S(a, \epsilon)$. If k is missing, then the default value is 2. ♣

Classes of relations, semantically defined. From the semantic point of view we are interested in studying the following properties of a relation $f \in \mathbb{R}\mathrm{el}_S(a,b)$:

P1. f is *total*: $(\forall i \in [|a|])(\exists j \in [|b|]) : (i,j) \in f$;
P2. f is *surjective*: $(\forall j \in [|b|])(\exists i \in [|a|]) : (i,j) \in f$;
P3. f is *univocal*: $\forall i \in [|a|],\ j,k \in [|b|] : [(i,j) \in f$ and $(i,k) \in f \Rightarrow j = k]$;
P4. f is *injective*: $\forall k \in [|b|], i,j \in [|a|] : [(i,k) \in f$ and $(j,k) \in f \Rightarrow i = j]$.

These properties are very independent and their combinations give sixteen types of finite relations for which we shall find presentations by equations.

Classes of relations, syntactically defined. Passing to the syntactical level, as we shall see soon, every relation $f \in \mathbb{R}\mathrm{el}_S(a,b)$ may be represented in the following form

$$(*) \qquad f = (\textstyle\sum_{j\in[|a|]} \wedge^{a_j}_{m_j}) \cdot f_2 \cdot (\sum_{i\in[|b|]} \vee^{n_i}_{b_i})$$

where, moreover, f_2 is a bijection, hence it may be represented by a symocat ground term. This shows that NA operators are complete for representing all relations.

Table 3.1. xy-restrictions for branching constants

(a) all $m_j = 1$	(α) all $n_i = 1$
(b) all $m_j \leq 1$	(β) all $n_1 \leq 1$
(c) all $m_j \geq 1$	(γ) all $n_i \geq 1$
(d) arbitrary m_j	(δ) arbitrary n_i

We may syntactically define 16 classes of finite relations xy-$\mathbb{R}\mathrm{el}_S$ by imposing certain restrictions on the branching indices m_j and n_i used in representation (*). These restrictions are shown in Table 3.1.

Notation 3.1. (xy-$\mathbb{R}\mathrm{el}_S$) If $x \in \{a,b,c,d\}$ and $y \in \{\alpha,\beta,\gamma,\delta\}$ are two restrictions as in Table 3.1, then xy-$\mathbb{R}\mathrm{el}_S$ denotes the set of all relations having *at least* one representation (*) satisfying restrictions x and y.

Restrictions xy give natural classes of relations xy-$\mathbb{R}\mathrm{el}_S$ which coincide with the classes of relations defined by the properties P1–P4 defined above. The classes of relations under consideration may be ordered using the relation of inclusion. The resulting ordered set is presented by the Hasse diagram in Fig. 3.1.

Intuitive acronyms for xy-$\mathbb{R}\mathrm{el}_S$. Fig. 3.1 contains more intuitive acronyms for xy-$\mathbb{R}\mathrm{el}_S$, e.g, $\mathbb{B}\mathrm{i}_S$ for $a\alpha$-$\mathbb{R}\mathrm{el}_S$ (bijections), $\mathbb{I}\mathrm{n}_S$ for $a\beta$-$\mathbb{R}\mathrm{el}_S$ (injections), $\mathbb{S}\mathrm{ur}_S$ for $a\gamma$-$\mathbb{R}\mathrm{el}_S$ (surjections), $\mathbb{F}\mathrm{n}_S$ for $a\delta$-$\mathbb{R}\mathrm{el}_S$ (functions), $\mathbb{P}\mathrm{fn}_S$ for $b\delta$-$\mathbb{R}\mathrm{el}_S$ (partially defined functions), $\mathbb{R}\mathrm{el}_S$ for $d\delta$-$\mathbb{R}\mathrm{el}_S$ (relations). $\diamond$

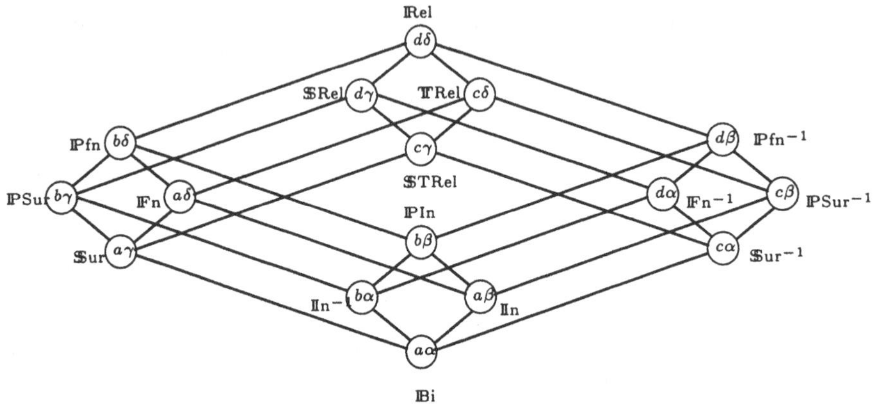

Fig. 3.1. Hasse diagram for classes of xy-relations

Axiomatization results. Roughly speaking, the main result of the first part of this chapter is the following:

> *Axioms for branching constants, acyclic case*
>
> A correct (i.e., sound and complete) axiomatization for relations consists of the following axioms for branching constants: *A1. associativity, A2. commutativity, A3. commutation*, and *A4. idempotency.*

The meaning of the axioms is the following. Associativity assures that arrows in a branching point may be collected. Commutativity implies that there is no difference among the arrows reaching a branching point. Commutation allows us to permute identification with ramification branching constants in a sequential composition. Finally, idempotency says that a k ramification followed by a k identification is identity, provided $k \geq 1$.

This axiomatization result uses general branching constants and it may be specialized for all classes of relations considered so far.

xy-symocats. In more detail, the axiomatization result shows how xy-relations may be mapped in enriched symocats. These symocats, called *xy-symocats*, use xy-branching constants and satisfy the axioms in Table 3.2 written with these constants.

♠ The axioms of type A1–4 are particular instances of the corresponding above stated properties: associativity, commutativity, commutation, and idempotency. The remaining axioms of type A5 (of which A5e–h are sometimes superfluous) show how the branching constants behave when the interface is refined. ♣

Some comments. (1) Why do we need axioms for branching constants? The main point is that a branching point is not, say, a test or something similar. A test is just a network cell. We want to provide a default behaviour in branching points and this is what the axioms do.

Table 3.2. The standard axioms for (angelic) branching constants

III. Axioms for (angelic) branching constants, without feedback		IV. Feedback on (angelic) branching constants	
A1a	$\wedge^a \cdot (\wedge^a \star \mathsf{I}_a) = \wedge^a \cdot (\mathsf{I}_a \star \wedge^a)$	R8	$\wedge^a \uparrow^a = \top_a$
A1b	$\wedge^a \cdot (\perp^a \star \mathsf{I}_a) = \mathsf{I}_a$	R9	$\vee_a \uparrow^a = \perp^a$
A1c	$(\vee_a \star \mathsf{I}_a) \cdot \vee_a = (\mathsf{I}_a \star \vee_a) \cdot \vee_a$	R10	$[(\mathsf{I}_a \star \wedge^a) \cdot ({}^a\mathsf{X}^a \star \mathsf{I}_a)$
A1d	$(\top_a \star \mathsf{I}_a) \cdot \vee_a = \mathsf{I}_a$		$\cdot (\mathsf{I}_a \star \vee_a)] \uparrow^a = \mathsf{I}_a$
A2a	$\wedge^a \cdot {}^a\mathsf{X}^a = \wedge^a$		
A2b	${}^a\mathsf{X}^a \cdot \vee_a = \vee_a$		
A3a	$\top_a \cdot \perp^a = \mathsf{I}_\epsilon$		
A3b	$\top_a \cdot \wedge^a = \top_a \star \top_a$		
A3c	$\vee_a \cdot \perp^a = \perp^a \star \perp^a$		
A3d	$\vee_a \cdot \wedge^a =$		
	$(\wedge^a \star \wedge^a) \cdot (\mathsf{I}_a \star {}^a\mathsf{X}^a \star \mathsf{I}_a) \cdot (\vee_a \star \vee_a)$		
A4	$\wedge^a \cdot \vee_a = \mathsf{I}_a$		
A5a	$\perp^{a \star b} = \perp^a \star \perp^b$		
A5b	$\wedge^{a \star b} = (\wedge^a \star \wedge^b) \cdot (\mathsf{I}_a \star {}^a\mathsf{X}^b \star \mathsf{I}_b)$		
A5c	$\top_{a \star b} = \top_a \star \top_b$		
A5d	$\vee_{a \star b} = (\mathsf{I}_a \star {}^b\mathsf{X}^a \star \mathsf{I}_b) \cdot (\vee_a \star \vee_b)$		
[A5e	$\perp^\epsilon = \mathsf{I}_\epsilon$		
A5f	$\wedge^\epsilon = \mathsf{I}_\epsilon$		
A5g	$\top_\epsilon = \mathsf{I}_\epsilon$		
A5h	$\vee_\epsilon = \mathsf{I}_\epsilon$]		

(2) At a more semantic level, there are two basic models for the behaviour in a branching point. One is similar, say, to the behaviour of a car at a road crossing; the car is "atomic," hence it has to go on a unique forthcoming way. The other model inherits the behaviour of water in a pipe bifurcation; in this case the water goes simultaneously on all forthcoming ways. One model is additive, the other multiplicative. One is corpuscular, the other continuum. ◊

EXERCISES AND PROBLEMS (SEC. 3.1)

1. What restriction corresponds to partial surjective functions? What relations are in $c\beta$-$\mathbb{R}$el? (05)
2. Describe in detail all classes xy-$\mathbb{R}$el, for $x \in \{a, b, c, d\}$, $y \in \{\alpha, \beta, \gamma, \delta\}$. (08)

3.2 Angelic vs. demonic operators

In this section we briefly look at certain demonic relation operators. More on this may be found in Appendix C.

This section may be skipped if the reader is not interested in the demonic calculi.

Some graphical algorithms may facilitate the computations in the algebraic calculi of relations with angelic, forward demonic, backward demonic, or two-way demonic operators. In general, a relational term $f : m \to n$, written with NA operations and constants, may be represented as a graph

$$G(f) = (I, O, Int; E)$$

where $I = \{in_1, \ldots, in_m\}$ (resp. $O = \{out_1, \ldots, out_n\}$; resp. Int) is the set of input (resp. output; resp. internal) vertices of $G(f)$, and E is the set of edges of $G(f)$. We suppose that I, O and Int are disjoint sets. An example is given in Fig. 3.2; the formal definition is given in the following paragraphs.

The critical NA operations on relations having *different* meanings within the demonic approaches are *composition* and *feedback.* Let $\cdot_z/\uparrow_z$ denote a pair of such operations. Their definition is obtained using the following rules.

In the *angelic* case an input and an output of a relational term f are connected in the resulting relation iff there exists a path in $G(f)$ connecting them. On the opposite extreme, in the *two-way demonic* case a whole connected component containing an input or output vertex in $G(f)$ is destroyed if that vertex belongs to a maximal path which does not connect an input with an output. The *forward demonic* case is in between and in this case all the paths containing an output are destroyed if that vertex belongs to a maximal path which fails to start from an input. The situations with *backward demonic* operators is similar, but now the paths containing an input are deleted if this input belongs to a maximal path which fails to reach an output.

Graphical representation of relational terms: A relation $f \subseteq [m] \times [n]$ is graphically represented using m input vertices $in_1, \ldots, in_m$ and n output vertices $out_1, \ldots, out_n$ (and no internal vertices) by

$$G(f) = (\{in_1, \ldots, in_m\}, \{out_1, \ldots, out_n\}, \emptyset; \{(in_i, out_j) : (i,j) \in f\})$$

For instance, the constants may be represented as follows:

$$G(\mathsf{I}_1) = (\{in_1\}, \{out_1\}, \emptyset; \{(in_1, out_1)\});$$
$$G({}^1\mathsf{X}^1) = (\{in_1, in_2\}, \{out_1, out_2\}, \emptyset; \{(in_1, out_2), (in_2, out_1)\});$$
$$G(\wedge_k^1) = (\{in_1\}, \{out_1, \ldots, out_k\}, \emptyset; \{(in_1, out_1), \ldots, (in_1, out_k)\});$$
$$G(\vee_1^k) = (\{in_1, \ldots, in_k\}, \{out_1\}, \emptyset; \{(in_1, out_1), \ldots, (in_k, out_1)\})$$

Let $t : m \to n$ be a relational term over $\mathsf{I}, \mathsf{X}, \perp, \wedge, \top, \vee, \star, \cdot_z, \uparrow_z$, where $\cdot_z/\uparrow_z$ generically denotes a pair of sequential composition/feedback operators used in one of the angelic, forward demonic, backward demonic, or two-way demonic calculus. Then t is represented as a graph $G(t)$ starting from the graphical representation of the constants and using the following rules for the operators:

—$G(f \star g)$, for $f : m \to n$, $g : p \to q$, is obtained renaming the labels of the input (resp. output) vertices in $G(g)$ into $in_{m+1}, \ldots, in_{m+p}$ (resp. into $out_{n+1}, \ldots, out_{n+q}$) and making the disjoint union with $G(f)$.

—$G(f \cdot_z g)$ is the graph obtained identifying the outputs of $G(f)$ with the inputs of $G(g)$; these vertices become internal vertices in $G(f \cdot g)$.
—$G(f \uparrow_z^1)$ for $f : m+1 \to n+1$ is obtained from $G(f)$ by adding an edge from out_{n+1} to in_{m+1} and transforming in_{m+1} and out_{n+1} into internal vertices.

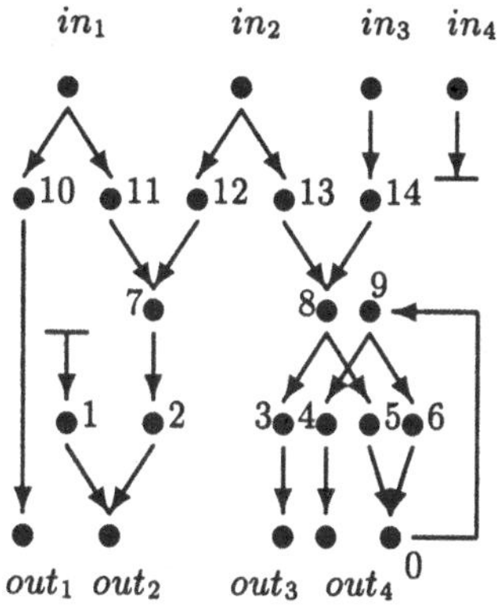

Fig. 3.2. The graphical representation of a relational term

♡ While the graphical representation in Fig. 3.2 has some similarities with the standard graphical representation of relations, actually it is *different*. The main point is that in the current representation the intermediary vertices are kept within the graph, while in the previous examples they were immaterial, the final outcome being a representation without internal vertices. ◇

Definitions. A few definitions simplify the analysis to follow. A maximal path in $G(f)$ is call *i-disconnected* if it fails to start with an input. Furthermore, an i-disconnected path is either infinite (say, it is *looping*) or starts from an internal vertex with no incoming edges (say, it is *not looping*). Similarly, an *o-disconnected* path is a maximal path in $G(f)$ not ending in an output vertex. Finally, a *connected component* in $G(f)$ is just a connected component in the graph obtained making undirected all the edges of $G(f)$. (Hence, the connected component which contains a vertex v consists of all the vertices which may be reached by paths starting from v and using the arrows in both directions.) ◇

Algorithm: Suppose $t : m \to n$ is a relational term and $G(t)$ is the corresponding graph obtained using the above rules. Then:
—the resulting relation $\mathbf{a}(G(f))$ computed in the *angelic* case is

$$\mathbf{a}(G(f)) = \{(in_i, out_j) : \text{there is a finite path } in_i \to \ldots \to out_j \text{ in } G(f)\}$$

—the resulting relation $\mathbf{fd}(G(f))$ computed in the *forward demonic* case is $\mathbf{a}(G_o(f))$, where $G_o(f)$ is obtaining from $G(f)$ by deleting all i-disconnected paths,
—the resulting relation $\mathbf{bd}(G(f))$ computed in the *backward demonic* case is

$\mathbf{a}(G_i(f))$, where $G_i(f)$ is obtaining from $G(f)$ by deleting all o-disconnected paths,
—finally, the resulting relation $\mathbf{fbd}(G(f))$ computed in the *two-way demonic* case is $\mathbf{a}(G_{io}(f))$, where $G_{io}(f)$ is the graph obtaining from $G(f)$ by iteratively deleting all i-disconnected or o-disconnected paths as long as this is possible.

♡ It may be seen that $G_{io}(f)$ is obtained by removing a connected component in $G(f)$ if it contains an i- or o-disconnected path. ◇

Example 3.2. For an illustration of the algorithm we consider the relational term

$$f = (\wedge^1 \star \wedge^1 \star \mathsf{I}_1 \star \bot^1) \cdot (\mathsf{I}_1 \star \vee_1 \cdot (\top_1 \star \mathsf{I}_1) \cdot \vee_1 \star \vee_1 \cdot [(\wedge^2 \cdot (\mathsf{I}_2 \star \vee_1)) \uparrow^1]).$$

Its graphical representation $G(f)$ is given in Fig. 3.2.

Using the algorithm one can easily compute the resulting relation obtained in various calculi, namely:

angelic calculus	$\{(in_1, out_1), (in_1, out_2), (in_2, out_2), (in_2, out_3), (in_2, out_4), (in_3, out_3), (in_3, out_4)\}$
forward demonic calculus	$\{(in_1, out_1), (in_2, out_3), (in_3, out_3)\}$
backward demonic calculus	$\{(in_1, out_1), (in_1, out_2)\}$
two-way demonic calculus	$\emptyset$

A few explanations may be useful. To compute the result obtained in the forward demonic case first we detect the maximal i-disconnected paths

$1 \to out_2$	(not looping, i-disconnected)
$\ldots \to \underline{6 \to 0 \to 9} \to \underline{6 \to 0 \to 9} \to 4 \to out_4$	(looping, i-disconnected)

Hence all the tuples obtained in the angelic case which contain out_2 or out_4 are deleted and the result follows.

Similarly for the backward demonic case.

In the two-way demonic case, in a first step we delete the maximal i- or o-disconnected paths

$1 \to out_2$	(not looping, i-disconnected)
$\ldots \to \underline{6 \to 0 \to 9} \to \underline{6 \to 0 \to 9} \to 4 \to out_4$	(looping, i-disconnected)
$in_2 \to 13 \to 8 \to 5 \to \underline{0 \to 9 \to 6} \to \underline{0 \to 9 \to 6} \to \ldots$	(looping, o-disconnected)
$in_3 \to 14 \to 8 \to 5 \to \underline{0 \to 9 \to 6} \to \underline{0 \to 9 \to 6} \to \ldots$	(looping o-disconnected)

At the second step, new maximal disconnected paths occur

$3 \to out_3$	(not looping, i-disconnected)
$in_1 \to 11 \to 7 \to 2$	(not looping, o-disconnected)

which are deleted. Finally, at the third step, the path

$10 \to out_1$ (not looping, i-disconnected)

is deleted and the empty relation is obtained.

The adaptation of axiomatization from the angelic case to the demonic cases is presented in Appendix C.

EXERCISES AND PROBLEMS (SEC. 3.2)

1. Give explicit definitions for composition and feedback for all demonic versions. (08)
2. Show that the BNA axioms are valid for demonic relational models. What more axioms from Table 3.2 are valid? (10)

3.3 Semantic models: II. NA structure

Cantorian models. Branching constants have the following meaning in the additive model AddRel(D):

- *Additive Ramification:* ${}_a{\bullet\!\!<}_k : a \to ka$ is defined by
  ${}_a{\bullet\!\!<}_k = \{(x, i.x) : i \in [k], x \in D_a\}$
- *Additive Identification:* ${}_k{>\!\!\bullet}_a : ka \to a$ is defined by
  ${}_k{>\!\!\bullet}_a = \{(x, y) : \text{ if } x \text{ is } i.x_i, \text{ then } y = x_i\}$

Cartesian models. The branching constants have the following meaning in the multiplicative model MultRel(D):

- *Multiplicative Ramification:* $\curlywedge_k^a : a \to a^k$ is defined by
  $\curlywedge_k^a = \{(x, (x, \ldots, x)) : x \in D^a\}$
- *Multiplicative Identification:* $\curlyvee_a^k : a^k \to a$ is defined by
  $\curlyvee_a^k = \{((x, \ldots, x), x) : x \in D^a\}$

Notice that $\curlyvee_a^k$ is a partial function: if the input contains two components with different values, then the output is not defined.

EXERCISES AND PROBLEMS (SEC. 3.3)

1. Check that both models AddRel(D) and MultRel(D) satisfy the axioms in Table 3.2. (08)

3.4 Normal form for relations

A normal form for representing relations is introduced in this section. As a byproduct, it shows that our setting suffices for representing all finite relations. The main technical result describes the relation between the normal form representation of a relation and an arbitrary (quasi-nf) representation of that relation.

A *quasi normal form* (or *quasi-nf*) is a term of the following kind: $(\wedge_{k_1}^{a_1} \star \ldots \star \wedge_{k_m}^{a_m}) \cdot bij \cdot (\vee_{b_1}^{l_1} \star \ldots \star \vee_{b_m}^{l_m})$, bij being an abstract bijection, i.e., a ground symocat term.

We start with an example which illustrates how a normal form representation of a relation may be obtained. Let $S \supset \{s, t, u, v\}$ and f be the relation

$$f = \{(1,3), (1,1), (4,1), (4,5), (3,2), (5,2)\} \in \mathbb{R}\mathrm{el}_S(a, b)$$

where $a = s \star t \star v \star s \star v$ and $b = s \star v \star s \star u \star s$, see Fig. 3.3.(a).

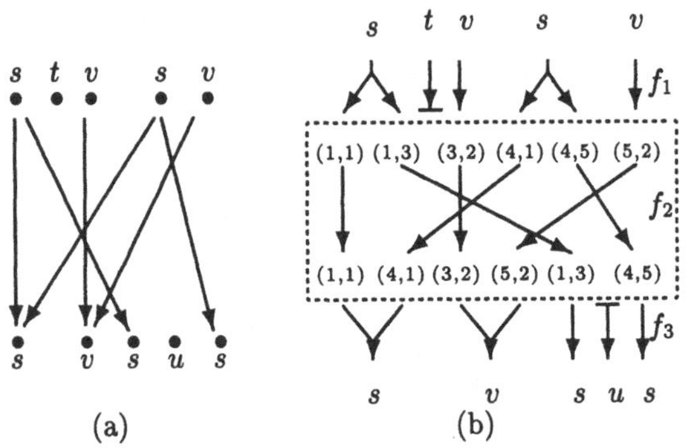

Fig. 3.3. An arbitrary relation (a) and its normal form (b)
Notice that the lexicographical order on pairs is used on the top of the normal form picture, while the anti-lexicographical order is used on the bottom part; the edges of bijection f_2 connect equal pairs.

Then f may be represented as in Fig. 3.3.(b) using the following method:

Procedure for normal form representations of relations

(1) Attach a sort to every arrow $(j, i) \in f$, namely a_j (equal with b_i).
(2) Order the pairs (j, i) in f with respect to the lexicographical order

$$(*) \qquad (j, i) \prec (j', i') \iff j < j' \text{ or } (j = j' \text{ and } i < i')$$

to get the upper side of the rectangle f_2.
(3) Order the pairs (j, i) in f with respect to the anti-lexicographical order

$$(**) \qquad (j, i) \prec^a (j', i') \iff i < i' \text{ or } (i = i' \text{ and } j < j')$$

to get the lower side of the rectangle f_2.
(4) Finally: f_1 is the opposite of the projection on the first component $p(j, i) = j$; the arrows of f_2 are obtained connecting each pair (j, i) of the upper side to the pair (j, i) of the lower side; and f_3 corresponds to the projection on the second component $q(j, i) = i$.

Consequently, we get a quasi-nf decomposition of f as $f_1 \cdot f_2 \cdot f_3$, where

$f_1 = \wedge_2^s \star \wedge_0^t \star \wedge_1^v \star \wedge_2^s \star \wedge_1^v$, f_2 is a bijection $(1_s 5_s 3_v 2_s 6_s 4_v)$, and
$f_3 = \vee_s^2 \star \vee_v^2 \star \vee_s^1 \star \vee_u^0 \star \vee_s^1$.

A quasi-nf representation $f_1 \cdot f_2 \cdot f_3$ of a relation f is not unique due to the following reasons:

- the pairs in f corresponding to a branching cone may be arranged in another linear order, so f_2 is not unique;
- more than one path may connect an input j with an output i, hence the branching degrees of the ramification or identification constants may vary.

More general, quasi-nf representations. For instance, the relation that has been described above may also be represented by the quasi-nf term $(\wedge_3^s \star \wedge_0^t \star \wedge_2^v \star \wedge_3^s \star \wedge_1^v)(3_s 7_s 1_s 5_v 6_v 9_s 2_s 8_s 4_v)(\vee_s^3 \star \vee_v^3 \star \vee_s^1 \star \vee_u^0 \star \vee_s^2)$. In this representation, two paths connect the pair $(3,2)$. At the same time, the order of the arrows is arbitrary; for instance, in the branching point of the first input, the outgoing arrows are not increasingly ordered. $\diamond$

More constraints to such representations are necessary to get the unique normal form representation corresponding to the procedure. They are stated in the theorem below. Before this we explain a notation:

$k + [n]$ denotes $\{k+1, \ldots, k+n\}$ (recall that $[n]$ denotes $\{1, \ldots, n\}$)

Theorem 3.3. *Every relation $f : a \to b$ may be written in a unique way as $f_1 \cdot f_2 \cdot f_3$, where*

(i) $f_1 = \wedge_{m_1}^{a_1} \star \ldots \star \wedge_{m_{|a|}}^{a_{|a|}}$, *$f_2$ is sorted bijection, and* $f_3 = \vee_{b_1}^{n_1} \star \ldots \star \vee_{b_{|b|}}^{n_{|b|}}$;

(ii) *for every $(j,i) \in f$ there exists a unique pair $(k,k') \in f_2$ such that $(j,k) \in f_1$ and $(k',i) \in f_3$;*

(iii) *the restrictions of f_2 on the sets $|m_1 + \ldots + m_{j-1}| + [m_j]$, $j \in [|a|]$ are increasing functions and the corestrictions of f_2 to the sets $|n_1 + \ldots + n_{i-1}| + [n_i]$, $i \in [|b|]$ are increasing (partial) functions.*

PROOF: The representation obtained using the above procedure satisfies (i)–(iii). Moreover, this decomposition of f is the unique one satisfying (i)–(iii). Indeed, if $f_1' \cdot f_2' \cdot f_3'$ is another decomposition of f fulfilling conditions (i)–(iii), then: by (i) there are certain numbers m_j' and n_i' such that $f_1' = \wedge_{m_1'}^{a_1} \star \ldots \star \wedge_{m_{|a|}'}^{a_{|a|}}$ and $f_3' = \vee_{b_1}^{n_1'} \star \ldots \star \vee_{b_{|b|}}^{n_{|b|}'}$; by (ii) $m_j' = card\{i : (j,i) \in f\} = m_j, \ \forall j \in [|a|]$ and $n_i' = card\{j : (j,i) \in f\} = n_i, \ \forall i \in [|b|]$; finally, with (iii) $f_2' = f_2$, hence the decomposition is unique. $\square$

Condition (iii) of this theorem is rather important in what follows. There will be many references to it, so let us give it a name: a bijection f_2 satisfying condition (iii) of Theorem 3.3 is said to be *naturally ordered.*

Convention. In order to shorten the formulae we write $\sum_{i\in[n]} a_i$ instead of $a_1\star\ldots\star a_n$. Since the juxtaposition "$\star$" is not commutative, the notation may be meaningless. To avoid this ambiguity we use the convention that the indices are taken from linearly ordered sets and juxtaposition is done in the increasing order of indices. When nothing else is said and the indices are natural numbers the default order in "less than." ◇

In the rest of this section we are looking for a connection between an arbitrary quasi-nf representation of $f : a \to b$ and the normal form representation. For the normal form representation of f we keep the notation used in the statement of Theorem 3.3.

Lemma 3.4. *If $f = f_1' \cdot f_2' \cdot f_3'$ is a quasi-nf representation of $f : a \to b$, where $f_1' = \sum_{j\in[|a|]} \wedge_{m_j'}^{a_j}$ and $f_3' = \sum_{i\in[|b|]} \vee_{b_i}^{n_i'}$, then there exist bijections $h_j : m_j'a_j \to m_j'a_j$ and $g_i : n_i'b_i \to n_i'b_i$ such that $f = f_1' \cdot f_2'' \cdot f_3'$ with $f_2'' = (\sum_{j\in[|a|]} h_j) \cdot f_2' \cdot (\sum_{i\in[|b|]} g_i)$ a naturally ordered bijection.*

PROOF: Using permutations in each cone of a ramification constant we transform the given representation into one satisfying the first part of (iii) in Theorem 3.3. Then with permutation in each cone of the identification constants the second part of (iii) in Theorem 3.3 is satisfied. (The result obtained after the first step is preserved by the second step.) □

The proof is illustrated in Fig. 3.3. An arbitrary representation of the relation in Fig. 3.3(a) is drawn in (c). Then in (c1) and (c2) are drawn the steps towards a representation (d) with a naturally ordered bijection. The passing from (c) to (c1) consists of the introduction of some bijections in the branching cones of the ramifications to increasingly order the outgoing arrows. E.g., for the first input the arrows go to outputs (3,7,1) of the bijection f_2'; when h_1 is introduced, the corresponding arrows go to outputs (1,3,7). The passing from (c1) to (c2) is similar, but now with respect to the identification cones of the output. E.g., for the first output the arrows which reach its identification cone come from inputs (1,6,2) of the bijection $(\Sigma h_j)f_2'$; when g_1 is introduced, the corresponding arrows come from inputs (1,2,6). The passing from (c2) to (d) consists of a simple redrawing of bijection $(\Sigma h_j)f_2'(\Sigma g_i)$.

The importance of the "naturally ordered" property of the involved bijection comes from the following fact: if a quasi-nf representation $f = f_1' \cdot f_2' \cdot f_3'$ has a naturally ordered f_2', then all the edges of f_2' belonging to paths connecting two elements j and i are parallel and grouped one near the other. In brief, decomposition $f = f_1' \cdot f_2' \cdot f_3'$ has *the same shape* as the normal form decomposition.

♡ This shows that f_2' may be obtained from f_2 by a multiplication of edges. For example, in Fig. 3.4 we have $f_2'' = (1_2 5_1 3_2 2_1 6_2 4_1)$ while $f_2 = (1_1 5_1 3_1 2_1 6_1 4_1)$. ◇

Notation 3.5. ($f_\prec$, $f_\prec^{j,i}$; $f_{\prec^a}$, $f_{\prec^a}^{j,i}$) We denote by $f_\prec$ the set of $\prec$-ordered pairs in f and by $f_\prec^{j,i}$ the number of the pair (j,i) in $f_\prec$, counted in increasing order. $f_{\prec^a}$ and $f_{\prec^a}^{j,i}$ are similarly defined using the anti-lexicographical order $\prec^a$.

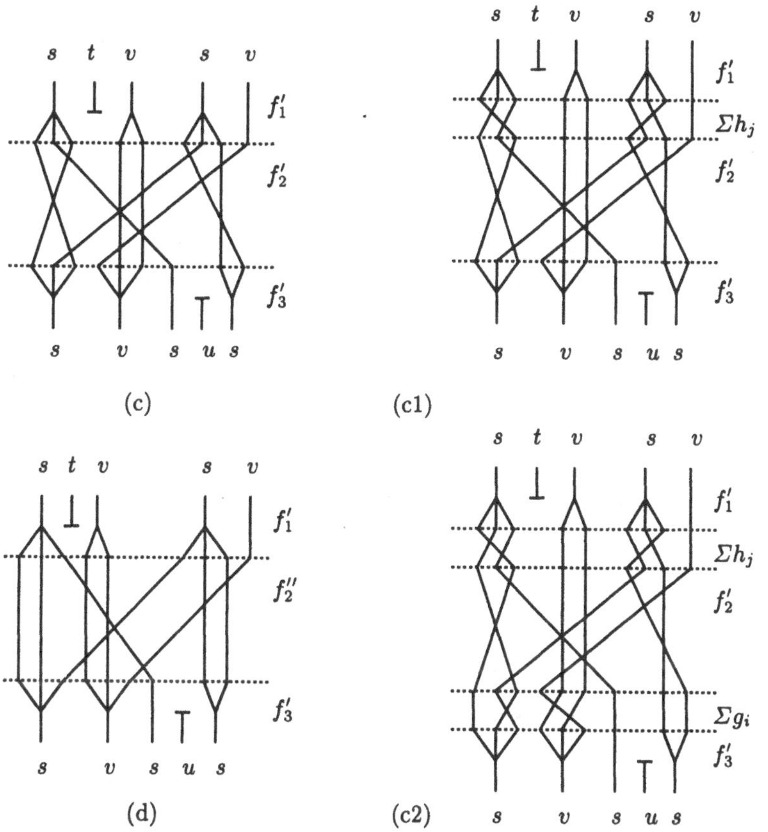

Fig. 3.4. Normalizing representaions

♡ For instance, in the case of our starting example $f_{\prec}^{3,2} = 3$, $f_{\prec}^{5,2} = 6$, $f_{\prec^a}^{5,2} = 4$, etc. With this notation the definition of f_2 used in the normal form representation of f becomes simpler: $f_2(f_{\prec}^{j,i}) = f_{\prec^a}^{j,i}$. ◇

With the bracket notation f_2 is written as

$$f_2 = (\dots (f_{\prec^a}^{j,i})_{a_j} \dots)$$

where the generic term is indexed by (j,i) and this pair varies from left to right increasingly in $f_{\prec}$.

♡ For our example we reobtain the representation $f_2 = (1_1 5_1 3_1 2_1 6_1 4_1)$. ◇

We have got the following result which relates a quasi-nf representation using a naturally ordered bijection and the normal form representation.

Proposition 3.6. *Let $f = f_1 \cdot f_2 \cdot f_3$ be the normal form representation of $f : a \to b$ and $f = f_1' \cdot f_2' \cdot f_3'$ another quasi-nf representation of f with $f_1' = \sum_{j\in[|a|]} \wedge_{m_j'}^{a_j}$, f_2' naturally ordered, and $f_3' = \sum_{i\in[|b|]} \vee_{b_i}^{n_i'}$.*

If $p_{j,i} = card\{(k,k') \in f'_2 : (j,k) \in f'_1 \text{ and } (k',i) \in f'_3\}$, *then*

$$f'_2 = (\dots (f^{j,i}_{\prec^a})_{p_{j,i} a_j} \dots)$$

where the generic term is indexed by (j,i) *and this pair varies from left to right increasingly in* $f_\prec$. □

This shows f'_2 is obtained from f_2 multiplying the edge corresponding to the pair (j,i) by $p_{j,i}$ times. The identity in Proposition 2.5 for f_2 and $g_{j,i} : a_{j,i} \to b_{j,i}$ for $(j,i) \in f$ becomes:

$$\Big(\sum_{(j,i)\in f_\prec} g_{j,i}\Big) \cdot (\dots (f^{j,i}_{\prec^a})_{b_{j,i}} \dots) = (\dots (f^{j,i}_{\prec^a})_{a_{j,i}} \dots) \cdot \Big(\sum_{(j,i)\in f_{\prec^a}} g_{j,i}\Big)$$

EXERCISES AND PROBLEMS (SEC. 3.4)

1. Give a formal presentation of the transformations in Fig. 3.3. (08)

3.5 Axioms for relations

In this section an axiomatization for relations with general branching constants (and angelic relation operators) is presented. It consists of the associativity, commutativity, idempotency, and commutation axioms for the branching constants.

In this section we present an axiomatization for relations using the general branching constants $\wedge^a_k$ and $\vee^k_a$. The result may be specialized for xy-relations, where $x \in \{a,b,c,d\}$ and $y \in \{\alpha,\beta,\gamma,\delta\}$ are two restrictions on the branching degrees of identification and ramification as in Table 3.1.

Roughly speaking the axiomatization consists of the following axioms for branching constants:

associativity, commutativity, idempotency, and *commutation*

They are stated in a precise form in Table 3.3. The commutation axiom is by far the most complicated one. In order to express it we introduce the following family of bijections.

Notation 3.7. $\varphi^s_{m,n} : m(ns) \to n(ms)$ is the bijection given by

- $\varphi^s_{0,n} = \varphi^s_{m,0} = \mathsf{I}_0$;
- $\varphi^s_{m,n}(m(i-1)+j) = n(j-1)+i$ if $m \geq 1, n \geq 1,\ i \in [m]$, and $j \in [n]$

Table 3.3. Axioms for general branching constants.
$F : \mathbb{B}\text{is}_S \to B$ is a morphism of symocats with $F(s) = a$ for an $s \in S$.

A0	$\vee_a^1 = \wedge_1^a = \mathsf{I}_a$	
A1	The branching constants are *associative*	
	$\wedge_m^a \cdot (\sum_{j\in[m]} \wedge_{m_j}^a) = \wedge_{\Sigma_{j\in[m]} m_j}^a$	if m and all m_j satisfy x
	$(\sum_{i\in[n]} \vee_a^{n_i}) \cdot \vee_a^n = \vee_a^{\Sigma_{i\in[n]} n_i}$	if n and all n_i satisfy y
A2	The branching constants are *commutative*	
	$\wedge_m^a \cdot F(f) = \wedge_m^a$	for every bijection $f : ms \to ns$, if m satisfies x
	$F(f) \cdot \vee_a^n = \vee_a^n$	for every bijection $f : ns \to ns$, if n satisfies y
A3	*Commutation of identification and ramification* constants	
	$\vee_a^n \cdot \wedge_m^a = (\sum_{i\in[n]} \wedge_m^a) \cdot F(\varphi_{n,m}^s) \cdot (\sum_{i\in[m]} \vee_a^n)$	if m satisfies x and n satisfies y
A4	*Idempotency*	
	$\wedge_n^a \cdot \vee_a^n = \mathsf{I}_a$	if $n \geq 1$ and n satisfies both restrictions x and y

♡ The intuitive meaning of the bijection $\varphi_{m,n}^s$ is as a natural permutation which rearranges m groups of n elements in n groups of m elements. ◇

The meaning of the axioms is the following:

(1) Associativity assures that arrows in a branching point may be collected. Hence successive ramifications may be replaced with a unique one. (This is a property which, in the given form, is questionable in the demonic view. For instance, in the backward-demonic case, a ramification point with a sink on one of its branches, is equivalent to a sink (followed by the corresponding source). E.g., $\wedge_2(\wedge_2 \star \wedge_0)$ is equal to $\wedge_0\vee_2^0$. In this case the "collection" of the arrows corresponding to the monoidal operation of juxtaposition $\star$ is not their sum $2+0$, but a sort of multiplication 2×0 which models the fact that a zero branching degree destroys the other arrows of the branching point.)

(2) Commutativity implies that there are no differences between the arrows reaching a branching point.

(3) The commutation axiom allows us to permute an identification followed by a ramification constant in a sequential composite.

(4) Finally, the idempotency axiom says that a k ramification followed by a k identification is an identity, provided $k \geq 1$.

Notice that these axioms are valid for both additive and multiplicative interpretations of the branching constants.

Definition 3.8. (xy-object) An object a in a symocat B is an *xy-object* if arrows $\wedge_m^a \in B(a, ma)$ and $\vee_a^n \in B(na, a)$ are given for m satisfying x and n satisfying y such that conditions A0–A4 in Table 3.3 hold.

Particular xy-objects. We describe how concrete relations are mapped in such abstract structures. In the current preliminary setting we need such a branching structure only for the objects in B which are in the image of the map $h : S \to \mathcal{O}(B)$ which generates relation mapping. We do not require a uniform branching structure on the target category B. This will come later. However, notice that in xy-$\mathbb{R}\mathrm{el}_S$ we have a uniform branching structure (every object is an xy-object). ◇

The theorem which maps xy-relations in symocats with such xy-objects follows. It is rather long and technical, so it may be useful to present informally the structure of the proof.

Proof structure. (1) The mapping is defined using arbitrary quasi-nf representations of xy-relations. The correctness of the mapping's definition is based on the results of the previous section, more precisely on the analysis of the relationship between different quasi-nf representations for xy-relations. See Fig. 3.4. With associativity and commutativity axioms A1–A2 for branching constants the image in B of an arbitrary quasi-nf representation is reduced to the shape of the normal form representation. Then apply idempotency axiom A4 to get its equality to the image in B of the normal form representation.

(2) The proof of the commutation of the mapping with the symocat operations is basically reduced to the commutation with composition. (The other is trivial.) A representation is written schematically $\wedge b \vee$ ("b" means "bijection"). What we need here is a procedure to reduce a composite of two representations $(\wedge b \vee) \cdot (\wedge b \vee)$ to a representation. The procedure steps are

$$\wedge\ b\ \vee \wedge\ b\ \vee \quad \to \quad \wedge\ b\ \wedge\ \vee\ b\ \vee \quad \to \quad \wedge \wedge\ b\,b\ \vee \vee \quad \to \quad \wedge\ b\ \vee$$

where in the first step axiom A3 is used. Such a transformation is valid for abstract xy-relations in B, but also for concrete xy-relations in $\mathbb{R}\mathrm{el}_S$. This is used to show the mapping commutes with composition. ◇

Theorem 3.9. *Let xy be a restriction and B a symocat. If $h : S^* \to \mathcal{O}(B)$ is a function which maps every $s \in S$ to an xy-object $h(s)$ in B, then there exists a unique morphism of symocats $H : xy$-$\mathbb{R}\mathrm{el}_S \to B$ such that for $s \in S$: (1) $H(s) = h(s)$; (2) $H(\wedge_m^s) = \wedge_m^{h(s)}$, for m satisfying x; and (3) $H(\vee_s^n) = \vee_{h(s)}^n$, for n satisfying restriction y.*

PROOF: First of all we have to provide a definition for H. We know how to map bijections in symocats. Indeed, by Theorem 2.3 there exists a unique morphism of symocats $F : \mathbb{B}\mathrm{i}_S \to B$ whose restriction to objects agrees with h. This may be extended to xy-relations as follows.

For every xy-relation $f \in xy$-$\mathbb{R}\mathrm{el}_S(a, b)$ and every quasi-nf representation of it

$$f = (\Sigma_{j \in [|a|]} \wedge_{m'_j}^{a_j}) \cdot f'_2 \cdot (\Sigma_{i \in [|b|]} \vee_{b_i}^{n'_i})$$

such that f'_2 is in $\mathbb{B}\mathrm{i}_S$, all m'_j-s obey x and all n'_i-s obey y

we define $H(f) := E$, where

$$E = (\Sigma_{j\in[|a|]}\wedge^{h(a_j)}_{m'_j}) \cdot F(f'_2) \cdot (\Sigma_{i\in[|b|]}\vee^{n'_i}_{h(b_i)})$$

The task now is to show the definition is correct, i.e., the result does not depend on the particular representation we choose for f. It is completed with a proof that E is equal to the expression corresponding to the normal form representation of f.

To fix the notation, let the normal form of f be $(\Sigma_{j\in[|a|]}\wedge^{a_j}_{m_j}) \cdot f_2 \cdot (\Sigma_{i\in[|b|]}\vee^{n_i}_{b_i})$ and take another quasi-nf representation $f = f'_1 \cdot f'_2 \cdot f'_3$, where $f' = \Sigma_{j\in[|a|]}\wedge^{a_j}_{m'_j}$ and $f'_3 = \Sigma_{i\in[|b|]}\vee^{n'_i}_{b_i}$ (that is, as in the proposed definition for H).

First we bring the former arbitrary representation to the shape of the normal form. Using Lemma 3.4 we find bijections $h_j : m'_j a_j \to m'_j a_j$ and $g_i : n'_i b_i \to n'_i b_i$ such that

$$f = f'_1 \cdot f''_2 \cdot f'_3$$

with $f''_2 = (\Sigma_{j\in[|a|]}h_j) \cdot f'_2 \cdot (\Sigma_{i\in[|b|]}g_i)$ a naturally ordered bijection. Using A2 it follows that

$$\begin{aligned} E &= (\Sigma_{j\in[|a|]}\wedge^{h(a_j)}_{m'_j} \cdot F(h_j)) \cdot F(f'_2) \cdot (\Sigma_{i\in[|b|]}F(g_i) \cdot \vee^{n'_i}_{h(b_i)}) \\ &= (\Sigma_{j\in[|a|]}\wedge^{h(a_j)}_{m'_j}) \cdot F(f''_2) \cdot (\Sigma_{i\in[|b|]}\vee^{n'_i}_{h(b_i)}) \end{aligned}$$

Denote

$$p_{j,i} = card\{(k,k') \in f''_2 : (j,k) \in f'_1 \text{ and } (k',i) \in f'_3\}$$

If $(j,i) \in f$, then $1 \le p_{j,i} \le min\{m'_j, n'_i\}$; so $p_{j,i}$ obeys restrictions x and y. By Proposition 3.6

$$\begin{aligned} E &= (\Sigma_{j\in[|a|]}\wedge^{h(a_j)}_{m_j} \cdot (\Sigma_{i\in f(j)}\wedge^{h(a_j)}_{p_{j,i}})) \cdot F(f''_2) \cdot (\Sigma_{i\in[|b|]}(\Sigma_{j\in f^{-1}(i)}\vee^{p_{j,i}}_{b_i}) \cdot \vee^{n_i}_{h(b_i)}) \\ &= (\Sigma_{j\in[|a|]}\wedge^{h(a_j)}_{m_j}) \cdot (\Sigma_{(j,i)\in f_\prec}\wedge^{h(a_j)}_{p_{j,i}}) \cdot (\ldots (f^{j,i}_{\prec^a})_{p_{j,i}h(a_j)} \ldots) \\ &\quad \cdot (\Sigma_{(j,i)\in f_{\prec^a}}\vee^{p_{j,i}}_{h(b_i)}) \cdot (\Sigma_{i\in[|b|]}\vee^{n_i}_{h(b_i)}) \end{aligned}$$

Next we eliminate multiple paths between input–output pairs. To this end we shift the redundant $\wedge_{p_{j,i}}$ along the edges of the bijection. Using Proposition 2.5 (in the form that has been written below Proposition 3.6) we get

$$(\Sigma_{(j,i)\in f_\prec}\wedge^{h(a_j)}_{p_{j,i}}) \cdot (\ldots (f^{j,i}_{\prec^a})_{p_{j,i}h(a_j)} \ldots) = F(f_2) \cdot (\Sigma_{(j,i)\in f_{\prec^a}}\wedge^{h(a_j)}_{p_{j,i}})$$

Replace this in the above expression to get

$$E = (\Sigma_{j\in[|a|]}\wedge^{h(a_j)}_{m_j}) \cdot F(f_2) \cdot (\Sigma_{(j,i)\in f_{\prec^a}} \wedge^{h(a_j)}_{p_{j,i}} \cdot \vee^{p_{j,i}}_{h(b_i)}) \cdot (\Sigma_{i\in[|b|]}\vee^{n_i}_{h(b_i)})$$

Finally, f being a sorted relation, $a_j = b_i$, for every $(j,i) \in f$; hence by A4 we get

$$E = (\Sigma_{j\in[|a|]}\wedge^{h(a_j)}_{m_j}) \cdot F(f_2) \cdot (\Sigma_{i\in[|b|]}\vee^{n_i}_{h(b_i)})$$

This shows the definition of H does not depend on the particular quasi-nf representation of f one starts with.

The second part of the proof is to show H is a morphism. With A0, $H(f) = F(f)$ for a bijection f, hence H extends F. It is also easy to see that $H(\wedge^s_m) = \wedge^{h(s)}_m$ for an m satisfying restriction x and $H(\vee^n_s) = \vee^n_{h(s)}$ for an n satisfying y.

Next, we show h commutes with the symocat operations. In order to prove $H(f \cdot g) = H(f) \cdot H(g)$ for $a \xrightarrow{f} b \xrightarrow{g} c$ we use the normal form representation of f with the notations in Theorem 3.3 and the normal form representation of g, say, $g_1 \cdot g_2 \cdot g_3$, where $g_1 = \Sigma_{i\in[|b|]}\wedge^{b_i}_{m'_i}$, g_2 is bijection, and $g_3 = \Sigma_{k\in[|c|]}\vee^{n'_k}_{c_k}$. Then

$$\begin{aligned} H(f) \cdot H(g) &= (\Sigma_{j\in[|a|]}\wedge^{h(a_j)}_{m_j}) \cdot F(f_2) \cdot (\Sigma_{i\in[|b|]}\vee^{n_i}_{h(b_i)}) \\ &\quad \cdot (\Sigma_{i\in[|b|]}\wedge^{h(b_i)}_{m'_i}) \cdot F(g_2) \cdot (\Sigma_{k\in[|c|]}\vee^{n'_k}_{h(c_k)}) \end{aligned}$$

Step $\wedge\ b\ \vee \wedge\ b \vee \rightarrow \wedge\ b\ \wedge \vee\ b\ \vee$: apply A3 to get

$$\begin{aligned} &(\Sigma_{i\in[|b|]}\vee^{n_i}_{h(b_i)}) \cdot (\Sigma_{i\in[|b|]}\wedge^{h(b_i)}_{m'_i}) \\ &= \ (\Sigma_{(j,i)\in f_{\prec^a}}\wedge^{h(b_i)}_{m'_i}) \cdot (\Sigma_{i\in[|b|]}F(\varphi^{b_i}_{n_i,m'_i})) \cdot (\Sigma_{(i,k)\in g_{\prec}}\vee^{n_i}_{h(b_i)}) \end{aligned}$$

Step $\wedge\ b\ \wedge \vee\ b\ \vee \rightarrow \wedge\wedge\ b\ b\ \vee\vee$: replace this in the above identity and use Proposition 2.5 (as in the proof of the definition correctness) to get

$$\begin{aligned} &H(f) \cdot H(g) \\ &= \ (\Sigma_{j\in[|a|]}\wedge^{h(a_j)}_{m_j}) \cdot (\Sigma_{(j,i)\in f_{\prec}}\wedge^{h(b_i)}_{m'_i}) \\ &\quad \cdot (\ldots (f^{j,i}_{\prec^a})_{m'_i h(b_i)} \ldots) \cdot (\Sigma_{i\in[|b|]}F(\varphi^{b_i}_{n_i,m'_i})) \cdot (\ldots (g^{i,k}_{\prec^a})_{n_i h(b_i)} \ldots) \\ &\quad \cdot (\Sigma_{(i,k)\in g_{\prec}}\vee^{n_i}_{h(b_i)}) \cdot (\Sigma_{k\in[|c|]}\vee^{n'_k}_{h(c_k)}) \end{aligned}$$

Step $\wedge\wedge\ b\ b\ \vee\vee \rightarrow \wedge\ b\ \vee$: since $a_j = b_i$ for all $(j,i) \in f$ and $b_i = c_k$ for all $(i,k) \in g$, with A1 we get

$$\begin{aligned} (*)\quad &H(f) \cdot H(g) \\ &= \ (\Sigma_{j\in[|a|]}\wedge^{h(a_j)}_{\Sigma_{\{i:\,(j,i)\in f\}}m'_i}) \\ &\quad \cdot F([(\ldots (f^{j,i}_{\prec^a})_{m'_i b_i} \ldots) \cdot (\Sigma_{i\in[|b|]}\varphi^{b_i}_{n_i,m'_i}) \cdot (\ldots (g^{i,k}_{\prec^a})_{n_i b_i} \ldots)]) \\ &\quad \cdot (\Sigma_{k\in[|c|]}\vee^{\Sigma_{\{i:\,(i,k)\in g\}}n_i}_{h(c_k)}) \end{aligned}$$

By a case analysis for $x \in \{a, b, c, d\}$, it is easy to see that all the sums above $\Sigma_{\{i:\,(j,i)\in f\}}m'_i$ satisfy restriction x. Similarly, all the sums $\Sigma_{\{i:\,(i,k)\in g\}}n_i$ satisfy restriction y.

We first observe that the above deduction is valid in every xy-symocat, hence also in xy-$\mathbb{R}\mathrm{el}_S$. Applying (*) when h is the identity on objects, in which case the extensions F and H are identities, we get a representation of $f \cdot g$.

Now we return to the general case. From the above observation and the definition of H it follows that the right hand side of (*) represents an expression equal to $H(f \cdot g)$. Hence, $H(f \cdot g) = H(f) \cdot H(g)$.

The proof of the commutation with juxtaposition is fairly simple and it is omitted. □

Mapping based on normal forms. Why not defining the extension using normal forms only? Indeed, at first sight it seems to be easier. But the problems occur when one composes such normal forms. The composite is no longer in normal form, hence the same machinery to reduce the result to the normal form is necessary. ◇

3.6 Simplification

In this section we describe an instance of the above axiomatization result using branching constants of branching degree 0 and 2 only. This way an equivalent axiomatic system for relations with more, but simpler axioms is obtained.

The system of axioms we have described for relations is conceptually simple. Indeed, we have 4 basic axioms, which are fairly easy to understand. However, axiom A3 is quite complex and cannot be easily verified. It will be better if we can simplify it. It is the purpose of this section to show that this can be done.

The particular branching constants we are considering here are those with branching degree 0 or 2. In $\mathbb{R}\mathrm{el}_S$ they have the following meaning:

- IDENTIFICATION: $\mathsf{V}_a = \{(i,i) : i \in [|a|]\} \cup \{(|a|+i, i) : i \in [|a|]\} : a \star a \to a$;
- SOURCE: $\mathsf{T}_a = \emptyset : \epsilon \to a$, where ϵ is the empty word;
- SINK: $\perp^a = \emptyset : a \to \epsilon$;
- RAMIFICATION: $\wedge^a = \{(i,i) : i \in [|a|]\} \cup \{(i, |a|+i) : i \in [|a|]\} : a \to a \star a$.

The general branching constants $\mathsf{V}_a^n \in \mathbb{R}\mathrm{el}_S(na, a)$ and $\wedge_n^a \in \mathbb{R}\mathrm{el}_S(a, na)$ are derived from these particular branching constants with the following inductive definitions:

$$\mathsf{V}_a^0 = \mathsf{T}_a \qquad\qquad \wedge_0^a = \perp^a$$
$$\mathsf{V}_a^{n+1} = (\mathsf{V}_a^n \star \mathsf{I}_a) \cdot \mathsf{V}_a \qquad\qquad \wedge_{n+1}^a = \wedge^a \cdot (\wedge_n^a \star \mathsf{I}_a).$$

These rules will be used to define general branching constants in arbitrary symocats starting with their particular cases of branching degree 0 and 2. Note that $\mathsf{V}_a^1 = \wedge_1^a = \mathsf{I}_a$, $\mathsf{V}_a^2 = \mathsf{V}_a$ and $\wedge_2^a = \wedge^a$.

A set of axioms for these constants is presented in Table 3.2. The branching constants satisfy the axioms of type A1 to A4 and axioms of type A5.

♠ These latter axioms show how the vectorial constants may be expressed in terms of scalar ones ("vectorial" means $\mathsf{V}_a, \mathsf{T}_a, \perp^a, \wedge^a$ for $|a| \geq 0$, while "scalar" is the same, but for $|a| = 1$). Note that the axioms B6 and B8 are of the same type. ♣

The identities in Table 3.2 hold in $\mathbb{R}\mathrm{el}_S$.

The aim of this section is to simplify as much as possible conditions A1–A4. We prove that axioms of type A1 to A4 in Table 3.2, which are particular instances of the conditions A1–A4, are enough to imply the validity of A1–A4 in all cases.

If $x = d$ (resp. $y = \delta$) the first step in the inductive definition is $\wedge_0^a = \perp^a$ (resp. $\vee_a^0 = \top_a$). In this case one may use axiom A7 (resp. A3) in order to prove A0. In the other cases A0 follows from the definition: that is, $\wedge_1^a = \mathsf{I}_a$ (resp. $\vee_a^1 = \mathsf{I}_a$) is the basis of the inductive definition in the case $x = c$ (resp. $y = \gamma$). Let us note that A0 implies $\wedge_2^a = \wedge^a$ and $\vee_a^2 = \vee_a$.

Roughly speaking there are 60 variants (4 conditions × 15 cases, i.e., the cases xy with $xy \neq a\alpha$). The study is reduced to 7 cases using the following simplifications:

a) Duality. The duality we are considering here is the process of changing of statements with respect to the following rules: (it formalizes the idea of changing the sense of the arrows)

$$f \cdot g \text{ is replaced by } g \cdot f, \quad {}^a\mathsf{X}^b \text{ is replaced by } {}^b\mathsf{X}^a,$$
$$\wedge_n^a \text{ is replaced by } \vee_a^n \qquad \vee_a^n \text{ is replaced by } \wedge_n^a,$$
$$f \star g \text{ and } \mathsf{I}_a \text{ remain unchanged.}$$

The symocat structure is self-dual. The second identity in A1 (resp. A2) is dual with the first one. By duality restrictions a, b, c, d are interchanged with $\alpha, \beta, \gamma, \delta$, respectively. So, we reduce the study to 9 cases: $a\beta$, $a\gamma$, $a\delta$, $b\beta$, $b\gamma$, $b\delta$, $c\gamma$, $c\delta$, $d\delta$. E.g., by duality a result about $b\gamma$ produces a result about $c\beta$.

b) No mixed constants. If an identity in condition Ai contains only constants $\wedge_m$ (resp. $\vee^n$), then its study may be reduced to the cases $x\alpha$ (resp. ay); this reduction may be applied to conditions A1 and A2.

c) Fixed branching degree. If an identity in a condition Ai has all the occurrences of the constants $\wedge_m$ (resp. $\vee^n$) to the same power m (resp. n), then its study in the case dy (resp. $x\delta$) may be reduced to its study in the cases by and cy (resp. $x\beta$ and $x\gamma$); this reduction applies to A2, A3 and A4.

Analysis of A1. By a) and b) we reduce the study of A1 to the study of the second identity in A1 in cases $a\beta$, $a\gamma$ and $a\delta$. Since for $n \in \{0, 1\}$ the second identity in A1 holds, the case $a\beta$ is over. Hence it remains to prove *the second identity in* A1 *in cases* $a\gamma$ *and* $a\delta$.

Analysis of A2. As in the above analysis of A1 the study is reduced to the second identity in cases $a\gamma$ and $a\delta$. By reduction c), case $a\delta$ is reduced to cases $a\beta$ and $a\gamma$. Hence we still have to prove *the second identity in* A2 *in case* $a\gamma$.

Analysis of A3. In the case $m = 1$, A3 holds. By duality we reduce the study to six cases: $b\beta$, $b\gamma$, $b\delta$, $c\gamma$, $c\delta$, $d\delta$. Finally, using reduction c) the problem is reduced to the study of A3 *in cases* $b\beta, b\gamma$ *and* $c\gamma$.

Analysis of A4. Since A4 holds when $x \in \{a, b\}$ or $y \in \{\alpha, \beta\}$ it remains to study four cases: $c\gamma, c\delta, d\gamma, d\delta$. By reduction c) the proof is reduced to the study of A4 *in case* $c\gamma$.

The reduced set of cases to be proved are collected and proved in the lemma to follow.

Lemma 3.10. *(A0–A4 from their simplified versions in Table 3.2)*

1. *The second identity in A1 in case* $a\gamma$ *follows from A1c.*
2. *The second identity in A1 in case* $a\delta$ *follows from A1c, A2b and A1d.*
3. *The second identity in A2 in case* $a\gamma$ *follows from A1c and A2b.*
4. *A3 in case* $b\beta$ *follows from A3a.*
5. *A3 in case* $b\gamma$ *follows from A3c.*
6. *A3 in case* $c\gamma$ *follows from A3d.*
7. *A4 in case* $c\gamma$ *follows from A4.*

PROOF: 1. *Second identity in A1, case* $a\gamma$. For $n = 1$ the equality obviously holds. For $n = 2$ we have to prove that

$$(\mathsf{V}_a^i \star \mathsf{V}_a^j) \cdot \mathsf{V}_a = \mathsf{V}_a^{i+j}$$

for every $i, j \geq 1$. We prove this by induction on j. The case $j = 1$ follows by definition. For the inductive step using: definition in $=^1$, axiom A1c in $=^2$, inductive hypothesis in $=^3$, and definition in $=^4$ we get

$$\begin{aligned}(\mathsf{V}_a^i \star \mathsf{V}_a^{j+1}) \cdot \mathsf{V}_a &=^1 (\mathsf{V}_a^i \star \mathsf{V}_a^j \star \mathsf{I}_a) \cdot (\mathsf{I}_a \star \mathsf{V}_a) \cdot \mathsf{V}_a \\ &=^2 (\mathsf{V}_a^i \star \mathsf{V}_a^j \star \mathsf{I}_a) \cdot (\mathsf{V}_a \star \mathsf{I}_a) \cdot \mathsf{V}_a \\ &=^3 (\mathsf{V}_a^{i+j} \star \mathsf{I}_a) \cdot \mathsf{V}_a \\ &=^4 \mathsf{V}_a^{i+j+1}\end{aligned}$$

For $n \geq 3$ using: definition in $=^1$, inductive hypothesis in $=^2$, and step $n = 2$ in $=^3$ we get

$$\begin{aligned}(\Sigma_{i\in[n]}\mathsf{V}_a^{k_i}) \cdot \mathsf{V}_a^n &=^1 (\Sigma_{i\in[n-1]}\, \mathsf{V}_a^{k_i} \star \mathsf{V}_a^{k_n}) \cdot (\mathsf{V}_a^{n-1} \star \mathsf{I}_a) \cdot \mathsf{V}_a \\ &=^2 (\mathsf{V}_a^{\Sigma_{i\in[n-1]}k_i} \star \mathsf{V}_a^{k_n}) \cdot \mathsf{V}_a \\ &=^3 \mathsf{V}_a^{\Sigma_{i\in[n]}k_i}\end{aligned}$$

2. *Second identity in A1, case* $a\delta$. Using axiom A2b in A1d we get $(\mathsf{I}_a \star \mathsf{T}_a) \cdot \mathsf{V}_a = \mathsf{I}_a$. Now the proof of this case is similar to the proof of the above one, with the difference that each induction starts with 0. For $n = 0$ it is clear (the empty juxtaposition is I_ϵ). In the case $n = 2$, the starting step $j = 0$ is implied by

$$(\mathsf{V}_a^i \star \mathsf{V}_a^\epsilon) \cdot \mathsf{V}_a = (\mathsf{V}_a^i \star \mathsf{I}_\epsilon) \cdot (\mathsf{I}_a \star \mathsf{T}_a) \cdot \mathsf{V}_a = \mathsf{V}_a^i$$

3. *Second identity in A2, case aγ.* Every bijection $f \in \mathbb{B}\mathrm{i}_S(ns, ns)$ may be written as a composition of bijections of the type $\mathsf{I}_{is} \star {}^s\mathsf{X}^s \star \mathsf{I}_{(n-i-2)s}$, where $0 \leq i \leq n-2$. Hence it is enough to prove the statement in the lemma for this type of bijection. This follows using: the first statement 1 of the present lemma in $=^1$; axiom A2b in $=^2$; and again step 1 of this lemma in $=^2$ in

$$\begin{aligned}
&(\mathsf{I}_{ia} \star {}^a\mathsf{X}^a \star \mathsf{I}_{(n-i-2)a}) \cdot \mathsf{V}_a^n \\
&=^1 \quad (\mathsf{I}_{ia} \star {}^a\mathsf{X}^a \star \mathsf{I}_{(n-i-2)a}) \cdot (\mathsf{I}_{ia} \star \mathsf{V}_a \star \mathsf{I}_{(n-i-2)a}) \cdot \mathsf{V}_a^{n-1} \\
&=^2 \quad (\mathsf{I}_{ia} \star \mathsf{V}_a \star \mathsf{I}_{(n-i-2)a}) \cdot \mathsf{V}_a^{n-1} \\
&=^3 \quad \mathsf{V}_a^n
\end{aligned}$$

4. *A3, case bβ.* In the case $n = 1$ or $m = 1$, property A3 obviously holds. The case left open $m = 0 = n$ follows from axiom A3a.

5. *A3, case bγ.* The case $m = 1$ clearly holds. In the case $m = 0$, by induction on n it follows that

$$\begin{aligned}
\mathsf{V}_a^{n+1} \cdot \bot^a &=^1 (\mathsf{V}_a^n \star \mathsf{I}_a) \cdot \mathsf{V}_a \cdot \bot^a \\
&=^2 (\mathsf{V}_a^n \star \mathsf{I}_a) \cdot (\bot^a \star \bot^a) \\
&=^3 \Sigma_{i\in[n]} \bot^a \star \bot^a \\
&=^4 \Sigma_{i\in[n+1]} \bot^a
\end{aligned}$$

where we used axiom A3c in $=^2$ and inductive hypothesis in $=^3$.

6. *A3, case cγ.* We prove A3 using an induction on $n+m$, for pairs (n, m) with $n \geq 2$ and $m \geq 2$. For convenience we repeat the definition of $\varphi_{n,m}^s$:

- $\varphi_{0,n}^s = \varphi_{m,0}^s = \mathsf{I}_0$;
- $\varphi_{m,n}^s(m(i-1)+j) = n(j-1)+i$ if $m \geq 1, n \geq 1$, and $i \in [m],\ j \in [n]$

The case (2,2) follows from axiom A3d. The case $(2, m+1)$ may be reduced to the case $(2, m)$ as follows (use A3d in $=^2$ and the inductive hypothesis for $(2, m)$ in $=^3$)

$$\begin{aligned}
&\mathsf{V}_a^2 \cdot \wedge_{m+1}^a \\
&=^1 \mathsf{V}_a\ \wedge^a\ (\wedge_m^a \star \mathsf{I}_a) \\
&=^2 (\wedge^a \star \wedge^a)\ (\mathsf{I}_a \star {}^a\mathsf{X}^a \star \mathsf{I}_a)\ (\mathsf{V}_a \star \mathsf{V}_a)\ (\wedge_m^a \star \mathsf{I}_a) \\
&=^3 (\wedge^a \star \wedge^a)\ (\mathsf{I}_a \star {}^a\mathsf{X}^a \star \mathsf{I}_a)\ [(\wedge_m^a \star \wedge_m^a)\ F(\varphi_{2,m}^s)\ (\Sigma_{i\in[m]}\mathsf{V}_a) \star \mathsf{V}_a] \\
&=^4 (\wedge^a \star \wedge^a)\ (\wedge_m^a \star {}^a\mathsf{X}^a(\wedge_m^a \star \mathsf{I}_a) \star \mathsf{I}_a)\ (F(\varphi_{2,m}^s) \star \mathsf{I}_{2a})\ (\Sigma_{i\in[m+1]}\mathsf{V}_a) \\
&=^5 (\wedge^a \star \wedge^a)\ (\wedge_m^a \star \mathsf{I}_a \star \wedge_m^a \star \mathsf{I}_a)\ (\mathsf{I}_{ma} \star {}^a\mathsf{X}^{ma} \star \mathsf{I}_a)\ (F(\varphi_{2,m}^s) \star \mathsf{I}_{2a})\ (\Sigma_{i\in[m+1]}\mathsf{V}_a) \\
&=^6 (\wedge_{m+1}^a \star \wedge_{m+1}^a)\ F((\mathsf{I}_{ms} \star {}^s\mathsf{X}^{ms} \star \mathsf{I}_s)\ (\varphi_{2,m}^s) \star \mathsf{I}_{2s}))\ (\Sigma_{i\in[m+1]}\mathsf{V}_a) \\
&=^7 (\Sigma_{i\in[2]}\wedge_{m+1}^a)\ F(\varphi_{2,m+1}^s)\ (\Sigma_{i\in[m+1]}\mathsf{V}_a)
\end{aligned}$$

Hence the identity is valid for $n = 2$ and arbitrary m.

Finally, the case $(n+1,m)$ with $n>2$ may be reduced to the cases $(2,m)$ and (n,m) as follows (use inductive hypothesis for $(2,m)$ in $=^2$ and for (n,m) in $=^3$)

$$
\begin{aligned}
&\mathsf{V}_a^{n+1}\cdot\Lambda_m^a\\
&=^1 \quad (\mathsf{V}_a^n\star \mathsf{I}_a)\ \mathsf{V}_a\ \Lambda_m^a\\
&=^2 \quad (\mathsf{V}_a^n\star \mathsf{I}_a)\ (\Lambda_m^a\star\Lambda_m^a)\ F(\varphi_{2,m}^s)\ (\Sigma_{i\in[m]}\mathsf{V}_a)\\
&=^3 \quad [(\Sigma_{i\in[n]}\Lambda_m^a)\ F(\varphi_{n,m}^s)\ (\Sigma_{i\in[m]}\mathsf{V}_a^n)\ \star\ \Lambda_m^a]\ F(\varphi_{2,m}^s)\ (\Sigma_{i\in[m]}\mathsf{V}_a)\\
&=^4 \quad (\Sigma_{i\in[n+1]}\Lambda_m^a)\ (F(\varphi_{n,m}^s)\star \mathsf{I}_{ma})\ (\Sigma_{i\in[m]}\ \mathsf{V}_a^n\ \star \mathsf{I}_{ma})\ F(\varphi_{2,m}^s)\ (\Sigma_{i\in[m]}\mathsf{V}_a)
\end{aligned}
$$

The proof is finished if we show that

$$(*)\quad (F(\varphi_{n,m}^s)\star \mathsf{I}_{ma})\cdot(\Sigma_{i\in[m]}\ \mathsf{V}_a^n\ \star \mathsf{I}_{ma})\cdot F(\varphi_{2,m}^s) = F(\varphi_{n+1,m}^s)\cdot(\Sigma_{i\in[m]}(\mathsf{V}_a^n\star \mathsf{I}_a))$$

To this end it is useful to use an auxiliary function ψ related to φ. It appears when one tries to write an inductive definition for φ as

$$(a)\quad \varphi_{n+1,m}^s = (\varphi_{n,m}^s\star \mathsf{I}_{ms})\psi_{n,m}^s$$

Before the spelling of the definition for ψ we note that now $(*)$ is reduced to

$$(**)\quad (\Sigma_{i\in[m]}\ \mathsf{V}_a^n\ \star \mathsf{I}_{ma})\cdot F(\varphi_{2,m}^s) = F(\psi_{n,m}^s)\cdot(\Sigma_{i\in[m]}(\mathsf{V}_a^n\star \mathsf{I}_a))$$

The intuitive meaning of the recurrent definition for φ is the following. To rearrange $n+1$ groups of m elements in m groups of $n+1$ elements, first one rearranges the left-most n groups of m elements in m groups of n elements and then inserts one element from the remaining $n+1$-th group in each group obtained so far. $\psi_{n,m}$ does the second part of the job. Its formal definition is

$$
\begin{aligned}
&\psi\in \mathbb{B}\mathrm{i}_S(m(ns)+ms,\ m((n+1)s)),\ m,n\geq 1 \text{ is defined by}\\
&\psi_m(in+k)=i(n+1)+k, \quad \text{if } 0\leq i<m \text{ and } k\in[n] \text{ and}\\
&\psi_m(mn+i)=i(n+1), \qquad \text{if } i\in[m]
\end{aligned}
$$

First note that

b) $\psi_1=\mathsf{I}_{(n+1)s}$
c) $\varphi_{2,m+1}^s=(\mathsf{I}_{ms}\star{}^s\mathsf{X}^{ms}\star \mathsf{I}_s)\cdot(\varphi_{2,m}^s\star \mathsf{I}_{2s})$ and
d) $(\mathsf{I}_{m(ns)}\star{}^{ns}\mathsf{X}^{ms}\star \mathsf{I}_s)\cdot(\psi_m\star \mathsf{I}_{(n+1)s})=\psi_{m+1}$

The proof of $(**)$ follows by induction on m. The inductive step looks as follows (use the inductive hypothesis $(**)$ for m in $=^3$)

$$
\begin{aligned}
&(\Sigma_{i\in[m+1]}\ \mathsf{V}_a^n\star \mathsf{I}_{(m+1)a})\cdot F(\varphi_{2,m+1}^s)\\
&=^1 \quad [\Sigma_{i\in[m]}\ \mathsf{V}_a^n\star(\mathsf{V}_a^n\star \mathsf{I}_{ma})\cdot{}^a\mathsf{X}^{ma}\star \mathsf{I}_a]\ (F(\varphi_{2,m}^s)\star \mathsf{I}_{2a})\\
&=^2 \quad (\mathsf{I}_{m(na)}\star{}^{na}\mathsf{X}^{ma}\star \mathsf{I}_a)\ [(\Sigma_{i\in[m]}\ \mathsf{V}_a^n\star \mathsf{I}_{ma})\ F(\varphi_{2,m}^s)\ \star\ \mathsf{V}_a^n\ \star\ \mathsf{I}_a]\\
&=^3 \quad F(\ (\mathsf{I}_{m(ns)}\star{}^{ns}\mathsf{X}^{ms}\star \mathsf{I}_s)(\psi_m\star \mathsf{I}_{(n+1)s})\ )\cdot(\Sigma_{i\in[m+1]}(\mathsf{V}_m^n\star \mathsf{I}_a))\\
&=^4 \quad F(\psi_{m+1})\cdot(\Sigma_{i\in[m+1]}(\mathsf{V}_m^n\star \mathsf{I}_a))
\end{aligned}
$$

7. *A4, case cγ*. By induction on m we get

$$\Lambda_{m+1}^a\cdot\mathsf{V}_a^{m+1}=\Lambda^a\cdot(\Lambda_m^a\star \mathsf{I}_a)\cdot(\mathsf{V}_a^m\star \mathsf{I}_a)\cdot\mathsf{V}_a=\Lambda^a\cdot(\mathsf{I}_a\star \mathsf{I}_a)\cdot\mathsf{V}_a=\mathsf{I}_a. \qquad \square$$

EXERCISES AND PROBLEMS (SEC. 3.6)

1. Show that the following rewriting system is confluent and terminating, providing a presentation for finite functions ($\mathbb{P}$fn), using atomic constants ${}^1\mathsf{X}^1, \mathsf{V}_1, \mathsf{T}_1$, only;
   1. $(\mathsf{V}_1 \star \mathsf{I}_1)\mathsf{V}_1 \to (\mathsf{I}_1 \star \mathsf{V}_1)\mathsf{V}_1$
   2. $(\mathsf{T}_1 \star \mathsf{I}_1)\mathsf{V}_1 \to \mathsf{I}_1$
   3. $(\mathsf{I}_1 \star \mathsf{T}_1)\mathsf{V}_1 \to \mathsf{I}_1$
   4. ${}^1\mathsf{X}^1 \cdot {}^1\mathsf{X}^1 \to \mathsf{I}_2$
   5. $({}^1\mathsf{X}^1 \star \mathsf{I}_1)(\mathsf{I}_1 \star {}^1\mathsf{X}^1)({}^1\mathsf{X}^1 \star \mathsf{I}_1) \to (\mathsf{I}_1 \star {}^1\mathsf{X}^1)({}^1\mathsf{X}^1 \star \mathsf{I}_1)(\mathsf{I}_1 \star {}^1\mathsf{X}^1)$
   6. $(\mathsf{V}_1 \star \mathsf{I}_1){}^1\mathsf{X}^1 \to (\mathsf{I}_1 \star {}^1\mathsf{X}^1)({}^1\mathsf{X}^1 \star \mathsf{I}_1)(\mathsf{I}_1 \star \mathsf{V}_1)$
   7. $({}^1\mathsf{X}^1 \star \mathsf{I}_1)(\mathsf{I}_1 \star {}^1\mathsf{X}^1)(\mathsf{V}_1 \star \mathsf{I}_1) \to (\mathsf{I}_1 \star \mathsf{V}_1){}^1\mathsf{X}^1$
   8. $({}^1\mathsf{X}^1 \star \mathsf{I}_1)(\mathsf{I}_1 \star \mathsf{V}_1){}^1\mathsf{X}^1 \to (\mathsf{I}_1 \star {}^1\mathsf{X}^1)(\mathsf{V}_1 \star \mathsf{I}_1)$
   9. $(\mathsf{T}_1 \star \mathsf{I}_1){}^1\mathsf{X}^1 \to \mathsf{I}_1 \star \mathsf{T}_1$
   10. $(\mathsf{I}_1 \star \mathsf{T}_1){}^1\mathsf{X}^1 \to \mathsf{T}_1 \star \mathsf{I}_1$
   11. ${}^1\mathsf{X}^1\mathsf{V}_1 \to \mathsf{V}_1$
   12. $({}^1\mathsf{X}^1 \star \mathsf{I}_1)(\mathsf{I}_1 \star \mathsf{V}_1)\mathsf{V}_1 \to (\mathsf{I}_1 \star \mathsf{V}_1)\mathsf{V}_1$. (18)
2. Extend the above rewriting system to the case of all relations. (**)

3.7 Relations in xy-symocats

We collect, and extend, here the results of the previous two sections on mapping relations in xy-symocats. To this end the interface refinement rules are considered.

Definition 3.11. An *xy-symocat* is a symocat where every object in B is an xy-object and the corresponding axioms A5 of type xy in Table 3.2 hold.

♠ In detail, axioms A5c,g have to be satisfied when $y \in \{\beta, \delta\}$, axioms A5d,h when $y \in \{\gamma, \delta\}$, axioms A5a,e when $x \in \{b, d\}$, and axioms A5b,f when $x \in \{c, d\}$. ♣

A *morphism of xy-symocats* $H : B \to B'$ is a morphism of symocats that preserves the additional constants.

♠ That is, for every object a in B the following relations are valid:
$H(\mathsf{T}_a) = \mathsf{T}_{H(a)}$ when $y \in \{\beta, \delta\}$; $H(\mathsf{V}_a) = \mathsf{V}_{H(a)}$ when $y \in \{\gamma, \delta\}$;
$H(\perp^a) = \perp^{H(a)}$ when $x \in \{b, d\}$; $H(\wedge^a) = \wedge^{H(a)}$ when $x \in \{c, d\}$. ♣

Overlapping. Notice that with this general definition a symocat may be identified with an $a\alpha$-symocat. The commuting conditions in the above definition of xy-symocat morphisms are equivalent to the ones defined with general branching constants. ♢

Redundancy of axioms A5e–h. A comment of axioms A5e–h is maybe necessary. In the full setting $d\delta$ they are not necessary, more or less as was the case with the symocat rule ${}^{a}\mathsf{X}^{\epsilon} = \mathsf{I}_a$. E.g., $\wedge^{\epsilon} = \wedge^{\epsilon\star\epsilon} = (\wedge^{\epsilon}\star\wedge^{\epsilon})(\mathsf{I}_{\epsilon}\star{}^{\epsilon}\mathsf{X}^{\epsilon}\star\mathsf{I}_{\epsilon}) = \wedge^{\epsilon}\star\wedge^{\epsilon}$; then we can compose with $\vee_{\epsilon}$ to get $\mathsf{I}_{\epsilon} = \wedge^{\epsilon}\cdot\vee_{\epsilon} = (\wedge^{\epsilon}\star\wedge^{\epsilon})(\vee_{\epsilon}\star\mathsf{I}_{\epsilon}) = \mathsf{I}_{\epsilon}\star\wedge^{\epsilon} = \wedge^{\epsilon}$. Next, $\perp^{\epsilon} = \perp^{\epsilon\star\epsilon} = \perp^{\epsilon}\star\perp^{\epsilon}$; composing with $\top_{\epsilon}$ one gets $\mathsf{I}_{\epsilon} = \top_{\epsilon}\perp^{\epsilon} = (\top_{\epsilon}\star\mathsf{I}_{\epsilon})(\perp^{\epsilon}\star\perp^{\epsilon}) = \mathsf{I}_{\epsilon}\star\perp^{\epsilon} = \perp^{\epsilon}$. The other cases are similar. ◊

Now we may formulate the main result of this first part of the chapter.

Theorem 3.12. *Let $x \in \{a,b,c,d\}$ and $y \in \{\alpha,\beta,\gamma,\delta\}$ be two restrictions. If B is an xy-symocat, then for every function $h : S \to \mathcal{O}(B)$ there exists a unique morphism of xy-symocats*

$$H : xy\text{-}\mathbb{R}\mathrm{el}_S \to B$$

extending h, i.e., satisfying the condition $H(s) = h(s)$, $\forall s \in S$.

PROOF: This result is based on Theorem 3.9. The only fact still open is the commutation with the constants $\wedge^a$, $\perp^a$, $\vee_a$ and $\top_a$ in the case a is an arbitrary word in $S^\star$, not only a letter. This follows by induction on the length of a. The case $|a| = 0$ is trivial. The inductive step may be proved as follows: for $a \in S^*$ and $s \in S$

$$\begin{aligned}
H(\wedge^{a\star s}) &=^1 H((\wedge^a\star\wedge^s)\cdot(\mathsf{I}_a\star{}^a\mathsf{X}^s\star\mathsf{I}_s)) \\
&=^2 (H(\wedge^a)\star H(\wedge^s))\cdot(\mathsf{I}_{h(a)}\star{}^{h(a)}\mathsf{X}^{h(s)}\star\mathsf{I}_{h(s)}) \\
&=^3 (\wedge^{h(a)}\star\wedge^{h(s)})\cdot(\mathsf{I}_{h(a)}\star{}^{h(a)}\mathsf{X}^{h(s)}\star\mathsf{I}_{h(s)}) \\
&=^4 \wedge^{h(a)\star h(s)} \\
&=^5 \wedge^{h(a\star s)}
\end{aligned}$$

where we have applied an axiom of type A5 in $=^4$. For the other constants the proof is similar. □

Finally, consider the case of a fixed monoid. Denote by xy-**SyMoCat**$_M$ the category of all xy-symocats having the monoid M as the monoid of the objects and whose morphisms are morphisms of xy-symocats which preserve the objects.

Corollary 3.13. *xy-$\mathbb{R}$el$_S$ is an initial object in the category xy-**SyMoCat**$_{S^\star}$, for $x \in \{a,b,c,d\}$ and $y \in \{\alpha,\beta,\gamma,\delta\}$.* □

Notice that the mapping H is an embedding (i.e., injective) when the target symocat is $\mathsf{AddRel}(D)$, but not in the case of $\mathsf{MultRel}(D)$.

EXERCISES AND PROBLEMS (SEC. 3.7)

1. Single out the set of constants and axioms used to axiomatize the following particular classes of relations: surjective functions; partial injective functions; surjective and total relations; total relations; and partial functions. (04)
2. Are axioms A5e–h superfluous for all restrictions xy? (**)

3.8 Relations in xy-symocats with feedback

In this section we extend the previous axiomatization to the cyclic setting. Only 9 classes of relations are closed to feedback. Algebraic presentations for these classes as enriched BNAs (i.e., xy-symocats with feedback) are given.

Classes of relations closed to feedback. Not all sixteen classes of previously described relations xy-$\mathbb{R}$el are closed to feedback. E.g., the feedback operation applied to a surjective function may produce a partial surjective function. Only 9 classes are closed with respect to feedback - see Fig. 3.5.

Relations closed to feedback. The expanded list of the classes which are closed to feedback may be useful. They are: $\mathbb{B}$i (bijections), $\mathbb{I}$n (injections), $\mathbb{I}\mathrm{n}^{-1}$ (converses on injections), $\mathbb{P}$Sur (partial surjections), $\mathbb{P}\mathrm{Sur}^{-1}$ (converses of partial surjections), $\mathbb{P}$fn (partial functions), $\mathbb{P}\mathrm{fn}^{-1}$ (converses of partial functions), and $\mathbb{R}$el (relations) ◇

The resulting algebras are BNAs over xy-symocats and such that the feedback acts on the additional constants as is shown in axioms R8–10 of Table 3.2.

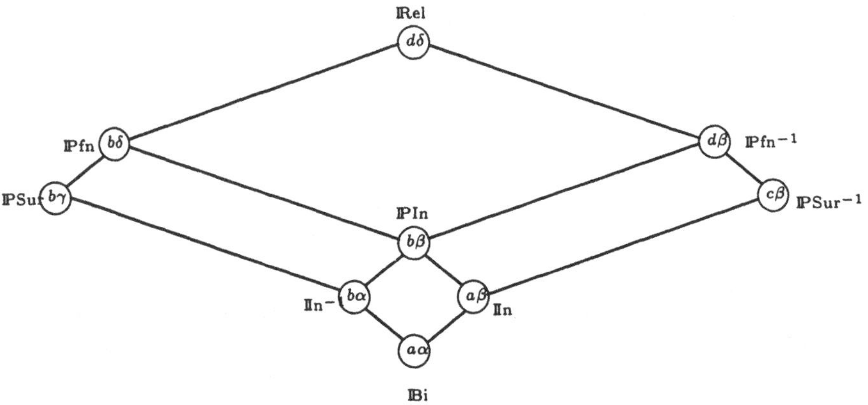

Fig. 3.5. Relations closed to feedback

For a restriction xy with $x \in \{a, b, c, d\}$ and $y \in \{\alpha, \beta, \gamma, \delta\}$ the class of relations xy-$\mathbb{R}\mathrm{el}_S$ is closed under feedback iff $x = b$ or $y = \beta$ or $xy \in \{a\alpha, d\delta\}$. Indeed, in $\mathbb{R}\mathrm{el}_S$ $\wedge^s \uparrow^s = \top_s$ and $\vee_s \uparrow^s = \perp^s$, for $s \in S$. Hence the classes xy-$\mathbb{R}\mathrm{el}_S$ with $xy \in \{a\gamma, a\delta, c\alpha, c\gamma, c\delta, d\alpha, d\gamma\}$ are not closed under feedback. The remaining cases are those collected in the above statement and in these cases it is easy to verify the closure to feedback.

xy-symocat with feedback. The xy-symocat with feedback to be defined below is a key algebraic structure of the calculus of flownomials. It extends the BNA structure with weak axioms for the branching constants.

Definition 3.14. (xy-symocat with feedback) Suppose xy is a restriction such that $x = b$ or $y = \beta$ or $xy \in \{a\alpha, d\delta\}$. A structure T is an *xy-symocat with feedback* iff

(1) T is an xy-symocat;
(2) T is a BNA; and
(3) axioms R8–10 of Table 3.2 corresponding to xy hold in T.

(The last condition means: axiom R8 when $xy \in \{b\gamma, b\delta, d\delta\}$, axiom R9 when $xy \in \{c\beta, d\beta, d\delta\}$, and axiom R10 when $xy = d\delta$.)

A *morphism of xy-symocats with feedback* is a morphism of xy-symocats preserving the feedback operation. (If h is the corresponding monoid morphism, then this condition is $H(f \uparrow^a) = H(f) \uparrow^{h(a)}$.)

Overlapping. There is a slight overlapping of terminology here. In the case $a\alpha$ the resulted structure of $a\alpha$-symocat with feedback coincides with the already familiar BNA structure. $\diamond$

As we said, this is a basic algebraic structure used in the network algebra approach. It allows us to extend the axiomatization for networks modulo graph isomorphism to the full network algebra setting. That is, branching constants are present, but they are used in a *weak* way. Roughly speaking, this means they are not allowed to duplicate, discard, etc. general network cells, but they are allowed to do such things for other (branching) constants.

Theorem 3.15. *(mapping relations, cyclic case)*
If xy is a restriction with ($x = b$ or $y = \beta$ or $xy \in \{a\alpha, d\delta\}$) and B is an xy-symocat with feedback over a monoid M, then for every morphism of monoids $h : S^\star \to M$ there exists a unique morphism of xy-symocats with feedback $H : xy\text{-}\mathbb{R}\mathrm{el}_S \to B$ which extends h.

PROOF: Let xy be a restriction such that $x = b$ or $y = \beta$ or $xy \in \{a\alpha, d\delta\}$. We have to prove that the (unique) morphism of xy-symocats $H : xy\text{-}\mathbb{R}\mathrm{el}_S \to B$ which extends $h : S \to \mathcal{O}(B)$ also commutes with the feedback operation.

Let $f \in xy\text{-}\mathbb{R}\mathrm{el}_S(a \star s, b \star s)$ with $s \in S$ be a relation. Suppose its normal form is:

$$f = (\Sigma_{j\in[|a|]} \wedge^{a_j}_{m_j} \star \wedge^{s}_{m}) \cdot g \cdot (\Sigma_{i\in[|b|]} \vee^{n_i}_{b_i} \star \vee^{n}_{s})$$

Using axiom R7 we may separate the parts of $\wedge$ and $\vee$ which are not in the scope of feedback. This reduces the problem to f of the following form

$$f = (\mathsf{I}_a \star \wedge^{s}_{m}) \cdot g \cdot (\mathsf{I}_b \star \vee^{n}_{s})$$

Formally, if $g' = (\mathsf{I}_{\Sigma_{j\in[|a|]} m_j a_j} \star \wedge^{s}_{m}) \cdot g \cdot (\mathsf{I}_{\Sigma_{i\in[|b|]} n_i b_i} \star \vee^{n}_{s})$, then $f \uparrow^s = (\Sigma_{j\in[|a|]} \wedge^{a_j}_{m_j}) \cdot g' \uparrow^s \cdot (\Sigma_{i\in[|b|]} \vee^{n_i}_{b_i})$. In B we have $H(f \uparrow^s) = (\Sigma_{j\in[|a|]} \wedge^{h(a_j)}_{m_j}) \cdot H(g' \uparrow^s) \cdot (\Sigma_{i\in[|b|]} \vee^{n_i}_{h(b_i)})$; by R7, $H(f) \uparrow^{h(s)} = (\Sigma_{j\in[|a|]} \wedge^{h(a_j)}_{m_j}) \cdot H(g') \uparrow^{h(s)} \cdot (\Sigma_{i\in[|b|]} \vee^{n_i}_{h(b_i)})$.

Since f is a normal form representation, g is normally ordered. Hence $g \in \mathbb{B}\mathrm{i}_S(a\star ms, b\star ns)$ is such that

$$(\mathsf{T}_a \star \mathsf{I}_{ms}) \cdot g \cdot (\perp^b \star \mathsf{I}_{ns}) \in \{\perp^{ms} \cdot \mathsf{T}_{ns}, \quad \perp^{(m-1)s} \cdot \mathsf{T}_{(n-1)s} \star \mathsf{I}_s\}$$

Indeed, either the rightmost input and output pins of f are disconnected, or are connected. Due to normal ordering of g in the latter case this connection should be in the rightmost part of the g. This explain the form of g's restriction stated above.

Now we prove the result. The proof is separated by cases according to the values of the branching degree numbers m and n.

1. Case $m = 0$ (when $x \in \{b, d\}$). $f = (\mathsf{I}_a \star \perp^s) \cdot g \cdot (\mathsf{I}_b \star \mathsf{V}^n_s)$ and $f \uparrow^s = g \cdot (\mathsf{I}_b \star \perp^{ns})$

$$\begin{aligned} H(f) \uparrow^{h(s)} &= [(\mathsf{I}_{h(a)} \star \perp^{h(s)})\ H(g)\ (\mathsf{I}_{h(b)} \star \mathsf{V}^n_{h(s)})] \uparrow^{h(s)} \\ &= [H(g)\ (\mathsf{I}_{h(b)} \star \mathsf{V}^n_{h(s)} \cdot \perp^{h(s)})] \uparrow^0 && \text{R3} \\ &= H(g)\ (\mathsf{I}_{h(b)} \star \perp^{nh(s)}) \\ &= H(g\ (\mathsf{I}_b \star \perp^{ns})) \\ &= H(f \uparrow^s) \end{aligned}$$

2. Case $n = 0$ (when $y \in \{\beta, \delta\}$). Similar to case 1.

3. Case $m \geq 1$ and $n \geq 1$. We can use bijections to permute the input and output ports which are not in the scope of feedback to get simpler forms. There are two possibilities:

31. Case when $(\mathsf{T}_a \star \mathsf{I}_{ms}) \cdot g \cdot (\perp^b \star \mathsf{I}_{ns}) = \perp^{ms} \cdot \mathsf{T}_{ns}$. In this case there are bijections $u \in \mathbb{B}\mathrm{i}_S(a, c\star ns)$ and $v \in \mathbb{B}\mathrm{i}_S(c\star ms, b)$ for a $c \in S^\star$ such that

$$g = (u \star \mathsf{I}_{ms})\ (\mathsf{I}_c \star {}^{ns}\mathsf{X}^{ms})\ (v \star \mathsf{I}_{ns})$$

(u is used to collect all the inputs in a going to the last ns outputs just before the last ms inputs in a; in addition, they are collected the same order as they were in the output. Similarly for v.)

It follows that

$$\begin{aligned} f &= (u \star \wedge^s_m)\ (\mathsf{I}_c \star {}^{ns}\mathsf{X}^{ms})\ (v \star \mathsf{V}^n_s) \\ &= (u \star \mathsf{I}_s)\ [\mathsf{I}_c \star (\mathsf{I}_{ns} \star \wedge^s_m) \cdot {}^{ns}\mathsf{X}^{ms} \cdot (\mathsf{I}_{ms} \star \mathsf{V}^n_s)]\ (v \star \mathsf{I}_s) \\ &= (u \star \mathsf{I}_s)\ [\mathsf{I}_c \star (\mathsf{V}^n_s \star \mathsf{I}_s) \cdot {}^s\mathsf{X}^s \cdot (\wedge^s_m \star \mathsf{I}_s)]\ (v \star \mathsf{I}_s) \\ &= [u(\mathsf{I}_c \star \mathsf{V}^n_s) \star \mathsf{I}_s]\ (\mathsf{I}_c \star {}^s\mathsf{X}^s)\ [(\mathsf{I}_c \star \wedge^s_m)v \star \mathsf{I}_s] \end{aligned}$$

hence $f \uparrow^s = u(\mathsf{I}_c \star \mathsf{V}^n_s) \cdot (\mathsf{I}_c \star \wedge^s_m)v$. In B one gets

$$\begin{aligned} H(f) \uparrow^{h(s)} &= H(u(\mathsf{I}_c \star \mathsf{V}^n_s)) \cdot [\mathsf{I}_{h(c)} \star {}^{h(s)}\mathsf{X}^{h(s)}] \uparrow^{h(s)} \cdot H((\mathsf{I}_c \star \wedge^s_m)v) \\ &= H(u(\mathsf{I}_c \star \mathsf{V}^n_s) \cdot (\mathsf{I}_c \star \wedge^s_m)v) && \text{R7} \\ &= H(f \uparrow^s) \end{aligned}$$

32. Case when $(\top_a \star \mathsf{I}_{ms}) \cdot g \cdot (\perp^b \star \mathsf{I}_{ns}) = \perp^{(m-1)s} \cdot \top_{(n-1)s} \star \mathsf{I}_s$.

321. Sub-case $m = 1$ or $n = 1$. First note that $g = g' \star \mathsf{I}_s$ with $g' \in \mathbb{B}\mathrm{i}_S(a\star(m-1)s,\ b\star(n-1)s)$.

3211. The case $m = 1$ and $n = 1$ is trivial.

3212. If $m = 1$ and $n > 1$ (case $y \in \{\gamma, \delta\}$), then $f = (g' \star \mathsf{I}_s) \cdot (\mathsf{I}_b \star \vee_s^n) = (g'\ (\mathsf{I}_b \star \vee_s^{n-1}) \star \mathsf{I}_s) \cdot (\mathsf{I}_b \star \vee_s)$ hence $f \uparrow^s = g'\ (\mathsf{I}_b \star \vee_s^{n-1}) \cdot (\mathsf{I}_b \star \perp^s)$. In B one get

$$\begin{aligned} & H(f) \uparrow^{h(s)} \\ = \ & H(g'\ (\mathsf{I}_b \star \vee_s^{n-1})) \cdot (\mathsf{I}_{h(b)} \star \vee_{h(s)}) \uparrow^{h(s)} \\ = \ & H(g'\ (\mathsf{I}_b \star \vee_s^{n-1})) \cdot (\mathsf{I}_{h(b)} \star \perp^{h(s)}) \qquad \text{R8} \\ = \ & H(g'\ (\mathsf{I}_b \star \vee_s^{n-1}) \cdot (\mathsf{I}_b \star \perp^s)) \\ = \ & H(f \uparrow^s) \end{aligned}$$

3212. The case $m > 1$ and $n = 1$ (when $x \in \{c, d\}$) is similar.

322. Sub-case $m > 1$ and $n > 1$ (when $xy = d\delta$). In this case there are bijections $u \in \mathbb{B}\mathrm{i}_S(a,\ c\star(n-1)s)$ and $v \in \mathbb{B}\mathrm{i}_S(c\star(m-1)s, b)$ for a $c \in S^\star$, such that

$$g = (u \star \mathsf{I}_{ms}) \cdot (\mathsf{I}_c \star {}^{(n-1)s}\mathsf{X}^{(m-1)s} \star \mathsf{I}_s) \cdot (v \star \mathsf{I}_{ns})$$

It follows that

$$\begin{aligned} f \ = \ & (u \star \wedge_m^s) \cdot (\mathsf{I}_c \star {}^{(n-1)s}\mathsf{X}^{(m-1)s} \star \mathsf{I}_s) \cdot (v \star \vee_s^n) \\ = \ & (u \star \mathsf{I}_s) \cdot [\ \mathsf{I}_c \star (\mathsf{I}_{(n-1)s} \star \wedge^s(\wedge_{m-1}^s \star \mathsf{I}_s)) \cdot ({}^{(n-1)s}\mathsf{X}^{(m-1)s} \star \mathsf{I}_s) \\ & \cdot (\mathsf{I}_{(m-1)s} \star (\vee_s^{n-1} \star \mathsf{I}_s)\vee_s)\ ] \cdot (v \star \mathsf{I}_s) \\ = \ & (u \star \mathsf{I}_s) \cdot [\mathsf{I}_c \star (\vee_s^{n-1} \star \wedge^s) \cdot ({}^s\mathsf{X}^s \star \mathsf{I}_s) \cdot (\wedge_{m-1}^s \star \vee_s)] \cdot (v \star \mathsf{I}_s) \\ = \ & [u(\mathsf{I}_c \star \vee_s^{n-1}) \star \mathsf{I}_s] \cdot [\mathsf{I}_c \star (\mathsf{I}_s \star \wedge^s)\ ({}^s\mathsf{X}^s \star \mathsf{I}_s)\ (\mathsf{I}_s \star \vee_s)] \\ & \cdot [(\mathsf{I}_c \star \wedge_{m-1}^s)v \star \mathsf{I}_s] \end{aligned}$$

hence $f \uparrow^s = u(\mathsf{I}_c \star \vee_s^{n-1})(\mathsf{I}_c \star \wedge_{m-1}^s)v$. In B one gets

$$\begin{aligned} & H(f) \uparrow^{h(s)} \\ = \ & H(u(\mathsf{I}_c \star \vee_s^{n-1})) \\ & \cdot [\ \mathsf{I}_{h(c)} \star ((\mathsf{I}_{h(s)} \star \wedge^{h(s)}) \cdot ({}^{h(s)}\mathsf{X}^{h(s)} \star \mathsf{I}_{h(s)}) \cdot (\mathsf{I}_{h(s)} \star \vee_{h(s)})) \uparrow^{h(s)}\ ] \\ & \cdot H((\mathsf{I}_c \star \wedge_{m-1}^s)v) \\ = \ & H(u(\mathsf{I}_c \star \vee_s^{n-1})) \cdot H((\mathsf{I}_c \star \wedge_{m-1}^s)v) \qquad \text{R10} \\ = \ & H(u(\mathsf{I}_c \star \vee_s^{n-1}) \cdot (\mathsf{I}_c \star \wedge_{m-1}^s)v) \\ = \ & H(f \uparrow^s) \end{aligned}$$

This shows that H commutes with feedback operation, hence it is a xy-symocat with feedback morphism. In addition, the morphism H is unique. (This follows from the similar property of F and the definition of H.) □

3.9 Networks with branching constants

In this section a network algebra (NA) axiomatization for networks modulo graph-isomorphism is presented. In other words, branching constants are present in the signature. Moreover, graph-isomorphism is considered up to a transformation of the branching constants according to the rules they satisfy in $\mathbb{R}el_S$. The corresponding algebraic structures are the *xy-symocats with feedback* introduced in the previous section.

The basic BNA model for networks modulo isomorphism was described in Chapter 2. One may use arbitrary relations there, but at the semantic level, in the connecting theory. At the syntactic level only bijections were allowed.

Here we extend the algebra to the full NA syntax for the calculus of flownomials

$$\star \quad \cdot \quad \uparrow^a \quad {}^a\mathsf{X}^b \quad \wedge_k^a \quad \vee_a^k$$

by adding axiomatizations for $\wedge_k$ and $\vee^k$. We use the basic results in the previous sections which give presentations for certain classes of relations as xy-symocats and Theorem 3.15 that lifts those axiomatizations to the cyclic setting. We get the following results

Theorem 3.16. *(Correctness theorem for NA)*
If xy be a restriction with $x \in \{a, b, c, d\}$ and $y \in \{\alpha, \beta, \gamma, \delta\}$ such that $x = b$ or $y = \beta$ or $xy \in \{a\alpha, d\delta\}$ and T an xy-symocat with feedback, then
(Soundness) $[X, T]_{a\alpha}$ is an xy-symocat with feedback.
(Completeness) The rules of xy-symocat with feedback are complete for networks with xy-constants modulo isomorphism. □

Comments, problems, bibliographic remarks

Problems. The development of practical tools to support the computation with relations and flownomials is an important issue of concern for further work.

Efficient representations. Our representation is like the Roman representation for natural numbers. In this algebraic setting certain large but regular relations may be easily handled. For instance, short representations may be found, e.g.,

$$(\vee_{343}^{524} \star \wedge_{10^4}^{10}) \cdot (\wedge_{50^5}^{43} \star {}^{300}\mathsf{X}^1 \star \vee_{101}^{99}) : 179742 \to 13437500402$$

and operations may be easily computed, but using general block constants. Find a sort of representation for finite relations similar to the Arabic representation for natural numbers.

Implementations. To make the calculus more effective, efficient implementations for working with relations and flownomials have to be developed. One possibility may be to use the CafeOBJ tools [DF98], eventually combined with Lafont's results on associated rewriting systems [Laf92]. A direct implementation using Java for a related calculus was developed by Jeffrey; see [Jef97].

Notes The results of the first part of the chapter, including Theorems 3.9 and 3.12, are based on [CŞ91, CŞ94]. In their presentation, we have followed the slightly improved version presented in Chapter A of [Şte91].

The key result of the second part of the chapter is Theorem 3.15, giving the mapping of relations in enriched symocats with feedback; it is taken from [CŞ88b]. The presentation follows Chapter C, Sections 1 of [Şte91].

For demonic calculi see, e.g., [DBS$^+$95], [dNS95] and [DMN97]; the adaptation of the current axiomatization to the demonic calculi is presented in [BŞ98]; it will be described in Appendix C.

Exercise 3.6.1 is from [Laf92].

Enriched symocats were used in [Gib95] to handle acyclic graphs in a functional programming setting; in [BGM99] they are used to get a unifying framework for representing concurrent and distributed systems. An extension of the "relations are enriched symocats" view to "planar relations" is presented in [Mol88] and used to model VLSI design of computer circuits.

4. Network behaviour

Some call it evolution and others call it God.
W.H. Carruth

We continue the study of the branching structure of networks. The aim of the present chapter is to introduce certain basic algebraic structures that may be used to capture a broad class of intuitive notions of network behaviour in a precise mathematical framework. The study of particular network behaviours is postponed to later chapters.

For acyclic setting three basic critical structures are used to describe the behaviour of general many-inputs/many-outputs networks. (1) The first one is called *matrix theory*; it uses ordinary one-input/one-output elements; general networks are modeled by *matrices* of such elements. (2) The second one is called *algebraic theory*; it uses ranked (e.g., many-inputs/single-output) elements; general networks are described by *tuples* of such elements. (3) The last case is the general one, already familiar to us. It is the symocat setting where one uses variables for such general networks; no a priori reduction to "simpler" elements is supposed.

A pleasant property is that matrix and algebraic theories may be easily defined as particular xy-symocats, namely as xy-symocats which satisfy certain *strong commutation axioms.*

Next, we pass to the cyclic setting. Here *xy-flow* structures are defined. These parametric structures represent our basic proposal for axiomatization of various types of network behaviour. Their definition is based on an *enzymatic rule* for the identification of cyclic networks, relative to a class of xy-relations. In many cases this rule is equivalent with *simulation* via the corresponding class of relations.

Finally, we may classify the standard algebraic structures spread throughout the book with three parameters

$$\boxed{\text{BNA} + x_1y_1\text{-weak} + x_2y_2\text{-strong} + x_3y_3\text{-enzymatic rule}}$$

where $x_1y_1 \geq x_2y_2 \geq x_3y_3$ ("$\geq$" means "more branching constants"). The first restriction shows what branching constants may be used, the second which branching constants strongly commute with arbitrary arrows, and the third for which branching constants the enzymatic rule may be applied.

4.1 Strong xy-symocats (xy-strong rules)

In this section *strong* xy-symocats are introduced; they are xy-symocats where the xy-branching constants commute with arbitrary elements. A few basic results are presented, too.

Table 4.1. The xy-strong axioms ($f : a \to b$).

$C_{a\beta}$	$\top_a \cdot f = \top_b$	$C_{b\alpha}$	$f \cdot \bot^b = \bot^a$
$C_{a\gamma}$	$\vee_a \cdot f = (f \star f) \cdot \vee_b$	$C_{c\alpha}$	$f \cdot \wedge^b = \wedge^a \cdot (f \star f)$

Definition 4.1. Let xy be a restriction with $x \in \{a, b, c, d\}, y \in \{\alpha, \beta, \gamma, \delta\}$. An xy-symocat is *strong* if the corresponding xy-strong axioms in Table 4.1 hold.

Commutation via general branching constants. These strong commutation conditions may be stated with general branching constants as follows:

$$\vee_a^k \cdot f = (kf) \cdot \vee_b^k \quad \text{and} \quad f \cdot \wedge_k^b = \wedge_k^a \cdot (kf)$$

Here we used our notational convention: kf denotes k times juxtaposition of f.

The axioms in Table 4.1 are strong in the following sense: in an arbitrary xy-symocat only their restriction to terms f that are abstract xy-relations is required. (Notice that B6 and B9 are of the same type showing the commutation of the constants I_a and ${}^a\mathsf{X}^b$ with arbitrary arrows.) ◇

Recall that an *abstract xy-relation* is the image of an xy-relation in an xy-symocat. The following simply inferred property is of fundamental importance in the following.

Proposition 4.2. *(commuting abstract xy-relations with arbitrary arrows)* *Let B be a strong xy-symocat, φ an xy-relation in xy-$\mathrm{IRel}_S(a, b)$, and i, o : xy-$\mathrm{IRel}_S \to B$ two morphisms of xy-symocats.*

If $f_j \in B(i(a_j), o(a_j))$ and $g_k \in B(i(b_k), o(b_k))$ are such that $f_j = g_k$ whenever $(j, k) \in \varphi$, then

$$(f_1 \star \ldots \star f_{|a|}) \cdot o(\varphi) = i(\varphi) \cdot (g_1 \star \ldots \star g_{|b|})$$

PROOF: Denote by $Z \subseteq xy$-IRel_S the substructure given by those ϕ which obey the property in the statement of the proposition for all f_j, g_k. Since B is xy-strong the constants in $\top_a$, $\vee_a$, $\bot^a$, $\wedge^a$ which obey the xy-restriction are included in Z. In addition, Z is closed under juxtaposition and composition. Hence it is equal to xy-IRel_S.

Below we give some details for the closure to composition (the closure to juxtaposition is fairly easy). Let $\phi \in xy$-$\mathrm{IRel}_S(a, b)$ be of the form $\phi = \sigma \cdot \tau$ with $\sigma \in Z(a, c)$ and $\tau \in Z(c, b)$. Suppose f_j, $j \in [|a|]$ and g_k, $k \in [|b|]$ are as in the statement of the proposition. Let us further consider some arrows $h_r \in B(i(c_r), o(c_r))$, $r \in [|c|]$ defined by

$$h_r = \begin{cases} f_j & \text{if } (j,r) \in \sigma \\ g_k & \text{if } (r,k) \in \tau \\ \text{arbitrary} & \text{otherwise} \end{cases}$$

The first two cases may overlap, but, nevertheless, the definition is correct; indeed, if $(j,r) \in \sigma$ and $(r,k) \in \tau$, then $(j,k) \in \sigma \cdot \tau = \phi$, hence $f_j = g_k$. As σ and τ are in Z we have $(f_1 \star \ldots \star f_{|a|}) \cdot o(\sigma) = i(\sigma) \cdot (h_1 \star \ldots \star h_{|c|})$ and $(h_1 \star \ldots \star h_{|c|}) \cdot o(\tau) = i(\tau) \cdot (g_1 \star \ldots \star g_{|b|})$. These identities clearly imply $(f_1 \star \ldots \star f_{|a|}) \cdot o(\phi) = i(\phi) \cdot (g_1 \star \ldots \star g_{|b|})$, hence $\phi \in Z$. □

The following heuristics rule is obtained from this result.

Commuting relations with network cells
In a strong xy-symocat one can shift cells along edges of a xy-relation accompanied by the duplication or deletion of cells according to the branching structure of the relation.

4.2 Algebraic theories

In this section we introduce *algebraic theories.* They are rather old and well known mathematical objects. In the NA context they naturally occur as $d\alpha$-strong symocats. In the dual case, $a\delta$-strong symocats are called *coalgebraic theories.*

General introduction. Before the definition, a few comments may be useful. Algebraic theories have been used for a long time. They were formally introduced by Lawvere in the 1960s to capture the essential structure of classical *algebraic theories*, e.g., for groups, modules, etc. In this view one is interested in the full theory, in considering all derived operations, not only in the operations used to define the given structure.

A usual algebraic operation has many inputs, but a unique output. Derived operations are obtained as combinations of such elements, hence tree or term structures form basic models for them. A general many-inputs/many-outputs object is reduced to a *tuple* of such elements.

We may fairly name the algebraic theory world as a *tree world.* Each field where trees are important may be a target for algebraic theory applications. Included are: algebraic specifications for data-types, (deterministic) data-flow networks, semantics of circuits, etc.

Notice that the tree-world is asymmetric. The dual *coalgebraic* view was widely used to get algebraic semantics for flowchart programs. Recently there is a renewed interest in developing such coalgebraic methods, especially related to object-oriented programming studies, or with concurrent process-algebra like models. ◇

Definition 4.3. ((co)algebraic theory) An *algebraic theory* is a strong $d\alpha$-symocat. A *coalgebraic theory* is a strong $a\delta$-symocat.

Consequently, algebraic theories have ramification constants $\wedge_k$ satisfying $f \cdot \wedge_k^b = \wedge_k^b \cdot (kf)$ (and the usual $d\alpha$-symocat axioms); coalgebraic theories have constants $\vee_k$ satisfying $\vee_a^k \cdot f = (kf) \cdot \vee_a^k$ (and the usual $a\delta$-symocat axioms).

♡ The interested reader may check that this definition of the algebraic theories coincides with the classical one, presented e.g., in [Law63, Elg75]. In our symocat context we need the proof of one implication: *in an algebraic theory every arrow with multiple inputs may be uniquely decomposed as a tuple of single input arrows.* ◇

Proposition 4.4. *If B is a coalgebraic theory, then for every arrow $f \in B(a \star b, c)$ there exists a unique pair of arrows $f_1 \in B(a, c)$ and $f_2 \in B(b, c)$ such that $f = (f_1 \star f_2) \cdot \mathsf{V}_c$.*

PROOF: For $f \in B(a \star b, c)$ take $f_1 = (\mathsf{I}_a \star \mathsf{T}_b) \cdot f$ and $f_2 = (\mathsf{T}_a \star \mathsf{I}_a) \cdot f$. Then obviously,

$$\begin{aligned}
(f_1 \star f_2) \cdot \mathsf{V}_c &= [(\mathsf{I}_a \star \mathsf{T}_b)f \star (\mathsf{T}_a \star \mathsf{I}_b) \cdot f] \cdot \mathsf{V}_c && \text{definition}\\
&= (\mathsf{I}_a \star \mathsf{T}_b \star \mathsf{T}_a \star \mathsf{I}_b) \cdot (f \star f) \cdot \mathsf{V}_c\\
&= (\mathsf{I}_a \star \mathsf{T}_b \star \mathsf{T}_a \star \mathsf{I}_b) \cdot \mathsf{V}_{a \star b} \cdot f && \mathrm{C}_{a\gamma}\\
&= f
\end{aligned}$$

Next we show the pair f_1, f_2 is unique. If $g_1 \in B(a, c)$ and $g_2 \in B(b, c)$ are arrows satisfying condition $f = (g_1 \star g_2) \cdot \mathsf{V}_c$, then

$$\begin{aligned}
f_1 &= (\mathsf{I}_a \star \mathsf{T}_b) \cdot f\\
&= (\mathsf{I}_a \star \mathsf{T}_b) \cdot (g_1 \star g_2) \cdot \mathsf{V}_c && \text{hypothesis}\\
&= (g_1 \star \mathsf{T}_c) \cdot \mathsf{V}_c && \mathrm{C}_{a\beta}\\
&= g_1 \cdot (\mathsf{I}_c \star \mathsf{T}_c) \cdot \mathsf{V}_c\\
&= g_1
\end{aligned}$$

In a similar way one may prove that $f_2 = g_2$. This shows that the pair f_1, f_2 is unique. □

In this context f is called the *cotupling* of f_1 and f_2. The standard notation used for cotupling is $f = \langle f_1, f_2 \rangle$. In the dual case of algebraic theories the name is *tuple* and the standard notation is $f = (f_1, f_2)$.

4.3 Matrix theories

Here matrix theories are introduced. They are simultaneously algebraic and coalgebraic theories. As before, they naturally occur as strong symocats, namely satisfying the $a\delta$ and $d\alpha$ strong commutation axioms.

Matrix theories (strong $a\delta$- and $d\alpha$-symocats). In a matrix theory one may reduce each arrow with multiple inputs and multiple outputs to arrows with one input and one output. A one-input/one-output arrow is an *ordinary element.* Then matrices over ordinary elements may be used to describe the behaviour of general arrows.

Monadic world. The view here is of a *monadic world*. There is a unique and homogeneous universe. Every complicated relational structure of a stratified world should be described, analyzed and understood using combinations of elements from this unique universe. As we shall see in the chapter on finite automata, this monadic world view may produce complicated combinatorial problems. Compare, for instance, the BNA rule $f \uparrow^{m\star n} = (f \uparrow^n) \uparrow^n$ with its complicated equivalent form R3 used in Kleene/Conway theories. Also process algebra inherited this monadic world view and it suffers the same combinatorial problems with its "interleaving semantics." ◇

Definition 4.5. A *matrix theory* is a strong $a\delta$- and $d\alpha$-symocat. It is *idempotent* if $\wedge_k^a \cdot \vee_a^k = \mathsf{l}_a, k \geq 1$ is valid (i.e., axiom A4 in Table 3.2).

♠ Axioms A3a, A3b, A3c, and A3d follow from $\mathsf{C}_{b\alpha}$, $\mathsf{C}_{a\beta}$, $\mathsf{C}_{a\gamma}$, and $\mathsf{C}_{c\alpha}$, respectively. Hence all the acyclic axioms in Table 3.2 are valid in a matrix theory with one possible exception only: axiom A4. As we have seen, if A4 holds, the matrix theory is called idempotent. ♣

It is easy to see that this definition coincides with the classical one. An implication follows from Proposition 4.6, below.

Proposition 4.6. *Let B be a matrix theory. For every arrow $f \in B(a \star b, c \star d)$ there exists a unique 4-uple of arrows $f_{11} \in B(a,c)$, $f_{12} \in B(a,d)$, $f_{21} \in B(b,c)$, and $f_{22} \in B(b,d)$ such that*

$$[\wedge^a \cdot (f_{11} \star f_{12}) \star \wedge^b \cdot (f_{21} \star f_{22})] \cdot \vee_{c\star d} = f$$

The arrow of the left-hand-side of the above equality is equal to the arrow $\wedge^{a\star b} \cdot [(f_{11} \star f_{21}) \cdot \vee_c \star (f_{12} \star f_{22}) \cdot \vee_d]$ and it is simply written as a matrix

$$\begin{pmatrix} f_{11} & f_{12} \\ f_{21} & f_{22} \end{pmatrix}$$

PROOF: For an $f \in B(a \star b, c \star d)$ by Proposition 4.4 there is a unique pair $\langle f_1, f_2 \rangle$ with $f_1 \in B(a, c \star d)$, $f_2 \in B(b, c \star d)$ such that

$$(f_1 \star f_2) \cdot \vee_{c\star d} = f$$

Now we can apply the dual version of Proposition 4.4 for f_1 (resp. f_2) to get the unique pair (f_{11}, f_{12}) with $f_{11} \in B(a,c)$, $f_{12} \in B(a,d)$ (resp. (f_{21}, f_{22}) with $f_{21} \in B(b,c)$, $f_{22} \in B(b,d)$) such that

$$f_1 = \wedge^a \cdot (f_{11} \star f_{12}) \quad (\text{resp. } f_2 = \wedge^b \cdot (f_{21} \star f_{22}))$$

Hence the desired identity was found.

For the uniqueness part, if g_{ij}, $i,j \in [2]$ satisfy the same identity and have the same sources and targets as the arrows f_{ij} defined above, then from the uniqueness part provided by Proposition 4.4 it follows that $\wedge^a \cdot (g_{11} \star g_{12}) = f_1$

and $\wedge^b \cdot (g_{21} \star g_{22}) = f_2$. Next, we apply the uniqueness given by the dual of Proposition 4.4. Hence, $g_{11} = f_{11}$ and $g_{12} = f_{12}$ (resp. $g_{21} = f_{21}$ and $g_{22} = f_{22}$).

Finally, the equality of both expressions that define $\begin{pmatrix} f_{11} & f_{12} \\ f_{21} & f_{22} \end{pmatrix}$ follows by permuting (with B9) the middle terms f_{12} and f_{21} and applying axioms A5b and A5d. □

The components of the matrix associated with an arrow $f \in B(a \star b, c \star d)$ may be explicitly written as

$$f = \begin{pmatrix} (\mathsf{I}_a \star \top_b) \cdot f \cdot (\mathsf{I}_c \star \perp^d) & (\mathsf{I}_a \star \top_b) \cdot f \cdot (\perp^c \star \mathsf{I}_d) \\ (\top_a \star \mathsf{I}_b) \cdot f \cdot (\mathsf{I}_c \star \perp^d) & (\top_a \star \mathsf{I}_b) \cdot f \cdot (\perp^c \star \mathsf{I}_d) \end{pmatrix}$$

Matrix theory as the theory of matrices over a semiring. In this subsection we look at the structure induced by the matrix theory axioms on its components. It turn out that this is the well-known "semiring structure." Hence a matrix theory may be seen as a classical theory of matrices over a semiring.

Let B be a matrix theory over $\mathbb{N}$. On the set $B(1,1)$ we introduce the operations

$$f + g = \wedge^1 \cdot (f \star g) \cdot \vee_1; \quad f \cdot g = f \cdot g; \quad 0 = \perp^1 \cdot \top_1; \quad \text{and } 1 = \mathsf{I}_1.$$

This way we get a semiring $(B(1,1), +, \cdot, 0, 1)$. Conversely, given a semiring S the usual matrices with elements in S may be structured as a matrix theory.

We give some details below. A *semiring* $S = (S, +, \cdot, 0, 1)$ is a set S endowed with two binary operations $+, \cdot$ and constants 0,1 such that: $(S, +, 0)$ is a commutative monoid, $(S, \cdot, 1)$ is a monoid, and the distributive laws

$$a \cdot (b + c) = (a \cdot b) + (a \cdot c),$$

$$(a + b) \cdot c = (a \cdot c) + (b \cdot c)$$

and the zero law

$$0 \cdot a = 0 = a \cdot 0$$

hold, for every $a, b, c \in S$. A semiring is called *idempotent* if $a + a = a$, for every $a \in S$.

An easy computation shows that $\mathcal{S}(B) = (B(1,1), +, \cdot, 0, 1)$ is a semiring whenever B is a matrix theory. For example, one may use the strong commuting axioms to prove the left distributivity law as follows:

$$\begin{aligned} f \cdot (g + h) &= f \cdot [\wedge^1 \cdot (g \star h) \cdot \vee_1] \\ &= \wedge^1 \cdot (f \star f) \cdot (g \star h) \cdot \vee_1 && \mathsf{C}_{c\alpha} \\ &= \wedge^1 \cdot (f \cdot g \star f \cdot h) \cdot \vee_1 \\ &= (f \cdot g) + (f \cdot h) \end{aligned}$$

Actually, the strong commuting axioms match these special semiring rules as follows:

$\mathsf{C}_{a\gamma}$ — left distributivity
$\mathsf{C}_{c\alpha}$ — right distributivity
$\mathsf{C}_{a\beta}$ — left zero rule
$\mathsf{C}_{b\alpha}$ — right zero rule.

Conversely, if $(S,+,\cdot,0,1)$ is a semiring, then one may construct the theory $\mathcal{M}(S)$ of matrices over S in the usual way:

$$\mathcal{M}(S)(m,n) = \{A\colon\ A \text{ is an } m\times n \text{ matrix with entries in } S\}, \quad \text{for } m,n\geq 0$$

(In the case $m = 0$ by convention $\mathcal{M}(S)(m,n)$ has a unique element, namely the "matrix" with 0 rows and n columns; similarly for $n = 0$.)

In $\mathcal{M}(S)$ the following operations and constants may be introduced:

– $A\star B = \begin{pmatrix} A & 0 \\ 0 & B \end{pmatrix}$,

– $A\cdot B =$ "the usual product of matrices A and B",

– $\mathsf{I}_n = \begin{pmatrix} 1 & \dots & 0 \\ \vdots & \ddots & \vdots \\ 0 & \dots & 1 \end{pmatrix}$, ${}^m\mathsf{X}^n = \begin{pmatrix} 0 & \mathsf{I}_m \\ \mathsf{I}_n & 0 \end{pmatrix}$, $\mathsf{V}_n = \begin{pmatrix} \mathsf{I}_n \\ \mathsf{I}_n \end{pmatrix}$, $\wedge^n = (\mathsf{I}_n \ \ \mathsf{I}_n)$, $\mathsf{T}_n =$ "the unique element in $\mathcal{M}(S)(0,n)$," $\perp_n =$ "the unique element in $\mathcal{M}(S)(n,0)$."

An easy computation shows that this way one gets a matrix theory (in the NA sense) and transformations $\mathcal{M}$ and $\mathcal{S}$ are mutually inverse. Consequently,

Proposition 4.7. *A structure is a matrix theory over* $\mathbb{N}$, *in the given network algebra sense, iff it is a usual theory of matrices over a semiring.* □

From this result it follows that a strong $d\delta$-symocat (actually being a matrix theory satisfying axiom A4) is equivalent with the usual theory of matrices over an idempotent semiring, provided the underlining monoid is $\mathbb{N}$.

EXERCISES AND PROBLEMS (SEC. 4.3)

1. Fill in the full proofs for the above result, i.e., (1) $\mathcal{S}(B)$ is a semiring, whenever B is a matrix theory; (2) if S is a semiring, than $\mathcal{M}(S)$ is a matrix theory. (07)

4.4 Enzymatic rule (xy-enzymatic rules)

In this section we describe the enzymatic rule. This is of crucial importance in the development of network algebra. A simple proof of its validity in the standard additive relational model $\mathsf{AddRel}(D)$ is presented.

What is the enzymatic rule?

We begin with a look at the equivalence relations generated by certain local commutation rules described in Table 4.2 applied to simple variables only.

♡ They are "local," in the sense that only the commutation of the abstract xy-relations with variables is required. ◇

Table 4.2. Commutation of the branching constants with variables ($x : a \to b$)

$\mathsf{C}_{a\beta}$-var	$\mathsf{T}_a \cdot x = \mathsf{T}_b$	$\mathsf{C}_{b\alpha}$-var	$x \cdot \bot^b = \bot^a$
$\mathsf{C}_{a\gamma}$-var	$\mathsf{V}_a \cdot x = (x \star x) \cdot \mathsf{V}_b$	$\mathsf{C}_{c\alpha}$-var	$x \cdot \wedge^b = \wedge^a \cdot (x \star x)$

Let $\equiv_{xy}$ denotes the congruence relation generated by the rules in Table 4.2 corresponding to xy. These congruences give natural equivalence relations on *acyclic networks*. E.g.,

$\equiv_{a\beta}$ captures accessibility,
$\equiv_{a\gamma}$ captures reduction,
$\equiv_{a\delta}$ captures minimization (deterministic case),
etc.

The question is:

Do the analogous simple statements work for cyclic networks?

The answer is generally *not*! The reason is the impossibility to use C_{xy}-var in cycles. To overcome this difficulty we combine C_{xy}-var with an additional identification rule, the "enzymatic rule." It is defined below.

Definition 4.8. (enzymatic rule)
For $f : m \star p \to n \star p$, $g : m \star q \to n \star q$, and $y : p \to q$ consider the implication

$$(Enz_{y,f,g}) \qquad f \cdot (\mathsf{I}_m \star y) = (\mathsf{I}_m \star y) \cdot g \quad \Rightarrow \quad f \uparrow^p = g \uparrow^q$$

We say, T *satisfies* Enz_E, for a class E of arrows in T, if it satisfies $\mathrm{Enz}_{y,f,g}$ for y in E and arbitrary f, g.

Example: inaccessible cells. In an acyclic network one may use $\mathsf{C}_{a\beta}$-var to throw away the inaccessible cells. The procedure is easy: (1) start with the cells without incoming connections and apply $\mathsf{C}_{a\beta}$-var; (2) repeat this as long as possible; (3) finally you will get the accessible part of the network.

Try to see how $\mathsf{C}_{a\beta}$-var may be used to eliminate the inaccessible cells of a cyclic network. It may be the case that all these inaccessible cells, from the network inputs' point of view, are in a cycle. Then, there is *no point* where you may start to apply the elimination procedure using $\mathsf{C}_{a\beta}$-var. The enzymatic rule helps you to cut the cycle and to apply the elimination procedure thereafter. See [CŞ92], [Şte94] for more details. (However, notice that in case $a\beta$ if moreover the stronger coalgebraic rules are valid (that is, the structure is even $a\delta$ strong), then the full nonaccessible part may be eliminated at once, as one can see from the proof of Proposition 5.13.) ◊

It is useful to have a similar definition relative to an arbitrary equivalence, not only for equality. The natural extension is the following. For an equivalence relation $\sim$ on T denote

$$(Enz^{\sim}_{y,f,g}) \qquad f \cdot (\mathsf{I}_m \star y) \sim (\mathsf{I}_m \star y) \cdot g \quad \Rightarrow \quad f \uparrow^p \sim g \uparrow^q$$

We say $\sim$ *satisfies* Enz_E *in* T, where E is a set of arrows in T, if it satisfies $\mathrm{Enz}^{\sim}_{y,f,g}$ for y in E and arbitrary f, g.

A technical result. Here we prove a simple, but very useful fact, which may simplify the verification of the enzymatic rule. It applies to an enzyme which may be decomposed as a juxtaposition of other enzymes and reduces the verification task to these latter enzymes.

Lemma 4.9. *If the enzymatic axiom holds for two arrows, then it also holds for their juxtaposition. Put formally,*

$$Enz_{\{u\}} \ \& \ Enz_{\{u'\}} \Rightarrow Enz_{\{u\star u'\}}$$

PROOF: Suppose $u : p \to q$ and $u' : p' \to q'$. Take arbitrary $f : m\star p\star p' \to n\star p\star p'$ and $g : m \star q \star q' \to n \star q \star q'$ such that the premise of the implication in the enzymatic rule holds, i.e.,

$$f \cdot (\mathsf{I}_n \star u \star u') = (\mathsf{I}_m \star u \star u') \cdot g$$

Write this identity as $[f(\mathsf{I}_n \star u \star \mathsf{I}_{p'})] \cdot (\mathsf{I}_{n\star q} \star u') = (\mathsf{I}_{m\star p} \star u') \cdot [(\mathsf{I}_m \star u \star \mathsf{I}_{q'})g]$ and apply $\mathrm{Enz}_{\{u'\}}$. One gets

$$[f(\mathsf{I}_n \star u \star \mathsf{I}_{p'})] \uparrow^{p'} = [(\mathsf{I}_m \star u \star \mathsf{I}_{q'})g] \uparrow^{q'}$$

With axiom R1 we may write this identity as $(f \uparrow^{p'}) \cdot (\mathsf{I}_n \star u) = (\mathsf{I}_m \star u) \cdot (g \uparrow^{q'})$ and apply $\mathrm{Enz}_{\{u\}}$. One gets $(f \uparrow^{p'}) \uparrow^{p} = (g \uparrow^{q'}) \uparrow^{q}$ or equivalently, $f \uparrow^{p\star p'} = g \uparrow^{q\star q'}$.

□

Validity of $\mathbf{Enz}_E$ in AddRel(D). A semantic justification of the enzymatic rule is inserted here. The enzymatic rule is valid in the additive relational models, even in a very strong sense: each arrow satisfies the enzymatic rule. This is strong because usually we require the validity of the enzymatic rule for arrows represented by ground relational terms only.

Proposition 4.10. *In* $\mathsf{AddRel}(D)$ *the enzymatic rule holds for every arrow.*

PROOF: Suppose $f \in \mathsf{AddRel}(D)(m+p, n+p)$, $g \in \mathsf{AddRel}(D)(m+q, n+q)$ and $y \in \mathsf{AddRel}(D)(p, q)$ fulfil the condition

$$f \cdot (\mathsf{I}_n \oplus y) = (\mathsf{I}_m \oplus y) \cdot g$$

Write f and g as matrices

$$f = \begin{pmatrix} f_{11} & f_{12} \\ f_{21} & f_{22} \end{pmatrix} \quad \text{and} \quad g = \begin{pmatrix} g_{11} & g_{12} \\ g_{21} & g_{22} \end{pmatrix}$$

with f_{11}, $g_{11} : m \to n$. Then equality $f \cdot (\mathsf{I}_n \oplus y) = (\mathsf{I}_m \oplus y) \cdot g$ gives

$$\begin{aligned} f_{11} &= g_{11} \\ f_{12} \cdot y &= g_{12} \\ f_{21} &= y \cdot g_{21} \\ f_{22} \cdot y &= y \cdot g_{22} \end{aligned}$$

From the definition of the feedback in $\mathsf{AddRel}(D)$ one gets

$$f \uparrow^p = f_{11} \cup f_{12} \cdot f_{22}^* \cdot f_{21}$$

where $f_{22}^* = \mathsf{I}_p \cup f_{22} \cup f_{22}^2 \ \dots$. Similarly,

$$g \uparrow^q = g_{11} \cup g_{12} \cdot g_{22}^* \cdot g_{21}$$

Using the relationship between f_{ij} and g_{ij} one gets

$$\begin{aligned} g \uparrow^q &= f_{11} \cup f_{12} \cdot y \, (\mathsf{I}_q \cup g_{22} \cup g_{22}^2 \ \dots) \cdot g_{21} \\ &= f_{11} \cup f_{12} \cdot (y \cup y \cdot g_{22} \cup y \cdot g_{22}^2 \ \dots) \cdot g_{21} \\ &= f_{11} \cup f_{12} \cdot (y \cup f_{22} \cdot y \cup f_{22}^2 \cdot y \ \dots) \cdot g_{21} \\ &= f_{11} \cup f_{12} \cdot (\mathsf{I}_p \cup f_{22} \cup f_{22}^2 \ \dots) \, y \cdot g_{21} \\ &= f_{11} \cup f_{12} \cdot f_{22}^* \cdot f_{21} \\ &= f \uparrow^p \end{aligned}$$

This completes the proof. □

An enzymatic, reactive process for network equivalence. One may imagine an analogue of a physical device to check network equivalence as follows: suppose one has a "recipient" where he may introduce a large amount of copies of the network f we are starting with and one copy for each possible enzyme. The reaction is the one already explained: the feedback is cut, an enzyme is attached and the explained transformation takes place; if the enzyme is appropriate, then this transformation process ends liberating the enzyme in the reaction, hence it may participate again in this game; if not, then an inert constituent is obtained. After some time, the solution will contain significant parts from all equivalent networks. (This explain the word "enzymes" used to describe the simulation relations: they may be used to start and preserve the reaction; finally, a small amount of enzyme is enough for a large, potentially unbounded, quantity of network(s) to be transformed.)

However, the reaction requires more attention in the following sense: do we allow for the newly produced networks to participate to the reaction as well, or not? If the answer is "not," then only the one-step equivalent networks may be identified, which is not too interesting. If the answer is "yes," then one may hope to have significant parts from all equivalent networks present in the solution. The problem now is that it may be the case that *one* enzyme is blocked due to an inappropriate linkage to a current network, but still useful for some other ones. A possible solution to this situation may be to mix the one-step and iterative processes as follows: one has to provide more rooms for reaction, each one with the corresponding enzymes; each room is used for one network only; the results of the reaction in a room are separated on classes and forwarded to the appropriate new rooms.

This may be seen as a rather speculative process; it may be interesting for certain massively parallel machines, but we have no claim on its effectiveness or efficiency.

4.5 Strong axioms: from cells to networks

We show that the presence of the enzymatic rule assures the lifting of the strong axioms from connections to networks.

The local validity of xy-strong axioms on X and T does not automatically extend to all networks in $[X,T]_{a\alpha}$. However, the extension is possible in the presence of the enzymatic rule.

Proposition 4.11. *Let T be a BNA which is also a strong xy-symocat. If $\sim$ is a congruence on* $\mathsf{FlowExp}[X,T]$ *which: (1) contains $\xrightarrow{a\alpha}$; (2) contains identifications C_{xy}-var; and (3) satisfies Enz_{xy}, then* $\mathsf{FlowExp}[X,T]/_\sim$ *is a BNA and a strong xy-symocat, too.*

PROOF: Since $\xrightarrow{a\alpha} \subseteq \sim$, the quotient $[X,T]/_\sim$ is also a quotient of $[X,T]_{a\alpha}$. But the latter is a BNA, hence the former is also a BNA. (Use the hypothesis that $\sim$ is a congruence.)

Below we check the strong commutation conditions. Suppose $F = [(\mathsf{I}_m \star \underline{x}) \cdot f] \uparrow^{i(\underline{x})} \in [X,T](m,n)$. As usual, $\underline{x}$ denotes a juxtaposition of variables in X.

- For $\mathrm{C}_{a\beta}$ in $[X,T]_{a\alpha}$, first notice that $\mathsf{T}_m \cdot F = [(\mathsf{I}_\epsilon \star \underline{x}) \cdot (\mathsf{T}_m \star \mathsf{I}_{o(\underline{x})}) \cdot f] \uparrow^{i(\underline{x})}$. Then

$$\begin{array}{lll}
(\mathsf{I}_\epsilon \star \mathsf{T}_{i(\underline{x})}) \cdot [(\mathsf{I}_\epsilon \star \underline{x})\ (\mathsf{T}_m \star \mathsf{I}_{o(\underline{x})})\ f] & & \\
\sim\ (\mathsf{I}_\epsilon \star \mathsf{T}_{o(\underline{x})})\ (\mathsf{T}_m \star \mathsf{I}_{o(\underline{x})}) \cdot f & & \mathrm{C}_{a\beta}\text{-var} \\
=\ \mathsf{T}_{m\star o(\underline{x})} \cdot f & & \\
=\ \mathsf{T}_{n\star i(\underline{x})} & & \mathrm{C}_{a\beta} \text{ in } T \\
=\ \mathsf{T}_n \cdot (\mathsf{I}_n \star \mathsf{T}_{i(\underline{x})}) & &
\end{array}$$

Applying $\mathrm{Enz}_{a\beta}$ we get $[(\mathsf{I}_\epsilon \star \underline{x})\ (\mathsf{T}_m \star \mathsf{I}_{o(\underline{x})})\ f] \uparrow^{i(\underline{x})} \sim \mathsf{T}_n \uparrow^\epsilon$ hence $\mathsf{T}_m \cdot F \sim \mathsf{T}_n$.

- For $\mathrm{C}_{a\gamma}$ in $[X,T]_{a\alpha}$, first notice that $(F \star F) \cdot \mathsf{V}_n = g \uparrow^{i(\underline{x})\star i(\underline{x})}$, where

$$g = (\mathsf{I}_{m\star m} \star \underline{x} \star \underline{x}) \cdot (\mathsf{I}_m \star {}^m\mathsf{X}^{o(\underline{x})} \star \mathsf{I}_{o(\underline{x})}) \cdot (f \star f) \cdot (\mathsf{I}_n \star {}^{i(\underline{x})}\mathsf{X}^n \star \mathsf{I}_{i(\underline{x})}) \cdot (\mathsf{V}_n \star \mathsf{I}_{i(\underline{x})\star i(\underline{x})})$$

Then

$$\begin{array}{ll}
g \cdot (\mathsf{I}_n \star \mathsf{V}_{i(\underline{x})}) & \\
= (\mathsf{I}_{2m} \star \underline{x} \star \underline{x}) \cdot (\mathsf{I}_m \star {}^m\mathsf{X}^{o(\underline{x})} \star \mathsf{I}_{o(\underline{x})}) \cdot (f \star f) \cdot \mathsf{V}_{n\star i(\underline{x})} & \\
= (\mathsf{I}_{2m} \star \underline{x} \star \underline{x}) \cdot (\mathsf{I}_m \star {}^m\mathsf{X}^{o(\underline{x})} \star \mathsf{I}_{o(\underline{x})}) \cdot \mathsf{V}_{m\star o(\underline{x})} \cdot f & \mathrm{C}_{a\gamma} \text{ in } T \\
= (\mathsf{I}_{2m} \star \underline{x} \star \underline{x}) \cdot (\mathsf{V}_m \star \mathsf{V}_{o(\underline{x})}) \cdot f & \\
= (\mathsf{I}_{2m} \star (\underline{x} \star \underline{x}) \cdot \mathsf{V}_{o(\underline{x})}) \cdot (\mathsf{V}_m \star \mathsf{I}_{o(\underline{x})}) \cdot f & \\
\sim (\mathsf{I}_{2m} \star \mathsf{V}_{i(\underline{x})}) \cdot [(\mathsf{I}_{2m} \star \underline{x}) \cdot (\mathsf{V}_m \star \mathsf{I}_{o(\underline{x})}) \cdot f] & \mathrm{C}_{a\gamma}\text{-var}
\end{array}$$

Applying $\mathrm{Enz}_{a\gamma}$ we get

$$\begin{aligned} g \uparrow^{i(\underline{x}) * i(\underline{x})} &\sim [(\mathsf{I}_{2m} \star \underline{x}) \cdot (\mathsf{V}_m \star \mathsf{I}_{o(\underline{x})})\ f] \uparrow^{i(\underline{x})} \\ &= \mathsf{V}_m \cdot F \end{aligned}$$

hence $\mathsf{V}_m \cdot F \sim (F \star F) \cdot \mathsf{V}_n$.

- The proof of $F \cdot \perp^n \sim \perp^m$ is dual to the first case.
- The proof of $F \cdot \wedge^n \sim \wedge^m \cdot (F \star F)$ is dual to the second case. □

4.6 xy-flows

The basic algebraic structure, called xy-flow, is introduced here. Its instances are used to axiomatize networks behaviours. In the $a\alpha$-case it coincides with BNA. Important cases are also $a\delta$, $b\delta$, and $d\delta$, as we will see in the next two chapters.

The basic algebras for the axiomatization of the different notions of network behaviour used in the network algebra context are obtained adding the strong axioms and the enzymatic rules to xy-symocats with feedback axioms. Recall the basic scheme

BNA + x_1y_1-weak + x_2y_2-strong + x_3y_3-enzymatic rule

In general 3 restrictions are to be fixed in order to have an appropriate algebraic setting: (1) one for branching constants (what branching constants are to be used), (2) one for the strong commuting rules, and finally (3) one for xy-relations to be used as enzymes in the enzymatic rule. The choices are clearly ordered: $x_1y_1 \geq x_2y_2 \geq x_3y_3$ ("$\geq$" means more branching constants).

♡ It's clear that $x_1y_1 \geq x_2y_2$ and $x_1y_1 \geq x_3y_3$. Why $x_2y_2 \geq x_3y_3$? It seems meaningless to study the enzymatic rule in a context where the corresponding strong rules are not valid. Actually the enzymatic rule was introduced to allow for the use of the strong axioms in a cyclic setting. ◇

In xy-flows there is a *unique* choice xy, with $x \in \{a, b, c, d\}, y \in \{\alpha, \beta, \gamma, \delta\}$. Then $x_2y_2 = x_3y_3 = xy$. In some cases xy-relations are not closed to feedback. Therefore x_1y_1 may contain more branching constants; actually x_1y_1 is the closure of xy to feedback; for the new branching constants the strong and enzymatic rules are not required.

Before stating xy-flow definition it is necessary to introduce a partial order relation on restrictions. The set $\{a, b, c, d\}$ is ordered by using the relation $\prec_L$ given by $a \prec_L b \prec_L d$, $a \prec_L c \prec_L d$, $\neg(b \prec_L c)$ and $\neg(c \prec_L b)$. In a similar way the relation $\prec_G$ on the Greek letters $\{\alpha, \beta, \gamma, \delta\}$ is defined. Finally $xy \prec x'y'$ if $x \prec_L x'$ and $y \prec_G y'$.

♡ This partial order on restrictions coincides with the partial order provided by the inclusion relation induced by the Hasse diagram in Fig. 3.1. ◇

The closure of a restriction to feedback may be defined now as follows. For xy with $x \in \{a, b, c, d\}$ and $y \in \{\alpha, \beta, \gamma, \delta\}$, let x^+y^+ be the least restriction greater than xy (i.e., $x \prec_L x^+$ and $y \prec_G y^+$) and closed to feedback (i.e., satisfying the condition: $x^+ = b$ or $y^+ = \beta$ or $x^+y^+ \in \{a\alpha, d\delta\}$).

Definition 4.12. (xy-flow) An *xy-flow* is a

(1) x^+y^+-symocat with feedback, which is also
(2) a strong xy-symocat and
(3) satisfies the enzymatic rule for xy-relations.

An *xy-flow morphism* is a morphism of xy-symocats which preserves feedback.

Comments. (1) The passage from xy to x^+y^+ is made adding constants of type $\top$ or $\perp$, eventually. In the definition of xy-flow morphisms the commutation with these constants was not required. This follows from axioms R8 and R9 of the symocats with feedback definition.
(2) With this general definition we have a new small overlap: $a\alpha$-flow and BNA means the same thing. There are no branching constants in the $a\alpha$ case, hence no new axioms are to be added to the BNA structure to get a $a\alpha$-flow. ◇

4.7 Semantic models: III. xy-flow structure

The additive relational model AddRel(D) is a $d\delta$-flow. This follows from already proved results, in particular using Proposition 4.10. This is the stronger structure from the hierarchy of xy-flow algebras.

For the multiplicative relational model MultRel(D), the situation is more complicated, as we did not provide a simple iterative definition for the feedback yet; and it is this iterative definition which is the crucial fact on which the proof of the enzymatic axiom is based. The situation will be clarified in Chapters 7 and 10, where two different iterative mechanisms for defining the multiplicative feedback will be described. The first one is based on the fundamental duality of the additive and multiplicative networks described in Section 7.7; this uses classes, and is fully dual to the additive case. The second possibility is to use a temporal order on data, i.e., streams; in such a case, for the particular type of deterministic data-flow networks described by (continuous) stream processing functions, a $d\alpha$-flow structure may be found.

4.8 Simulation

Simulation via arbitrary finite relations is introduced here; it may be seen as the result of applying the strong commuting axioms and the enzymatic rule to network representations (nf-pairs).

An iterative process for generating $\sim_{xy}$ is described in Appendix D, Prop. V.12, but it is complicated. It is fairly complex, indeed. It is not easy to analyze, understand, and use the resulting congruence. It produces $\sim_{xy}$ using a doubly iterative process: one for generating equivalences, the other to close these equivalences to the enzymatic implication rule.

To simplify it we use "simulation," a combination of strong axioms and a bit of enzymatic rule. It may be considered as the first step of the closure of C_{xy}-var to Enz_{xy}. This is a transformation on the normal form flownomial expressions that inherits the form of an enzymatic process. The rôle of enzymes is played by xy-arrows.

Enzymatic transformation of normal form flownomial expressions.

Enzymatic transformation

- Take a network represented by the normal form flownomial expression $F = [(\mathsf{l}_m \star x_1 \star \ldots \star x_k) \cdot f] \uparrow^{i(x_1)\star\ldots\star i(x_k)}$.
- Introduce an enzyme (perturbation) $r : \alpha \to i(x_1) \star \ldots \star i(x_k)$ on the feedback on the top of the variables/atoms and cut the feedback.
- Move the enzyme through the network trying to pass it over the variables (using C_{xy}-var rules), then over the connections, and finally displace it along the feedback.
- If the enzyme is reproduced at the end of the transformation, then take out the enzyme, reconnect the resulted feedback, and declare the resulted network F' equivalent to the initial one.

This process may be applied in the *dual* form, as well, i.e., starting with an enzyme $r : i(x_1) \star \ldots \star i(x_k) \to \alpha$ and moving the enzyme on the opposite sense: cut feedback and insert the enzyme, displace it along feedback, commute with connections, then with cells, take-out the enzyme, and finally, reconnect the feedback.

This, perhaps strange, transformation method of flownomials proves to be very useful. Its meaning depends on the type of the enzymes. It is presented in Table 4.3 in some particular, but typical, cases.

Example 4.13. Two examples are given in Figs. 4.8 and 4.2. The legend of the first figure (4.8) describes in some detail the enzymatic mechanism. It uses a particular enzyme $\wedge_2$, but it is typical for all relations which represent converse of surjective functions. As a result of this process, some cells may be identified if they have the same label and coherent input connections; in the particular case illustrated in the figure the two cells labeled by x are identified. The second figure describes a similar process, but now using an enzyme which is an injective function.

We may now formally define the simulation relation. Let T be an xy-symocat over a monoid M. For an X-sorted xy-relation $u : a \to b$, denote by $i(u) \in T(i(a),\ i(b))$ its extension to inputs, namely the image of u via the unique morphism of xy-symocats $H : \mathbb{R}\mathrm{el}_X \to T$ that extends $i : X \to M$. Similarly, the extension of u to outputs is denoted by $o(u)$.

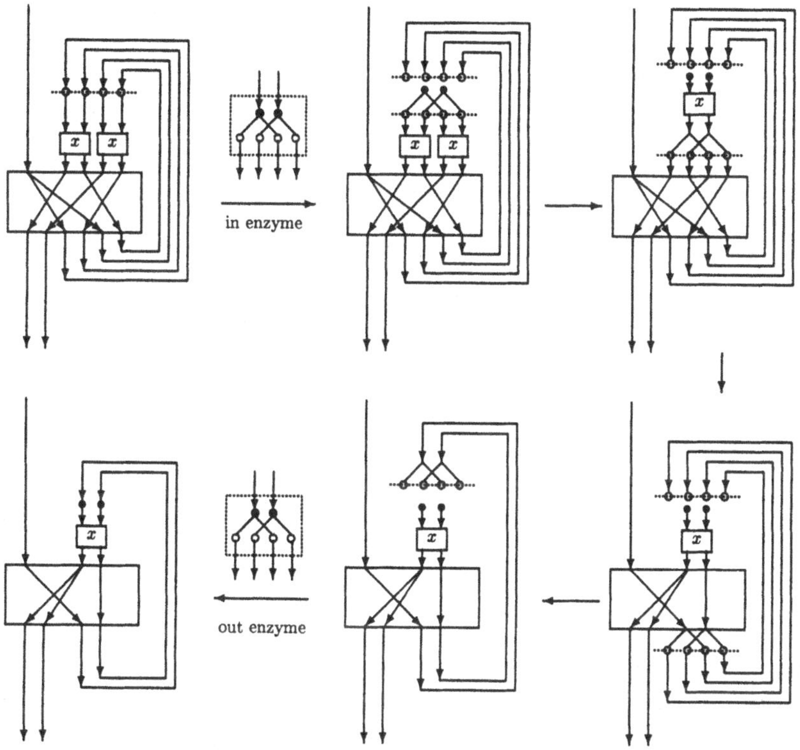

Fig. 4.1. Simulation via converses of surjections (case $c\alpha$)
Comments: In this example simulation via a converse of surjection is illustrated by an enzymatic-like process. The feedback is cut on the top of variable x's and the enzyme $\wedge_2^2$ is introduced. By the strong commuting rule we have $\wedge_2^2(x \star x) = x\wedge_2^2$, hence the enzyme passes over the variable cells. In general, the enzyme may be transformed during this process, but it was not the case here. Then, the enzyme $\wedge_2^2$ is shifted over the connections, i.e., we decompose the resulting relation $\wedge_2^2 \cdot f$ as $g \cdot \wedge_2^2$ to reobtain the original enzyme. Finally, this enzyme is shifted along the feedback, thrown away, and the resulting feedback is reconnected.

Table 4.3. The meaning of simulation

Enzyme u	The meaning of simulation		
$a\alpha$: bijection	$F \xrightarrow{a\alpha}_u F'$	$\Leftrightarrow$	GR(F) =GR(F') (i.e., both flownomial expressions F and F' represent the same network; the bijection u gives the passage from the linear order in F to the one in F')
$a\beta$: injection	$F \xrightarrow{a\beta}_u F'$	$\Leftrightarrow$	GR(F') has a subnetwork isomorphic with GR(F) and – in addition – has an inaccessible part (i.e., a part with no incoming arrows from this copy of GR(F); this additional part corresponds to the complement of $Im(u)$)
$a\gamma$: surjection	$F \xrightarrow{a\gamma}_u F'$	$\Leftrightarrow$	GR(F') may be obtained from GR(F) by identifying vertices that have common labels and their outgoing connections match (two vertices x and x' are identified iff $u(x) = u(x')$)
$b\alpha$: converse of injection	$F \xrightarrow{b\alpha}_u F'$	$\Leftrightarrow$	GR(F) has a subnetwork isomorphic with GR(F') and – in addition – has a non-coaccessible part (i.e., a part with no outgoing arrows to this copy of GR(F'); this additional part corresponds to the complement of $Dom(u)$)
$c\alpha$: converse of surjection	$F \xrightarrow{c\alpha}_u F'$	$\Leftrightarrow$	GR(F) may be obtained from GR(F') by identifying vertices that have common labels and their incoming connections match (two vertices x and x' in F' are identified iff $\exists v$ in F: $(v, x) \in u$ and $(v, x') \in u$)

Definition 4.14. (simulation) Let $F = [(\mathsf{l}_m \star x_1 \star \ldots \star x_k) \cdot f] \uparrow^{i(x_1)\star\ldots\star i(x_k)}$ and $F = [(\mathsf{l}_m \star x'_1 \star \ldots \star x'_{k'}) \cdot f'] \uparrow^{i(x'_1)\star\ldots\star i(x'_{k'})}$ be two normal form flownomial expressions. We say a relation between vertices $u \in \mathbb{R}\text{el}(k, k')$ is a *simulation from F to F'*, and write $F \rightarrow_u F'$, or $F \xrightarrow{u} F'$, iff

$$\text{(i)}\quad (j, j') \in u \;\Rightarrow\; x_j = x'_{j'} \text{ and}$$

$$\text{(ii)}\quad f \cdot (\mathsf{l}_n \star i(u)) = (\mathsf{l}_m \star o(u)) \cdot f'$$

$F \xrightarrow{xy} F'$ means $F \rightarrow_u F'$ for an abstract xy-relation u.

♡ Condition (i) shows that relation u preserves the variables, hence $u \in \mathbb{R}\text{el}_X(x_1\star\ldots\star x_k,\ x'_1\star\ldots\star x'_{k'})$. The second condition (ii) gives a correlation between the connections of F and F'. ◇

The following simple result enlightens our previous statement that simulation may be seen as a combination of strong commuting axioms and enzymatic rule.

Proposition 4.15. *(enzymatic and strong rules imply simulation)*
Let $\sim$ be a congruence on $[X, T]$ which: (1) contains $\xrightarrow{a\alpha}$, (2) contains C_{xy}-var, and (3) satisfies Enz_{xy}. Then $\sim$ contains simulation via xy-relations.

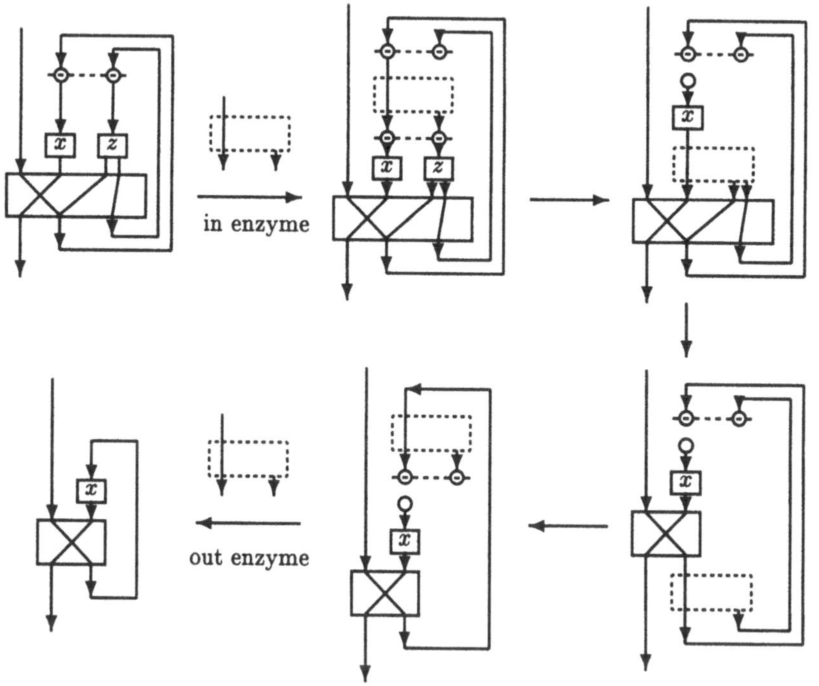

Fig. 4.2. Simulation via injections (case $a\beta$)

PROOF: Let $[(\mathsf{I}_m\star\underline{x})\cdot f]\uparrow^{i(\underline{x})} \to_u [(\mathsf{I}_m\star\underline{x}')\cdot f']\uparrow^{i(\underline{x}')}$, where $\underline{x} = x_1\star\ldots\star x_k$, $\underline{x}' = x_1'\star\ldots\star x_{k'}'$, and $u \in xy\text{-}\mathbb{R}\mathrm{el}(k,k')$. Then

$$\begin{aligned}
[(\mathsf{I}_m\star\underline{x})f]\cdot(\mathsf{I}_n\star i(u)) &= (\mathsf{I}_m\star\underline{x})\ (\mathsf{I}_m\star o(u))\cdot f' \\
&= [\mathsf{I}_m\star(x_1\star\ldots\star x_k)\cdot o(u)]\cdot f' \\
&\sim [\mathsf{I}_m\star i(u)\cdot(x_1'\star\ldots\star x_{k'}')]\cdot f' \\
&= (\mathsf{I}_m\star i(u))\cdot[(\mathsf{I}_m\star\underline{x}')f']
\end{aligned}$$

(The third step follows by a proof similar to the one of Proposition 4.2, but now applied to $\sim$ and C_{xy}-var.)

Hence one may apply Enz_{xy} to get $[(\mathsf{I}_m\star\underline{x})f]\uparrow^{i(\underline{x})}\sim[(\mathsf{I}_m\star\underline{x}')f']\uparrow^{i(\underline{x}')}$. □

Simulation is *compatible* with the operations in $[X,T]$. Indeed, if T is a BNA and $F_1\to_u F_2$ and $F_1'\to_{u'} F_2'$ are two simulations in $[X,T]$, then: (a) $F_1\star F_1'\to_{u\star u'} F_2\star F_2'$, (b) $F_1\cdot F_1'\to_{u\star u'} F_2\cdot F_2'$, and (c) $F_1\uparrow^p\to_u F_2\uparrow^p$, whenever the operations have sense. (From all BNA axioms for feedback only the axiom R1 is used here.) Consequently the simulation relation of is reflexive, transitive, and compatible with the operations. However, it is not symmetric if $xy \neq a\alpha$. Hence:

Proposition 4.16. *The congruence relation* $\overset{xy}{\Longleftrightarrow}$ *generated by simulation* $\overset{xy}{\longrightarrow}$ *is*

$$\overset{xy}{\Longleftrightarrow} = (\overset{xy}{\longrightarrow} \cup \overset{xy}{\longleftarrow})^+$$

where $\overset{xy}{\longleftarrow}$ *denotes* $(\overset{xy}{\longrightarrow})^{-1}$ *and* $^+$ *denotes the transitive closure.* □

Notation 4.17. $[X,T]_{xy}$ denotes $[X,T]/\overset{xy}{\Longleftrightarrow}$.

In the proof of Theorem 4.11 the enzymatic rule was applied only once. The result still holds when the enzymatic rule is replaced by simulation. We get the following result.

Proposition 4.18. *If* T *is a BNA and a strong xy-symocat, then* $[X,T]_{xy}$ *is also a BNA and a strong xy-symocat.* □

EXERCISES AND PROBLEMS (SEC. 4.8)

1. (i) Find the normal form representations for the networks $E = (f \uparrow^1 \star f \uparrow^1) \cdot \vee_1^2$ and $F = [(\mathsf{I}_1 \star {}^2\mathsf{X}^1) \cdot (f \star f) \cdot (\mathsf{I}_1 \star {}^1\mathsf{X}^1 \star \mathsf{I}_1)] \uparrow^2$, where $2 \overset{f}{\longrightarrow} 2$.
(ii) Are the corresponding normal forms similar?
(iii) If the answer is positive, which is the minimal class of relations that does that job? (10)
2. Take different restrictions, say $b\beta, a\delta, c\gamma, d\beta$, etc. and find the meaning of a simulation via that sort of relation. (15)
3. Show that simulation is compatible with the BNA operations. (09)
4. Fill in the details for the proof of Proposition 4.18. (08)

4.9 Enzymatic rule: from connections to networks

In this section we describe certain general conditions which assure that the enzymatic rule is preserved by passing from connection to (equivalent) networks.

We know that simulation is implied by the collection of strong commutation rules and the enzymatic rule; see Proposition 4.15. The above proposition (Proposition 4.18) shows that a part of the converse of this implication holds, i.e., $[X,T]_{xy}$ is a strong xy-symocat. An important question is to see when the other part of the implication holds, i.e.,

does Enz_{xy} *hold in* $[X,T]_{xy}$*?*

This requires us to see when the generative process for closing a relation to the enzymatic rule stops in one step. A partial answer is given in Lemma 4.20 below.

Reasons for testing Enz_{xy} in $[X,T]_{xy}$. One reason for which this question is important is the following. Semantically, the use of simulation is more appropriate, while syntactically, the approach using C_{xy} + Enz_E is better. Indeed:

(1) The basic xy-flow structure is defined using C_{xy} + Enz_E. A definition of xy-flow structure based on a rule which reflects simulation equivalence is possible, but leads to very messy definitions. In brief: C_{xy} + Enz_E are the simplest forms of the basic properties used by simulation.

(2) But, for reasoning about equivalent networks, simulation is more appropriate than C_{xy} + Enz_E. E.g., the congruence relation generated by simulation $\stackrel{xy}{\Longleftrightarrow}$ is obtained with a unique iterative process; this is more complicated for C_{xy} + Enz_E, see Proposition V.12; moreover, in many particular cases, including the ones to be analyzed in detail in the next two chapters, we will arrive at certain short and beautiful characterization theorems for $\stackrel{xy}{\Longleftrightarrow}$.

A mixture of these approaches occurs when one tries to prove that the algebraic structure is preserved by passing from connections to equivalent networks. This is of crucial importance for higher order networks, namely when one wants to iterate the construction, say, as in $[Y,[X,T]_{x'y'}]_{xy}$. The algebraic rules are described in terms of C_{xy} + Enz_E, while network equivalence uses simulation. All these comments reinforce our desire to know when these two approaches are equivalent. ◇

Lemma 4.19. *($\rightarrow$ preserves enzymatic rule) If $Enz_{\{j\}}$ holds in T, then simulation $\rightarrow$ obeys $Enz_{\{j\}}$ in $[X,T]$. More precisely, the implications*

$$(a)\quad F\cdot(\mathsf{I}_n\star j)\stackrel{u}{\longrightarrow}(\mathsf{I}_m\star j)\cdot G \;\Rightarrow\; F\uparrow^p\stackrel{u}{\longrightarrow}G\uparrow^q$$
$$(b)\quad F\cdot(\mathsf{I}_n\star j)\stackrel{u}{\longleftarrow}(\mathsf{I}_m\star j)\cdot G \;\Rightarrow\; F\uparrow^p\stackrel{u}{\longleftarrow}G\uparrow^q$$

hold for every $F\in[X,T](m\star p,n\star p)$, $G\in[X,T](m\star q,n\star q)$ and $j\in T(p,q)$.

PROOF: a) Suppose $F=(\underline{x},f)$ and $G=(\underline{y},g)$ with $\underline{x},\underline{y}\in X^\star$. The simulation gives $f\cdot(\mathsf{I}_n\star j\star i(u))=(\mathsf{I}_m\star j\star o(u))\cdot g$. This implies:

$$\begin{aligned}&[(\mathsf{I}_m\star{}^{o(\underline{x})}\mathsf{X}^p)\cdot f\cdot(\mathsf{I}_n\star{}^p\mathsf{X}^{i(\underline{x})})\cdot(\mathsf{I}_n\star i(u)\star\mathsf{I}_p)]\cdot(\mathsf{I}_{n\star i(\underline{y})}\star j)\\ =\;&(\mathsf{I}_{m\star o(\underline{x})}\star j)\cdot[(\mathsf{I}_m\star o(u)\star\mathsf{I}_q)\cdot(\mathsf{I}_m\star{}^{o(\underline{y})}\mathsf{X}^q)\cdot g\cdot(\mathsf{I}_n\star{}^q\mathsf{X}^{i(\underline{y})})]\end{aligned}$$

Applying $Enz_{\{j\}}$ in T we get

$$\begin{aligned}&[(\mathsf{I}_m\star{}^{o(\underline{x})}\mathsf{X}^p)\cdot f\cdot(\mathsf{I}_n\star{}^p\mathsf{X}^{i(\underline{x})})]\uparrow^p\cdot(\mathsf{I}_n\star i(u))\\ =\;&(\mathsf{I}_m\star o(u))\cdot[(\mathsf{I}_m\star{}^{o(\underline{y})}\mathsf{X}^q)\cdot g\cdot(\mathsf{I}_n\star{}^q\mathsf{X}^{i(\underline{y})})]\uparrow^q\end{aligned}$$

This shows that $F\uparrow^p\stackrel{u}{\longrightarrow}G\uparrow^q$.

(b) The proof is similar. □

Lemma 4.20. *Let A, B, C, and D be certain restrictions in $\{a,b,c,d\}\times\{\alpha,\beta,\gamma,\delta\}$, $j\in T(p,q)$ an arrow, and $F\in[X,T](m\star p,n\star p)$, $G\in[X,T](m\star q,n\star q)$ two nf-pairs. Suppose moreover the following conditions hold:*

(1) $F\cdot(\mathsf{I}_n\star j) \overset{B}{\longleftarrow} \circ \overset{C}{\longrightarrow} \circ \overset{D}{\longleftarrow} (\mathsf{I}_m\star j)\cdot G$ *and* $A \supseteq B \cup C \cup D$

(2) $F\cdot(\mathsf{I}_n\star j) \overset{u}{\longleftarrow} H_1 \quad\Rightarrow\quad \exists H_1' : H_1 = H_1'\cdot(\mathsf{I}_n\star j)\ \&\ F \overset{u}{\longleftarrow} H_1'$

for u obeying restriction B and arbitrary H_1

(3) $H_2 \overset{u}{\longleftarrow} (\mathsf{I}_m\star j)\cdot G \quad\Rightarrow\quad \exists H_2' : H_2 = (\mathsf{I}_m\star j)\cdot H_2'\ \&\ H_2' \overset{u}{\longleftarrow} G$

for u obeying restriction D and arbitrary H_2

Then

if $Enz_{\{j\}}$ *holds in* T, *then* $Enz_{j,F,G}$ *holds in* $[X,T]_A$

PROOF: Using (1) we get

$$F\cdot(\mathsf{I}_n\star j) \overset{B}{\longleftarrow} H_1 \overset{C}{\longrightarrow} H_2 \overset{D}{\longleftarrow} (\mathsf{I}_m\star j)\cdot G$$

Using (2) for the left simulation we get a pair H_1' such that $H_1 = H_1'\cdot(\mathsf{I}_n\star j)$ and $F \overset{B}{\longleftarrow} H_1'$; simulation is compatible with feedback, hence $F\uparrow^p \overset{B}{\longleftarrow} H_1'\uparrow^p$. In a similar way, using (3) for the right simulation we get a pair H_2' such that $H_2 = (\mathsf{I}_m\star j)\cdot H_2'$ and $H_2' \overset{D}{\longleftarrow} G$. Hence $H_2'\uparrow^q \overset{D}{\longleftarrow} G\uparrow^q$.
Now the middle simulation may be written as $H_1'\cdot(\mathsf{I}_n\star j) \overset{C}{\longrightarrow} (\mathsf{I}_m\star j)\cdot H_2'$. One may apply Lemma 4.19(a) to get $H_1'\uparrow^p \overset{C}{\longrightarrow} H_2'\uparrow^q$. All these observations show that

$$F\uparrow^p \overset{B}{\longleftarrow} H_1'\uparrow^p \overset{C}{\longrightarrow} H_2'\uparrow^q \overset{D}{\longleftarrow} G\uparrow^q$$

Consequently, $F\uparrow^p \overset{A}{\Longleftrightarrow} G\uparrow^q$, or $F = G$ in $[X,T]_A$. □

4.10 Duality: I. Reversing arrows

Duality is a very important and complex phenomena; the aim of the present short section is very modest: we describe how simulation behaves when the sense of arrows is reversed.

We first recall the duality studied in the acyclic case (Section 3.6). It is the process of transformation of statements with respect to the following rules (it formalizes the idea of changing the sense of the arrows)

$f\cdot g$ is replaced by $g\cdot f$,
${}^a\mathsf{X}^b$ is replaced by ${}^b\mathsf{X}^a$,
$\wedge_n^a$ is replaced by $\vee_a^n$ $\qquad$ $\vee_a^n$ is replaced by $\wedge_n^a$,
$f\star g$ and I_a remain unchanged.

Next, we extend it to the cyclic case adding the following duality rule:

$$(f\uparrow^p)^\circ = (f^\circ)\uparrow^p, \quad \text{for } f: m\star p \to n\star p$$

The dual expression corresponding to a flownomial expression $F = [(\mathsf{I}_m \star x) \cdot f] \uparrow^{i(x)} : m \to n$ is

$$F^o : n \to m \quad \text{in} \quad (\mathsf{FlowExp}[X^o, T^o],\ \overline{\star},\ \bar{\cdot},\ \overline{\uparrow},\ \bar{\mathsf{I}},\ \overline{\mathsf{X}})$$

where

- $X^o(m,n) = \{x^o : x \in X(n,m)\}$ and
- $(T^o,\ \overline{\star},\ \bar{\cdot},\ \overline{\uparrow},\ \bar{\mathsf{I}},\ \overline{\mathsf{X}})$ is the dual theory associated with T.

  The first condition means in X^o the functions $i^o,\ o^o : X^o \to M$ which specify the inputs and the outputs, respectively, are obtained by interchanging the ones of X, i.e. $i^o(x^o) = o(x)$ and $o^o(x^o) = i(x)$.

The duality rules may be applied to produce the following result:

$$\begin{aligned} F^o &= (((\mathsf{I}_m \star x) \cdot f) \uparrow^{i(x)})^o \\ &= ((\mathsf{I}_m \star x) \cdot f)^o)\ \overline{\uparrow}^{(i(x))^o} \\ &= (f^o \mathbin{\bar{\cdot}} (\bar{\mathsf{I}}_m \mathbin{\overline{\star}} x^o))\ \overline{\uparrow}^{o^o(x^o)} \\ &= ((\bar{\mathsf{I}}_n \mathbin{\overline{\star}} x^o) \mathbin{\bar{\cdot}} f^o)\ \overline{\uparrow}^{i^o(x^o)} \qquad \text{by R3} \end{aligned}$$

Written for pairs, the duality means

$$(x,\ f)^o = (x^o,\ f^o)$$

Finally, let us see what we get applying the duality machinery to a simulation $F = (x,\ f) \to_u G = (y,\ g)$. The simulation means $f \cdot (\mathsf{I}_n \star i(u)) = (\mathsf{I}_m \star o(u)) \cdot g$. By duality this gives $(\bar{\mathsf{I}}_n \mathbin{\overline{\star}} i^o(u^o)) \mathbin{\bar{\cdot}} f^o = g^o \mathbin{\bar{\cdot}} (\bar{\mathsf{I}}_m \mathbin{\overline{\star}} o^o(u^o))$. Hence $F^o = (x^o,\ f^o)\ {}_{u^o}\!\leftarrow G^o = (y^o,\ g^o)$.

It is clear that if u obeys a restriction xy, the u^o obeys the dual restriction $(xy)^o$. To conclude,

Fact 4.21. *By duality a simulation: (1) changes its sense and (2) its restriction is replaced by the dual one.* □

Comments, problems, bibliographic remarks

Comments. One of the most important ingredients used in the network algebra approach is the enzymatic rule and the companion simulation relation. It has striking similarities with the classical operator equivalence in linear algebra.

Philosophy behind the enzymatic rule. The enzymatic rule clearly has to do with the possibility to change a cyclic system. How can one make a cyclic system evolving? We come close to the enzymatic rule. Suppose a "perturbation" produces a "mutation," cutting the circular auto-reproduction of a system. If the transformation is such that the change matches the system, in the sense that it is able to close the circle, to let the system still exist as a working circular system, then the transformation is valid. Of course, the result depends on the type of perturbation that was used. But it seems to be the unique possibility to change a run-time system: *one has to work on the wave.*

On simulation relation. The current simulation relation *preserves* the behaviour, while usually the "simulation" term is used to express the relationship between two systems where the behaviour of one is *part* of the behaviour of the other. For instance, in process algebra the simulation is seen as a "half" of the standard bisimulation relation, hence the behaviour of one process is only a part of the other; a precise relationship between simulation and bisimulation is stated in Chapter 5. The simulation relation used in testing practical systems also has such a restricting meaning; indeed, the testing scenarios are just a few, if possibly typical, running possibilities of the tested system. In the same vein, our simulation is distinct from Eilenberg's automata simulation used in [Eil74].

Notes. Matrix theories are classical structures; their relationship with algebraic theories is singled out in [Elg76a].

Algebraic theories are introduced in [Law63] and widely used in the study of the semantics of programs, e.g., in [Elg75], [GTWW77], [TWW79], [MA86], [Man92] and [BÉ93b].

The term "symmetric strict monoidal category" is used in [Lan71]; similar structures, called "x-categories," were studied by Hotz [Hot65] and followers. It is quite popular to view algebraic or matrix theories as enriched symocats fulfilling the strong commuting rules. Symocats are frequently used in the field of flowchart theories; a pioneering paper is [ES82]. In logic, a symocats are used in the study of the algebra of proofs in [Sza80]; the passage from matrix theories to symocats corresponds to the passage from classical logic to linear logic, a step made in [Gir87]; a lot of works on linear logic use these structures. Symocats appears in other algebraic studies related to various kind of nets, e.g., in [Mol88] and [MM90].

The second part of the chapter is more technical and it is based on Chapter C of [Şte91]. The enzymatic rule via general functions is used in [AM80], but was called "functoriality rule," a name which was also used in our previous papers. In an implicit way, the enzymatic rule for functions appears in the axiomatization of iteration theories in [Ési80]; see also [BÉ93b]. The simulation relation we are using here has its roots in the corresponding equivalence relations used in [Gog74] and [Elg77]; a kind of simulation via functions is used in these papers. Its current formal definition appeared in [Şte87a, Şte87b, CŞ90a], based on a formal representation of flowcharts/networks. In a more abstract setting, simulation is also studied in [CŞ92].

The xy-flow structure is parametric and captures a lot of relevant algebraic structures used in computer science, ranging from BNAs to regular algebras; a list of important cases was included in this book's introduction; more comments on particular xy-flows are also included in the bibliographic notes of the forthcoming chapters.

5. Elgot theories

> Those who write clearly have readers; those who write obscurely have commentators.
>
> Albert Camus

The *one-way behaviour* we are studying here is "half" of the standard input–output behaviour. The latter is defined in terms of input–output paths, while the former uses all finite and infinite paths connecting one of the network interfaces. Actually, there are two dual versions of the one-way behaviour: the *input behaviour*, which considers all paths starting from the network's inputs, and the *output behaviour*, which takes into account all the paths arriving at the network's outputs.

The input behaviour may be equivalently described using the regular trees obtained by network *unfolding* starting from the inputs. For instance, this type of network behaviour is relevant for (deterministic) flowchart schemes.

Elgot theory is the short name for the already defined $a\delta$-flow. In the general view of the previous chapter *it is an ($b\delta$-weak, $a\delta$-strong, $a\delta$-enzymatic) structure.*

The main result of this chapter shows that *Elgot theory axioms are correct for networks modulo this unfolding equivalence.* The proof is given in the abstract setting, namely in the case the theory for connections is an arbitrary Elgot theory. It heavily depends on a structural theorem for characterizing the equivalence relation generated by simulation via functions; subsequently, this theorem is based on network minimization. The soundness part follows from the preservation of the Elgot theory structure by passing from connections to equivalent networks. The completeness part holds since the minimization procedure may be modeled within an Elgot theory.

A short account of the dual *output behaviour* is presented, too. It fits with the natural behaviour of (deterministic) data-flow networks.

The chapter ends with a study of the *input behaviour of nondeterministic networks.* Two networks with single-input/single-output atoms have the same input behaviour iff they are *bisimilar.* The main axiomatization result is then just a combination of the above theorem for $a\delta$-flows and the axiomatization of networks with arbitrary $d\delta$-branching constants. This setting is relevant for process algebra, as one may describe nondeterministic sequential processes modulo bisimulation.

5.1 Input behaviour; regular trees

In this section regular trees are briefly introduced as a way to model the input behaviour. The equivalent presentation of the input behaviour in terms of minimal networks is described too, with a special emphasis on the algebraic counterpart of the minimization procedure.

Regular trees. We are working here in a coalgebraic theory; hence the network cells may be specified by cotuples of single-entry elements. Then, one may consider the complete unfolding of a network starting from a certain input. A, possibly infinite, tree is obtained; it is called a *regular (or finite index) tree*. The *input behaviour* of the network is the cotuple of the regular trees corresponding to the network inputs.

Finite-index trees. The "regularity," or "finite-index" property corresponds to the following fact: such a tree *has a finite number of different subtrees*. E.g., the subtrees obtained within the unfolding procedure that correspond to a specific network cell are equal; consequently, the number of different subtrees in the resulting tree does not exceed the number of network cells, hence it is finite. ◇

While this description of the input behaviour by unfolding trees is useful at the intuitive level, it fails to be useful in computation. Actually, for handling infinite trees, one definitely needs to represent them with finite descriptions. Then, network algebra representations (or something similar, e.g., finite systems of equations) have to be used. A second critic refers to the case of abstract networks: a naive unfolding procedure is difficult to describe in this abstract case. Both points are naturally solved by passing from regular trees to $\stackrel{a\delta}{\Longleftrightarrow}$-minimal networks.

One-way minimization and its algebraic counterpart. As we said, the main result of this chapter shows that Elgot theory axioms give a correct axiomatization of deterministic networks modulo unfolding equivalence. An important technical result on which this result is based shows that this equivalence is formally captured as the equivalence relation generated by simulation via functions $\stackrel{a\delta}{\Longleftrightarrow}$.

Proof structure, via minimization. In the proof of these results minimal deterministic networks play a very important role; actually the Structural Theorem for $\stackrel{a\delta}{\Longleftrightarrow}$ is based on network minimization. The frame of the proof is the following: (1) each network may be minimized using transformations as in the example below; (2) these transformations are modeled by Elgot theory axioms; (3) minimal networks are unique up to isomorphism; and (4) minimal (concrete) networks are in bijective correspondence with the unfolding trees. ◇

The interplay between Elgot theory axioms and minimization is perhaps easier to explain for the dual case; in that case familiar systems of equations may be used to represent networks, considered with, e.g., their data-flow interpretation. The axiomatic system is given by the dual Elgot theory axioms. Recall that besides the graph isomorphism axioms this axiomatic system has two new ingredients:

(1) the strong commuting axioms $C_{b\alpha}$ and $C_{c\alpha}$ and
(2) the enzymatic axiom $Enz_{d\alpha}$ (i.e., the enzymatic rule applied to abstract conversions of functions; they are specified using the branching constants $\wedge$ and $\perp$ and the acyclic BNA signature).

Of these new axioms, $Enz_{d\alpha}$ is, by far, the most complicated.

Example 5.1. We show here how the enzymatic rule may be used to model network minimization. To help the intuition we will give in parallel the meaning of the transformations using network representation via systems of equations. In this latter representation, the network behaviour is the restriction of the system's (usually least fix-point) solution to outputs. In the examples below the outputs are y_1 and y_2, the input is x_1, the other variables being auxiliary.

(N1)	(N2)	(N3)	(N4)
$y_1 = v$	$y_1 = v$	$y_1 = f(x_1, t)$	$y_1 = f(x_1, w)$
$y_2 = f(x_1, z)$	$y_2 = f(x_1, z)$	$y_2 = f(x_1, z)$	$y_2 = f(x_1, w)$
$v = f(x_1, t)$	$v = f(x_1, t)$	$z = g(x_1, t)$	$w = g(x_1, w)$
$z = g(x_1, t)$	$z = g(x_1, t)$	$t = g(x_1, z)$	
$t = g(x_1, z)$	$t = g(x_1, z)$		
$u = h(v, u)$			

The system of equations we are starting with is N1. A network algebra representation of N1 is (the procedure to get $F1$ is explained below)

$$F1 = \{([2,4,6,8][1,10][3,9][5,7][11]) \cdot (\mathsf{l}_1 \otimes f \otimes f \otimes g \otimes g \otimes h)\} \uparrow^4_\otimes$$

The sorts are omitted, being irrelevant here; the particular NA expression above looks like a dual normal form (i.e., some connections on top, followed by a simple linear term), but this is not relevant, too.

(1) Network algebra is a variable-free framework (with respect to interface pins), hence there are some differences in passing arguments to operations which have to be clarified. First, the function symbols of the right-hand side terms in N1 correspond to the terms in $\mathsf{l}_1 \otimes f \otimes f \otimes g \otimes g \otimes h$. Next, all the interface pins in this term $\mathsf{l}_1 \otimes f \otimes f \otimes g \otimes g \otimes h$ are counted from left to right. E.g., the pin corresponding to the 2nd input of the 2nd occurrence of g has number 9. Then, the numbers in $F1$ correspond to the occurrences of the variables x_1, v, z, t, u in terms, as follows: the 1st bracket [...] to x_1, the 2nd to v, and so on. To have an example again, variable z occurs in the 2nd input of the 1st f (position 3) and in the 2nd input of the 2nd g (position 9), hence the 3rd square bracket in $F1$ is $[3, 9]$. Actually, the first part of $F1$ is just a relation in the bracket representation.

(2) Now we can pass to the enzymatic rule analysis. First, we consider the case of *converses of injective functions*. In the system-of-equations representation of networks this rule allows us to *delete* some equations of the system, provided they define variables that are not used in the remaining equations. (Of course, the equations corresponding to the outputs cannot be deleted.)

In the present example u and the corresponding equation may be deleted. What NA rules may be used to capture this phenomenon? We show that the enzymatic rule

$\mathrm{Enz}_{b\alpha}$, together with the strong rule $\mathrm{C}_{b\alpha}$, do the job. Indeed, using $h \cdot \perp^1 = \perp^2$ we may delete the term h, then its incoming connections 10 and 11 as follows:

$$
\begin{aligned}
&([2,4,6,8][1,10][3,9][5,7][11]) \cdot (\mathsf{I}_1 \otimes f \otimes f \otimes g \otimes g \otimes h) \cdot (\mathsf{I}_2 \otimes \underline{\mathsf{I}_3 \otimes \perp^1})\\
&= ([2,4,6,8][1][3,9][5,7][]) \cdot (\mathsf{I}_1 \otimes f \otimes f \otimes g \otimes g)\\
&= (\mathsf{I}_1 \otimes \underline{\mathsf{I}_3 \otimes \perp^1}) \cdot ([2,4,6,8][1][3,9][5,7]) \cdot (\mathsf{I}_1 \otimes f \otimes f \otimes g \otimes g)
\end{aligned}
$$

Hence, we are in a situation where the enzymatic rule can be applied: we have started with the enzyme $\mathsf{I}_3 \otimes \perp^1$ at the bottom feedback interface, then we made some transformations, and finally we re-obtained the enzyme at the top feedback interface. The enzymatic rule $\mathrm{Enz}_{b\alpha}$ may be applied, hence $F1$ is equivalent to

$$F2 = \{(([2,4,6,8][1][3,9][5,7]) \cdot (\mathsf{I}_1 \otimes f \otimes f \otimes g \otimes g)\} \uparrow^3_\otimes$$

This $F2$ corresponds the second system N2.

(3) Next, a graph-isomorphic transformation allows us to eliminate the *renaming* of variable v as y_1. We get an equivalent network expression that corresponds to N3:

$$F3 = \{(([1,3,5,7][4,8][2,6]) \cdot (f \otimes f \otimes g \otimes g)\} \uparrow^2_\otimes$$

(Notice that the first f in this expression corresponds to the 2nd f of the previous formula describing $F2$.)

(4) Next, let's go to the meaning of the enzymatic axiom for *converses of surjective functions*, represented using $\wedge$ and the acyclic BNA signature. In this case, we may *identify* certain variables, provided the right-hand-side terms in their defining equations become equal after that identification. In the present example, we may identify z and t, and rewrite them as a new variable w, since after this identification the corresponding terms $g(x_1, t)$ and $g(x_1, z)$ become equal, namely $g(x_1, w)$.

This is important, so let us repeat how the rule is applied, in slightly more general terms. We make a guess that we can separate the variables in certain disjoint classes and the variables in a class may be identified. The right-hand-side terms of the equations corresponding to identified variables may *not* be equal! However, these terms should be consistent with the proposed identification, namely a unique term has to be obtained when their variables are identified according to the guess.

Again, we have to look for some algebraic rules that allow us to model this phenomenon. The strong commuting rule $\mathrm{C}_{c\alpha}$ and the enzymatic rule $\mathrm{Enz}_{c\alpha}$ do the job. Indeed, using $\mathrm{C}_{c\alpha}$ one gets $\wedge^2 \cdot (g \otimes g) = g \cdot \wedge^1$. Hence,

$$
\begin{aligned}
&(\mathsf{I}_1 \otimes \underline{\wedge^1}) \cdot ([1,3,5,7][4,8][2,6]) \cdot (f \otimes f \otimes g \otimes g)\\
&= ([1,3,5,7][2,4,6,8]) \cdot (f \otimes f \otimes g \otimes g)\\
&= ([1,3,5][2,4,6]) \cdot (\mathsf{I}_4 \otimes \wedge^2) \cdot (f \otimes f \otimes g \otimes g)\\
&= ([1,3,5][2,4,6]) \cdot (f \otimes f \otimes g) \cdot (\mathsf{I}_2 \otimes \underline{\wedge^1})
\end{aligned}
$$

Hence, we are again in a situation where the enzymatic rule can be applied: we have started with the enzyme $\wedge^1$ on the top feedback interface, then we made some transformations, and finally we re-obtained the enzyme on the bottom feedback interface. By $\mathrm{Enz}_{c\alpha}$, $F3$ is equivalent to

$$F4 = \{(([1,3,5][2,4,6]) \cdot (f \otimes f \otimes g)\} \uparrow^1_\otimes$$

This is a network algebra representation of the 4th system of equations N4.

(5) Notice that till now the two occurrences of f couldn't be identified, as they had different connections inside the feedback part. Now this may be done, but this is a simple acyclic transformation. Anyways, we get the form

$$F5 = \{\wedge^2 \cdot (\mathsf{I}_2 \otimes g)\} \uparrow^1_\otimes \cdot f \cdot \wedge^1$$

corresponding to, perhaps the simplest representation of this behaviour, $y_1 = y_2 = f(x_1, w)$ where $w = g(x_1, w)$.

Theorem 5.2. *The algebra $R[X]$ of regular trees over X is isomorphic to the algebra of $\overset{a\delta}{\Longleftrightarrow}$-minimal networks $[X, \mathbb{P}\mathrm{fn}]_{a\delta}$.* □

For a sketch of the proof see for example [Şte87a].

EXERCISES AND PROBLEMS (SEC. 5.1)

1. Prove that $\{([2,4,6,8][1][3,9][5,7]) \cdot (\mathsf{I}_1 \otimes f \otimes f \otimes g \otimes g)\} \uparrow^3_\otimes$ $= \{([1,3,5,7][4,8][2,6]) \cdot (f \otimes f \otimes g \otimes g)\} \uparrow^2_\otimes$ is valid in a $d\beta$-symocat with feedback. (This shows the transformation in step (3) of the example above is correct in a dual Elgot theory.) (07)

5.2 Elgot theories ($a\delta$-flows)

The definition of Elgot theory was already given in the previous chapter. In general NA terms, it is a ($b\delta$-weak, $a\delta$-strong, $a\delta$-enzymatic) structure. That means, it is a BNA with (1) $b\delta$-branching constants ($b\delta$-symocat with feedback), of which (2) $a\delta$-constants strongly commutes with arbitrary arrows, and (3) the same $a\delta$-constants generate relations for which the enzymatic rule is valid.

Syntactic calculus. An algebraic calculus for this kind of network behaviour is the $a\delta$-Flow-Calculus, an instance of the general class of xy-Flow-Calculi defined below.

Definition 5.3. (xy-Flow-Calculus; xy-flownomials)
Let xy be a restriction, i.e., $x \in \{a, b, c, d\}$ and $y \in \{\alpha, \beta, \gamma, \delta\}$.

xy-Flow-Calculus is the calculus obtained from BNA with xy-branching constants adding: (1) C_{xy} and (2) Enz_{xy}. An *xy-flownomial over X and T* is a class of $\sim_{xy}$-equivalent flownomial expressions in $\mathsf{FlowExp}[X, T]$.

ⓘ Recall that: C_{xy} is the strong commutation conditions of xy-branching constants with arbitrary arrows; Enz_{xy} is the enzymatic rule for the corresponding abstract xy-relations; $\sim_{xy}$ is the congruence relation with property Enz_{xy} generated on $\mathsf{FlowExp}[X, T]$ by C_{xy}-var, i.e., by the commutation of xy-branching constants with variables. ⓘ

Iteration theories. A closely related algebraic structure is the *iteration theory* presented in [BÉ93b]. In this algebraic structure, the enzymatic (implication) rule for functions is replaced by an equation, the so called Esik commutative axiom. The idea is to replace the premise of the implication rule by a mechanism of generating all situations (occurring in regular trees) leading to such valid premises. To have an example, notice that $f(x,x,y,t)$ and $f(x,y,y,t)$ become equal under the identification $x = y$; this situation is modeled using a more general term $f(a,b,c,t)$ from which the previous case follows by using the identification $a = x,\ b = x,\ c = y,\ t = t$ and $a = x,\ b = y,\ c = y,\ t = t$, respectively, both consistent with the identification $a = b = c,\ t = t$.

This mechanism has a tight syntactic extent: for instance, one cannot model the situation where in a semantical domain one has different f and g such that $f(x,x,y,t)$ and $g(x,y,y,t)$ become equal under the identification $x = y$.

Exercises and Problems (sec. 5.2)

1. Show that ω-continuous theories (cf., [TWW79, BÉ93b]) are Elgot theories. (08)

5.3 Structural Theorem, case $a\delta$

The structural theorem, proved in this section, gives a characterization for $\overset{a\delta}{\Longleftrightarrow}$, the equivalence relation generated by simulation via functions. Roughly speaking, it says that two networks are $\overset{a\delta}{\Longleftrightarrow}$-equivalent iff by deletion of non-accessible cells and identification of certain similar cells, both networks may be reduced to the same form.

The core of the calculus of flownomials for the current setting is provided by a structural theorem for the equivalence relation generated by simulation via functions. Combined with the completeness theorem, it will give an effective procedure to decide when two flownomial expressions are equivalent in the current one-way behavioural setting.

General presentation. Recall from the previous chapter that: $\overset{a\delta}{\longrightarrow}$ denotes simulation via functions; $\overset{a\delta}{\Longleftrightarrow}$ denotes the equivalence relation generated by $\overset{a\delta}{\longrightarrow}$; if the connecting theory T is a BNA, actually $\overset{a\delta}{\Longleftrightarrow}$ is even a congruence on $[X,T]$. The Structural Theorem we shall prove is

$$\overset{a\delta}{\Longleftrightarrow} = \overset{a\gamma}{\longrightarrow} \cdot \overset{a\beta}{\longleftarrow} \cdot \overset{a\beta}{\longrightarrow} \cdot \overset{a\gamma}{\longleftarrow}$$

This means that two networks are $\overset{a\delta}{\Longleftrightarrow}$-equivalent iff by

- reduction $\stackrel{a\gamma}{\longrightarrow}$ (i.e., by identification of vertices that have the same label and whose outgoing connections become identical after identification) and
- deletion of inaccessible vertices $\stackrel{a\beta}{\longleftarrow}$ (i.e., reduction of those vertices which are not reached by paths starting from the network inputs)

both networks may be brought to the same form.

The connecting theory T is supposed to be an $a\delta$-strong $b\delta$-symocat with feedback.

Needed axioms. The context for proving this theorem is that of a coalgebraic theory T with a feedback operation. The feedback axioms are not too important here. However, we have to use the *congruence* relation generated by simulation; consequently, certain axioms for feedback are to be imposed. BNA axioms suffice to prove the result is a congruence, see Proposition 4.16. ◊

Commuting lemmas. The proof of the Structural Theorem is based on certain commuting relations between the elementary simulations $\stackrel{a\beta}{\longrightarrow}$, $\stackrel{a\gamma}{\longrightarrow}$ and their converses. They are displayed in Table 5.1.

How to use Table 5.1. The elementary simulations are ordered by raws and columns according to their order in the Structural Theorem $\stackrel{a\gamma}{\longrightarrow} \cdot \stackrel{a\beta}{\longleftarrow} \cdot \stackrel{a\beta}{\longrightarrow} \cdot \stackrel{a\gamma}{\longleftarrow}$. The commuting lemmas are used in the following context. We want to show that an elementary simulation occurring on the right of this chain may be eliminated. To this end, it is commuted with the elementary reductions till it reaches its mate. This explain why only the commutations in the lower-left part of Table 5.1 are necessary. Indeed all elementary simulations have to pass over the right one in the chain $\stackrel{a\gamma}{\longleftarrow}$, hence the full 4th low line is necessary; next, all elementary simulations, except for $\stackrel{a\gamma}{\longleftarrow}$, have to pass over the second right one $\stackrel{a\beta}{\longrightarrow}$, hence the commutations in the 3rd low line are necessary; and so on. ◊

Table 5.1. Is $x \cdot y \subseteq y \cdot x$? (case $a\delta$)

x \ y	$\stackrel{a\gamma}{\longrightarrow}$	$\stackrel{a\beta}{\longleftarrow}$	$\stackrel{a\beta}{\longrightarrow}$	$\stackrel{a\gamma}{\longleftarrow}$
$\stackrel{a\gamma}{\longrightarrow}$	Obvious			
$\stackrel{a\beta}{\longleftarrow}$	5.7	Obvious		
$\stackrel{a\beta}{\longrightarrow}$	5.4	5.6	Obvious	
$\stackrel{a\gamma}{\longleftarrow}$	5.9	5.5	5.8	Obvious

To fix a notation, we suppose that the networks we are using here are in a fixed $[X,T](m,n)$, for some $m,n \in M$.

Injection, then surjection. The first commuting lemma, to be proved below, has an additional importance. It not only shows the required commutation, but it also proves that simulation via functions $\xrightarrow{a\delta}$ may be reduced to elementary simulations via surjections $\xrightarrow{a\gamma}$ and injections $\xrightarrow{a\beta}$. ◊

Lemma 5.4. $\xrightarrow{a\beta}\cdot\xrightarrow{a\gamma}\ \subseteq\ \xrightarrow{a\delta}\ \subseteq\ \xrightarrow{a\gamma}\cdot\xrightarrow{a\beta}$, *when T is a $b\delta$-symocat.*

PROOF: The first inclusion is obviously true; indeed, both $\xrightarrow{a\beta}$ and $\xrightarrow{a\gamma}$ are included in $\xrightarrow{a\delta}$ and the latter relation is transitive.

For the second inclusion, suppose $F' = (x', f') \xrightarrow{a\delta}_u F'' = (x'', f'')$.

This means: (1) $f'\ (\mathsf{I}_n \star i(u)) = (\mathsf{I}_m \star o(u))\ f''$.

Function $u \in a\delta\text{-}\mathbb{R}\mathrm{el}_X(x', x'')$ may be written as a composite of a surjection and an injection, say

$$u = u_s \cdot u_i$$

with $u_s \in a\gamma\text{-}\mathbb{R}\mathrm{el}_X(x', x)$ and $u_i \in a\beta\text{-}\mathbb{R}\mathrm{el}_X(x, x'')$. Then there exists a function $v_s \in a\delta\text{-}\mathbb{R}\mathrm{el}_X(x, x')$ such that $v_s \cdot u_s = \mathsf{I}_x$ and an (eventually partial) function $v_i : x'' \to x$ such that $u_i \cdot v_i = \mathsf{I}_x$.

(Actually, the function v_i is partial in the case x'' contains some sorts that does not occur in x. Since our relations are sorted, in such a case one cannot make v_i total, even if the actual value on a pin with such an exotic sort does not matter.)

Take the network $F = (x, f)$ with $f = (\mathsf{I}_m \star o(v_s))\ f'\ (\mathsf{I}_n \star i(u_s))$.

We show that $F' \longrightarrow_{u_s} F \longrightarrow_{u_i} F''$.

Indeed, the left relation gives a simulation by

$$\begin{array}{rcll}
(\mathsf{I}_m \star o(u_s)) \cdot f & = & (\mathsf{I}_m \star o(u_s\ v_s))\ f'\ (\mathsf{I}_n \star i(u_s)) & \text{def } f\\
& = & (\mathsf{I}_m \star o(u_s\ v_s))\ f'\ (\mathsf{I}_n \star i(u_s\ u_i\ v_i)) & \\
& = & (\mathsf{I}_m \star o(u_s\ v_s))\ f'\ (\mathsf{I}_n \star i(u))\ (\mathsf{I}_n \star i(v_i)) & \\
& = & (\mathsf{I}_m \star o(u_s\ v_s\ u))\ f''\ (\mathsf{I}_n \star i(v_i)) & \text{by (1)}\\
& = & (\mathsf{I}_m \star o(u_s\ u_i))\ f''\ (\mathsf{I}_n \star i(v_i)) & \\
& = & f' \cdot (\mathsf{I}_n \star i(u\ v_i)) & \text{by (1)}\\
& = & f' \cdot (\mathsf{I}_n \star i(u_s)) &
\end{array}$$

and the right one by

$$\begin{array}{rcll}
f \cdot (\mathsf{I}_n \star i(u_i)) & = & (\mathsf{I}_m \star o(v_s))\ f'\ (\mathsf{I}_n \star i(u_s\ u_i)) & \text{def } f\\
& = & (\mathsf{I}_m \star o(u_i)) \cdot f'' & \text{by (1)}
\end{array}$$

□

Using converse relations we can reverse the sense of arrows in the statement of the above lemma. The following result is then obtained.

Lemma 5.5. $\xleftarrow{a\gamma}\cdot\xleftarrow{a\beta}\ \subseteq\ \xleftarrow{a\delta}\ \subseteq\ \xleftarrow{a\beta}\cdot\xleftarrow{a\gamma}$, *when T is a $b\delta$-symocat.* □

Injection, then against injection. The next lemma is on commutation of simulations via injections. It has to do with the situation where two networks F' and F'' are subnetworks in a given network F and, moreover, their complementary parts in F are non-accessible from these subnetworks, respectively. Then, one can show that their common part in F is a subnetwork in both, having a non-accessible complementary part, too. ◇

Lemma 5.6. $\overset{a\beta}{\longrightarrow} \cdot \overset{a\beta}{\longleftarrow} \subseteq \overset{a\beta}{\longleftarrow} \cdot \overset{a\beta}{\longrightarrow}$*, when T is a $b\delta$-symocat.*

PROOF: Suppose $F' \overset{a\beta}{\longrightarrow}_{u'} F \;{}_{u''}\!\overset{a\beta}{\longleftarrow} F''$.

Using isomorphic representations for F', F and F'' we may suppose that

$$F' = (a \star b,\ f'),\quad F = (a \star b \star c \star d,\ f),\ F'' = (b \star c,\ f'')$$
$$u' = \mathsf{I}_{a\star b} \star \top_{c\star d} \text{ and } u'' = \top_a \star \mathsf{I}_{b\star c} \star \top_d.$$

Pause: isomorphic representations. Since this wording "using isomorphic representations we may suppose that [...]" will be used quite often in the following, we present it in more detail here.

(1) We may permute the cells of F' (in our current case as $a \star b$, where in b one puts the cells which meet in F cells from F'' via u' and u''). If the lemma holds for this new network obtained from F', then it will hold for the original F', too. Indeed, one simply has to use the simulation via the bijection which connects both versions of F' to pass from one statement of the lemma to the other. (Use $\overset{xy}{\longrightarrow} \cdot \overset{a\alpha}{\longrightarrow} = \overset{xy}{\longrightarrow}$, for all xy.)

(2) The same transformation may be applied to F'' (in the current case the cells having common images in F with those in F' are put in the front and in the same order b as in F').

(3) Finally, the cells in F may be permuted and put in the specified order (in our case, first the ones from F', but not from F''; next the ones from both F' and F''; then the ones from F'', but not from F'; and finally, the remaining ones; moreover, within each of the first 3 groups, the order is the same in F and the corresponding F' and/or F''). The simulation with the bijection which relates the new and the old versions of F is then included in both $\overset{a\beta}{\longrightarrow}$ and $\overset{a\beta}{\longleftarrow}$.

Hence by such transformations we have an equivalent case to prove. But now, while the relationship between the networks replacing F', F, F'' is similar to the starting one, the simulating relations u' and u'' have the desired new form. ◇

The simulations show that

$$\text{(1)}\qquad f' \cdot (\mathsf{I}_n \star i(\mathsf{I}_{a\star b} \star \top_{c\star d})) = (\mathsf{I}_m \star o(\mathsf{I}_{a\star b} \star \top_{c\star d})) \cdot f \quad \text{and}$$
$$\text{(2)}\qquad f'' \cdot (\mathsf{I}_n \star i(\top_a \star \mathsf{I}_{b\star c} \star \top_d)) = (\mathsf{I}_m \star o(\top_a \star \mathsf{I}_{b\star c} \star \top_d)) \cdot f$$

Take the network $\overline{F} = (b, \overline{f})$ with

$$\overline{f} = (\mathsf{I}_m \star o(\top_a \star \mathsf{I}_b \star \top_{c\star d})) \cdot f \cdot (\mathsf{I}_n \star i(\perp^a \star \mathsf{I}_b \star \perp^{c\star d}))$$

We show that $F' \;{}_{\top_a \star \mathsf{I}_b}\!\overset{a\beta}{\longleftarrow} \overline{F} \overset{a\beta}{\longrightarrow}_{\mathsf{I}_b \star \top_c} F''$.

Indeed, in the left part we have a simulation since using the definition of $\overline{f}$ in $=^1$ below, (2) in $=^2$, again (2) in $=^4$, and (1) in $=^5$ we get

$$\begin{aligned}
&\overline{f} \cdot (\mathsf{I}_n \star i(\mathsf{T}_a \star \mathsf{I}_b))\\
&=^1 (\mathsf{I}_m \star o(\mathsf{T}_a \star \mathsf{I}_b \star \mathsf{T}_{c\star d})) \cdot f \cdot (\mathsf{I}_n \star i(\bot^a \cdot \mathsf{T}_a \star \mathsf{I}_b \star \bot^{c\star d}))\\
&=^2 (\mathsf{I}_m \star o(\mathsf{I}_b \star \mathsf{T}_c)) \cdot f'' \cdot (\mathsf{I}_n \star i(\mathsf{T}_a \star \mathsf{I}_{b\star c} \star \mathsf{T}_d) \cdot i(\bot^a \cdot \mathsf{T}_a \star \mathsf{I}_b \star \bot^{c\star d}))\\
&=^3 (\mathsf{I}_m \star o(\mathsf{I}_b \star \mathsf{T}_c)) \cdot f'' \cdot (\mathsf{I}_n \star i(\mathsf{T}_a \star \mathsf{I}_{b\star c} \star \mathsf{T}_d) \cdot i(\mathsf{I}_a \star \mathsf{I}_b \star \bot^{c\star d}))\\
&=^4 (\mathsf{I}_m \star o(\mathsf{T}_a \star \mathsf{I}_b \star \mathsf{T}_{c\star d})) \cdot f \cdot (\mathsf{I}_n \star i(\mathsf{I}_{a\star b} \star \bot^{c\star d}))\\
&=^5 (\mathsf{I}_m \star o(\mathsf{T}_a \star \mathsf{I}_b)) \cdot f' \cdot (\mathsf{I}_n \star i(\mathsf{I}_{a\star b} \star \mathsf{T}_{c\star d}) \cdot i(\mathsf{I}_{a\star b} \star \bot^{c\star d}))\\
&=^6 (\mathsf{I}_m \star o(\mathsf{T}_a \star \mathsf{I}_b)) \cdot f'
\end{aligned}$$

In a similar way one may prove that in the right part we have a simulation, hence the lemma is proved. □

Against injection, then surjection. The next lemma starts with the situation where a network F is a subnetwork in F', with a non-accessible complementary part, and, on the other hand, may be reduced to F'' by cell identification. To commute the order of simulations, we need a new network obtained from F'' adding the complementary part of F within F'. This combination may be done if the stronger coalgebraic rules are accepted. ◊

Lemma 5.7. $\stackrel{a\beta}{\longleftarrow} \cdot \stackrel{a\gamma}{\longrightarrow} \subseteq \stackrel{a\gamma}{\longrightarrow} \cdot \stackrel{a\beta}{\longleftarrow}$, *when T is an $a\delta$-strong $b\delta$-symocat.*

PROOF: Let $F' \;{}_{u'}\!\stackrel{a\beta}{\longleftarrow} F \stackrel{a\gamma}{\longrightarrow}_{u''} F''$.

Using an isomorphic copy of F' we may suppose $F' = (a \star b,\ f')$, $F = (a,\ f)$ and $u' = \mathsf{I}_a \star \mathsf{T}_b$. Let $F'' = (c,\ f'')$. The simulations give

$$\begin{aligned}
&(1) \quad f \cdot (\mathsf{I}_n \star i(\mathsf{I}_a \star \mathsf{T}_b)) = (\mathsf{I}_m \star o(\mathsf{I}_a \star \mathsf{T}_b)) \cdot f' \qquad \text{and}\\
&(2) \quad f \cdot (\mathsf{I}_n \star i(u'')) = (\mathsf{I}_m \star o(u'')) \cdot f''.
\end{aligned}$$

Take the network $\overline{F} = (c \star b,\ \overline{f})$, where

$$\overline{f} = (f''\ (\mathsf{I}_n \star i(\mathsf{I}_c \star \mathsf{T}_b))\ \star\ (\mathsf{T}_m \star o(\mathsf{T}_a \star \mathsf{I}_b))\ f'\ (\mathsf{I}_n \star i(u'' \star \mathsf{I}_b))) \cdot \mathsf{V}_{n \star i(c\star b)}.$$

We show that $F' \stackrel{a\gamma}{\longrightarrow}_{u''\star \mathsf{I}_b} \overline{F}\ {}_{\mathsf{I}_c \star \mathsf{T}_b}\!\stackrel{a\beta}{\longleftarrow} F''$.

Indeed, for the left part we may compute, using the definition of $\overline{f}$ in $=^1$, (2) in $=^2$, (1) in $=^3$, and $\mathsf{C}_{a\gamma}$ in $=^4$, as follows:

$$\begin{aligned}
&(\mathsf{I}_m \star o(u'' \star \mathsf{I}_b)) \cdot \overline{f}\\
&=^1 [(\mathsf{I}_m \star o(u''))\ f''\ (\mathsf{I}_n \star i(\mathsf{I}_c \star \mathsf{T}_b)) \star (\mathsf{T}_m \star o(\mathsf{T}_a \star \mathsf{I}_b))\ f'\ (\mathsf{I}_n \star i(u'' \star \mathsf{I}_b))] \cdot \mathsf{V}_{n\star i(c\star b)}\\
&=^2 [f\ (\mathsf{I}_n \star i(u'' \star \mathsf{T}_b)) \star (\mathsf{T}_m \star o(\mathsf{T}_a \star \mathsf{I}_b))\ f'\ (\mathsf{I}_n \star i(u'' \star \mathsf{I}_b))] \cdot \mathsf{V}_{n\star i(c\star b)}\\
&=^3 [(\mathsf{I}_m \star o(\mathsf{I}_a \star \mathsf{T}_b))f'(\mathsf{I}_n \star i(u'' \star \mathsf{I}_b)) \star (\mathsf{T}_m \star o(\mathsf{T}_a \star \mathsf{I}_b))f'(\mathsf{I}_n \star i(u'' \star \mathsf{I}_b))] \cdot \mathsf{V}_{n\star i(c\star b)}\\
&=^4 f' \cdot (\mathsf{I}_n \star i(u'' \star \mathsf{T}_b))
\end{aligned}$$

and for the right part, using the definition of $\overline{f}$ in $=^1$ and $\mathsf{C}_{a\beta}$ in $=^2$, the computation is

$$\begin{aligned}
&(\mathsf{I}_m \star o(\mathsf{I}_c \star \mathsf{T}_b)) \cdot \overline{f} \\
&=^1 [f'' \; (\mathsf{I}_n \star i(\mathsf{I}_c \star \mathsf{T}_b)) \star (\mathsf{T}_m \star o(\mathsf{T}_a \star \mathsf{T}_b)) \; f' \; (\mathsf{I}_n \star i(u'' \star \mathsf{I}_b))] \cdot \mathsf{V}_{n \star i(c \star b)} \\
&=^2 (f'' \; (\mathsf{I}_n \star i(\mathsf{I}_c \star \mathsf{T}_b))
\end{aligned} \qquad \square$$

Using converse relations we solve the commuting problem for one more case.

Lemma 5.8. $\overset{a\gamma}{\longleftarrow} \cdot \overset{a\beta}{\longrightarrow} \;\subseteq\; \overset{a\beta}{\longrightarrow} \cdot \overset{a\gamma}{\longleftarrow}$, *when T is an $a\delta$-strong $b\delta$-symocat.* $\square$

Against surjection, then surjection. Finally, we have to commute two reductions which identify the cells of a given network F, say, via surjections u' and u'', respectively. This is quite complicated and it requires us to work with the components corresponding to the network's cells. Hence the strong coalgebraic theory rules are necessary. Roughly speaking, the proof says that by considering the identification generated by both u' and u'' a unique network may be obtained; then the partial identification given by u' and u'', respectively, may be continued by further identifications to reach this final common network. $\diamondsuit$

Lemma 5.9. $\overset{a\gamma}{\longleftarrow} \cdot \overset{a\gamma}{\longrightarrow} \;\subseteq\; \overset{a\gamma}{\longrightarrow} \cdot \overset{a\gamma}{\longleftarrow}$, *when T is an $a\delta$-strong $b\delta$-symocat.*

PROOF: Suppose $F' \;{}_{u'}\!\overset{a\gamma}{\longleftarrow} F \overset{a\gamma}{\longrightarrow}_{u''} F''$ where $F' = (x',\ f')$, $F = (x,\ f)$ and $F'' = (x'',\ f'')$. It follows that

$$\begin{array}{lll}
(1) & f \cdot (\mathsf{I}_n \star i(u')) = (\mathsf{I}_m \star o(u')) \cdot f' & \text{and} \\
(2) & f \cdot (\mathsf{I}_n \star i(u'')) = (\mathsf{I}_m \star o(u'')) \cdot f''. &
\end{array}$$

For a function $g : p \to q$ we denote its kernel by

$$Ker(g) = \{(j,k) : \ j,k \in [p] \text{ and } g(j) = g(k)\}$$

Let $\sim$ be the least equivalence relation on $[|x|]$ which contain both $Ker(u')$ and $Ker(u'')$. It may be constructively introduced as

$$j \sim k \quad \Leftrightarrow \quad \left\{ \begin{array}{l}
\text{there exists a sequence of elements in } [|x|] \text{ denoted by} \\
n_1, \ldots, n_r \text{ with } n_1 = j \text{ and } n_r = k, \text{ such that} \\
(n_s, n_{s+1}) \in Ker(u') \;\cup\; Ker(u''), \text{ for every } s < r
\end{array} \right.$$

Relation $\sim$ inherits from $Ker(u')$ and $Ker(u'')$ the property that it does not identify elements $j,k \in [|x|]$ with $x_j \neq x_k$. Hence $\sim$ may be represented as

$$\sim \;=\; Ker(u)$$

for a surjection $u \in a\gamma\text{-}\mathrm{IRel}_X(x, \overline{x})$.

Let $z' : x' \to \overline{x}$ and $z'' : x'' \to \overline{x}$ denote the induced multi-sorted surjections which satisfy the condition $u' \cdot z' = u = u'' \cdot z''$.

For $j \in [|x|]$ the component of f corresponding to the outputs of x_j is denoted by f_j, i.e., $f_j = (\top_{m\star o(x_1\star\ldots\star x_{j-1})} \star \mathsf{I}_{o(x_j)} \star \top_{o(x_{j+1}\star\ldots\star x_{|x|})}) \cdot f$.
The simulations u' and u'' show that

$$\begin{array}{lll}(j,k) \in Ker(u') & \Rightarrow & f_j \cdot (\mathsf{I}_n \star i(u')) = f_k \cdot (\mathsf{I}_n \star i(u')) \quad \text{and} \\ (j,k) \in Ker(u'') & \Rightarrow & f_j \cdot (\mathsf{I}_n \star i(u'')) = f_k \cdot (\mathsf{I}_n \star i(u''))\end{array}$$

respectively. Using $u' \cdot z' = u = u'' \cdot z''$ we get

$$(j,k) \in Ker(u') \cup Ker(u'') \quad \Rightarrow \quad f_j \cdot (\mathsf{I}_n \star i(u)) = f_k \cdot (\mathsf{I}_n \star i(u))$$

and this implication may easily be extended to $(j,k) \in Ker(u)$ using the above constructive definition of $Ker(u)$ (equal to $\sim$). This shows that

$$(*) \quad f \cdot (\mathsf{I}_n \star i(u)) = (\mathsf{I}_m \star o(u\ v)) \cdot f \cdot (\mathsf{I}_n \star i(u)), \quad \text{for every } v \text{ with } v \cdot u = \mathsf{I}_{\overline{x}}.$$

(Roughly speaking, this property shows that if one interchanges the connecting components of f corresponding to cells in the same equivalence class of $\sim$ (this is the effect of using $u\ v$ in the right-hand-side term), then the resulting connections does not change the equivalence classes of the target cells.)

Take the network $\overline{F} = (\overline{x}, \overline{f})$ with $\overline{f} = (\mathsf{I}_m \star o(v)) \cdot f \cdot (\mathsf{I}_n \star i(u))$, where $v \in a\beta\text{-}\mathbb{R}\mathrm{el}_X(\overline{x}, x)$ is an arbitrary right inverse for u, i.e., $v \cdot u = \mathsf{I}_{\overline{x}}$.

(By (*), all choices for v have the same effect. Hence the resulting network $\overline{F}$ is unique, i.e., it does not depend on the choice of v.)

With this definition of $\overline{F}$ we may prove that $F' \xrightarrow{a\gamma}_{z'} \overline{F}\ {}_{z''}\!\xleftarrow{a\gamma} F''$.
For the left simulation, if $v' \in a\beta\text{-}\mathbb{R}\mathrm{el}_X(x', x)$ is such that $v' \cdot u' = \mathsf{I}_{x'}$, then

$$\begin{array}{lll} f' \cdot (\mathsf{I}_m \star i(z')) & & \\ \quad = (\mathsf{I}_m \star o(v'\ u'))\ f'\ (\mathsf{I}_n \star i(z')) & & \\ \quad = (\mathsf{I}_m \star o(v'))\ f\ (\mathsf{I}_n \star i(u'\ z') & & \text{by (1)} \\ \quad = (\mathsf{I}_m \star o(v'))\ f\ (\mathsf{I}_n \star i(u)) & & \\ \quad = (\mathsf{I}_m \star o(v'\ u\ v))\ f\ (\mathsf{I}_n \star i(u)) & & \text{by } (*) \\ \quad = (\mathsf{I}_m \star o(z'))\ (\mathsf{I}_m \star o(v))\ f\ (\mathsf{I}_n \star i(u)) & & \text{by } v'u = v'u'z' = z' \\ \quad = (\mathsf{I}_m \star o(z')) \cdot \overline{f} & & \text{def } \overline{f} \end{array}$$

For the other simulation the proof is similar. □

Structural Theorem. The just proved lemmas may be used to get the structural theorem. Adding BNA axioms to their hypotheses, one realizes that the framework where all these lemmas are valid is that of Căzănescu–Ungureanu theories. This explains the condition in the theorem.

Theorem 5.10. *(Structural Theorem for $\overset{a\delta}{\Longleftrightarrow}$) If T is a Căzănescu–Ungureanu theory, then the following decomposition holds in $[X,T]$:*

$$\overset{a\delta}{\Longleftrightarrow} = \overset{a\gamma}{\longrightarrow} \cdot \overset{a\beta}{\longleftarrow} \cdot \overset{a\beta}{\longrightarrow} \cdot \overset{a\gamma}{\longleftarrow}$$

PROOF: Two networks are related by $\overset{a\delta}{\Longleftrightarrow}$ iff they are connected by a chain of simulations $\overset{a\delta}{\longrightarrow}$ or $\overset{a\delta}{\longleftarrow}$. From Lemmas 5.4 and 5.5 it follows that $\overset{a\delta}{\longrightarrow} \subseteq \overset{a\gamma}{\longrightarrow} \cdot \overset{a\beta}{\longrightarrow}$ and $\overset{a\delta}{\longleftarrow} \subseteq \overset{a\gamma}{\longleftarrow} \cdot \overset{a\beta}{\longleftarrow}$, respectively. Moreover, $\overset{a\gamma}{\longrightarrow}$, $\overset{a\beta}{\longrightarrow}$, $\overset{a\gamma}{\longleftarrow}$ and $\overset{a\beta}{\longleftarrow}$ are reflexive and transitive relations. Hence, inserting such arbitrary simulations, we may arrive at the following relation

$$F \overset{a\delta}{\Longleftrightarrow} F' \text{ iff } \exists n \geq 0 \colon F\ \rho^n F'$$

where $\rho = \overset{a\gamma}{\longrightarrow} \cdot \overset{a\beta}{\longleftarrow} \cdot \overset{a\beta}{\longrightarrow} \cdot \overset{a\gamma}{\longleftarrow}$. Using the commuting lemmas 5.4–5.9, it follows that all the relations $\rho \cdot \overset{a\gamma}{\longrightarrow}$, $\rho \cdot \overset{a\beta}{\longrightarrow}$, $\rho \cdot \overset{a\gamma}{\longleftarrow}$ and $\rho \cdot \overset{a\beta}{\longleftarrow}$ are included in ρ, hence ρ is a transitive relation. This gives $\overset{a\delta}{\Longleftrightarrow} = \rho$. □

Minimization with respect to the input behaviour. The above theorem provides a procedure for minimizing abstract networks with respect to the input behaviour. Before its presentation a few definitions are necessary.

A network (x, f) is $\sim$-*minimal*, for an equivalence on networks $\sim$, if there is no $(x', f') \sim (x, f)$ with $|x'| < |x|$. The relations $\overset{a\gamma}{\longrightarrow}$ and $\overset{a\beta}{\longleftarrow}$ are *reductions*, i.e., if $(x, f) \overset{a\gamma}{\longrightarrow} (x', f')$ or $(x, f) \overset{a\beta}{\longleftarrow} (x', f')$ then $|x'| < |x|$. The relation $\overset{a\alpha}{\longrightarrow}$ is also called *isomorphism*.

Now the following easy corollaries may be obtained.

Corollary 5.11. *A network F is $\overset{a\delta}{\Longleftrightarrow}$-minimal if and only if ($F \overset{a\gamma}{\longrightarrow} F'$ and $F \overset{a\beta}{\longleftarrow} F'$) implies $F \overset{a\alpha}{\longrightarrow} F'$.* □

Corollary 5.12. *Two $\overset{a\delta}{\Longleftrightarrow}$-minimal and equivalent networks are isomorphic.*

PROOF: The proof is simple. Suppose F' and F'' are $\overset{a\delta}{\Longleftrightarrow}$-minimal and equivalent networks. By the Structural Theorem 5.10 there exists F_1, F_2, F_3 such that $F' \overset{a\gamma}{\longrightarrow} F_1 \overset{a\beta}{\longleftarrow} F_2 \overset{a\beta}{\longrightarrow} F_3 \overset{a\gamma}{\longleftarrow} F''$. But F' is minimal, hence the first two simulations actually are isomorphisms. Next, F'' is minimal, hence the last two simulations are isomorphism. Thus F' and F'' are isomorphic. □

In order to get the minimal network with respect to the input behaviour one has to throw away *all* non-accessible cells and then to use the *maximal* possible identification.

EXERCISES AND PROBLEMS (SEC. 5.3)

1. When is $\overset{a\beta}{\longrightarrow} \cdot \overset{a\gamma}{\longleftarrow} \subseteq \overset{a\gamma}{\longleftarrow} \cdot \overset{a\beta}{\longrightarrow}$ valid (see also Lemma 6.18)? (10)
2. Study the other commutations left open in Table 5.1. (15)

3. May you find, perhaps by adding some mild conditions to Căzănescu–Ungureanu theory axioms, other useful characterizations for $\overset{a\delta}{\Longleftrightarrow}$? (*Hint:* suggestions for possible decompositions are $\overset{a\delta}{\Longleftrightarrow} = \overset{a\beta}{\longleftarrow} \cdot \overset{a\gamma}{\longrightarrow} \cdot \overset{a\gamma}{\longleftarrow} \cdot \overset{a\beta}{\longrightarrow}$ or $\overset{a\delta}{\Longleftrightarrow} = \overset{a\delta}{\longleftarrow} \cdot \overset{a\delta}{\longrightarrow}$.) (15)

5.4 Soundness for $a\delta$-flow

Here we prove that the $a\delta$-flow structure is preserved by passing from connections to $\overset{a\delta}{\Longleftrightarrow}$-equivalent networks. The proof mainly uses the Structural Theorem 5.10 to lift the enzymatic rule $\mathrm{Enz}_{a\delta}$.

The enzymatic rule for injections within a coalgebraic theory. Here we simplify a bit the enzymatic axiom in this particular case. The enzymatic rule for injections comes for free, provided the acyclic structure is stronger, namely a coalgebraic theory.

Proposition 5.13. *$Enz_{a\beta}$ holds in a Căzănescu–Ungureanu theory.*

PROOF: Using permutations every $a\beta$-relation may be reduced to a juxtaposition $u \star \mathsf{T}_q$ where u is an $a\alpha$-relation. By Lemma 4.9, Enz_u & $\mathrm{Enz}_v \Rightarrow \mathrm{Enz}_{u\star v}$. Hence it is enough to show that $\mathrm{Enz}_{\{\mathsf{T}_q\}}$ holds for an arbitrary q.

Suppose $f : m \to n$ and $g : m \star q \to n \star q$ are such that

$$(*) \qquad f \cdot (\mathsf{I}_n \star \mathsf{T}_q) = (\mathsf{I}_m \star \mathsf{T}_q) \cdot g$$

Then

$$\begin{array}{lll}
g \uparrow^q & & \\
= & ((\mathsf{I}_m \star \mathsf{T}_q \star \mathsf{T}_m \star \mathsf{I}_q) \cdot \mathsf{V}_{m\star q} \cdot g) \uparrow^q & \\
= & (((\mathsf{I}_m \star \mathsf{T}_q)\ g \star (\mathsf{T}_m \star \mathsf{I}_q)\ g) \cdot \mathsf{V}_{n\star q}) \uparrow^q & \text{by } \mathrm{C}_{a\gamma} \\
= & ((f\ (\mathsf{I}_n \star \mathsf{T}_q) \star (\mathsf{T}_m \star \mathsf{I}_q)\ g)) \cdot \mathsf{V}_{n\star q}) \uparrow^q & \text{by } (*) \\
= & ((f \star \mathsf{I}_q) \cdot (\mathsf{I}_n \star (\mathsf{T}_m \star \mathsf{I}_q)\ g) \cdot (\mathsf{I}_n \star \mathsf{T}_q \star \mathsf{I}_{n\star q}) \cdot \mathsf{V}_{n\star q}) \uparrow^q & \\
= & f \cdot ((\mathsf{I}_n \star (\mathsf{T}_m \star \mathsf{I}_q)\ g) \cdot (\mathsf{V}_n \star \mathsf{I}_q)) \uparrow^q & \\
= & f \cdot (\mathsf{I}_n \star \mathsf{T}_m \cdot (g \uparrow^q)) \cdot \mathsf{V}_n & \\
= & f \cdot (\mathsf{I}_n \star \mathsf{T}_n) \cdot \mathsf{V}_n & \text{by } \mathrm{C}_{a\beta} \\
= & f & \\
= & f \uparrow^0 &
\end{array}$$

The proof is complete. □

Lifting $\mathrm{Enz}_{a\delta}$ from connections to $\overset{a\delta}{\Longleftrightarrow}$-equivalent networks. Now the announced lifting of the enzymatic rule from connections to equivalent networks, in the current Elgot theory setting, may be proved.

Theorem 5.14. *If T is an Elgot theory, then $Enz_{a\delta}$ holds in $[X,T]_{a\delta}$.*

PROOF: Proposition 4.18 shows that $[X,T]_{a\delta}$ is a Căzănescu–Ungureanu theory. By the above proposition it obeys $\mathrm{Enz}_{\{\mathsf{T}_q\}}$. Up to a composition with some bijections, every abstract $a\delta$-relation is a juxtaposition of elements of the type $\vee_p^k$, $k \geq 0$. By using Lemma 4.9 the problem is finally reduced to the verification of Enz_u for $u = \vee_p^k$ and $k \geq 2$.

Let $F \in [X,T](m \star kp, n \star kp)$ and $G \in [X,T](m \star p, n \star p)$ be such that

$$F \cdot (\mathsf{I}_n \star \vee_p^k) \overset{a\delta}{\Longleftrightarrow} (\mathsf{I}_m \star \vee_p^k) \cdot G$$

We may suppose that $G = (\underline{x}, g)$ is an $\overset{a\delta}{\Longleftrightarrow}$-minimal pair.

It follows that $(\mathsf{I}_m \star \vee_p^k) \cdot G$ is also minimal. (Indeed, suppose the opposite, i.e., $(\mathsf{I}_m \star \vee_p^k) \cdot G$ is not minimal. Then, it has an effective reduction $(\mathsf{I}_m \star \vee_p^k) \cdot G \overset{a\gamma}{\longrightarrow} H$ or $(\mathsf{I}_m \star \vee_p^k) \cdot G \overset{a\beta}{\longleftarrow} H$ for an $H = (y, h)$ with $|y| < |x|$. By a left composition with $\mathsf{I}_m \star (\mathsf{I}_p \star \mathsf{T}_{(k-1)p})$ we get an effective reduction of G. But this is impossible as G is minimal.)

By the Structural Theorem we get

$$F \cdot (\mathsf{I}_n \star \vee_p^k) \overset{a\gamma}{\longrightarrow} \cdot \overset{a\beta}{\longleftarrow} (\mathsf{I}_m \star \vee_p^k) \cdot G$$

Now the desired equivalence

$$F \uparrow^{kp} \overset{a\delta}{\Longleftrightarrow} G \uparrow^{p}$$

follows from Lemma 4.20 applied for $A = a\delta$, $B = a\alpha$, $C = a\gamma$ and $D = a\beta$.

The conditions of Lemma 4.20 are valid. Condition (3) of that lemma is maybe slightly more difficult to prove. We check it below. Recall that it is:

$$H \;{}_u\!\overset{a\beta}{\longleftarrow} (\mathsf{I}_m \star \vee_p^k) \cdot G \;\Rightarrow\; \exists H': \quad H = (\mathsf{I}_m \star \vee_p^k) \cdot H' \;\&\; H' \;{}_u\!\overset{a\beta}{\longleftarrow} G$$

If $H = (\underline{y}, h)$, the simulation shows that $(\mathsf{I}_{m\star kp} \star o(u))\, h = (\mathsf{I}_m \star \vee_p^k \star \mathsf{I}_{o(\underline{x})})\, g\, (\mathsf{I}_{n\star p} \star i(u))$, hence $(\mathsf{T}_m \star \mathsf{I}_{kp} \star \mathsf{T}_{o(\underline{y})})\, h = \vee_p^k\, g'$ where $g' = (\mathsf{T}_m \star \mathsf{I}_p \star \mathsf{T}_{o(\underline{x})})\, g\, (\mathsf{I}_{n\star p} \star i(u))$.

Now, take the pair $H' = (\underline{y}, h')$, where $h' = [(\mathsf{I}_m \star \mathsf{T}_{kp\star o(\underline{y})})\, h \star g' \star (\mathsf{T}_{m\star kp} \star \mathsf{I}_{o(\underline{y})})\, h]\, \vee^3_{n\star p\star i(\underline{y})}$. Then it is obvious that $(\mathsf{I}_m \star \vee_p^k)\, H' = H$. Moreover, $H' \;{}_u\!\leftarrow G$ follows by

$$\begin{aligned}
&(\mathsf{I}_{m\star kp} \star o(u)) \cdot [(\mathsf{I}_m \star \mathsf{T}_{kp\star o(\underline{y})})\, h \star g' \star (\mathsf{T}_{m\star kp} \star \mathsf{I}_{o(\underline{y})})\, h] \cdot \vee^3_{n\star p\star i(\underline{y})} \\
&= [(\mathsf{I}_m \star \mathsf{T}_{kp\star o(\underline{y})})\, h \star (\mathsf{T}_m \star \mathsf{I}_p \star \mathsf{T}_{o(\underline{x})})\, g\, (\mathsf{I}_{n\star p} \star i(u)) \star (\mathsf{T}_{m\star kp} \star o(u))\, h] \\
&\quad \cdot \vee^3_{n\star p\star i(\underline{y})} \\
&= [(\mathsf{I}_m \star \mathsf{T}_{kp\star o(\underline{x})})\, g\, (\mathsf{I}_{n\star p} \star i(u)) \star (\mathsf{T}_m \star \mathsf{I}_p \star \mathsf{T}_{o(\underline{x})})\, g\, (\mathsf{I}_{n\star p} \star i(u)) \\
&\quad \star (\mathsf{T}_{m\star p} \star \mathsf{I}_{o(\underline{x})})\, g\, (\mathsf{I}_{n\star p} \star i(u))] \cdot \vee^3_{n\star p\star i(\underline{y})} \\
&= g \cdot (\mathsf{I}_{n\star p} \star i(u))
\end{aligned}$$

The proof is complete. □

Corollary 5.15. $\overset{a\delta}{\Longleftrightarrow} = \sim_{xy}$

PROOF: It follows by Proposition 4.15 and Theorem 5.14. □

♠ This shows that the equivalence relation generated by simulation via $a\delta$-relations coincides with the equivalence relation generated by commutations $C_{a\delta}$-var in the class of congruences satisfying the enzymatic rule for $a\delta$-relations. ♣

Lifting Elgot theory structure from connections to networks. Finally, combining Theorem 5.14 and Proposition 4.18 we get the following theorem which shows that the Elgot theory structure is preserved when one passes from the support theory of connections to the classes of $a\delta$-similar pairs.

Theorem 5.16. *$[X, T]_{a\delta}$ is an Elgot theory, whenever T is an Elgot theory.* □

Soundness Theorem. The above theorems (Theorems 5.16 and 5.2) show that the Elgot theory axioms are sound with respect to input behaviour in the case of concrete deterministic networks over $\mathbb{P}$fn.

The following reasoning motivates this statement. The elements in $[X, \mathbb{P}\text{fn}]_{a\delta}$ may be represented as $\overset{a\delta}{\Longleftrightarrow}$-minimal networks, up to an isomorphism. But in this particular case of $\mathbb{P}$fn these minimal networks are isomorphic with the regular trees, hence they represent the input behaviour. On the other hand, the Elgot theory axioms are valid in $\mathbb{P}$fn. By the general theorems above they are lifted to $[X, \mathbb{P}\text{fn}]_{a\delta}$, hence they are sound with respect to the input behaviour.

EXERCISES AND PROBLEMS (SEC. 5.4)

1. Give a proof for Proposition 5.13 using graphical transformations. (05)

5.5 Completeness for $a\delta$-flow

In order to get the meaning of the $a\delta$-flownomials we combine Corollary 5.4 and Theorem 5.16 with the observation that the morphism $\varphi^{\#}$ in Theorem 2.27 obtained in the particular case when $Q = R_X$ and when $\varphi_{\mathbb{P}\text{fn}}$ and φ_X are the natural embeddings of $\mathbb{P}$fn and X into R_X is the unfolding of networks. Hence:

Theorem 5.17. *($a\delta$-flownomials = input behaviours of networks) The algebras* $\mathsf{FlowExp}[X, \mathbb{P}\text{fn}]_{a\delta}$, *$[X, \mathbb{P}\text{fn}]_{a\delta}$ and $R[X]$ are isomorphic.* □

Corollary 5.18. *The following conditions are equivalent for two flownomial expressions E and E' in* $\mathsf{FlowExp}[X, \mathbb{P}\text{fn}]$*:*

1. $E \sim_{a\delta} E'$

2. $\mathsf{nf}(E) \overset{a\delta}{\Longleftrightarrow} \mathsf{nf}(E')$

3. *The networks associated with E and E' unfold into the same cotuple of trees.*

4. *The computation processes denoted by E and E' have the same finite and infinite computation sequences.*

□

Hence the $a\delta$-flow axioms are complete for proving that networks with the same input behaviour are equal.

5.6 Working with $a\delta$-flownomials

Theorem 5.19. *(universality of $a\delta$-flownomials) Let T be an Elgot theory.*

(i) Algebra $\mathsf{FlowExp}[X,T]_{a\delta}$ *of $a\delta$-flownomials is an Elgot theory and it is isomorphic to the algebra* $[X,T]_{a\delta}$ *of the classes of* $\overset{a\delta}{\Longleftrightarrow}$*-equivalent networks.*

(ii) There exist two embeddings $E_X^{a\delta}$ *and* $E_T^{a\delta}$ *of X and T into* $[X,T]_{a\delta}$*, respectively, where* $E_X^{a\delta}$ *is a function and* $E_T^{a\delta}$ *is a morphism of Elgot theory, such that for every Elgot theory Q and every pair* (φ_X, φ_T)*, where* $\varphi_X : X \to Q$ *is a function and* $\varphi_T : T \to Q$ *is a morphism of Elgot theory, there exists a unique morphism of Elgot theory* $\varphi^\# : [X,T]_{a\delta} \to Q$ *such that* $E_X^{a\delta} \cdot \varphi^\# = \varphi_X$ *and* $E_T^{a\delta} \cdot \varphi^\# = \varphi_T$. □

$\mathbb{P}\mathrm{fn}$ is the initial Elgot theory in the category of Elgot theories. By applying the above theorem in this particular case $T = \mathbb{P}\mathrm{fn}$ we get the following corollary.

Corollary 5.20. *(free Elgot theory)* $[X, \mathbb{P}\mathrm{fn}]_{a\delta}$ *is the Elgot theory freely generated by X.* □

5.7 Output behaviour

All the results given in this chapter hold in the dual case, i.e., replacing $a\delta$ by $d\alpha$. In this dual form, the behaviour of a multiple-entries/multiple-exits variable may be specified by a tuple of multiple-entries/single-exit ones. After such a transformation the resulting network may be unfolded getting a (regular) tree for every output, but now the unfolding is towards the inputs. The resulting tuple of regular trees gives the *output behaviour* of the network. (As we already indicated in the first section, this setting fits with the interpretation of the deterministic data-flow networks.)

5.8 Bisimulation: two-way simulation

We prove here that, on transition systems, bisimulation is the equivalence relation generated by simulation via functions. The proof entirely rests on simple rules of the calculus of relations.

Let X be a set of atomic actions. A *transition system over X with multiple entries and multiple exits* $F : m \xrightarrow{k} n$ consists of three interfaces: m, k, n (m specifies the input vertices, k the internal ones, and m the output vertices) and a transition relation $f \subseteq (m \star k) \times X \times (n \star k)$.

Transition systems as networks. How may these transition systems be integrated in the general network algebra view? One possibility is the following. The cells are trivial and are identified with the system's vertices. The cell connections are specified by the transition relation. These transition relations have most of the properties of matrix theories, but fail to have identities. To solve this problem one may simply add a special ϵ action with some natural properties and a matrix theory is obtained, so apparently there are no complications! It is one subtle point which complicates things: the parallel operator, actually not used in the definition of simple transition systems. Transition systems are mainly used in process algebra, where the parallel operator is essential. If ϵ is added, then also its behaviour with respect to the (interleaving-based) parallel operator has to be specified, and this is not so simple, or natural. Indeed, in parallel with a different process ϵ may prevent the current process for making any transition: the current process may idle forever, an undesirable feature, indeed. For this reason, usually the transition systems studied in the literature are without empty steps. $\diamond$

Definition 5.21. Two transition systems $F : m \xrightarrow{k} n$ and $F' : m \xrightarrow{k'} n$ are *similar* via a function $\phi : k \to k'$ if

$$(\mathsf{I}_m \star \phi)f' = f(\mathsf{I}_n \star \phi)$$

We briefly write this as $F \to_\phi F'$ (or $F \xrightarrow{a\delta}_\phi F'$, to emphasize the functional properties of ϕ). Two such transition systems F and F' are *bisimilar* via a relation $\rho : k \to k'$ if

$$(\mathsf{I}_m \star \rho)f' \subseteq f(\mathsf{I}_n \star \rho)$$
$$(\mathsf{I}_m \star \rho^{-1})f \subseteq f'(\mathsf{I}_n \star \rho^{-1})$$

This relation is briefly denoted by $F \leftrightarrow_\rho F'$.

Notice that simulation $\xrightarrow{a\delta}$ is a reflexive-transitive, but not a symmetric relation, whilst bisimulation $\leftrightarrow$ is even an equivalence relation.

♠ This simulation relation *is nothing else than* the standard NA simulation via functions $\xrightarrow{a\delta}$, adapted to the present context of transition systems. It is *different* from the process algebra simulation relation; this usually considers half of the conditions from the bisimulation definition; then, two processes may be process algebra similar in both directions, but still not bisimilar; an example is described in Exercise 5.8.1.

One more important fact has to be emphasized. The definition of simulation uses an equation, while bisimulation is defined using two inclusions. This has important side-effects when one tries to lift these definitions to a general abstract setting, where the equational style is much easier to use. ♣

The following simple fact may be proved, being of fundamental importance in linking bisimulation by simulation.

Proposition 5.22. *Simulation via a function coincides with bisimulation via a function.*

PROOF: The proof is reduced to the following observation. For two transition systems $F : m \xrightarrow{k} n$ and $F' : m \xrightarrow{k'} n$ and a function $\phi : k \to k'$ the following equivalence holds:

$$(i)\ (\mathsf{I}_m \star \phi^{-1})f \subseteq f'(\mathsf{I}_n \star \phi^{-1}) \quad \Longleftrightarrow \quad (ii)\ f(\mathsf{I}_n \star \phi) \subseteq (\mathsf{I}_m \star \phi)f'$$

Indeed, since ϕ is a function, $\phi \cdot \phi^{-1} \supseteq \mathsf{I}_k$ (i.e., ϕ is a total relation) and $\phi^{-1} \cdot \phi \subseteq \mathsf{I}_{k'}$ (i.e., ϕ is a univocal relation). The "$\Rightarrow$" part of the implication follows easily: compose (i) on the left with $\mathsf{I}_m \star \phi$ and on the right with $\mathsf{I}_n \star \phi$; then

$$(\mathsf{I}_m \star \phi\phi^{-1}) \cdot f \cdot (\mathsf{I}_n \star \phi) \subseteq (\mathsf{I}_m \star \phi) \cdot f' \cdot (\mathsf{I}_n \star \phi^{-1}\phi)$$

and, by the above properties of ϕ, this implies (ii). The other part of the implication "$\Leftarrow$" is proved in a similar way. □

Now, in order to pass from bisimilar transition systems to chains of similar ones we need a characterization of bisimulation in terms of bisimulation via functions. The result given here is based on a construction of a common refinement of two bisimilar transition systems.

Lemma 5.23. *(interpolation property) Let $F' : m \xrightarrow{k'} n$ and $F'' : m \xrightarrow{k''} n$ be two transition systems and $\rho \subseteq k' \times k''$ a relation such that $F' \leftrightarrow_\rho F''$.*
Then there exists a transition system $F : m \xrightarrow{k} n$ and two functions $\phi : k \to k'$ and $\psi : k \to k''$ such that $F \leftrightarrow_\phi F'$ and $F \leftrightarrow_\psi F''$.

PROOF: Take a decomposition of ρ as $\phi^{-1} \cdot \psi$ for two functions $\phi : k \to k'$ and $\psi : k \to k''$. By definition $F' \leftrightarrow_\rho F''$ gives

$$(i)\ (\mathsf{I}_m \star \phi^{-1}\psi) \cdot f'' \subseteq f' \cdot (\mathsf{I}_n \star \phi^{-1}\psi) \text{ and}$$
$$(ii)\ (\mathsf{I}_m \star \psi^{-1}\phi) \cdot f' \subseteq f'' \cdot (\mathsf{I}_n \star \psi^{-1}\phi)$$

Take the transition system $F : m \xrightarrow{k} n$ given by the transition relation

$$f = [(\mathsf{I}_m \star \phi) \cdot f' \cdot (\mathsf{I}_n \star \phi^{-1})] \cap [(\mathsf{I}_m \star \psi) \cdot f'' \cdot (\mathsf{I}_n \star \psi^{-1})]$$

Then $F \leftrightarrow_\phi F'$ and $T \leftrightarrow_\psi T''$.

We prove the first statement only. (The second is similar). Indeed, for $F \leftrightarrow_\phi F'$ one inclusion is straighforward (use the inclusion of f in the first term of the above intersection and the functional property of ϕ, hence $\phi^{-1}\phi \subseteq \mathsf{I}_{k'}$)

$$\begin{aligned}(\mathsf{I}_m \star \phi^{-1}) \cdot f &\subseteq (\mathsf{I}_m \star \phi^{-1}\phi) \cdot f' \cdot (\mathsf{I}_n \star \phi^{-1}) \\ &\subseteq f' \cdot (\mathsf{I}_n \star \phi^{-1})\end{aligned}$$

For the second one we use (*ii*). By a left composition of (*ii*) with $\mathsf{I}_m \star \psi$ we get

$$L := (\mathsf{I}_m \star \psi\psi^{-1}\phi) \cdot f' \subseteq (\mathsf{I}_m \star \psi) \cdot f'' \cdot (\mathsf{I}_n \star \psi^{-1}\phi) =: R$$

First, $\mathsf{I}_k \subseteq \psi\psi^{-1}$ (as ψ is a total relation), hence $(\mathsf{I}_m \star \phi) \cdot f' \subseteq L$.

Next, notice that $(\mathsf{I}_m \star \phi) \cdot f' \subseteq L \subseteq (\mathsf{I}_m \star \psi) \cdot f'' \cdot (\mathsf{I}_n \star \psi^{-1}\phi)$ shows that the image of the relation $(\mathsf{I}_m \star \phi) \cdot f'$ is included in $A := [n] \cup \{n + j : j \in Im(\phi)\}$, where $Im(\phi)$ denotes the image of ϕ. Hence this relation is not changed when it is composed on the right with $\mathsf{I}_n \star \phi^{-1}\phi$ (i.e., with the partial identity restricted to A); formally we get

$$(\mathsf{I}_m \star \phi) \cdot f' = (\mathsf{I}_m \star \phi) \cdot f' \cdot (\mathsf{I}_n \star \phi^{-1}\phi)$$

Combined with $(\mathsf{I}_m \star \phi) \cdot f' \subseteq L \subseteq (\mathsf{I}_m \star \psi) \cdot f'' \cdot (\mathsf{I}_n \star \psi^{-1}\phi)$ this implies

$$(\mathsf{I}_m \star \phi) \cdot f' \subseteq [(\mathsf{I}_m \star \phi) \cdot f' \cdot (\mathsf{I}_n \star \phi^{-1}\phi)] \cap [(\mathsf{I}_m \star \psi) \cdot f'' \cdot (\mathsf{I}_n \star \psi^{-1}\phi)]$$

But the right-hand-side term of this inclusion is just $f \cdot (\mathsf{I}_n \star \phi)$, hence

$$(\mathsf{I}_m \star \phi) \cdot f' \subseteq f \cdot (\mathsf{I}_n \star \phi)$$

Both inclusions have been proved, hence $F \leftrightarrow_\phi F'$, indeed. As we said, the other case is similar, hence the interpolation property is proved. □

Theorem 5.24. *Bisimulation is the equivalence relation generated by simulation via functions. More precisely,* $\leftrightarrow \; = \; \stackrel{a\delta}{\longleftarrow} \cdot \stackrel{a\delta}{\longrightarrow}$.

PROOF: From the lemmas above. □

♡ As we have seen in this chapter, simulation is an useful tool to speak about minimization. One may give an alternative proof of this theorem by replacing Lemma 5.23 with another one using minimization, but the characterization is less direct, i.e., instead of two-way simulation one gets $\leftrightarrow \; = \; \stackrel{a\gamma}{\longrightarrow} \cdot \stackrel{a\beta}{\longleftarrow} \cdot \stackrel{a\beta}{\longrightarrow} \cdot \stackrel{a\gamma}{\longleftarrow}$, similar to the Structural Theorem. ◇

The result of this section may be checked for many other transition systems (e.g., with ϵ transitions, with silent actions τ, higher order bisimulations, etc.), provided appropriate algebraic models are introduced to describe their transition relations. This way one may hope to get a unique, abstract notion of bisimulation, parametrized by the "connecting theory" of transition relations, and not tens of separate bisimulation relations.

EXERCISES AND PROBLEMS (SEC. 5.8)

1. Say that two transition systems are *PA (process-algebra) similar* via ϕ if the 2nd inclusion from the definition of bisimulation is valid. Show that the transition systems $T_1 = a(b+c)+ab$ and $T_2 = ac+a(c+b)$ (represented as PA terms) are PA similar via functions in both directions, but not bisimilar. (08)

5.9 Milner theories

An axiomatization for nondeterministic networks with single-entry/single-exit cells modulo bisimulation is presented here. The involved algebraic structure is a Milner theory, i.e., an Elgot theory over a $d\delta$-symocat with feedback.

Axiomatizing bisimilar networks. The words "deterministic" and "nondeterministic" should be used with some care in the present calculus. We have an abstract calculus and we may freely use any kind of connecting theories, providing the underlining hypotheses are satisfied. For example, the results in the previous chapter (except for the meaning, which was given in the particular case $T = \mathbb{P}\mathrm{fn}$) still hold if the connecting theory is $\mathbb{R}\mathrm{el}$. To clarify the matters, we use the same convention as in Chapter 3:

Deterministic vs. nondeterministic network calculi
The words "deterministic" and "nondeterministic" refer to the syntactical aspects of the calculus. In the nondeterministic framework one may freely uses both types of branching constants $\wedge_k$ and $\vee^k$ at the syntactic level, while in the deterministic case only one type of constant of branching degree greater than 1 may be used.

Hence, what we have to do here is to add nondeterminism to the syntactic part, hence to use a $d\delta$-symocat with feedback.

Why is this simple instance of the general results of this chapter important? It is because of the important meaning of the $\overset{a\delta}{\Longleftrightarrow}$ equivalence in this case. Namely, as we have seen, $\overset{a\delta}{\Longleftrightarrow}$ coincides with bisimulation, whenever the support theory is $\mathbb{R}\mathrm{el}$ and only single-entry/single-exit cells are used. Now the main result of this chapter is a corollary of Theorem 3.16 and of the theorem of the previous chapter (Theorem 5.24).

Theorem 5.25. *If T is a $d\delta$-symocat with feedback, then:*
(Correctness) $[X,T]_{a\alpha}$ is a $d\delta$-symocat with feedback.
(Completeness) The rules of $a\delta$-flow and $d\delta$-symocat with feedback are complete for networks with single-entry/single-exit variables modulo bisimulation. □

Comments, problems, bibliographic remarks

Notes. Regular trees and their relationship with flowchart programs are studied in, e.g., [EBT78], [Gin79], [Ési80] and [Cou83]. Algebraic models for regular trees, called iterative or iteration theories, were proposed in [Elg75], [BEW80a] and [BEW80b]. An equational axiomatization of regular trees was obtained by Esik; see [Ési80], [BÉ93b], or [BÉ97] for more on his topics. Seen from the network algebra perspective, regular trees may be identified with flowcharts using (partial) functions as connecting relations.

Our basic algebraic structure is $a\delta$-flow, also called Elgot theory; it was formally introduced in [Şte87a]. The results of this chapter are dedicated to networks whose connections form an arbitrary Elgot theory, subsuming the case of regular trees above. They mainly come from [Şte87a]. (An extended abstract was presented in [Şte86a].) In particular, the key Structural Theorem 5.10 and axiomatization theorem (Theorem 5.19) are from [Şte87a]. However, they are presented here using feedback instead of iteration; in this presentation we have followed the presentation in Chapter D, Sections 1–3 of [Şte91].

Simulation via functions is a standard notion of graph homomorphism; it has been used in the study of flow diagram programs in, e.g., [Gog74], [Elg77] and [Şte87a]. The dual case of data-flow networks is based on the stream processing function model; see [Bro93], [BŞ94b] and [BŞ96].

Bisimulation is a standard equivalence used in the algebraic studies on concurrent processes. It was introduced in [Par81] in connection with Milner's work on concurrency [Mil80, Mil89]; see also [BK84, BW90a]. More on this may be found in Chapter 9. The main result of Section 5.8, showing that, in a nondeterministic setting, bisimulation is the equivalence relation generated by simulation via functions, is from [BŞ94a]. This result was also presented in [RT94], but within a more categorical framework; see also [BÉT93].

6. Kleene theories

Aim at the sun, and you may not reach it; but your arrow will fly far higher than if aimed at an object on a level with yourself.

Joel Hawes

This chapter is devoted to the study of the input–output (IO) behaviour of networks.

We recall the, by now familiar, general classification scheme for the basic algebraic structures. It uses three parameters (x_1y_1-weak, x_2y_2-strong, x_3y_3-enzymatic), where $x_1y_1 \geq x_2y_2 \geq x_3y_3$; the first restriction shows what branching constants may be used, the second which branching constants strongly commute with arbitrary arrows, and the third which branching constants may be used in the enzymatic rule.

First we handle the deterministic case. The connections for these networks include $b\delta$-relations (i.e., partial functions). An important fact is that the corresponding IO equivalence on networks is captured by the equivalence relation $\overset{b\delta}{\Longleftrightarrow}$ generated by simulation via these partial functions. Then, in order to get an axiomatization for deterministic IO behaviour of networks we shall use the Park theory (i.e., $b\delta$-flow); Park theory corresponds to the choice $(b\delta, b\delta, b\delta)$ for weak, strong, and enzymatic rules. The main result obtained here shows that Park theory axioms gives a correct (i.e., sound and complete) axiomatization for the deterministic IO behaviour of networks; for the soundness part a mild additional "identification-free" hypothesis is used.

The second part of this chapter is devoted to the nondeterministic case. It is shown that simulation via relations captures this nondeterministic IO behaviour. This is the key step in the proof of the completeness of the Kleene theory axioms for this equivalence on networks. Kleene theory (i.e., $d\delta$-flow) corresponds to the choice $(d\delta, d\delta, d\delta)$ for weak, strong, and enzymatic rules. The abstract soundness theorem (i.e., the preservation of Kleene theory structure by passing from connections to the classes of $\overset{b\delta}{\Longleftrightarrow}$-equivalent networks) is proved with some mild assumptions: the matrix theory should be "zero-sum-free" and "uniform divisible."

6.1 IO behaviour, deterministic case

The IO behaviour in this deterministic case is introduced here, both using regular trees and minimal networks.

The *deterministic IO (step-by-step) behaviour* of a network is the restriction of the input behaviour to the successful IO paths.

In this case the algebra $R[X]^\emptyset$ proves to be useful. This algebra is obtained from the algebra $R[X]$ of rational trees by the identification of the subtrees without outputs to the empty tree. It is clear that the morphisms in $R[X]^\emptyset$ are in a bijective correspondence with the IO behaviours of deterministic networks.

As we will see soon, it is also possible to use the $\stackrel{b\delta}{\Longleftrightarrow}$-minimal networks to have an equivalent definition for this type of behaviour.

6.2 Park theories ($b\delta$-flow)

"Park theory" is the equivalent, less esoteric, name for the $b\delta$-flow structure that was already given in the previous chapter. In general NA terms, it is a ($b\delta$-weak, $b\delta$-strong, $b\delta$-enzymatic) structure. That means, it is a BNA with (1) $b\delta$-branching constants ($b\delta$-symocat with feedback), of which (2) $b\delta$-constants strongly commutes with arbitrary arrows, and (3) the same $b\delta$-constants generate relations for which the enzymatic rule is valid.

♡ Spell out the definition of Park theories as extended Elgot theories. ◇

6.3 Structural Theorem, case $b\delta$

The structural theorem, proved in this section, gives a characterization for $\stackrel{b\delta}{\Longleftrightarrow}$, the equivalence relation generated by simulation via partial functions. Roughly speaking, it says that two networks are $\stackrel{b\delta}{\Longleftrightarrow}$-equivalent iff by (1) deletion of non-coaccessible vertices, (2) reduction, and (3) deletion of non-accessible vertices the associated networks may be brought to the same form.

General presentation. In this $b\delta$ case we use simulation via partially defined functions and we look for a theorem similar to Theorem 5.10, proved in the $a\delta$ case. Actually, we shall prove that

$$\stackrel{b\delta}{\Longleftrightarrow} = \stackrel{b\alpha}{\longrightarrow} \cdot \stackrel{a\gamma}{\longrightarrow} \cdot \stackrel{a\beta}{\longleftarrow} \cdot \stackrel{a\beta}{\longrightarrow} \cdot \stackrel{a\gamma}{\longleftarrow} \cdot \stackrel{b\alpha}{\longleftarrow} .$$

This means that two networks, represented as nf-pairs, are $\stackrel{b\delta}{\Longleftrightarrow}$-equivalent iff by (1) deletion of non-coaccessible vertices $\stackrel{b\alpha}{\longrightarrow}$ (i.e., deletion of those vertices

Table 6.1. Is $x \cdot y \subseteq y \cdot x$? (case $b\delta$)

$x \backslash y$	$\xrightarrow{b\alpha}$	$\xrightarrow{a\gamma}$	$\xleftarrow{a\beta}$	$\xrightarrow{a\beta}$	$\xleftarrow{a\gamma}$	$\xleftarrow{b\alpha}$
$\xrightarrow{b\alpha}$	Obvious					
$\xrightarrow{a\gamma}$	6.1	Obvious				
$\xleftarrow{a\beta}$	6.3	5.7	Obvious			
$\xrightarrow{a\beta}$	6.2	5.4	5.6	Obvious		
$\xleftarrow{a\gamma}$	(*)6.4	5.9	5.5	5.8	Obvious	
$\xleftarrow{b\alpha}$	5.6^o	$(*)6.4^{op}$	6.2^{op}	6.3^{op}	6.1^{op}	Obvious

for which no paths reaching the outputs do exist), (2) reduction $\xrightarrow{a\gamma}$, and (3) deletion of non-accessible vertices $\xleftarrow{a\beta}$, the associated networks may be brought to the same form.

The general framework we shall use here is given by a support theory T which is a $b\delta$-symocat equipped with a feedback operation. Sometimes we shall use the following *identification-free hypothesis*:

(identification-free) if $f \cdot (y_1 \star y_2) = g \star \top_q$ then $f = f \cdot (\mathsf{I}_n \star \perp^p \cdot \top_p)$, where $f \in T(m, n \star p)$, while $y_1 : n \to n'$ and $y_2 : p \to q$ are $a\gamma$-relations in T.

♡ Actually, this identification-free hypothesis says that if after some identifications of the outputs of f certain outputs become dummy, then these outputs were dummy before the identification, too. To get some confidence with this hypothesis we mention that if T is a matrix theory, then it is equivalent with the zero-sum-free condition, i.e., $x + y = 0 \Rightarrow x = y = 0$ in the associated semiring. ◇

Commuting lemmas. We start with the proof of some commutation lemmas for the elementary simulations that appear. Since in the characterization theorem we want to prove for $\xleftrightarrow{b\delta}$ the simulations $\xrightarrow{a\beta}$, $\xrightarrow{a\gamma}$ and their converses occur in the same order as in Theorem 5.10, the commutation lemmas used in the $a\delta$ case are still useful here. It remains to prove the validity of the commutations between $\xrightarrow{b\alpha}$ and $\xleftarrow{b\alpha}$ simulations and the other ones.

The commutation $\xleftarrow{b\alpha} \cdot \xrightarrow{b\alpha} \subseteq \xrightarrow{b\alpha} \cdot \xleftarrow{b\alpha}$ follows from Lemma 5.6 by duality, and the other ones in the bottom line of Table 6.1 are converses of some ones occurring in the left-hand side column of that table. Finally, four cases remain to be proved: the commutations of the simulations $\xrightarrow{a\gamma}$, $\xrightarrow{a\beta}$, $\xleftarrow{a\beta}$, and $\xleftarrow{a\gamma}$ with $\xrightarrow{b\alpha}$. (See Table 6.1.)

Lemma 6.1. $\xrightarrow{a\gamma} \cdot \xrightarrow{b\alpha} \subseteq \xrightarrow{b\alpha} \cdot \xrightarrow{a\gamma}$, *when T is an identification-free strong $b\delta$-symocat.*

PROOF: Let $F' \xrightarrow{a\gamma}_{u'} F \xrightarrow{b\alpha}_{u''} F''$.

By using isomorphic representations we may suppose that

$F' = (a \star b,\ f'), F = (c \star d,\ f), F'' = (c,\ f'')$ and
$u' = u_1 \star u_2$ (where $u_1 : a \to c$), $u'' = \mathsf{I}_c \star \perp^d$

The given simulations show that:

$$\text{(1)} \qquad f'\ (\mathsf{I}_n \star i(u_1 \star u_2)) = (\mathsf{I}_m \star o(u_1 \star u_2))\ f \qquad \text{and}$$
$$\text{(2)} \qquad f\ (\mathsf{I}_n \star i(\mathsf{I}_c \star \perp^d)) = (\mathsf{I}_m \star o(\mathsf{I}_c \star \perp^d))\ f''$$

Take the pair $\overline{F} = (a, \overline{f})$ where $\overline{f} = (\mathsf{I}_m \star o(\mathsf{I}_a \star \top_b))\ f'\ (\mathsf{I}_n \star i(\mathsf{I}_a \star \perp^b))$.

We show that $F' \xrightarrow{b\alpha}_{\mathsf{I}_a \star \perp^b} \overline{F} \xrightarrow{a\gamma}_{u_1} F''$.

For the right one, the simulation's conditions are easy to check:

$$\begin{aligned}
\overline{f} \cdot (\mathsf{I}_n \star i(u_1)) &= (\mathsf{I}_m \star o(\mathsf{I}_a \star \top_b))\ f'\ (\mathsf{I}_n \star i(u_1 \star \perp^b)) \\
&= (\mathsf{I}_n \star o(\mathsf{I}_a \star \top_b))\ f'\ (\mathsf{I}_n \star i(u_1 \star u_2 \perp^d)) \\
&= (\mathsf{I}_m \star o(u_1 \star \top_b \cdot u_2))\ f\ (\mathsf{I}_n \star i(\mathsf{I}_c \star \perp^d)) && \text{by (1)} \\
&= (\mathsf{I}_m \star o(u_1 \star \top_d \cdot \perp^d)) \cdot f'' && \text{by (2)} \\
&= (\mathsf{I}_m \star o(u_1)) \cdot f''
\end{aligned}$$

For the left one, first note that (1) and (2) implies

$$f' \cdot (\mathsf{I}_n \star i(u_1 \star \perp^b)) = (\mathsf{I}_m \star o(u_1 \star \perp^b)) \cdot f''$$

hence

$$\begin{aligned}
(\top_m \star o(\top_a \star \mathsf{I}_b)) \cdot f' \cdot (\mathsf{I}_n \star i(u_1 \star \perp^b)) &= \perp^{o(b)} (\top_m \star o(\top_c)) \cdot f'' \\
&= \perp^{o(b)} \star \top_{n \star i(c)} \quad \text{by } C_{a\beta}
\end{aligned}$$

Applying the identification-free hypothesis for surjection $\mathsf{I}_n \star i(u_1)$ one gets

$$\begin{aligned}
\text{(3)} \quad & (\top_m \star o(\top_a \star \mathsf{I}_b)) \cdot f' \cdot (\mathsf{I}_n \star i(\mathsf{I}_a \star \perp^b)) \\
&= [(\top_m \star o(\top_a \star \mathsf{I}_b))\ f'\ (\mathsf{I}_n \star i(\mathsf{I}_a \star \perp^b)] \cdot \perp^{n \star i(a)} \cdot \top_{n \star i(a)} \\
&= \perp^{o(b)} \cdot \top_{n \star i(a)} \qquad \text{by } C_{b\alpha}
\end{aligned}$$

Finally,

$$\begin{aligned}
& (\mathsf{I}_m \star o(\mathsf{I}_a \star \perp^b)) \cdot \overline{f} \\
&= (\mathsf{I}_m \star o(\mathsf{I}_a \star \perp^b \cdot \top_b))\ f'\ (\mathsf{I}_n \star i(\mathsf{I}_a \star \perp^b)) \\
&= [\ (\mathsf{I}_m \star o(\mathsf{I}_a \star \top_b))\ f'\ (\mathsf{I}_n \star i(\mathsf{I}_a \star \perp^b)) \\
&\quad \star (\top_m \star o(\top_a \star \underline{\perp^b \cdot \top_b} \cdot \mathsf{I}_b))\ f'\ (\mathsf{I}_n \star i(\mathsf{I}_a \star \perp^b))\ ] \cdot \vee_{n \star i(a)} \quad \text{by } C_{a\gamma}
\end{aligned}$$

Using (3) one may eliminate the underlined term and gets

$$(\mathsf{I}_m \star o(\mathsf{I}_a \star \perp^b)) \cdot \overline{f} \quad = \quad f' \cdot (\mathsf{I}_n \star i(\mathsf{I}_a \star \perp^b)) \qquad \text{by } C_{a\gamma}$$

This shows the simulation condition is verified in this case, too. □

Lemma 6.2. $\xrightarrow{a\beta} \cdot \xrightarrow{b\alpha} \;\subseteq\; \xrightarrow{b\alpha} \cdot \xrightarrow{a\beta}$, *when T is a $b\beta$-symocat.*

PROOF: Let $F' \xrightarrow{a\beta}_{u'} F \xrightarrow{b\alpha}_{u''} F''$.

Using isomorphic representations we may suppose that

$$F' = (a \star b,\ f'),\quad F = (a \star b \star c \star d,\ f),\quad F'' = (b \star c,\ f'') \quad \text{and}$$
$$u' = \mathsf{I}_{a\star b} \star \top_{c\star d},\quad u'' = \bot^a \star \mathsf{I}_{b\star c} \star \bot^d$$

The simulations show that:

$$(1) \qquad f' \cdot (\mathsf{I}_n \star i(\mathsf{I}_{a\star b} \star \top_{c\star d})) = (\mathsf{I}_m \star o(\mathsf{I}_{a\star b} \star \top_{c\star d})) \cdot f \quad \text{and}$$
$$(2) \qquad f \cdot (\mathsf{I}_n \star i(\bot^a \star \mathsf{I}_{b\star c} \star \bot^d)) = (\mathsf{I}_m \star o(\bot^a \star \mathsf{I}_{b\star c} \star \bot^d)) \cdot f''$$

Take $\overline{F} = (b, \overline{f})$ where $\overline{f} = (\mathsf{I}_m \star o(\top_a \star \mathsf{I}_b \star \top_{c\star d}))\ f\ (\mathsf{I}_n \star i(\bot^a \star \mathsf{I}_b \star \bot^{c\star d}))$.

We show that $F' \xrightarrow{b\alpha}_{\bot^a \star \mathsf{I}_b} \overline{F} \xrightarrow{a\beta}_{\mathsf{I}_b \star \top_c} F''$.

Indeed, for the left simulation we may compute

$$\begin{aligned}
&(\mathsf{I}_m \star o(\bot^a \star \mathsf{I}_b)) \cdot \overline{f} \\
&= (\mathsf{I}_m \star o(\bot^a \cdot \top_a \star \mathsf{I}_b \star \top_{c\star d}))\ f\ (\mathsf{I}_n \star i(\bot^a \star \mathsf{I}_b \star \bot^{c\star d})) \\
&= (\mathsf{I}_m \star o(\bot^a \cdot \top_a \cdot \bot^a \star \mathsf{I}_b \star \top_c \star \top_d \cdot \bot^d))\ f''\ (\mathsf{I}_n \star i(\mathsf{I}_b \star \bot^c)) && \text{by (2)} \\
&= (\mathsf{I}_m \star o(\mathsf{I}_{a\star b} \star \top_{c\star d}))\ f\ (\mathsf{I}_n \star i(\bot^a \star \mathsf{I}_b \star \bot^{c\star d})) && \text{by (2)} \\
&= f'\ (\mathsf{I}_n \star i(\bot^a \star \mathsf{I}_b \star \top_{c\star d} \cdot \bot^{c\star d})) && \text{by (1)} \\
&= f' \cdot (\mathsf{I}_n \star i(\bot^a \star \mathsf{I}_b))
\end{aligned}$$

The condition for the right simulation may be checked in a similar way. □

Lemma 6.3. $\xleftarrow{a\beta} \cdot \xrightarrow{b\alpha} \;\subseteq\; \xrightarrow{b\alpha} \cdot \xleftarrow{a\beta}$, *when T is an $a\delta$-strong $b\delta$-symocat.*

PROOF: Let $F'\ {}_{u'}\!\xleftarrow{a\beta} F \xrightarrow{b\alpha}_{u''} F''$.

Using isomorphic representations we may suppose that

$$F' = (a \star b \star c,\ f'),\quad F = (a \star b,\ f),\quad F'' = (a,\ f'') \quad \text{and}$$
$$u' = \mathsf{I}_{a\star b} \star \top_c,\quad u'' = \mathsf{I}_a \star \bot^b$$

The simulations show that:

$$(1) \qquad f \cdot (\mathsf{I}_n \star i(\mathsf{I}_{a\star b} \star \top_c)) = (\mathsf{I}_m \star o(\mathsf{I}_{a\star b} \star \top_c)) \cdot f' \quad \text{and}$$
$$(2) \qquad f \cdot (\mathsf{I}_n \star i(\mathsf{I}_a \star \bot^b)) = (\mathsf{I}_m \star o(\mathsf{I}_a \star \bot^b)) \cdot f''$$

Take $\overline{F} = (a \star c, \overline{f})$ where $\overline{f} = (\mathsf{I}_m \star o(\mathsf{I}_a \star \top_b \star \mathsf{I}_c))\ f'\ (\mathsf{I}_n \star i(\mathsf{I}_a \star \bot^b \star \mathsf{I}_c))$.

We show that $F' \xrightarrow{b\alpha}_{\mathsf{I}_a \star \bot^b \star \mathsf{I}_c} \overline{F}\ {}_{\mathsf{I}_a \star \top_c}\!\xleftarrow{a\beta} F''$.

For the right one, the simulation's conditions are easy to prove,

$$\begin{aligned}(\mathsf{I}_m \star o(\mathsf{I}_a \star \top_c)) \cdot \overline{f} &= (\mathsf{I}_m \star o(\mathsf{I}_a \star \top_{b\star c}))\ f'\ (\mathsf{I}_n \star i(\mathsf{I}_a \star \bot^b \star \mathsf{I}_c))\\ &= (\mathsf{I}_m \star o(\mathsf{I}_a \star \top_b))\ f\ (\mathsf{I}_n \star i(\mathsf{I}_a \star \bot^b \star \top_c)) && \text{by (1)}\\ &= (\mathsf{I}_m \star o(\mathsf{I}_a \star \top_b \cdot \bot^b))\ f''\ (\mathsf{I}_n \star i(\mathsf{I}_a \star \top_c)) && \text{by (2)}\\ &= f'' \cdot (\mathsf{I}_n \star o(\mathsf{I}_a \star \top_c))\end{aligned}$$

For the left one, as above one may show that

$$\begin{aligned}&(\top_m \star o(\top_a \star \mathsf{I}_b \star \top_c)) \cdot f' \cdot (\mathsf{I}_n \star i(\mathsf{I}_a \star \bot^b \star \mathsf{I}_c))\\ &= (\top_m \star o(\top_a \star \bot^b))\ f''\ (\mathsf{I}_n \star i(\mathsf{I}_a \star \top_c))\\ &= \bot^{o(b)} \cdot \top_{n\star i(a\star c)} && \mathrm{C}_{a\beta}\end{aligned}$$

hence

$$\begin{aligned}f' \cdot (\mathsf{I}_n \star i(\mathsf{I}_a \star \bot^b \star \mathsf{I}_c)) &= (\mathsf{I}_m \star o(\mathsf{I}_a \star \bot^b \cdot \top_b \star \mathsf{I}_c))\ f'\ (\mathsf{I}_n \star i(\mathsf{I}_a \star \bot^b \star \mathsf{I}_c))\\ &= (\mathsf{I}_m \star o(\mathsf{I}_a \star \bot^b \star \mathsf{I}_c)) \cdot \overline{f}\end{aligned}$$

by coalgebraic theory rules (i.e., working with the arrow's components). □

In general the inclusion $\overset{a\gamma}{\longleftarrow} \cdot \overset{b\alpha}{\longrightarrow}\ \subseteq\ \overset{b\alpha}{\longrightarrow} \cdot \overset{a\gamma}{\longleftarrow}$, does not hold. For example, if $x : 1 \to 1$ and $F,\ F',\ F'' : 0 \to 0$ are given by the pairs $F' = (x, \mathsf{I}_1)$, $F = (x\star x, \mathsf{V}_1 \star \top_1)$, $F'' = (x, \bot^1 \cdot \top_1)$, then $F' \overset{a\gamma}{\longleftarrow} F \overset{b\alpha}{\longrightarrow} F''$ but $(F', F'') \notin \overset{b\alpha}{\longrightarrow} \cdot \overset{a\gamma}{\longleftarrow}$.

Lemma 6.4. $\overset{a\gamma}{\longleftarrow} \cdot \overset{b\alpha}{\longrightarrow}\ \subseteq\ \overset{b\alpha}{\longrightarrow} \cdot \overset{a\gamma}{\longleftarrow} \cdot \overset{b\alpha}{\longleftarrow}$, *when T is an identification-free strong $b\delta$-symocat.*

PROOF: Let $F'\ {}_{u'}\!\overset{a\gamma}{\longleftarrow} F \overset{b\alpha}{\longrightarrow}_{u''} F''$.

Using isomorphisms we may order the variables in F as $a\star b$ such that $u'' = \mathsf{I}_a \star \bot^b$. Next, we order the variables in F' as $c\star e\star d$, where $\{|c|+1, \ldots, |c|+|e|\}$ is the intersection of the images via u' of the sets $\{1, \ldots, |a|\}$ and $\{|a|+1, \ldots, |a|+|b|\}$. Finally, we order again the variables in a and b corresponding to F as $a' \star a''$ and $b'' \star b'$ such that the preimage via u'^{-1} of $\{|c|+1, \ldots, |c|+|e|\}$ be $\{|a'|+1, \ldots, |a'|+|a''\star b''|\}$. Consequently, we may suppose the pairs are of the following type

$F' = (c\star e\star d,\ f')$, $F = (a' \star a'' \star b'' \star b',\ f)$, $F'' = (a' \star a'',\ f'')$ and
$u'' = \mathsf{I}_{a'\star a''} \star \bot^{b''\star b'}$, $u' = u_1 \star (u_2 \star u_3) \cdot \mathsf{V}_e \star u_4$,
where $u_1 : a' \to c$, $u_2 : a'' \to e$, $u_3 : b'' \to e$ and $u_4 : b' \to d$ are $a\gamma$-relations

The simulations show that:

(1) $f \cdot (\mathsf{I}_n \star i(u_1 \star (u_2 \star u_3)\ \mathsf{V}_e \star u_4)) = (\mathsf{I}_m \star o(u_1 \star (u_2 \star u_3)\ \mathsf{V}_e \star u_4)) \cdot f'$ and

(2) $f \cdot (\mathsf{I}_n \star i(\mathsf{I}_{a'\star a''} \star \bot^{b''\star b'})) = (\mathsf{I}_m \star o(\mathsf{I}_{a'\star a''} \star \bot^{b''\star b'})) \cdot f''$

Take the pairs $\overline{F} = (c, \overline{f})$ where $\overline{f} = (\mathsf{I}_m \star o(\mathsf{I}_c \star \top_{e\star d}))\ f'\ (\mathsf{I}_n \star i(\mathsf{I}_c \star \bot^{e\star d}))$ and $\tilde{F} = (a', \tilde{f})$ where $\tilde{f} = (\mathsf{I}_m \star o(\mathsf{I}_{a'} \star \top_{a''}))\ f''\ (\mathsf{I}_n \star i(\mathsf{I}_{a'} \star \bot^{a''}))$.

We show that $F' \xrightarrow{b\alpha}_{\mathsf{I}_c \star \bot^{e\star d}} \overline{F}\ {}_{u_1}\!\xleftarrow{a\gamma} \tilde{F}\ {}_{\mathsf{I}_{a'} \star \bot^{a''}}\!\xleftarrow{b\alpha} F''$.

The middle part is a simulation indeed, by

$$
\begin{aligned}
&(\mathsf{I}_m \star o(u_1)) \cdot \overline{f} \\
&\quad = (\mathsf{I}_m \star o(u_1 \star \top_{e\star d}))\ f'\ (\mathsf{I}_n \star i(\mathsf{I}_c \star \bot^{e\star d})) \\
&\quad = (\mathsf{I}_m \star o(\mathsf{I}_{a'} \star \top_{a''\star b''\star b'}))(\mathsf{I}_m \star o(u_1 \star (u_2 \star u_3) \vee_e \star u_4)) \cdot f'(\mathsf{I}_n \star i(\mathsf{I}_c \star \bot^{e\star d})) \\
&\quad = (\mathsf{I}_m \star o(\mathsf{I}_{a'} \star \top_{a''\star b''\star b'}))\ f\ (\mathsf{I}_n \star i(u_1 \star \bot^{a''\star b''\star b'})) && \text{by (1)} \\
&\quad = (\mathsf{I}_m \star o(\mathsf{I}_{a'} \star \top_{a''} \star \top_{b''\star b'} \cdot \bot^{b''\star b'}))\ f''\ (\mathsf{I}_n \star i(u_1 \star \bot^{a''})) && \text{by (2)} \\
&\quad = \tilde{f} \cdot (\mathsf{I}_n \star i(u_1))
\end{aligned}
$$

For the left part suppose that $v_1 : c \to a'$, $v_3 : e \to b''$ and $v_4 : d \to b'$ are left inverses of u_1, u_3 and u_4, respectively. Hence

$$
\begin{aligned}
&f' \cdot (\mathsf{I}_n \star i(\mathsf{I}_c \star \bot^{e\star d})) \\
&\quad = (\mathsf{I}_m \star o(v_1 u_1 \star (\top_{a''}\ u_2 \star v_3 u_3) \vee_e \star v_4 u_4))\ f'\ (\mathsf{I}_n \star i(\mathsf{I}_c \star \bot^{e\star d})) \\
&\quad = (\mathsf{I}_m \star o(v_1 \star \top_{a''} \star v_3 \star v_4))\ f\ (\mathsf{I}_n \star i(u_1 \star \bot^{a''\star b''\star b'})) && \text{by (1)} \\
&\quad = (\mathsf{I}_m \star o(v_1 \star \top_{a''} \star \bot^{e\star d}))\ f''\ (\mathsf{I}_n \star i(u_1 \star \bot^{a''})) && \text{by (2)}
\end{aligned}
$$

If we compose on left with $\mathsf{I}_m \star o(\mathsf{I}_c \star \bot^{e\star d} \cdot \top_{e\star d})$ then the last term remains unchanged. Hence,

$$
\begin{aligned}
(*) \quad f' \cdot (\mathsf{I}_n \star i(\mathsf{I}_c \star \bot^{e\star d})) &= (\mathsf{I}_m \star o(\mathsf{I}_c \star \bot^{e\star d} \cdot \top_{e\star d}))\ f'\ (\mathsf{I}_n \star i(\mathsf{I}_c \star \bot^{e\star d})) \\
&= (\mathsf{I}_m \star o(\mathsf{I}_c \star \bot^{e\star d}))\overline{f}
\end{aligned}
$$

Finally, for the right simulation, let us first note that

$$
\begin{aligned}
&(\top_m \star o(\top_{a'} \star \mathsf{I}_{a''})) \cdot f'' \cdot (\mathsf{I}_n \star i(u_1 \star \bot^{a''})) \\
&\quad = (\top_m \star o(\top_{a'} \star \mathsf{I}_{a''} \star \top_{b''\star b'} \cdot \bot^{b''\star b'}))\ f''\ (\mathsf{I}_n \star i(u_1 \star \bot^{a''})) \\
&\quad = (\top_m \star o(\top_{a'} \star \mathsf{I}_{a''} \star \top_{b''\star b'}))\ f\ (\mathsf{I}_n \star i(u_1 \star \bot^{a''\star b''\star b'})) && \text{by (2)} \\
&\quad = (\top_m \star o(\top_{a'} \star \mathsf{I}_{a''} \star \top_{b''\star b'}))\ f\ (\mathsf{I}_n \star i(u_1 \star [(u_2 \star u_3) \vee_e \star u_4]\bot^{e\star d})) \\
&\quad = (\top_m \star o(\top_{a'}\ u_1 \star (u_2 \star \top_{b''}\ u_3) \vee_e \star \top_{b'}\ u_4))\ f'\ (\mathsf{I}_n \star i(\mathsf{I}_c \star \bot^{e\star d})) && \text{by (1)} \\
&\quad = (\top_m \star o(\top_c \star u_2 \star \top_d))\ (\mathsf{I}_n \star o(\mathsf{I}_c \star \bot^{e\star d} \cdot \top_{e\star d}))\ f'\ (\mathsf{I}_n \star i(\mathsf{I}_c \star \bot^{e\star d})) && \text{by } (*) \\
&\quad = (\top_m \star o(\top_c \star \bot^{a''} \cdot \top_e \star \top_d))\ f'\ (\mathsf{I}_n \star i(\mathsf{I}_c \star \bot^{e\star d})) \\
&\quad = \bot^{o(a'')} \cdot \top_{n\star i(c)} && \mathrm{C}_{a\beta}
\end{aligned}
$$

Applying identification-free hypothesis for the $a\gamma$-relation $\mathsf{I}_n \star i(u_1)$ we get

$$
\begin{aligned}
(**) \quad &(\top_m \star o(\top_{a'} \star \mathsf{I}_{a''})) \cdot f'' \cdot (\mathsf{I}_n \star i(\mathsf{I}_{a'} \star \bot^{a''})) \\
&\quad = (\top_m \star o(\top_{a'} \star \mathsf{I}_{a''}))\ f''\ (\bot^n \cdot \top_n \star i(\bot^{a'} \cdot \top_{a'} \star \bot^{a''})) \\
&\quad = \bot^{o(a'')} \cdot \top_{n\star i(a')} && \text{by } \mathrm{C}_{b\alpha}
\end{aligned}
$$

Now the requested identity follows by

$$
\begin{aligned}
&(\mathsf{I}_m \star o(\mathsf{I}_{a'} \star \bot^{a''})) \cdot \tilde{f} \\
&\quad = (\mathsf{I}_m \star o(\mathsf{I}_{a'} \star \bot^{a''} \cdot \top_{a''}))\ f''\ (\mathsf{I}_n \star i(\mathsf{I}_{a'} \star \bot^{a''})) \\
&\quad = [(\mathsf{I}_m \star o(\mathsf{I}_{a'} \star \top_{a''}))\ f''\ (\mathsf{I}_n \star i(\mathsf{I}_{a'} \star \bot^{a''})) \\
&\qquad \star (\top_m \star o(\top_{a'} \star \bot^{a''} \cdot \top_{a''} \cdot \mathsf{I}_{a''}))\ f''\ (\mathsf{I}_n \star i(\mathsf{I}_{a'} \star \bot^{a''}))] \cdot \vee_{a \star i(a')} && \text{by } C_{a\gamma} \\
&\quad = [(\mathsf{I}_m \star o(\mathsf{I}_{a'} \star \top_{a''}))\ f''\ (\mathsf{I}_n \star i(\mathsf{I}_{a'} \star \bot^{a''})) \star \bot^{a''} \cdot \top_{n \star i(a')}] \cdot \vee_{a \star i(a')} && \text{by } C_{a\beta} \\
&\quad = [\ (\mathsf{I}_m \star o(\mathsf{I}_{a'} \star \top_{a''}))\ f''\ (\mathsf{I}_n \star i(\mathsf{I}_{a'} \star \bot^{a''})) \\
&\qquad \star (\top_m \star o(\top_{a'} \star \mathsf{I}_{a''}))\ f''\ (\mathsf{I}_n \star i(\mathsf{I}_{a'} \star \bot^{a''}))] \cdot \vee_{n \star i(a')} && \text{by } (**) \\
&\quad = f'' \cdot (\mathsf{I}_n \star i(\mathsf{I}_{a'} \star \bot_{a''})) && \text{by } C_{a\gamma}
\end{aligned}
$$

This completes the proof. □

Besides these commuting lemmas we need the following decomposition result showing that both relations $\xrightarrow{b\delta}$ and $\xrightarrow{a\beta} \cup \xrightarrow{a\gamma} \cup \xrightarrow{b\alpha}$ generate the same equivalence relation.

Lemma 6.5. $\xrightarrow{b\delta}\ \subseteq\ \xrightarrow{b\alpha} \cdot \xrightarrow{a\delta}$, *when T is an identification-free $a\delta$-strong $b\delta$-symocat.*

PROOF: Suppose $F' \xrightarrow{b\delta}_u F''$.

Using isomorphisms we may suppose that

$F' = (a \star b,\ f'),\ \ F'' = (c \star d,\ f'')$ and
$u = (\mathsf{I}_a \star \bot^b)\ v\ (\mathsf{I}_c \star \top_d)$ for an $a\gamma$-relation $v : a \to c$

This means

$$(1) \qquad f'(\mathsf{I}_n \star i((\mathsf{I}_a \star \bot^b)\ v\ (\mathsf{I}_c \star \top_d))) = (\mathsf{I}_m \star o((\mathsf{I}_a \star \bot^b)\ v\ (\mathsf{I}_c \star \top_d)))f''$$

Take the pair $\overline{F} = (a, \overline{f})$ where $\overline{f} = (\mathsf{I}_m \star o(\mathsf{I}_a \star \top_b)) \cdot f' \cdot (\mathsf{I}_n \star i(\mathsf{I}_a \star \bot^b))$.

We show that $F' \xrightarrow{b\alpha}_{\mathsf{I}_a \star \bot^b} \overline{F} \xrightarrow{a\delta}_{v(\mathsf{I}_c \star \top_d)} F''$.

For the right simulation it is easy to see that

$$
\begin{aligned}
&\overline{f} \cdot (\mathsf{I}_n \star i(v(\mathsf{I}_c \star \top_d))) \\
&\quad = \quad (\mathsf{I}_m \star o(\mathsf{I}_a \star \top_b))\ f'\ [\mathsf{I}_n \star i((\mathsf{I}_a \star \bot^b)\ v\ (\mathsf{I}_c \star \top_d))] \\
&\quad = \quad [\mathsf{I}_m \star o((\mathsf{I}_a \star \top_b \bot^b)\ v\ (\mathsf{I}_c \star \top_d))]\ f'' && \text{by (1)} \\
&\quad = \quad (\mathsf{I}_m \star o(v(\mathsf{I}_c \star \top_d))) \cdot f''
\end{aligned}
$$

For the left simulation, let us first note that

$$
\begin{aligned}
&[(\top_m \star o(\top_a \star \mathsf{I}_b)) \cdot f' \cdot (\mathsf{I}_n \star i(\mathsf{I}_a \star \bot^b))] \cdot (\mathsf{I}_n \star i(v)) \\
&\quad = (\top_m \star o(\top_a \star \mathsf{I}_b))\ f'\ [\mathsf{I}_n \star i((\mathsf{I}_a \star \bot^b)\ v\ (\mathsf{I}_c \star \top_d \cdot \bot^d))] \\
&\quad = [\top_m \star o((\top_a \star \bot^b)\ v\ (\mathsf{I}_c \star \top_d))]\ f''\ (\mathsf{I}_n \star i(\mathsf{I}_c \star \bot^d)) && \text{by (1)} \\
&\quad = \bot^{o(b)} \cdot \top_{m \star i(c \star d)} \cdot f''(\mathsf{I}_n \star i(\mathsf{I}_c \star \bot^d)) \\
&\quad = \bot^{o(b)} \cdot \top_{n \star i(c)} && \text{by } C_{a\beta}
\end{aligned}
$$

Applying identification-free hypothesis for surjection $\mathsf{I}_n \star i(v)$ we get

$$
\begin{aligned}
(*)\quad & (\mathsf{T}_m \star o(\mathsf{T}_a \star \mathsf{I}_b)) \cdot f' \cdot (\mathsf{I}_n \star i(\mathsf{I}_a \star \bot^b)) \\
& = [(\mathsf{T}_m \star o((\mathsf{T}_a \star \mathsf{I}_b))\ f'\ (\mathsf{I}_n \star i(\mathsf{I}_a \star \bot^b))] \cdot \bot^{n\star i(a)} \cdot \mathsf{T}_{n\star i(a)} \\
& = \bot^{o(b)} \cdot \mathsf{T}_{n\star i(a)} && \text{by } \mathrm{C}_{b\alpha}
\end{aligned}
$$

Hence

$$
\begin{aligned}
& (\mathsf{I}_m \star o(\mathsf{I}_a \star \bot^b)) \cdot \overline{f} \\
& = (\mathsf{I}_m \star o(\mathsf{I}_a \star \bot^b \mathsf{T}_b))\ f'\ (\mathsf{I}_n \star i(\mathsf{I}_a \star \bot^b)) \\
& = [(\mathsf{I}_m \star o(\mathsf{I}_a \star \mathsf{T}_b))\ f'\ (\mathsf{I}_n \star i(\mathsf{I}_a \star \bot^b)) \\
& \qquad \star (\mathsf{T}_m \star o(\mathsf{T}_a \star \bot^b \mathsf{T}_b))\ f'\ (\mathsf{I}_n \star i(\mathsf{I}_a \star \bot^b))] \cdot \mathsf{V}_{n\star i(a)} && \text{by } \mathrm{C}_{a\gamma} \\
& = [(\mathsf{I}_m \star o(\mathsf{I}_a \star \mathsf{T}_b))\ f'\ (\mathsf{I}_n \star i(\mathsf{I}_a \star \bot^b)) \star \bot^{o(b)} \cdot \mathsf{T}_{n\star i(a)}] \cdot \mathsf{V}_{n\star i(a)} && \text{by } \mathrm{C}_{a\beta} \\
& = [(\mathsf{I}_m \star o(\mathsf{I}_a \star \mathsf{T}_b))\ f'\ (\mathsf{I}_n \star i(\mathsf{I}_a \star \bot^b)) \\
& \qquad \star (\mathsf{T}_m \star o(\mathsf{T}_a \star \mathsf{I}_b))\ f'\ (\mathsf{I}_n \star i(\mathsf{I}_a \star \bot^b))] \cdot \mathsf{V}_{n\star i(a)} && \text{by } (*) \\
& = f' \cdot (\mathsf{I}_n \star i(\mathsf{I}_a \star \bot^b)) && \text{by } \mathrm{C}_{a\gamma}
\end{aligned}
$$

which certifies that the left part is a simulation, indeed. □

Using these lemmas we may now prove the following:

Theorem 6.6. *(Structural Theorem for $\overset{b\delta}{\Longleftrightarrow}$) If T is a strong identification-free $b\delta$-symocat with feedback, then the following decomposition holds in $[X,T]$:*

$$\overset{b\delta}{\Longleftrightarrow} = \xrightarrow{b\alpha} \cdot \xrightarrow{a\gamma} \cdot \xleftarrow{a\beta} \cdot \xrightarrow{a\beta} \cdot \xleftarrow{a\gamma} \cdot \xleftarrow{b\alpha}$$

Proof: Lemmas 5.4 and 6.5 show that both relations $\xrightarrow{b\delta}$ and $\xrightarrow{b\alpha} \cup \xrightarrow{a\gamma} \cup \xrightarrow{a\beta}$ generate the same equivalence relation.

Let $\rho = \xrightarrow{b\alpha} \cdot \xrightarrow{a\gamma} \cdot \xleftarrow{a\beta} \cdot \xrightarrow{a\beta} \cdot \xleftarrow{a\gamma} \cdot \xleftarrow{b\alpha}$. It is clear that

$$F' \overset{b\delta}{\Longleftrightarrow} F'' \quad \text{iff} \quad \exists n \geq 1 \text{ such that } F'\ \rho^n\ F''$$

All we still have to prove is that ρ is transitive, namely $\rho^n \subseteq \rho$.

From the commuting relations presented in Table 6.1 it follows that $\rho \cdot \xleftarrow{a\beta}$, $\rho \cdot \xrightarrow{a\beta}$, $\rho \cdot \xleftarrow{a\gamma}$ and $\rho \cdot \xleftarrow{b\alpha}$ are included in ρ. In the remaining cases it is neccessary to use the commutation given by Lemma 6.4 that provides a commutation, but it also adds an auxiliary term. Let us see what happens.

In one case it is easy,

$$
\begin{aligned}
& \rho \cdot \xrightarrow{b\alpha} \\
= \ & (\xrightarrow{b\alpha} \cdot \xrightarrow{a\gamma} \cdot \xleftarrow{a\beta} \cdot \xrightarrow{a\beta} \cdot \xleftarrow{a\gamma} \cdot \xleftarrow{b\alpha}) \cdot \xrightarrow{b\alpha} \\
\subseteq \ & \xrightarrow{b\alpha} \cdot \xrightarrow{a\gamma} \cdot \xleftarrow{a\beta} \cdot \xrightarrow{a\beta} \cdot (\xleftarrow{a\gamma} \cdot \xrightarrow{b\alpha}) \cdot \xleftarrow{b\alpha} && \text{by } 5.6^o \\
\subseteq \ & \xrightarrow{b\alpha} \cdot \xrightarrow{a\gamma} \cdot \xleftarrow{a\beta} \cdot \xrightarrow{a\beta} \cdot (\xrightarrow{b\alpha} \cdot \xleftarrow{a\gamma} \cdot \xleftarrow{b\alpha}) \cdot \xleftarrow{b\alpha} && \text{by } 6.4 \\
\subseteq \ & \rho && \text{by } 6.1\text{–}6.3
\end{aligned}
$$

For the inclusion $\rho \cdot \xrightarrow{a\gamma} \subseteq \rho$, suppose that $(F', F'') \in \rho \cdot \xrightarrow{a\gamma}$. It follows that there exists surjections u, u', u_1, u_1' such that

$$
\begin{array}{ll}
 & (F',\ F'') \in \xrightarrow{b\alpha} \cdot \xrightarrow{a\gamma} \cdot \xleftarrow{a\beta} \cdot \xrightarrow{a\beta} \cdot {}_{u'}\xleftarrow{a\gamma} \cdot \xleftarrow{b\alpha} \cdot \xrightarrow{a\gamma}_{u} \\
\Rightarrow & \{\text{by } 6.4^{op}\} \\
 & (F',\ F'') \in \xrightarrow{b\alpha} \cdot \xrightarrow{a\gamma} \cdot \xleftarrow{a\beta} \cdot \xrightarrow{a\beta} \cdot {}_{u'}\xleftarrow{a\gamma} \cdot \xrightarrow{b\alpha} \cdot \xrightarrow{a\gamma}_{u_1} \cdot \xleftarrow{b\alpha} \\
\Rightarrow & \{\text{by } 6.4\} \\
 & (F',\ F'') \in \xrightarrow{b\alpha} \cdot \xrightarrow{a\gamma} \cdot \xleftarrow{a\beta} \cdot \xrightarrow{a\beta} \cdot \xrightarrow{b\alpha} \cdot {}_{u_1'}\xleftarrow{a\gamma} \cdot \xleftarrow{b\alpha} \cdot \xrightarrow{a\gamma}_{u_1} \cdot \xleftarrow{b\alpha} \\
\Rightarrow & \{\text{by } 6.1\text{–}6.3\} \\
 & (F',\ F'') \in \xrightarrow{b\alpha} \cdot \xrightarrow{a\gamma} \cdot \xleftarrow{a\beta} \cdot \xrightarrow{a\beta} \cdot {}_{u_1'}\xleftarrow{a\gamma} \cdot \xleftarrow{b\alpha} \cdot \xrightarrow{a\gamma}_{u_1} \cdot \xleftarrow{b\alpha}
\end{array}
$$

It seems that the matters become more complicated. However, a fine analysis shows that we get a simplification. The simulation $\xrightarrow{b\alpha}$ between u' and u_1 (in the 2nd line above) shows that the length of the source of the new relation u_1 does not exceed the length of u'; moreover, if the lengths are equal, then the additional term $\xrightarrow{b\alpha}$ disappears (it becomes a simulation via an isomorphism). Similarly, in the 3rd line the source of the new relation u_1' either is strictly shorter in length than the one of u_1, or the additional term $\xleftarrow{b\alpha}$ disappears. But the source lengths are finite, hence they can not strictly decrease forever; hence after a finite number of such transformations we get

$$
\begin{array}{lll}
(F',\ F'') & \in & \xrightarrow{b\alpha} \cdot \xrightarrow{a\gamma} \cdot \xleftarrow{a\beta} \cdot \xrightarrow{a\beta} \cdot {}_{u_n'}\xleftarrow{a\gamma} \cdot \xrightarrow{a\gamma}_{u_n} \xleftarrow{b\alpha} \\
 & \subseteq & \rho \qquad\qquad\qquad\qquad\qquad\qquad \text{by } 5.4,\ 5.7,\ 5.9
\end{array}
$$

The theorem is proved. □

This theorem provides techniques to minimize (deterministic) networks with respect to the IO behaviour. In this type of minimization the following reductions are used: (1) $\xrightarrow{a\gamma}$ (identification of vertices having the same label and compatible outgoing connections), (2) $\xleftarrow{a\beta}$ (deletion of vertices with no incoming paths from the entries of the network) and (3) $\xrightarrow{b\alpha}$ (deletion of vertices with no outgoing paths to the exits of the network). We get the following corollaries.

Corollary 6.7. *A network F is $\overset{b\delta}{\Longleftrightarrow}$-minimal if and only if ($F \xrightarrow{a\gamma} F'$ or $F \xleftarrow{a\beta} F'$ or $F \xrightarrow{b\alpha} F'$) implies $F \xrightarrow{a\alpha} F'$.* □

Corollary 6.8. *Two $\overset{b\delta}{\Longleftrightarrow}$-equivalent and minimal networks are isomorphic.* □

Exercises and Problems (sec. 6.3)

1. When $\xrightarrow{b\alpha} \cdot \xleftarrow{a\gamma} \subseteq \xleftarrow{a\gamma} \cdot \xrightarrow{b\alpha}$ holds (see also Lemma 6.19)? (12)
2. Study the other commutations left open in Table 6.1. (- -)

6.4 Soundness for $b\delta$-flow

In this section we prove that, under the identification-free hypothesis, the Park theory structure may be lifted from connections to $\overset{b\delta}{\Longleftrightarrow}$-equivalent networks.

First, Theorem 6.6 allows us to prove that the equivalence relation generated by simulation via $b\delta$-relations fulfils the enzymatic axiom. Indeed, the following result hold.

Theorem 6.9. *If T is an identification-free Park theory, then $[X,T]_{b\delta}$ satisfies condition $Enz_{b\delta}$.*

PROOF: We use the same frame as in the proof of Theorem 5.14. Up to compositions with some bijections, every $b\delta$-relation is of the type $u \star \top_q$, with u a $b\gamma$-relation. As in Theorem 5.14 the problem is reduced to the verification of the axiom $\mathrm{Enz}_{b\gamma}$.

Let $u : p \to q$ be a $b\gamma$-relation and $F \in [X,T](m \star p, n \star p)$ and $G \in [X,T](m \star q, n \star q)$ be two pairs such that $F \cdot (\mathsf{I}_n \star u) \overset{b\delta}{\Longleftrightarrow} (\mathsf{I}_m \star u) \cdot G$.

We may suppose that $g = (\underline{x}, g)$ is a $\overset{b\delta}{\Longleftrightarrow}$-minimal pair. Then, it follows that $(\mathsf{I}_m \star u) \cdot G$ is a minimal, too. (The reason is as in the proof of 5.14 completed with the observation that there exists an $a\beta$-relation $v : q \to p$ such that $v \cdot u = \mathsf{I}_q$.)

By Theorem 6.6 it follows that $F \cdot (\mathsf{I}_n \star u) \overset{b\alpha}{\longrightarrow} \cdot \overset{a\gamma}{\longrightarrow} \cdot \overset{a\beta}{\longleftarrow} (\mathsf{I}_n \star u) \cdot G$.

The desired result follows from Lemma 4.20 applied for $A = b\delta$, $B = a\alpha$, $C = b\gamma$ and $D = a\beta$. (The verification of condition (3) in Lemma 4.20 in the case j is the $b\gamma$-relation u is similar to the verification done for the case $j = \mathsf{V}_p^k$ in the final part of the proof of Theorem 5.14.) □

Now, as in the previous case $a\delta$, from this theorem the following consequences may be inferred.

Corollary 6.10. *In the conditions of the above theorem, the congruence generated by simulation via $b\delta$-relations coincides with the congruence with property $Enz_{b\delta}$ generated by $C_{b\delta}$-var. Shortly, $\overset{b\delta}{\Longleftrightarrow} = \sim_{b\delta}$* □

Theorem 6.11. *If T is an identification-free Park theory, then $[X,T]_{b\delta}$ is a Park theory, too.* □

EXERCISES AND PROBLEMS (SEC. 6.4)

1. Is the identification-free condition necessary to prove that $[X,T]_{b\delta}$ is a Park theory? Check if this condition is preserved by passing from T to $[X,T]_{b\delta}$. (**)

6.5 Completeness for $b\delta$-flow

The issues related to the completness of the Park theory axioms are collected in this section. The main result is:

Theorem 6.12. *Algebras* $[X, \mathbb{P}\mathrm{fn}]_{b\delta}$ *and* $R[X]^{\emptyset}$ *are isomorphic.*

PROOF: Let us first note that $R[X]^{\emptyset}$ is an Park theory. Hence $b\delta$-flownomials may be correctly interpreted in this theory, i.e., if two networks in $[X, \mathbb{P}\mathrm{fn}]$ are equivalent using simulation via partial functions, then they have an equal interpretation in $R[X]^{\emptyset}$.

♠ The interpretation $\varphi^{\sharp} : \mathsf{FlowExp}[X, \mathbb{P}\mathrm{fn}] \to [X]^{\emptyset}$ acts as follows: the networks are unfolded and then all the subtrees without outputs are identified to the empty tree. Now, it is clear that, for two normal form expressions F_1 and F_2, if $F_1 \overset{b\delta}{\Longleftrightarrow} F_2$, then $\varphi^{\sharp}(F_1) = \varphi^{\sharp}(F_2)$. ♣

Next, we want to find out certain associated networks with the same input behaviour and to apply the completeness theorem for case $a\delta$. For $i \in [2]$, let F_i^{coacc} be the network obtained from F_i deleting the vertices with no outgoing paths through the network's outputs. It follows that $F_i \overset{b\alpha}{\longrightarrow} F_i^{coacc}$ and the unfolding of F_i^{coacc} is $\varphi^{\sharp}(F_i)$. Consequently the unfolding of F_1^{coacc} is equal to that of F_2^{coacc}.

By the previous completeness Theorem 5.2, we get $F_1^{coacc} \overset{a\delta}{\Longleftrightarrow} F_2^{coacc}$. Finally, this shows that $F_1 \overset{b\delta}{\Longleftrightarrow} F_2$, hence $\varphi^{\sharp}$ is injective. On the other hand, it is obvious that $\varphi^{\sharp}$ is surjective. Consequently, $\varphi^{\sharp}$ is an isomorphism, indeed. □

Using Corollary 6.10, we get the following result:

Corollary 6.13. *Algebras* $\mathsf{FlowExp}[X, \mathbb{P}\mathrm{fn}]_{b\delta}$, $[X, \mathbb{P}\mathrm{fn}]_{b\delta}$, *and* $R[X]^{\emptyset}$ *are isomorphic.*

Corollary 6.14. *The following conditions are equivalent for two flownomial expressions* E *and* E' *in* $\mathsf{FlowExp}[X, \mathbb{P}\mathrm{fn}]$:

1. $E \sim_{b\delta} E'$
2. $\mathsf{nf}(E) \overset{b\delta}{\Longleftrightarrow} \mathsf{nf}(E')$
3. *The trees obtained by unfolding the networks associated with E and E' are the same, up to the replacement of the empty tree for subtrees without outputs.*
4. *The networks associated with E and E' have the same successful computation sequences.* □

6.6 Working with $b\delta$-flownomials

Corollary 6.10 (and Theorem 5.19) gives us the following framework for working with $b\delta$-flownomials.

Theorem 6.15. *(universality of $b\delta$-flownomials) If T is an identification-free Park theory, then the statement obtained from that one of Theorem 5.19 by replacing everywhere $a\delta$ with $b\delta$ is valid.* □

Since $\mathbb{P}$fn is also an initial object in the category of Park theories, from the above theorem we get:

Corollary 6.16. *$[X,T]_{b\delta}$ is the Park theory freely generated by X.* □

6.7 IO behaviour, nondeterministic case

We give here some details for the formal definition of the IO behaviour of networks.

Formalization of the nondeterministic IO behaviour. The *nondeterministic IO behaviour* may be easier defined using an interpretation of the flownomial expressions in a particular matrix theory.

Nondeterministic IO behaviour

• First, if $X = \{X(m,n)\}_{m,n\in N}$ is a set of doubly-ranked variables, then we associate a new set of variables $\langle X\rangle$ consisting of $m \times n$ new variables $\langle i,x,j\rangle$, for each $x \in X(m,n)$ and $i \in [m]$, $j \in [n]$. (The new variable $\langle i,x,j\rangle$ denotes the network obtained restricting cell x to its i-th input and j-th output.)

• Next, we consider the theory of matrices $\mathcal{M}(\mathcal{P}(\langle X\rangle^*))$ over the language semiring $(\mathcal{P}(\langle X\rangle^*), \cup, \cdot, \emptyset, \{\lambda\})$; the elements in this semiring are subsets of words in $\langle X\rangle^*$; the operations are standard.

• Then, the unique morphism of matrix theories $\breve{\ } : \mathbb{R}\mathrm{el} \to \mathcal{M}(\mathcal{P}(\langle X\rangle^*))$ maps a relation, seen as 0-1 matrix, to the matrix obtained by replacing everywhere 0 by $\emptyset$ and 1 by $\{\lambda\}$.

• Finally, function ψ that interprets a variable $x \in X(m,n)$ as the matrix is

$$\psi(x) = \begin{pmatrix} \langle 1,x,1\rangle & \ldots & \langle 1,x,n\rangle \\ \vdots & & \vdots \\ \langle m,x,1\rangle & \ldots & \langle m,x,n\rangle \end{pmatrix} \in \mathcal{M}(\mathcal{P}(\langle X\rangle^*))(m,n)$$

• We claim that *the unique NA morphism*

$$\|\cdot\| : [X, \mathbb{R}\mathrm{el}] \to \mathcal{M}(\mathcal{P}(\langle X\rangle^*))$$

that extends $(\breve{\ }, \psi)$ produces the nondeterministic IO behaviour. To be more precise, for a network $F \in [X, \mathbb{R}\mathrm{el}](p,q)$, the component $F_{i,j}$ of the matrix $\|F\|$ consists of all the successful computation sequences of F starting from the network's i-th input and ending with its j-th output.

Sketch of proof. In order to prove this fact, let us suppose that

$$F = [(\mathsf{I}_p \star x_1 \star \ldots \star x_k) \cdot f] \uparrow^{m_1+\ldots+m_k}$$

where $x_i \in X(m_i, n_i)$, $\forall i \in [k]$. Write the matrix f as follows:

$$\begin{pmatrix} A & B_1 & \ldots & B_k \\ C_1 & D_{11} & \ldots & D_{1k} \\ \vdots & \vdots & & \vdots \\ C_k & D_{k1} & \ldots & D_{kk} \end{pmatrix}$$

where the entries of the matrix are matrices of the following dimensions:

$$\begin{array}{llll} A : p \to q & B_1 : p \to m_1 & \ldots & B_k : p \to m_k \\ C_1 : n_1 \to q & D_{11} : n_1 \to m_1 & \ldots & D_{1k} : n_1 \to m_k \\ \vdots & & & \\ C_k : n_k \to q & D_{k1} : n_k \to m_1 & \ldots & D_{kk} : n_k \to m_k \end{array}$$

With these notations we may write

$$\begin{aligned} \|F\| &= [(\mathsf{I}_p \star \psi(x_1) \star \ldots \star \psi(x_k)) \cdot \breve{f}] \uparrow^{m_1+\ldots+m_k} \\ &= \begin{pmatrix} \breve{A} & \breve{B}_1 & \ldots & \breve{B}_k \\ \psi(x_1)\breve{C}_1 & \psi(x_1)\breve{D}_{11} & \ldots & \psi(x_1)\breve{D}_{1k} \\ \vdots & \vdots & & \vdots \\ \psi(x_k)\breve{C}_k & \psi(x_k)\breve{D}_{k1} & \ldots & \psi(x_k)\breve{D}_{kk} \end{pmatrix} \uparrow^{m_1+\ldots+m_k} \\ &= \breve{A} \cup [\breve{B}_1 \ldots \breve{B}_k] \cdot \begin{pmatrix} \psi(x_1)\breve{D}_{11} & \ldots & \psi(x_1)\breve{D}_{1k} \\ \vdots & & \vdots \\ \psi(x_k)\breve{D}_{k1} & \ldots & \psi(x_k)\breve{D}_{kk} \end{pmatrix}^* \cdot \begin{pmatrix} \psi(x_1)\breve{C}_1 \\ \vdots \\ \psi(x_k)\breve{C}_k \end{pmatrix} \end{aligned}$$

Let us note that the matrices of the type $\breve{A}$, $\breve{B}$, $\breve{C}$ and $\breve{D}$ have all components of the forms $\emptyset$ or $\{\lambda\}$ and only the matrices $\psi(x_1), \ldots, \psi(x_k)$ have nontrivial components, more precisely of the form $\{\sigma\}$, with σ letter in $\langle X \rangle$.

Finally, an analysis of the language represented by this regular-like expression shows that the claim is true. $\diamond$

Exercises and Problems (sec. 6.7)

1. Show the language represented by the above regular-like expression consists of the input–output paths, indeed. (07)

6.8 Kleene theories ($d\delta$-flow)

"Kleene theory" is the equivalent name for $d\delta$-flow, a structure already defined in Chapter 4. In the general NA terms, it is a *($d\delta$-weak, $d\delta$-strong, $d\delta$-enzymatic)* *structure*. That means, it is a BNA with (1) $d\delta$-branching constants ($d\delta$-symocat with feedback), of which (2) $d\delta$-constants strongly commutes with arbitrary arrows, and (3) $d\delta$-constants generate relations for which the enzymatic rule is valid.

Notice that Kleene theory is the strongest structure among the structures defined using the general schemes with (xy-weak, $x'y'$-strong, $x''y''$-enzymatic) conditions. In contrast, BNA is the weakest.

6.9 Structural Theorem, case $d\delta$

The Structural Theorem proved in this section is different from the previous ones. It shows that two networks are $\overset{d\delta}{\Longleftrightarrow}$-equivalent if and only if, making abstraction of some non-accessible or non-coaccessible vertices, by partial unfolding the networks may be refined to a common extension; the unfolding is used in both directions, i.e., either towards the inputs or to the outputs.

General presentation. The Structural Theorems for $\overset{a\delta}{\Longleftrightarrow}$ and $\overset{b\delta}{\Longleftrightarrow}$ were based on minimization. A key property was the existence of a unique minimal network, up to an isomorphism. This idea cannot be used in the present nondeterministic case.

Nondeterministic minimization. The minimization of nondeterministic networks is not completely understood: there are no criteria for the characterization of their minimalness, no simple rules for minimization, etc. For example, the flownomials $F' = x \cdot (\vee \cdot x \cdot \wedge) \uparrow$ and $F'' = (\vee \cdot x \cdot \wedge) \uparrow \cdot\, x$ are $\overset{d\delta}{\Longleftrightarrow}$-minimal, equivalent, but not isomorphic. To see that they are equivalent, notice that both flownomials are equivalent with $F = x \wedge ((\vee x \wedge) \uparrow \star \mathsf{I}_1) \vee x$; indeed, $\mathsf{nf}(F) \overset{a\gamma}{\longrightarrow} \mathsf{nf}(F'')$ and $\mathsf{nf}(F) \overset{c\alpha}{\longleftarrow} \mathsf{nf}(F')$, where nf denotes the normal form application defined in Chapter 2. (The situation is similar as for nondeterministic automata; see the comments in Chapter 8.) ◊

To get a characterization theorem for $\overset{d\delta}{\Longleftrightarrow}$-equivalence we follow an opposite way: roughly speaking the Structural Theorem, to be proved in this section, shows that two networks are $\overset{d\delta}{\Longleftrightarrow}$-equivalent iff they may be refined to the same network.

♡ More precisely, two networks are $\overset{d\delta}{\Longleftrightarrow}$-equivalent if and only if, making abstraction of some non-accessible or non-coaccessible vertices, by partial unfolding the networks may be refined to a common extension (the unfolding is used in both directions, i.e., either towards the inputs or to the outputs). ◊

Actually, we shall prove that the following decomposition holds:

$$\overset{d\delta}{\Longleftrightarrow} = \overset{a\beta}{\longleftarrow} \cdot \overset{a\gamma}{\longleftarrow} \cdot \overset{b\alpha}{\longrightarrow} \cdot \overset{c\alpha}{\longrightarrow} \cdot \overset{a\gamma}{\longrightarrow} \cdot \overset{a\beta}{\longrightarrow} \cdot \overset{c\alpha}{\longleftarrow} \cdot \overset{b\alpha}{\longleftarrow}$$

The general setting is that of a matrix theory with a feedback theory. But for certain commuting lemmas we need some technical conditions. We suppose the monoid used to model network interfaces is $\mathbb{N}$ and the connecting theory is a matrix theory.

ⓘ As we have said in Proposition 4.7 such a theory may be seen as a theory of matrices over the semiring $(T(1,1), \cup, \cdot, 0, 1)$, where $f \cup g = \wedge^1 \cdot (f \star g) \cdot \vee_1$, $0 = \perp^1 \cdot \top_1$, $1 = \mathsf{I}_1$. Notice that composition in T coincides with the usual product of matrices. (The additive operation here is "$\cup$", called either union, or sum; in Chapter 4 it was denoted by "+". ⓘ

We start with a few definitions. A semiring $(S, \cup, \cdot, 0, 1)$ is *divisible* if for every $a_1, a_2, b_1, b_2 \in S$, if $a_1 \cup a_2 = b_1 \cup b_2$ then there exist some elements in S, denoted c_{ij}, for $i, j \in [2]$, such that $\cup_{j\in[2]} c_{ij} = a_i, \forall i \in [2]$ and $\cup_{i\in[2]} c_{ij} = b_j, \forall j \in [2]$. A semiring S is *zero-sum-free* if for every $a_1, a_2 \in S$: $a_1 \cup a_2 = 0 \Rightarrow a_1 = a_2 = 0$. From the implication that defines the divisibility condition one may deduce that the following more general implication holds:

(*) If I, J are *nonempty* finite sets, then for every family of elements $a_i \in S (i \in I)$, $b_j \in S (j \in J)$ such that $\cup_{i\in I} a_i = \cup_{j\in J} b_j$, there exist c_{ij} in S, for $i \in I$, $j \in J$, such that $\cup_{j\in J} c_{ij} = a_i, \forall i \in I$ and $\cup_{i\in I} c_{ij} = b_j, \forall j \in J$.

If we use the convention that the union over the empty set is zero, then the zero-sum-free condition is an extreme case of the implication (*), namely when $I = \{1, 2\}$ and $J = \emptyset$. So that

Fact 6.17. *In a zero-sum-free and divisible semiring the implication* (*) *holds for arbitrary finite sets I and J.* □

Finally, we say a *matrix theory T is divisible or zero-sum-free* if the associated semiring $(T(1,1), \cup, \cdot, 0, 1)$ has the corresponding property. It is clear how property (*) may be extended from ordinary elements to matrices. (The forthcoming Proposition 6.21 throws some light on the role of these conditions: an useful interpolation property may be proved, provided they are valid.)

Commuting lemmas. For proving the structural theorem in this case we need some more commuting lemmas. Their collection is presented in Table 6.2.

Lemma 6.18. $\xrightarrow{a\beta} \cdot \xleftarrow{a\gamma} \;\subseteq\; \xleftarrow{a\gamma} \cdot \xrightarrow{a\beta}$, *when T is an identification-free $b\delta$-symocat.*

PROOF: Suppose $F' \xrightarrow{a\beta}_{u'} F \;{}_{u''}\!\xleftarrow{a\gamma} F''$.

Using isomorphic representations we may suppose

$F' = (a,\ f')$, $F = (a \star b,\ f)$, $F'' = (c \star d,\ f'')$ and
$u' = \mathsf{I}_a \star \top_b$, $u'' = u_1 \star u_2$ (with $u_1 : c \to a$)

Table 6.2. Is $x \cdot y \subseteq y \cdot x$? (case $d\delta$) "Obv." means "Obvious". The index ^{op} (resp. ^{o}) means "opposite" (resp. "dual") version.

x \ y	$\overset{a\beta}{\longleftarrow}$	$\overset{a\gamma}{\longleftarrow}$	$\overset{b\alpha}{\longrightarrow}$	$\overset{c\alpha}{\longrightarrow}$	$\overset{a\gamma}{\longrightarrow}$	$\overset{a\beta}{\longrightarrow}$	$\overset{c\alpha}{\longleftarrow}$	$\overset{b\alpha}{\longleftarrow}$
$\overset{a\beta}{\longleftarrow}$	Obv.							
$\overset{a\gamma}{\longleftarrow}$	5.4^{op}	Obv.						
$\overset{b\alpha}{\longrightarrow}$	6.3^{o}	6.19	Obv.					
$\overset{c\alpha}{\longrightarrow}$	$(*)6.4^{o}$	$(*)6.24$	$5.4^{o,op}$	Obv.				
$\overset{a\gamma}{\longrightarrow}$	6.19^{op}	6.20	6.1	6.23	Obv.			
$\overset{a\beta}{\longrightarrow}$	5.6^{o}	6.18	6.2	$6.4^{o,op}$	5.4	Obv.		
$\overset{c\alpha}{\longleftarrow}$	6.1^{o}	6.23^{op}	$6.18^{o,op}$	6.20^{o}	$(*)6.24^{o,op}$	$6.19^{o,op}$	Obv.	
$\overset{b\alpha}{\longleftarrow}$	6.2^{op}	6.1^{op}	5.6^{o}	6.18^{o}	$(*)6.4^{op}$	6.3^{op}	5.4^{o}	Obv.

The simulations show that

(1) $f' \cdot (\mathsf{I}_n \star i(\mathsf{I}_a \star \top_b)) = (\mathsf{I}_m \star o(\mathsf{I}_a \star \top_b)) \cdot f$ and

(2) $f'' \cdot (\mathsf{I}_n \star i(u_1 \star u_2)) = (\mathsf{I}_m \star o(u_1 \star u_2)) \cdot f$

Take the pair $\overline{F} = (a, \overline{f})$ where $\overline{f} = (\mathsf{I}_m \star o(\mathsf{I}_c \star \top_d)) \cdot f'' \cdot (\mathsf{I}_n \star i(\mathsf{I}_c \star \bot^d))$.

We show that $F' \;{}_{u_1}\!\overset{a\gamma}{\longleftarrow} \overline{F} \overset{a\beta}{\longrightarrow}_{\mathsf{I}_c \star \top_d} F''$.

The left part follows by an easy computation. For the right one, first note that

$$\begin{aligned} &(\mathsf{I}_m \star o(\mathsf{I}_c \star \top_d)) \cdot f'' \cdot (\mathsf{I}_n \star i(u_1 \star u_2)) \\ &= (\mathsf{I}_m \star o(u_1 \star \top_b)) \cdot f && \text{by (2)} \\ &= (\mathsf{I}_m \star o(u_1)) \cdot f' \cdot (\mathsf{I}_n \star i(\mathsf{I}_a \star \top_b)) && \text{by (1)} \end{aligned}$$

hence the identification-free property may be applied for surjections u_1, u_2 to get $(\mathsf{I}_m \star o(\mathsf{I}_c \star \top_d)) \cdot f'' = (\mathsf{I}_m \star o(\mathsf{I}_c \star \top_d)) \cdot f'' \cdot (\mathsf{I}_n \star i(\mathsf{I}_a \star \bot^d \cdot \top_d))$.

It follows that

$$\begin{aligned} \overline{f} \cdot (\mathsf{I}_n \star i(\mathsf{I}_c \star \top_d)) &= (\mathsf{I}_m \star o(\mathsf{I}_c \star \top_d)) \cdot f'' \cdot (\mathsf{I}_n \star i(\mathsf{I}_a \star \bot^d \cdot \top_d)) \\ &= (\mathsf{I}_m \star o(\mathsf{I}_c \star \top_d)) \cdot f'' \end{aligned}$$ □

Lemma 6.19. $\overset{b\alpha}{\longrightarrow} \cdot \overset{a\gamma}{\longleftarrow} \;\subseteq\; \overset{a\gamma}{\longleftarrow} \cdot \overset{b\alpha}{\longrightarrow}$, *when T is a $d\alpha$-strong $d\gamma$-symocat.*

PROOF: Let $F' \overset{b\alpha}{\longrightarrow}_{u'} F \;{}_{u''}\!\overset{a\gamma}{\longleftarrow} F''$.

We may suppose that $F' = (a \star b,\ f')$, $F = (a,\ f)$, $F'' = (c,\ f'')$ and $u' = \mathsf{I}_a \star \bot^b$. The simulations give

(1) $f' \cdot (\mathsf{I}_n \star i(\mathsf{I}_a \star \bot^b)) = (\mathsf{I}_m \star o(\mathsf{I}_a \star \bot^b)) \cdot f$ and

(2) $f'' \cdot (\mathsf{I}_n \star i(u'')) = (\mathsf{I}_m \star o(u'')) \cdot f$

Take the pair $\overline{F} = (c \star b,\ \overline{f})$, where

$$\overline{f} = \wedge^{m \star o(c \star b)} \cdot [\ (\mathsf{I}_m \star o(\mathsf{I}_c \star \perp^b))\ f''\ \star\ (\mathsf{I}_m \star o(u'' \star \mathsf{I}_b))\ f'\ (\perp^n \star i(\perp^a \star \mathsf{I}_b))\ ]$$

We show that $F'\ {}_{u'' \star \mathsf{I}_b}\!\xleftarrow{a\gamma} \overline{F} \xrightarrow{b\alpha}_{\mathsf{I}_c \star \perp^b} F''$.

Indeed, for the right part one may see that

$$\begin{aligned}
&\overline{f} \cdot (\mathsf{I}_n \star i(\mathsf{I}_c \star \perp^b)) \\
&= \wedge^{m \star o(c \star b)} \cdot [\ (\mathsf{I}_m \star o(\mathsf{I}_c \star \perp^b))\ f''\ \star\ (\mathsf{I}_m \star o(u'' \star \mathsf{I}_b))\ f' \cdot \perp^{n \star i(a \star b)}\ ] \\
&= \wedge^{m \star o(c \star b)} \cdot [\ (\mathsf{I}_m \star o(\mathsf{I}_c \star \perp^b))\ f''\ \star\ \perp^{m \star o(c \star b)}\ ] \qquad \mathrm{C}_{b\alpha} \\
&= (\mathsf{I}_m \star o(\mathsf{I}_c \star \perp^b)) \cdot f''
\end{aligned}$$

For the other one the computation is

$$\begin{aligned}
&\overline{f} \cdot (\mathsf{I}_n \star i(u'' \star \mathsf{I}_b)) \\
&= \wedge^{m \star o(c \star b)}[(\mathsf{I}_m \star o(\mathsf{I}_c \star \perp^b))f''(\mathsf{I}_n \star i(u'')) \star (\mathsf{I}_m \star o(u'' \star \mathsf{I}_b))f'(\perp^n \star i(\perp^a \star \mathsf{I}_b))] \\
&= \wedge^{m \star o(c \star b)}[(\mathsf{I}_m \star o(u'' \star \perp^b))f\ \star (\mathsf{I}_m \star o(u'' \star \mathsf{I}_b))f'(\perp^n \star i(\perp^a \star \mathsf{I}_b))] \quad \text{by (2)} \\
&= \wedge^{m \star o(c \star b)}[(\mathsf{I}_m \star o(u'' \star \mathsf{I}_b))f'(\mathsf{I}_n \star i(\mathsf{I}_a \star \perp^b) \star (\mathsf{I}_m \star o(u'' \star \mathsf{I}_b))f'(\perp^n \star i(\perp^a \star \mathsf{I}_b))] \\
&\qquad \text{by (1)} \\
&= (\mathsf{I}_m \star o(u'' \star \mathsf{I}_b))\ f' \cdot \wedge^{n \star o(a \star b)} \cdot [\mathsf{I}_n \star i(\mathsf{I}_a \star \perp^b) \star \perp^n \star i(\perp^a \star \mathsf{I}_b)] \qquad \mathrm{C}_{c\alpha} \\
&= (\mathsf{I}_m \star o(u'' \star \mathsf{I}_b)) \cdot f'
\end{aligned}$$

□

The proof of the next results 6.20–6.23 may be found in [Şte87a].

Lemma 6.20. $\xrightarrow{a\gamma} \cdot \xleftarrow{a\gamma}\ \subseteq\ \xleftarrow{a\gamma} \cdot \xrightarrow{a\gamma}$, *whenever T is a divisible matrix theory.* □

Proposition 6.21. *(interpolation property) (1) If A and B are matrices over a zero-sum-free and divisible semiring and u and v are functions such that $A \cdot u^{-1} \cdot v = u^{-1} \cdot v \cdot B$, then there exists a matrix Z over S such that $A \cdot u^{-1} = u^{-1} \cdot Z$ and $Z \cdot v = v \cdot B$. (2) If the functions u and v are surjective, then only the divisiblity property of S is necessary for inferring the existence of Z.* □

Lemma 6.22. $\xrightarrow{d\delta}\ \subseteq\ \xrightarrow{b\alpha} \cdot \xrightarrow{c\alpha} \cdot \xrightarrow{a\gamma} \cdot \xrightarrow{a\beta}$, *when T is a zero-sum-free and divisible matrix theory.* □

Lemma 6.23. $\xrightarrow{a\gamma} \cdot \xrightarrow{c\alpha}\ \subseteq\ \xrightarrow{c\gamma}\ \subseteq\ \xrightarrow{c\alpha} \cdot \xrightarrow{a\gamma}$, *when T is a divisible matrix theory.* □

The last commutation to be studied is $\xrightarrow{c\alpha} \cdot \xleftarrow{a\gamma}\ \subseteq\ \xleftarrow{a\gamma} \cdot \xrightarrow{c\alpha}$. It does not hold in the general case. However, we will find a useful commutation which adds a parasite term. For proving this we need a condition stronger than divisibility, which we call "uniform divisibility".

A semiring $(S, \cup, \cdot, 0, 1)$ is *uniform divisible* if there is an operation $\cap : S \times S \to S$ fulfilling the following axioms:

$$\begin{array}{llll}
\text{(As)} & x \cap (y \cap z) = (x \cap y) \cap z & \text{(C)} & x \cap y = y \cap x \\
\text{(D)} & (x \cup y) \cap z = (x \cap z) \cup (y \cap z) & \text{(A)} & (x \cup y) \cap x = x
\end{array}$$

(Actually this means that $(S, \cup, \cap)$ is a lattice.) A matrix theory T is *uniform divisible* if its associated semiring is uniform divisible. The component-wise extension of "$\cap$" to matrices is denoted by the same sign.

Lemma 6.24. $\xrightarrow{c\alpha} \cdot \xleftarrow{a\gamma} \subseteq \xleftarrow{a\gamma} \cdot \xrightarrow{c\alpha} \cdot \xrightarrow{a\gamma}$, *when T is a uniform divisible matrix theory.* □

Identification-free vs zero-sum-free. Before collecting the results we made an observation: for a matrix theory the identification-free condition is equivalent to the zero-sum-free condition. Indeed, let T is a matrix theory. If $f = [\, A \;\; B \,] \in T(r, m+n)$ and $u : m \to p$, $v : n \to q$ are surjections, then the identification-free condition is $f \cdot (u \star v) = f' \cdot (\mathsf{I}_p \star \mathsf{T}_q) \;\Rightarrow\; f = f \cdot (\mathsf{I}_m \star \bot^n \cdot \mathsf{T}_n)$. It may be written as $[\, Au \;\; Bv \,] = [\, A' \;\; 0_{r,q} \,] \;\Rightarrow\; [\, A \;\; B \,] = [\, A \;\; 0_{r,n} \,]$, where we supposed $B : r \to n$. It follows that the identification-free condition is reduced to $Bv = 0 \;\Rightarrow\; B = 0$, for the surjection v. Using permutations we may suppose $v = \mathsf{V}_1^{n_1} \star \ldots \star \mathsf{V}_1^{n_q}$ and moreover, $q = 1$. In this case the last implication is $[\, b_1 \; \ldots \; b_k \,] \begin{bmatrix} 1 \\ \vdots \\ 1 \end{bmatrix} = 0 \;\Rightarrow\; [\, b_1 \; \ldots \; b_k \,] = 0$, or $b_1 \cup \ldots \cup b_k = 0 \;\Rightarrow\; b_1 = \ldots = b_k = 0$. This is just the zero-sum-free condition. ◇

Theorem 6.25. *(Structural Theorem, case $d\delta$) If T is a zero-sum-free and uniform divisible matrix theory with feedback, then the following decomposition holds in $[X, T]$:*

$$\overset{d\delta}{\Longleftrightarrow} = \xleftarrow{a\beta} \cdot \xleftarrow{a\gamma} \cdot \xrightarrow{b\alpha} \cdot \xrightarrow{c\alpha} \cdot \xrightarrow{a\gamma} \cdot \xrightarrow{a\beta} \cdot \xleftarrow{c\alpha} \cdot \xleftarrow{b\alpha}$$

PROOF: The inclusion "$\supset$" is obviously valid. For the other one, let ρ denotes the right-hand side term of the identity to prove. By Lemma 6.22 it follows that $\overset{d\delta}{\Longleftrightarrow} \subset \rho^+$, hence all we have to prove is that ρ is transitive. Furthermore, this is reduced to the verification that relations $\rho \cdot \xleftarrow{a\beta}$, $\rho \cdot \xleftarrow{a\gamma}$, $\rho \cdot \xrightarrow{b\alpha}$, $\rho \cdot \xrightarrow{c\alpha}$, $\rho \cdot \xrightarrow{a\gamma}$, $\rho \cdot \xrightarrow{a\beta}$, $\rho \cdot \xleftarrow{c\alpha}$, and $\rho \cdot \xleftarrow{b\alpha}$ are included in ρ.

All these inclusions, except for $\rho \cdot \xrightarrow{a\gamma}$, $\rho \cdot \xleftarrow{a\gamma}$, and $\rho \cdot \xleftarrow{a\beta}$, are easily validated by the commutations in Table 6.2. In the excepted cases, one has to use Lemmas 6.4, 6.24, and their duals. These lemmas give the necessary commutations, but they add some auxiliary terms. Fortunately, these terms may be eliminated using the commutations in Table 6.2, and the proof is finished. □

EXERCISES AND PROBLEMS (SEC. 6.9)

1. Show that uniform divisibility implies divisibility, but the converse is not true. Are there interesting divisible semirings which cannot be made uniform divisible? (- -)
2. Check in detail the last part of the proof of Theorem 6.25 (ρ is transitive). (05).

6.10 Soundness for $d\delta$-flow

The soundness theorem, in the form of the preservation of the Kleene theory structure by passing from connections to $\overset{d\delta}{\Longleftrightarrow}$-equivalent networks, is proved here. As in the previous cases, Theorem 6.25 allows us to prove that the equivalence relation generated by simulation via $b\delta$-relations fulfils the enzymatic axiom.

Theorem 6.26. *If T is a zero-sum-free and uniform divisible Kleene theory, then $[X,T]_{d\delta}$ satisfies condition $Enz_{d\delta}$.*

PROOF: Theorem 6.25 shows that $\overset{d\delta}{\Longleftrightarrow} = \overset{a\delta}{\longleftarrow} \cdot \overset{d\delta}{\longrightarrow} \cdot \overset{d\alpha}{\longleftarrow}$. Now, the theorem directly follows from Lemma 4.20 applied for $A = d\delta$, $B = a\delta$, $C = d\delta$ and $D = d\alpha$. (The conditions of that lemma are fulfilled: indeed, condition (2) in Lemma 4.20 is obviously satisfied – the proof is as in the final proof of Theorem 5.14; on the other hand, since $B^o = D$ condition (3) is dual to condition (2), therefore it is also valid.) □

Corollary 6.27. *Under the above conditions, the congruence generated by simulation via $d\delta$-relations coincides with the congruence with property $Enz_{d\delta}$ generated by $C_{d\delta}$-var. Shortly, $\overset{d\delta}{\Longleftrightarrow} = \sim_{d\delta}$.* □

Theorem 6.28. *If T is a zero-sum-free and uniform divisible Kleene theory, then $[X,T]_{d\delta}$ is a Kleene theory, too.* □

EXERCISES AND PROBLEMS (SEC. 6.10)

1. As in the previous case of Park theories it may be interesting to see whether the additional conditions (zero-sum-freeness and uniform divisibility) are necessary in order to prove that $[X,T]_{d\delta}$ is a Kleene theory and if these conditions are preserved by passing from T to $[X,T]_{d\delta}$. (**)

6.11 Completeness for $d\delta$-flow

The aim of this section is to show that the equivalence relation generated by simulation via relations captures the nondeterministic IO behaviour.

By Proposition 4.18, $[X,T]_{d\delta}$ is a matrix theory, hence we may suppose X has only one-input/one-output variables. Then, the completeness problem may be reduced to its version for one-input/one-output networks.

♡ It is easy to see why: (1) every network $F : p \to q$ is equivalent via $d\delta$-relations to the matrix $(F_{i,j})_{i\in[p],j\in[q]}$ of its components and (2) two networks $F, F' : p \to q$ have the same successful computation sequences iff for every $i \in [p], j \in [q]$ their components $F_{i,j}$ and $F'_{i,j}$ have the same set of successful computation sequences. ◇

A key idea is to use the translation of the notion of *deterministic automaton* in this setting. The translation gives the "semi-deterministic" condition below.

First, a notation. If the theory for network connections is a matrix theory, then we may define a *matricial representation of networks* as follows: for a pair $(x_1\star\ldots\star x_p, f)$ write the connection morphism $f \in T(m\star o(x_1)\star\ldots\star o(x_p)), n\star i(x_1)\star\ldots\star i(x_p))$ as

$$\left[\begin{array}{c|ccc} A & B_1 & \ldots & B_p \\ \hline C_1 & D_{11} & \ldots & D_{1p} \\ \vdots & \vdots & \ddots & \vdots \\ C_p & D_{p1} & \ldots & D_{pp} \end{array}\right] \quad \text{where} \quad \begin{array}{ll} A \in T(m,n) & \\ B_k \in T(m, i(x_k)) & \forall k \in [p] \\ C_j \in T(o(x_j), n) & \forall j \in [p] \\ D_{jk} \in T(o(x_j), i(x_k)) & \forall j \in [p],\ k \in [p] \end{array}$$

Bellow, $i(x) = o(x) = 1$ for all x. A network, written as

$$F = \begin{array}{c} \\ \\ \\ x_1 \\ \vdots \\ x_n \end{array} \left(\begin{array}{ccc|ccc} A_{11} & \ldots & A_{1q} & B_{11} & \ldots & B_{1n} \\ \vdots & & \vdots & \vdots & & \vdots \\ A_{p1} & \ldots & A_{pq} & B_{p1} & \ldots & B_{pn} \\ \hline C_{11} & \ldots & C_{1q} & D_{11} & \ldots & D_{1n} \\ \vdots & & \vdots & \vdots & & \vdots \\ C_{n1} & \ldots & C_{nq} & D_{n1} & \ldots & D_{nn} \end{array}\right) \in [X, \mathbb{R}\mathrm{el}](p,q)$$

is called *semi-deterministic* if

1. $\forall i \in [p],\ \forall j,k \in [n] \quad [j \neq k \wedge B_{ij} = B_{ik} = 1 \quad \Rightarrow \quad x_j \neq x_k]$ and
2. $\forall i \in [n],\ \forall j,k \in [n] \quad [j \neq k \wedge D_{ij} = D_{ik} = 1 \quad \Rightarrow \quad x_j \neq x_k]$.

♡ In other words, a semi-deterministic network still may be nondeterministic, but the possible multiple outgoing connecting edges point to vertices labeled with distinct variables or to distinct outputs.

From the network algebra point of view, with cells on vertices, the usual deterministic finite automata are still nondeterministic. More precisely, while the reaction of the automaton is unique to a given input action one still has a nondeterministic choice for the input action. To simulate a deterministic automaton with a fully deterministic network it requires us a few more transformations. ◇

Reduction of nondeterministic networks to semi-deterministic ones. The following result is an adaptation to the case of networks of a result of automata theory which states that a nondeterministic automaton is "similar" to the associated deterministic automaton obtained by the standard power-set construction. See the chapter on finite automata for more on this.

Lemma 6.29. *For every nondeterministic network $F \in [X, \mathbb{R}\mathrm{el}](p,q)$ there exists a semi-deterministic one $SD(F) \in [X, \mathbb{R}\mathrm{el}](p,q)$ and a relation u such that $SD(F) \xrightarrow{u} F$.* □

The proof of this lemma is omitted. It mimics the similar proof for passing from nondeterministic to deterministic automata (see Chapter 8); it is described in detail in [Şte94].

The next step consists of the transformation of a semi-deterministic network into a deterministic one having a nicely related behaviour.

Reduction of semi-deterministic networks to deterministic ones. In this subsection we briefly describe how a deterministic network may be associated with a semi-deterministic one.

We start with a linearly ordered set of variables $V = \{z_1, \ldots, z_r\} \subseteq X(1,1)$, say $z_1 \prec z_2 \prec \ldots \prec z_r$. Then we consider a new set of variables V_o consisting of a new variable $\overline{z}_i \in V_o(1, 1+r)$ for every $z_i \in V \subseteq X(1,1)$.

Let $F = [(\mathsf{I}_1 \star x_1 \star \ldots \star x_k) f] \uparrow^k \in [X, \mathbb{R}\mathrm{el}](1,1)$ be a semi-deterministic network and suppose V contains all the variables $x_1, \ldots, x_k$. Starting with the relation $f \in \mathbb{R}\mathrm{el}(1+k, 1+k)$ we consider the partial function $\overline{f} \in \mathbb{P}\mathrm{fn}((1+k)\times(1+r), 1+k)$ defined by the cotupling

$$\overline{f} = \langle \overline{f}_1, \overline{f}_2, \ldots, \overline{f}_{1+k} \rangle$$

where for $i \in [1+k]$ the partial function $\overline{f}_i \in \mathbb{P}\mathrm{fn}(1+r, 1+k)$ is defined as follows:

(1) $\overline{f}_i(1) =$ "**if** $(i,1) \in f$ **then** 1 **else** undefined," and
(2) $\overline{f}_i(1+j) =$ "**if** $(i, 1+l) \in f$ & $x_l = z_j$ **then** $1+l$ **else** undefined," for $j \in [r]$.

Since the network F is semi-deterministic, for every $i \in [1+k]$ and $j \in [r]$ there exists at most one l such that $(i, 1+l) \in f$ & $x_l = z_j$, hence f_i is a (partial) function, indeed.

Write f as a tupling $f = \langle f_1, f_2, \ldots, f_{1+k} \rangle$ with $f_i \in \mathbb{R}\mathrm{el}(1, 1+k)$. It follows that $\wedge_{1+r} \cdot \overline{f}_i = f_i$ (Indeed, in the definition of f_i only the pairs $(i, 1+l) \in f$ occur and all these pairs occur at least once in the definition of f_i, for $x_1, \ldots, x_k \in V$.) Consequently, $f = (\wedge_{1+r} \star \wedge_{1+r} \star \ldots \star \wedge_{1+r}) \cdot \overline{f}$. This implies,

$$\begin{aligned} F &= [(\mathsf{I}_1 \star x_1 \star \ldots \star x_k) \cdot f] \uparrow^k \\ &= [(\mathsf{I}_1 \star x_1 \star \ldots \star x_k) \cdot (\wedge_{1+r} \star \wedge_{1+r} \star \ldots \star \wedge_{1+r}) \cdot \overline{f}] \uparrow^k \\ &= \wedge_{1+r} \cdot [(\mathsf{I}_{1+r} \star x_1 \cdot \wedge_{1+r} \star \ldots \star x_k \cdot \wedge_{1+r}) \cdot \overline{f}] \uparrow^k \end{aligned}$$

Define $D_V(F) \in [V_o, \mathbb{P}\text{fn}](1+r,1)$ as $D_V(F) = [(\mathsf{I}_{1+r} \star \overline{x}_1 \star \ldots \star \overline{x}_k) \cdot \overline{f}] \uparrow^k$, where for $i \in [k]$, $\overline{x}_i = \overline{z}_{k_i} \in V_o(1, 1+r)$ whenever $x_i = z_{k_i} \in V$.

Now, if one looks to the successful computation sequences of $D_V(F)$, then one may see that

Proposition 6.30. *If F', $F'' \in [X, \mathbb{R}\text{el}](1,1)$ are semi-deterministic networks with $\|F'\| = \|F''\|$ and V is a finite, linearly ordered set containing all the variables that occur in F' or F'', then $\|D_V(F')\| = \|D_V(F'')\|$.* □

The original network F may be obtained from $D_V(F)$ by substituting $z_1 \cdot \wedge_{1+r}$ for $\overline{z}_l$, $\forall l \in [r]$; this substitution is denoted by $[\overline{z}_1/z_1 \cdot \wedge_{1+r}; \ \ldots \ ; \overline{z}_r/z_r \cdot \wedge_{1+r}]$. To be precise, $F = \wedge_{1+r} \cdot D_V(F)[\overline{z}_1/z_1 \cdot \wedge_{1+r} \ ; \ \ldots \ ; \overline{z}_r/z_r \cdot \wedge_{1+r}]$.

Lemma 6.31. *If $D_V(F') \to_u D_V(F'')$, then $F' \to_u F''$.*

PROOF: Let $F' = [(\mathsf{I}_1 \star x'_1 \star \ldots \star x'_{k'})\ f'] \uparrow^{k'}$ and $F'' = [(\mathsf{I}_1 \star x''_1 \star \ldots \star x''_{k''})\ f''] \uparrow^{k''}$ be two semi-deterministic networks and let $D_V(F') = [(\mathsf{I}_1 \star \overline{x}'_1 \star \ldots \star \overline{x}'_{k'})\overline{f}'] \uparrow^{k'}$ and $D_V(F'') = [(\mathsf{I}_1 \star \overline{x}''_1 \star \ldots \star \overline{x}''_{k''})\overline{f}''] \uparrow^{k''}$ be their deterministic versions obtained using the above procedure. The given simulation shows that: (1) $(i,j) \in u \Rightarrow \overline{x}'_i = \overline{x}''_j$ and (2) $\overline{f}' \cdot (\mathsf{I}_1 \star u) = (\mathsf{I}_{1+r} \star o(u)) \cdot \overline{f}''$.

Notice that: $f' \cdot (\mathsf{I}_1 \star u) = (\wedge_{1+r} \star \wedge_{1+r} \star \ldots \star \wedge_{1+r}) \cdot \overline{f} \cdot (\mathsf{I}_1 \star u) = [\wedge_{1+r} \star (\wedge_{1+r} \star \ldots \star \wedge_{1+r}) \cdot o(u)] \cdot \overline{f}'' = [\wedge_{1+r} \star u \cdot (\wedge_{1+r} \star \ldots \star \wedge_{1+r})] \cdot \overline{f}'' = (\mathsf{I}_1 + u) \cdot f''$. Combined with $(i,j) \in u \Rightarrow x'_i = x''_j$, this gets $F' \xrightarrow{u} F''$. □

Completeness theorem. The completeness of the Kleene theory axioms for capturing the (nondeterministic) IO behaviour is proved here. It is based on the above outlined results to reduce the problem to the completeness result of the previous case $b\delta$.

Theorem 6.32. *Two networks F', $F'' \in [X, \mathbb{R}\text{el}](p,q)$ have the same nondeterministic IO behaviour iff $F' \overset{d\delta}{\Longleftrightarrow} F''$.*

PROOF: Matrices having languages as their entries form a Kleene theory. This shows that two $\overset{d\delta}{\Longleftrightarrow}$-equivalent networks have the same IO behaviour.

For the converse implication, as we have mentioned in the beginning of this section, we may restrict our proof to the case $p = q = 1$. Suppose $\|F'\| = \|F''\|$.

As in Lemma 6.29 we construct the associated semi-deterministic networks $SD(F')$ and $SD(F'')$. But F' and $SD(F')$ are $d\delta$-similar (Lemma 6.29), hence they have the same behaviour $\|F'\| = \|SD(F'')\|$. Likely, $\|F''\| = \|SD(F'')\|$, so that $\|SD(F')\| = \|SD(F'')\|$. For an appropriate V we construct the deterministic networks $D_V(SD(F'))$ and $D_V(SD(F''))$. By Proposition 6.30 these networks have the same deterministic IO behaviour. By Theorem 6.12 it follows that $D_V(SD(F')) \overset{b\delta}{\Longleftrightarrow} D_V(SD(F''))$ and by Lemma 6.31 we get $SD(F') \overset{b\delta}{\Longleftrightarrow}$

$SD(F'')$. To conclude, we have got $F' \xrightarrow{d\delta} SD(F') \overset{b\delta}{\Longleftrightarrow} SD(F'') \xleftarrow{d\delta} F''$, hence $F' \overset{d\delta}{\Longleftrightarrow} F''$. □

Corollary 6.33. *The following conditions are equivalent for two flownomial expressions E and E' in* $\mathsf{FlowExp}[X, \mathbb{R}\mathrm{el}]$:

1. $E \sim_{d\delta} E'$
2. $\mathsf{nf}(E) \overset{d\delta}{\Longleftrightarrow} \mathsf{nf}(E')$
3. *The networks associated with E and E' have the same IO step-by-step behaviour.*
4. *The computation processes denoted by E and E' have the same successful computation sequences.* □

6.12 Working with $d\delta$-flownomials

The definitions for $d\delta$-Flow-Calculus and $d\delta$-flownomials appear as particular instances of the general definitions given in Chapter 5 (Definition 5.3). Theorems 2.27 and 6.28 and Corollary 6.27 give the following result.

Theorem 6.34. *(universality of $d\delta$-flownomials) If T is a zero-sum-free and uniform divisible Kleene theory, then the statement obtained from that one of Theorem 5.19 by replacing everywhere $a\delta$ by $d\delta$ is valid.* □

Since $\mathbb{R}\mathrm{el}$ is an initial object in the category of Kleene theories, the following corollary may be obtained.

Corollary 6.35. $[X, T]_{d\delta}$ *is the Kleene theory freely generated by X.* □

Comments, problems, bibliographic remarks

Notes. The first part of the chapter, dealing with input–output behaviour in the deterministic case, is based on Chapter D, Sections 4–6 of [Şte91]. The results were announced in [Şte87a] and [CŞ90a], but the detailed proofs were unpublished till now.

Except for the completeness part, the axiomatization result of the second part of the chapter is from [Şte87b]; this paper also contains a suggestion that simulation via relations capture the IO behaviour. The key point towards the completeness result is that the standard power-set construction of the deterministic automaton associated with a nondeterministic one may be modeled by simulation via relations; see [Koz91, BÉ93a]. Some more transformations were necessary to fit the

Table 6.3. The standard axioms for the calculus of flownomials

I. Axioms for symocats

B1	$f \star (g \star h) = (f \star g) \star h$
B2	$\mathsf{I}_\epsilon \star f = f = f \star \mathsf{I}_\epsilon$
B3	$f \cdot (g \cdot h) = (f \cdot g) \cdot h$
B4	$\mathsf{I}_a \cdot f = f = f \cdot \mathsf{I}_b$
B5	$(f \star f') \cdot (g \star g') = f \cdot g \star f' \cdot g'$
B6	$\mathsf{I}_a \star \mathsf{I}_b = \mathsf{I}_{a\star b}$
B7	${}^a\mathsf{X}^b \cdot {}^b\mathsf{X}^a = \mathsf{I}_{a\star b}$
B8	${}^a\mathsf{X}^{b\star c} = ({}^a\mathsf{X}^b \star \mathsf{I}_c) \cdot (\mathsf{I}_b \star {}^a\mathsf{X}^c)$
B9	$(f \star g) \cdot {}^c\mathsf{X}^d = {}^a\mathsf{X}^b \cdot (g \star f)$ for $f : a \to c,\quad g : b \to d$

II. Axioms for feedback

R1	$f \cdot (g \uparrow^c) \cdot h = ((f \star \mathsf{I}_c) \cdot g \cdot (h \star \mathsf{I}_c)) \uparrow^c$
R2	$f \star g \uparrow^c = (f \star g) \uparrow^c$
R3	$(f \cdot (\mathsf{I}_b \star g)) \uparrow^c = ((\mathsf{I}_a \star g) \cdot f) \uparrow^d$ for $f : a \star c \to b \star d,\quad g : d \to c$
R4	$f \uparrow^\epsilon = f$
R5	$(f \uparrow^b) \uparrow^a = f \uparrow^{a\star b}$
R6	$\mathsf{I}_a \uparrow^a = \mathsf{I}_\epsilon$
R7	${}^a\mathsf{X}^a \uparrow^a = \mathsf{I}_a$

III. Axioms for (angelic) branching constants, without feedback

A1a	$\wedge^a \cdot (\wedge^a \star \mathsf{I}_a) = \wedge^a \cdot (\mathsf{I}_a \star \wedge^a)$
A1b	$\wedge^a \cdot (\perp^a \star \mathsf{I}_a) = \mathsf{I}_a$
A1c	$(\vee_a \star \mathsf{I}_a) \cdot \vee_a = (\mathsf{I}_a \star \vee_a) \cdot \vee_a$
A1d	$(\top_a \star \mathsf{I}_a) \cdot \vee_a = \mathsf{I}_a$
A2a	$\wedge^a \cdot {}^a\mathsf{X}^a = \wedge^a$
A2b	${}^a\mathsf{X}^a \cdot \vee_a = \vee_a$
A3a	$\top_a \cdot \perp^a = \mathsf{I}_\epsilon$
A3b	$\top_a \cdot \wedge^a = \top_a \star \top_a$
A3c	$\vee_a \cdot \perp^a = \perp^a \star \perp^a$
A3d	$\vee_a \cdot \wedge^a =$ $(\wedge^a \star \wedge^a) \cdot (\mathsf{I}_a \star {}^a\mathsf{X}^a \star \mathsf{I}_a) \cdot (\vee_a \star \vee_a)$
A4	$\wedge^a \cdot \vee_a = \mathsf{I}_a$
A5a	$\perp^{a\star b} = \perp^a \star \perp^b$
A5b	$\wedge^{a\star b} = (\wedge^a \star \wedge^b) \cdot (\mathsf{I}_a \star {}^a\mathsf{X}^b \star \mathsf{I}_b)$
A5c	$\top_{a\star b} = \top_a \star \top_b$
A5d	$\vee_{a\star b} = (\mathsf{I}_a \star {}^b\mathsf{X}^a \star \mathsf{I}_b) \cdot (\vee_a \star \vee_b)$
[A5e	$\perp^\epsilon = \mathsf{I}_\epsilon$
A5f	$\wedge^\epsilon = \mathsf{I}_\epsilon$
A5g	$\top_\epsilon = \mathsf{I}_\epsilon$
A5h	$\vee_\epsilon = \mathsf{I}_\epsilon$]

IV. Feedback on (angelic) branching constants

R8	$\wedge^a \uparrow^a = \top_a$
R9	$\vee_a \uparrow^a = \perp^a$
R10	$[(\mathsf{I}_a \star \wedge^a) \cdot ({}^a\mathsf{X}^a \star \mathsf{I}_a) \cdot (\mathsf{I}_a \star \vee_a)] \uparrow^a = \mathsf{I}_a$

V. The strong axioms

Sa	$f \cdot \perp^b = \perp^a$
Sb	$f \cdot \wedge^b = \wedge^a \cdot (f \star f)$
Sc	$\top_a \cdot f = \top_b$
Sd	$\vee_a \cdot f = (f \star f) \cdot \vee_b$

VI. The enzymatic rule (E is a class of finite, abstract relations)

Enz_E: if for $f : a \star c \to b \star c$ and $g : a \star d \to b \star d$ there exists $y : c \to d$ in E such that $f \cdot (\mathsf{I}_b \star y) = (\mathsf{I}_a \star y) \cdot g$, then $f \uparrow^c = g \uparrow^d$

network algebra model, since from this point of view a deterministic automaton is still a nondeterministic network.

This axiomatization is closely related to that of regular algebras. An equational axiomatization is presented in [BÉ93a] using Esik's technique of replacing the enzymatic rule by an equation. A stronger result is presented in [Kro91] where two conjectures of Conway are proved using a different technique. More on regular algebras and related topics may be found in Chapter 8.

Part III

Algebraic theory of special networks

7. Flowchart schemes

> Reasoning is a bridge between young common sense and old common sense.
>
> Nichita Stănescu

In this chapter we give a brief presentation of the algebraic theory of flowchart schemes. While the chapter may be mainly read in an independent way, a full understanding will require some results included in the first part of the book.

After a short presentation of the classical structural "while programs" and of extended Elgot's representation, we describe our network algebra representation of flowchart programs. The main difference is that classically flowcharts were represented using iteration, while our presentation is based on feedback. Technically, the approaches are equivalent, provided the coalgebraic theory axioms are accepted.

A main topic of the chapter is Floyd–Hoare logic for program correctness. Roughly speaking, it is shown that Floyd–Hoare logic may be reduced to the equational logic of an appropriate algebraic structure. More concretely, it is shown that: (1) Floyd–Hoare logic assertions may be described by equations and, (2) it is proved that this logic is correct when it is interpreted in a BNA over a coalgebraic theory (Căzănescu–Ungureanu theory) that provides enough invariants.

We also describe certain duality results between AddRel (the additive relational model giving the standard semantics for flowchart schemes) and MultRel (the multiplicative relational network algebra model used to describe logical networks; these networks are interpreted as predicate transformers, on which Floyd–Hoare logic is based). We describe two such duality transformations, one based on Bainbridge duality, the other on weakest precondition technique.

Finally, a closer look to the relationship between feedback and iteration is included (a short description was already presented in Section 1.7). Several equivalent presentations for Căzănescu–Ungureanu theories are presented.

7.1 Structural programs

A brief presentation of while programs and of Elgot's extended setting is given here. More material is included in the "exercise" part of the section.

Structural programs via composition/if/while. Structural programs have been introduced in the 1960s and are very popular now. They are flowchart programs with one input and one exit built up from atomic statements, such as ***assignments*** and ***tests***, using ***sequential composition*** ";", IF_THEN_ELSE_ and WHILE_DO_. The name "while programs" is used for this class as well. Notice that the assignment blocks may be freely used, while the test blocks are allowed only in the restricted contexts provided by the IF and WHILE statements. Finally, a remark on notation: to simplify the presentation we describe the result in the one-sorted case, hence the monoid for interfaces is $\mathbb{N}$.

The while constructs are well known. They may be defined in terms of network algebra operations as follows (see Sections 1.1–1.5 for a brief presentation of network algebra setting):

$$F_1; F_2 = F_1 \cdot F_2$$
$$\text{IF } p \text{ THEN } F_1 \text{ ELSE } F_2 = p \cdot (F_1 \oplus F_2) {>\!\!\bullet}_1$$
$$\text{WHILE } p \text{ DO } F = \uparrow^1_{\oplus}({>\!\!\bullet}_1 \cdot p \cdot (F \oplus \mathsf{I}_1))$$

The atomic cells are: tests of the form $\boxed{p(y_1, \ldots, y_k)}$: $1 \to 2$ and assignments of the form $\boxed{y \leftarrow f(y_1, \ldots, y_k)}$: $1 \to 1$; here, p and f are terms built up using appropriate sets Σ and Π for function and test symbols, respectively.

The notation. A few explanations on notation are necessary. The set of symbols $\{\oplus, \cdot, \uparrow_{\oplus}, \mathsf{I}, \mathsf{X}, {\bullet\!\!<}_k, {}_k{>\!\!\bullet}\}$ denoting network algebra operations and constants are used here to replace the standard ones $\{\star, \cdot, \uparrow, \mathsf{I}, \mathsf{X}, \wedge_k, \vee^k\}$. The former are to be used for the current additive setting, while the latter were intended for a more abstract setting. ◊

The main result is given by the following expressiveness theorem, cf., [BJ66]:

Theorem 7.1. *Each program is equivalent to a structural program, provided* additional memory *can be used.*

PROOF: (hint) A program with arbitrary GOTO statements may be transformed into a structural program using additional memory to keep track of the labels and the control flow in the original program. □

Remark. This result, due to Böhm and Jacopini, had a rather broad impact, but it was slightly misunderstood. Thereafter, in less mathematical media, the slogan was to liberate programs from GOTO statements. However, one cannot completely do this; one only can *restrict* itself to the use of the GOTO statements in the forms provided by the while programs. The essence of (imperative) programs is the additive relational model. If no form of GOTO is admitted, then no tests can be used, too; and programs without tests will not be very useful. ◊

As a corollary, all functions that can be computed by flowchart programs may be computed by structural programs, as well. Actually, the model provided by flowchart programs is universal, i.e., all computable functions are representable by flowchart programs, see [DW81]. However, if one considers a *fixed* memory structure, then, by well known results of Ashkroft, Knuth, Kosaraju, Manna, etc., there are flowchart programs which are not equivalent with while programs.

Structural programs via composition/tupling/iteration. An important contribution to the development of the algebraic theory of flowchart schemes is due to Elgot. In 1975–76 Elgot published some seminal papers, e.g., [Elg75] and [Elg76b]. The class of "structural" programs is enlarged here to contain flowchart programs with one input, but arbitrary many outputs. The operations used by Elgot are *sequential composition* $\cdot$, *tupling* $\langle \_, \_ \rangle$ and *(scalar) iteration* †, using the same starting blocks (assignments and tests). The tupling and iteration operations are defined in terms of network algebra operations as follows (in the second case, F_1 and F_2 are supposed to have the same output type n; next, for the scalar iteration operation $F : 1 \to n$ the left-hand-side feedback is used):

$$\langle F_1, F_2 \rangle = (F_1 \oplus F_2) \succ\!\!\bullet_n$$
$$F^\dagger = \uparrow^1_\oplus (\succ\!\!\bullet_1 \cdot F)$$

Finally, the atomic cells are the same: tests $p : 1 \to 2$ and assignments $a : 1 \to 1$. Notice that both atomic assignment and test cells are freely used to generate programs in this setting.

This new algebraic setting is more expressive. Indeed, Elgot proved that

Theorem 7.2. *Each flowchart program is step-by-step equivalent with a program represented with sequential composition, tupling and (scalar) iteration. (No additional memory is used.)* □

However, the operations of Elgot are not expressive enough to *represent* programs themselves. Only their behavioural classes can be represented.

♡ The class of programs represented by such expressions have the syntactic property that all paths entering into a loop meet the loop in a unique point. E.g., $p[\uparrow^1_\oplus ((\succ\!\!\bullet_1 a \oplus \mathsf{I}_1) \succ\!\!\bullet_1 q)] : 1 \to 1$ cannot be represented ($a : 1 \to 1$; $p, q : 1 \to 2$). ◇

In order to represent all programs Elgot used a "vectorial" iteration $\uparrow^m_\oplus (\succ\!\!\bullet_m \cdot F)$, for $F : m \to m + n$ which cannot be obtained by a simple repetition of scalar iteration.

♡ Indeed, in order to relate vectorial and scalar iterations one has to use the complicated "pairing axiom" of Căzănescu–Ungureanu theories. For more on this point, see the comments in Section 1.7 on successive looping, or the results presented in the last section of the current chapter. ◇

Exercises and Problems (sec. 7.1)

1. Prove Theorem 7.1, i.e., any flowchart scheme may be transformed into an equivalent while scheme, provided adding memory variables is allowed. (12)

2. Consider the flowchart scheme (the blocks of type $p(y)$ are tests : $1 \to 2$, while the other blocks are assignments : $1 \to 1$)

   $$F = [\succ\!\bullet_1 \cdot \boxed{p_1(y)} \, (\boxed{y \leftarrow f_1(y)} \oplus \boxed{y \leftarrow f_2(y)} \cdot \boxed{p_2(y)}) \cdot (\succ\!\bullet_1 \oplus \boxed{y \leftarrow f_3(y)})] \uparrow^1_\oplus$$

   Find a while-program equivalent to F. Show that there is no while-program isomorphic to F. (12)

3. As before, find a while-program equivalent to (you may use extra-variables)

   $$F = [\succ\!\bullet_1 \cdot (\succ\!\bullet_1 \cdot \boxed{p_1(y)} \, (\mathsf{I}_1 \oplus \boxed{y \leftarrow f_1(y)})) \uparrow^1_\oplus \cdot \boxed{p_2(y)} \, (\boxed{y \leftarrow f_5(y)} \oplus \boxed{y \leftarrow f_2(y)} \cdot$$
   $$(\succ\!\bullet_1 \cdot \boxed{p_3(y)} \, (\mathsf{I}_1 \oplus \boxed{y \leftarrow f_3(y)})) \uparrow^1_\oplus \cdot \boxed{p_4(y)} \, (\boxed{y \leftarrow f_6(y)} \oplus \boxed{y \leftarrow f_4(y)}))] \uparrow^1_\oplus \cdot \succ\!\bullet_1$$

   (15)

4. Find a characterization of the flowchart programs which may be represented by Elgot operations up to an isomorphism. (12)

5. No flowchart program is equivalent to the recursive program: $z = F(a)$, where

   $F(y) \Leftarrow$ if $p(y)$ then $f(y)$ else $h(F(g_1(y)), F(g_2(y)))$

   Hence, recursion is more powerful than iteration, cf. Paterson and Hewitt. (*Hint*: any "equivalent flowchart" will require an infinite number of variables.) (15)

7.2 Flowchart representation

Our current network algebra representation of flowchart programs is briefly recalled here.

Structural programs via sum/composition/feedback. The most general setting is that of a program with many inputs and many exits. In this setting all flowchart programs may be represented starting with *atoms* (assignments and tests) and *connecting relations* and using the operations of *sum, sequential composition* and *(scalar) feedback.*

They are formally defined as *flownomial expressions*, i.e.,

$$F ::= F_1 \cdot F_2 \mid F_1 \oplus F_2 \mid F \uparrow^1_\oplus \mid p \mid a \mid \mathsf{I}_1 \mid {}^1_1\mathsf{X} \mid \succ\!\bullet_1 \mid \bullet\!\!-_1 \mid {}_1\!\!-\!\bullet$$

where the atomic cells are the same: tests $p : 1 \to 2$ and assignments $a : 1 \to 1$.

♡ Whether the feedback is allowed to be used in an endless cycle is a more delicate question; if this is the case, then $-\!\bullet$ has to be used to give meaning to stupid endless programs, e.g., ${}_1\!\!-\!\bullet = \succ\!\bullet_1 \uparrow^1_\oplus$. ◊

Theorem 7.3. *All flowchart programs may be represented by flownomial expressions starting from atomic blocks and connecting relations.* □

In the case of regular expressions each finite relation $\rho \subseteq [m] \times [n]$ is represented as a 0-1 matrix. However, in the more general symocat/graph-isomorphism setting the matrix building operator is not available. Hence the whole machinery of relation representation within symocats with branching constants has to be used. This machinery is based on the following constants: I_n, ${}^{n}_{m}\mathsf{X}$, ${}_{k}\!\succ\!\!\bullet_{n}$, ${}_{n}\!\bullet\!\!\prec_{k}$. (These constants model a trivial control flow; the flow of control is redirected, but the program state is not used at all; it is neither tested, nor modified.)

Semantic model. The standard semantic model for flowchart programs AddRel were already described in Sections 1.6 and in Chapters 2–6. In the notation of the current chapter it is $(\mathsf{AddRel}(D), \oplus, \cdot, \uparrow_{\oplus}, \mathsf{I}_n, {}^{n}_{m}\mathsf{X}, {}_{k}\!\succ\!\!\bullet_{n}, {}_{n}\!\bullet\!\!\prec_{k})$.

Example 7.4. A while program for $y = x^n$ may be easily described. J. Reynolds has shown that $y = x^n$ may be more efficiently computed by the following more general flowchart program, which has no isomorphic representation in the class of while programs:

$$init \cdot zero? \cdot (\mathsf{I}_1 \oplus \{{}_{2}\!\succ\!\!\bullet_{1} \cdot odd? \cdot [(dec \cdot zero? \oplus {}_{2}\!\succ\!\!\bullet_{1} \cdot div) \cdot (\mathsf{I}_1 \oplus {}^{1}_{1}\mathsf{X})] \uparrow\} \uparrow) \cdot {}_{2}\!\succ\!\!\bullet_{1}$$

It may be represented by "normal form" expressions, too. A normal form may be

$b1:$	*init*;	GOTO 2
$2:$	*zero?*;	GOTO 9, 3
$3,4:$	*odd?*;	GOTO 5, 6
$5:$	*dec*;	GOTO 8
$6,7:$	*div*;	GOTO 4
$8:$	*zero?*;	GOTO $e9$, 7
$e9:$		

or in a more conventional programming notation

[BEGIN 1:]	$y := 1;\ z := x;\ k := n;$ GOTO 2;
[2:]	IF $k = 0$ THEN GOTO 9 ELSE GOTO 3;
[3,4:]	IF '*k is odd*' THEN GOTO 5 ELSE GOTO 6;
[5:]	$y := y * z;\ k := k - 1;$ GOTO 8;
[6,7:]	$z := z * z;\ k := k/2;$ GOTO 4;
[8:]	IF '*k is zero*' THEN GOTO 9 ELSE GOTO 7;
[9:]	EXIT 1

The presence of multiple labels pointing to a statement requires some explanation. Actually, the current GOTO statement is somehow more restrictive, namely only *one* GOTO may point to a given statement label. Therefore, to provide multiple accesses to a statement one has to write explicitly a list with more such labels.

Exercises and Problems (sec. 7.2)

1. Give an interpretation for the above flowchart program into an appropriate $\mathsf{AddRel}(D)$ model and verify that it computes the function $y = x^n$. (08)

7.3 Floyd–Hoare logic

Floyd–Hoare logic was originally developed for while programs. We describe it here, but in an extended context that applies to all flowchart programs. Along the way, a presentation of the algebra of guards in a flowchart style is given, to replace the usual boolean algebra setting used for while programs.

Partial correctness. The proper algebraic setting to develop Floyd–Hoare logic is a BNA over an algebraic theory. The stronger algebraic theory rules are needed to model tests and programs, but no stronger than the BNA axioms for feedback are to be used. E.g., the enzymatic rule is not used here.

Let (1) Q be a set modeling program states, (2) $f : Q \to Q$ the partially defined function corresponding to a deterministic program, and (3) α, β two predicates on Q, represented as (total) functions $\alpha, \beta : Q \to \{\mathrm{T}, \mathrm{F}\}$. (As expected, T stands for *true* and F for *false*.) A *program correctness assertion* is a statement

$$\{\alpha\}f\{\beta\}$$

with the following intuitive meaning

$$\forall q \in Q \;\; [\alpha(q) = \mathrm{T} \;\&\; f(q) \text{ is defined} \quad \Rightarrow \quad \beta(f(q)) = \mathrm{T}]$$

In what follows we try to replace these assertion statements by *equations*. To this end we need an axiomatization of tests/guards in the style of flowchart schemes.

Guards. Let T be an algebraic theory. Define $\mathrm{T} := \mathsf{I}_1 \oplus \bullet\!\!-_1$ and $\mathrm{F} := \bullet\!\!-_1 \oplus \mathsf{I}_1$. (The "true" constants T is the direct connection between the input and the first output, while the "false" constants F has a similar property, but, in this case, the input is connected to the second output.)

A class $\Gamma \subseteq T(1,2)$, containing $\{\mathrm{T}, \mathrm{F}\}$ is *boolean* if
(1) every $\alpha \in \Gamma$ is *idempotent*, i.e., $\alpha \cdot {>\!\!\bullet}_1 = \mathsf{I}_1$;
(2) every $\alpha \in \Gamma$ is *diagonal*, i.e., $\alpha \cdot (\alpha \oplus \alpha) = \alpha \cdot (\mathrm{T} \oplus \mathrm{F})$;
(3) every two $\alpha, \beta \in \Gamma$ *commute*, i.e., $\alpha \cdot (\beta \oplus \beta) = \beta \cdot (\alpha \oplus \alpha) \cdot (\mathsf{I}_1 \oplus {}^1_1\mathsf{X} \oplus \mathsf{I}_1)$;
(4) Γ is *closed* to special compositions, i.e., $\alpha, \beta, \gamma \in \Gamma \;\Rightarrow\; \alpha(\beta \oplus \gamma) \cdot {>\!\!\bullet}_2 \in \Gamma$.

Meaning of the axioms. The first 3 axioms may be easily understood starting from the basic observation that a guard simply tests the current state for a certain logical property, *but this test has no side-effects*, i.e., the state is not changed after the test. Then, the axioms describe what happens when the branchings of a test are identified, or when a test is repeated, or when two successive tests are permuted. The last property (4) shows how more complex guards may be produced. (Recall that ${>\!\!\bullet}_2 : 4 \to 2$ maps 1,3 in 1 and 2,4 in 2.) ◇

One can easily see that: (1) $\{\mathrm{T}, \mathrm{F}\}$ is a boolean class; and (2) a boolean class Γ may be extended adding idempotent and diagonal morphisms which commutes with all morphisms in Γ.

A boolean class $\Gamma \subseteq T(1,2)$ may be seen as a *boolean algebra*. Indeed, for $\alpha, \beta \in \Gamma$ define

$$\begin{aligned} \alpha \text{ AND } \beta &= \alpha \cdot (\beta \oplus \mathrm{F}) \cdot \succ\!\bullet_2 \\ \alpha \text{ OR } \beta &= \alpha \cdot (\mathrm{T} \oplus \beta) \cdot \succ\!\bullet_2 \\ \text{NOT } \alpha &= \alpha \cdot {}^1_1\!\mathsf{X} \end{aligned}$$

. Then, if T is an algebraic theory and Γ is a boolean class in T, then $(\Gamma, \text{AND}, \text{OR}, \text{NOT}, \mathrm{T}, \mathrm{F})$ is a boolean algebra. Actually, every boolean algebra is isomorphic to a boolean class as above; see this section's exercises.

Let $\alpha_1, \ldots, \alpha_n \in \Gamma$, where $\Gamma \subseteq T(1,2)$ is a boolean class.

(1) An *n-guard* is a morphism

$$[\alpha_1, \ldots, \alpha_n] = (\alpha_1 \oplus \ldots \oplus \alpha_n) \cdot \varphi_{n,2}$$

where $\varphi_{n,2} \in \mathbb{R}\mathrm{el}(2n, 2n)$ is the bijection which maps $1, 2, 3, 4, \ldots, 2n-1, 2n$ into $1, n+1, 2, n+2, \ldots, n, n+n$, respectively.

ⓘ This $\varphi_{n,2}$ is a particular instance of the general constant $\varphi_{n,m}$ introduced in Chapter 3 used to naturally rearrange m groups of n elements in n groups of m elements; the effect is that all "true" branches of the guards are collected to the left. ⓘ

By $\Gamma_n \subseteq T(n, 2n)$ we denote the set of n-guards in T.

(2) For $\alpha = [\alpha_1, \ldots, \alpha_n] \in \Gamma_n$ and $\beta = [\beta_1, \ldots, \beta_p] \in \Gamma_p$ we denote by $[\alpha, \beta] \in \Gamma_{n+p}$ the *pair* guard

$$[\alpha_1, \ldots, \alpha_n, \beta_1, \ldots, \beta_p]$$

T_m denotes $[\mathrm{T}, \overset{m}{\ldots}, \mathrm{T}]$.

(3) Let $\alpha \in \Gamma_n$, $\beta \in \Gamma_p$ and $f \in T(n,p)$. The *validity* of a partial correctness assertion $\{\alpha\}f\{\beta\}$ in (T, Γ) is modeled by an equality:

$$\{\alpha\}f\{\beta\} \text{ is valid iff } \alpha \cdot (f \oplus \mathsf{I}_n) \cdot (\beta \oplus \mathsf{I}_n) = \alpha \cdot (f \oplus \mathsf{I}_n) \cdot (\mathrm{T}_p \oplus \mathsf{I}_n)$$

This may be read as follows: the assertion $\{\alpha\}f\{\beta\}$ is true if, after α and f, on the corresponding point in the program testing either β or T will lead to the same result. See the first part of the proof of Theorem 7.5 for a graphical illustration of this important definition.

(4) Finally, $\alpha \leq \beta$ means $\alpha(\beta \oplus \mathsf{I}_n) = \alpha(\mathrm{T}_n \oplus \mathsf{I}_n)$.

Floyd–Hoare logic deduction rules. Floyd–Hoare logic has 4 deduction rules:

(1) **Composition rule:** Let $m \xrightarrow{f} n \xrightarrow{g} p$ be programs and $m \xrightarrow{\alpha} 2m$, $n \xrightarrow{\beta} 2n$, $p \xrightarrow{\gamma} 2p$ guards. Then

$$\{\alpha\}f\{\beta\}, \ \{\beta\}g\{\gamma\} \vdash \{\alpha\}f \cdot g\{\gamma\}$$

" $\vdash$ " is to be read as "the left hand side premises imply the right-hand-side conclusions."

(2) **Sum rule:** Let $m \xrightarrow{f}, p \xrightarrow{g} q$ be programs and $m \xrightarrow{\alpha} 2m,\ n \xrightarrow{\beta} 2n,\ p \xrightarrow{\gamma} 2p,\ q \xrightarrow{\delta} 2q$ guards. Then

$$\{\alpha\}f\{\beta\},\ \ \{\gamma\}g\{\delta\} \vdash \{[\alpha,\gamma]\}(f \oplus g)\{[\beta,\delta]\}$$

(3) **Feedback rule:** Let $m+p \xrightarrow{f} n+p$ be a program and $m \xrightarrow{\alpha} 2m,\ n \xrightarrow{\beta} 2n,\ p \xrightarrow{\delta} 2p$ guards. Then

$$\{[\alpha,\delta]\}f\{[\beta,\delta]\} \vdash \{\alpha\}(f\uparrow^p)\{\beta\}$$

(4) **Rule for constants:** Let $\rho \in \mathbb{R}\mathrm{el}(m,n)$ be a representation for a trivial program having only GOTO statements and $\alpha = [\alpha_1,\ldots,\alpha_m] : m \to 2m,\ \ \beta = [\beta_1,\ldots,\beta_n] : n \to 2n$ guards. Then

$$\forall (i,j) \in \rho :\ \alpha_i \le \beta_j \vdash \{\alpha\}\rho\{\beta\}$$

Program expressions. These are simple flownomial expressions. For the sake of an independent reading of this chapter we briefly present their construction here. Let $X = \{X(m,n) | m,n \in \mathbb{N}\}$ be a doubly ranked family of sets; an element $x \in X(m,n)$ represents an atomic flowchart program with m inputs and n exits. Suppose, moreover, C is a class of particular connecting relations containing $\mathsf{I}_m, {}^n_m\mathsf{X}$ and some constants of type ${}_k{>\!\!\bullet}_n (k \ge 0)$, or ${}_n{\bullet\!\!<}_0$ and closed under sum and composition. Notice that no ${}_n{\bullet\!\!<}_k$ with $k \ge 2$ is used, hence we have "deterministic" programs.

The *program expressions over* X *and* C are inductively defined by:

(1) a variable $x \in X(m,n)$ is a program expression of type $: m \to n$;

(2) a constant in C is a program expression of an appropriate type, namely

$$\mathsf{I}_n : n \to n,\ \ {}^n_m\mathsf{X} : m+n \to n+m,\ \ {}_k{>\!\!\bullet}_n : kn \to n,\ \ {}_n{\bullet\!\!<}_0 : n \to 0$$

(3) if $E^1 : m \to n,\ E^2 : p \to q,\ E^3 : n \to p,\ E : m+p \to n+p$ are program expressions of the mentioned type, then

$$E^1 \oplus E^2 : m+p \to n+q,\ \ E^1 \cdot E^3 : m \to p,\ \ E\uparrow^p : m \to n$$

are program expressions of the mentioned type;

(4) all program expressions are obtained by a finite number of applications of the rules in (1)–(3).

In the case of usual flowchart programs $X(m,n)$ is empty, except for $X(1,1)$ and $X(1,2)$. Then, $X(1,1) = \Sigma$ is used to specify the "assignment" blocks, while $X(1,2) = \Pi$ is used for the "test" blocks. We simply describe this situation by stating that $X = \Sigma \cup \Pi$.

Guard expressions. The guards are special program expressions over X and C obtained using the atomic tests $\Pi \subseteq X(1,2)$, only. They are generated as follows:

(1) *1-guard expressions over* Π are: T, F, the elements in Π, and what can be further obtained using the composition rule $\alpha \cdot (\beta \oplus \gamma) \cdot \succ\!\!\bullet_2$, from 1-guard expressions α, β, γ;

(2) *n-guard expressions over* Π are defined by: l_0 is 0-guard expression; 1-guard expressions which were already defined; n-guard expressions are $[\alpha_1, \ldots, \alpha_n] = (\alpha_1, \ldots, \alpha_n) \cdot \varphi_{n,2}$, where $\alpha_1, \ldots, \alpha_n$ are 1-guard expressions and $\varphi_{n,2} : 2n \to 2n$ is the described bijection (it maps $1, 2, \ldots, 2n-1, 2n$ into $1, n+1, \ldots, n, 2n$, respectively).

Partial correctness expressions. Let the atomic blocks be represented by $X = \Sigma \cup \Pi$ and the program connections by C. A *partial correctness expression* (over Σ, Π, C) is an expression

$$\{\alpha\}\ f\ \{\beta\}$$

where $f : m \to n$ is a program expression over X and C and $\alpha : m \to 2m$, $\beta : n \to 2n$ are guard expressions over Π.

Interpreting partial correctness expressions. We start with a Căzănescu–Ungureanu theory T and certain interpretations $\phi_\Sigma : \Sigma \to T(1,1)$ and $\phi_\Pi : \Pi \to T(1,2)$ for atomic blocks (assignments and tests). These interpretations extend to a unique morphism from program expressions to T, i.e., $\phi : \mathsf{FlowExp}[\Sigma \cup \Pi, C](m,n) \to T(m,n)$, which commutes with the specified NA operations and constants. An interpretation ϕ is called *admissible* if $\phi(\Pi)$ generates a boolean class in T (by itself, it satisfies the guard axioms 1–3). Finally, the truth value of a partial correctness formula $\{\alpha\}f\{\beta\}$ is defined by

$$T \models \{\alpha\}f\{\beta\} \text{ iff the equation modeling } \{\phi(\alpha)\}\phi(f)\{\phi(\beta)\} \text{ holds in } T$$

Let $Val(T, \phi, \Sigma \cup \Pi)$ denotes this class of (T, ϕ)-valid formulae.

Axiomatic Floyd–Hoare calculus. Here we define the class of derived Floyd–Hoare logic formulas $FH(T, \phi, \Sigma \cup \Pi)$. As the notation suggests, the derivation depends on T, ϕ; more precisely, the starting axioms depend on these ingredients.

Axioms: (1) $\{\alpha\}\sigma\{\beta\}$ is axiom if $\{\phi(\alpha)\}\phi(\sigma)\{\phi(\beta)\}$ is valid in T (this mimics the semantics properties of the atomic operation σ);

(2) $\{\alpha\}\rho\{\beta\}$ is axiom, whenever $\alpha = [\alpha_1, \ldots, \alpha_m], \rho \in C(m,n), \beta = [\beta_1, \ldots, \beta_n]$ are such that for $(i,j) \in \rho$, $\phi(\alpha_i) \leq \phi(\beta_j)$ is valid in T (the guard implication in T is lifted at the syntactical level).

Derived expressions: The derived expressions are obtained as usual from axioms by a finite number of applications of Floyd–Hoare rules (for sum, composition, and feedback).

EXERCISES AND PROBLEMS (SEC. 7.3)

1. Show that an arrow a of a boolean class satisfies $a(a \oplus \mathsf{I}_1) = a(\mathrm{T} \oplus \mathsf{I}_1)$ and $a(\mathsf{I}_1 \oplus a) = a(\mathsf{I}_1 \oplus \mathrm{T})$. Show that a class of arrows satisfying the guard axioms 1–3 extends to a boolean class (take the closure to the composition operation described in axiom 4 of that definition). (08)
2. Let T be an algebraic theory. Show that $(\Gamma, \mathrm{AND}, \mathrm{OR}, \mathrm{NOT}, \mathrm{T}, \mathrm{F})$ is a boolean algebra, whenever Γ is a boolean class in T. (08)
3. Conversely, let $B = (B, +, \cdot, ', 1, 0)$ be a boolean algebra and $T = Pow(B)$ be the algebraic theory of functions over B, i.e., $Pow(B)(m,n) = \{f : B^n \to B^m\}$. Take the class $\Gamma(B) = \{if_b :\ b \in B\}$, where $if_b : B^2 \to B$ is the function $if_b(x,y) = (b \cdot x) + (b' \cdot y)$. Then $\Gamma(B)$ is a boolean class whose corresponding boolean algebra is isomorphic with B. (15)
4. Show that $\Gamma(B)$ above is a maximal boolean class in $Pow(B)$. (10)

7.4 Soundness of Floyd–Hoare logic

In this section we prove that Floyd–Hoare logic is sound when it is interpreted in a Căzănescu–Ungureanu theory.

For the proof of the soundness result we need a result of Chapter 5, namely Theorem 5.13; it states that the elimination of a non-accessible part of a flowchart is valid in Căzănescu–Ungureanu theories.

Theorem 7.5. *Floyd–Hoare logic is sound when it is interpreted in a Căzănescu–Ungureanu theory.*

PROOF: The Floyd–Hoare logic was reduced to an equational logic, hence we only have to prove that Floyd–Hoare logic rules are sound when they are interpreted in a Căzănescu–Ungureanu theory. We present here a graphical proof of the theorem.

Graphical proof. The main reason for choosing this graphical presentation of the proof is the complexity of the proof for the feedback rule. While possible, a fully formal proof will hide the, finally simple, underlying ideas of the proof. In teaching this result, I found that such a proof is more broadly understood by the students. The price to be paid was that sometimes the students have some difficulties to translate this reasoning into a fully formal proof. ◊

We start with the graphical description of the correctness assertions.

(1)

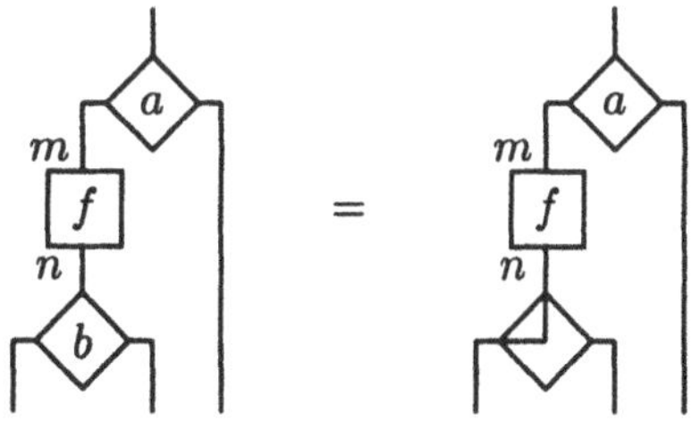

A correctness assertion $C = \{a\}f\{b\}$ is graphically represented as in this picture: if a is true, then b is true after f, hence cell b may be replaced by the corresponding "true" cell.

Summation: The check of the validity for the summation rule is simple and it is left for the reader.

Composition: For the composition rule the proof is described by the following transformations.

(2)

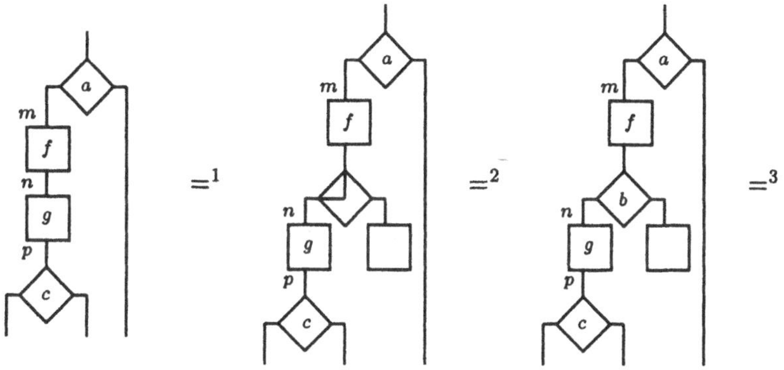

(3)

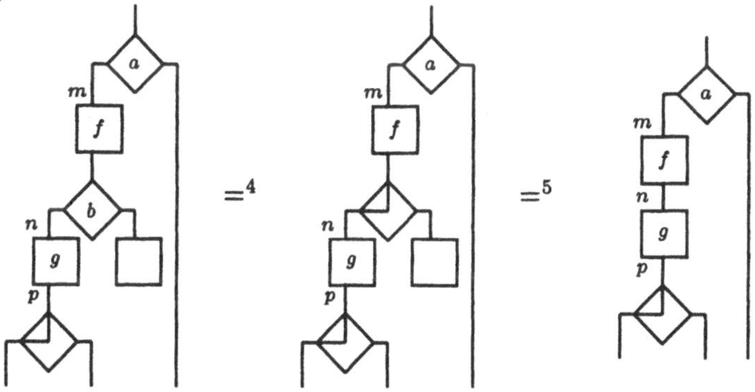

A few explanations may be useful. The hypothesis is that both assertions $C_1 = \{a\}f\{b\}$ and $C_2 = \{b\}g\{c\}$ are valid, hence one may draw pictures as in (1) above to represent their validity. Then,

— in $=^1$, a simple graph-isomorphic transformation is used;
— in $=^2$, the assertion C_1 is applied from right to left;
— in $=^3$ and $=^4$, one uses in turn C_2 and C_1 from left to right;
— finally, a graph-isomorphic transformation $=^5$, similar to the starting one, completes the proof.

Feedback: The verification of the soundness for the feedback rule is more tricky. Suppose that $C = \{a \oplus d\} f \{c \oplus d\}$ holds, hence

(4)

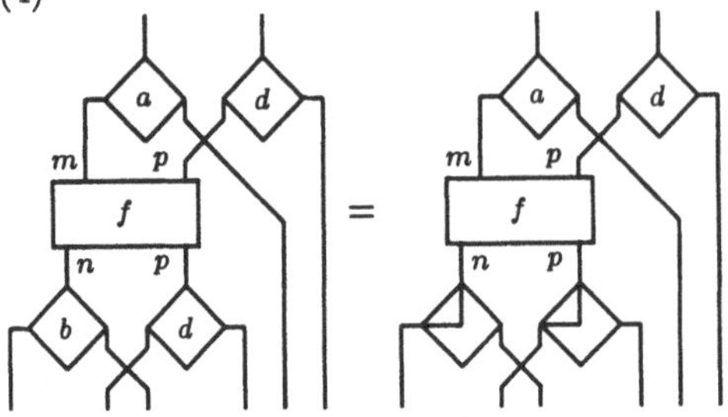

The proof is illustrated in the following pictures. It consists of 8 steps.

(5)

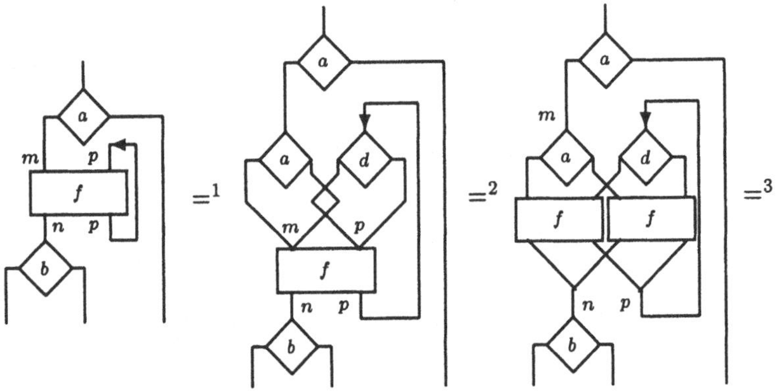

— in the 1st step one uses the fact that guards a and d are idempotents, hence they may be introduced to replace the identity before f as in the figure;

— in the 2nd step one applies the strong $a\gamma$-rule to duplicate the cell f (this rule is valid in a coalgebraic, hence Căzănescu–Ungureanu, theory);

— the 3rd step contains a simple graph-isomorphic transformation;

(6)

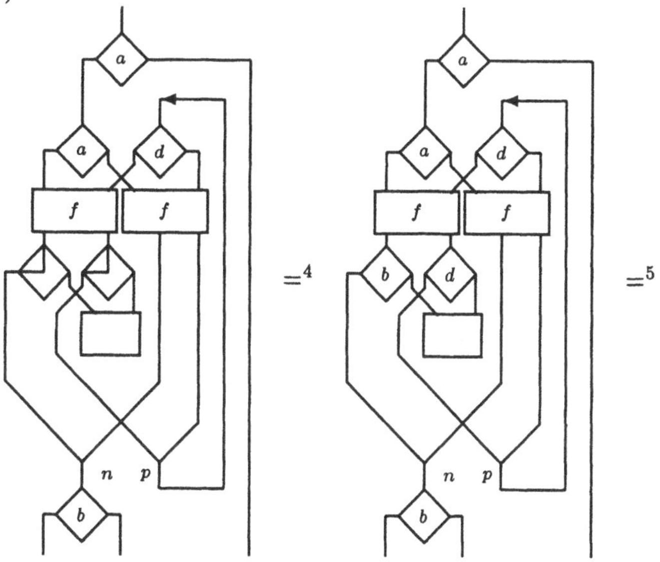

— the hypothesis C is used in the 4th step;
— a new graph-isomorphic transformation is used in the 5th step shifting the top guard d along the feedback;

(7)

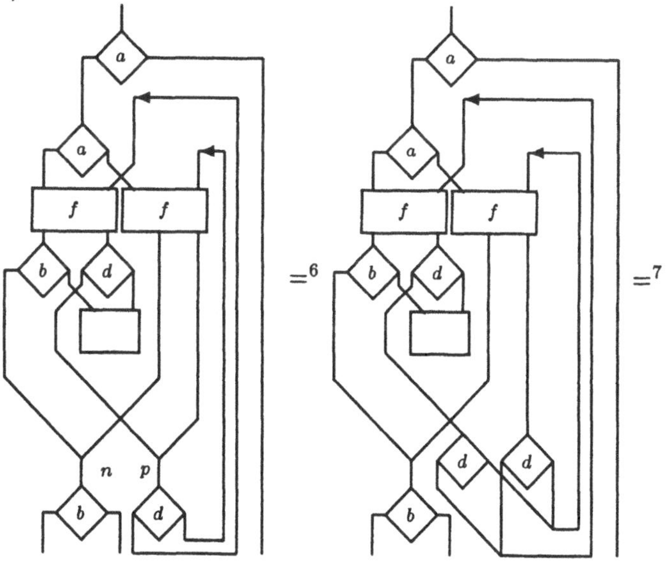

— in the 6th step one uses the strong $a\gamma$-rule to duplicate the bottom guard d;
— in the 7th step one uses the fact that guards a and d are diagonal;

(8)

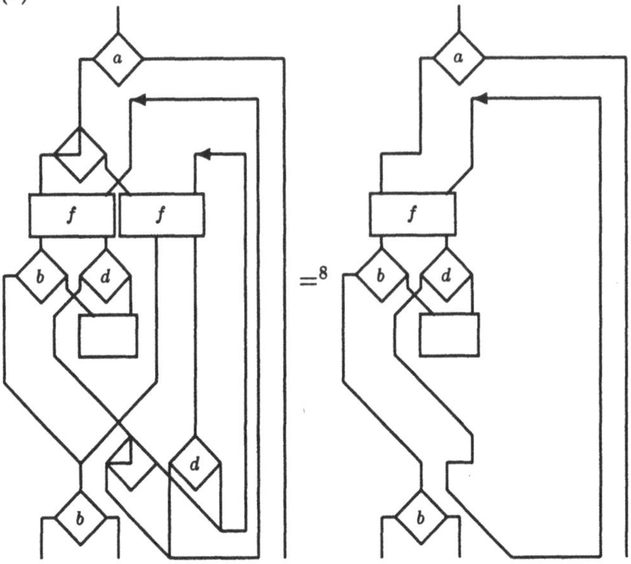

— finally, in the 8th step one uses the property that the inaccessible cells may be eliminated (in the beginning of this section we emphasized that this transformation is valid in a Căzănescu–Ungureanu theory) to get the final picture.

What was the purpose of this long transformation? We see that in the final flowchart one may replace the bottom b by the true T cell (indeed, b is diagonal). Now, if one starts the long transformation we just made with T instead of b, then the same result is obtained. Hence the feedback rule is valid.

Finally, the rule for the constants reduces to a simple transformation in coalgebraic theories. (It is based on the commutation property described in Proposition 4.2.) This completes the proof. □

Exercises and Problems (sec. 7.4)

1. Give a formal proof that Floyd–Hoare logic rules are valid when interpreted in Căzănescu–Ungureanu theories. (15)

7.5 Completeness of Floyd–Hoare logic

The result of this section is that Floyd–Hoare logic is complete when it is interpreted in a Căzănescu–Ungureanu theory, provided this theory contain invariants for the underlying programs. Moreover, this latter condition is necessary for the Floyd–Hoare logic be complete.

Let T be a Căzănescu–Ungureanu theory, Γ a collection of guards in T, and T_0 a collection of programs in T.

We say (T,Γ,T_0) *admits invariants* if for every $f \in T_0(m+p,n+p)$, $\alpha \in \Gamma(m,2m)$, $\beta \in \Gamma(n,2n)$

$$(T \models \{\alpha\}f \uparrow^p_\oplus \{\beta\}) \Rightarrow (\exists \delta \in \Gamma(p,2p).\ T \models \{[\alpha,\delta]\}f\{[\beta,\delta]\})$$

The name is self-explanatory: whenever an assertion about the feedback is valid, an invariant for the feedback loop may be found. A similar verbal translation may be given for the interpolation property to follow.

Composition may be reduced to sum, feedback, and transpositions, e.g., $f \cdot g = [(f \oplus g) \cdot {}^p_n\mathsf{X}] \uparrow^n_\oplus$ for $m \xrightarrow{f} n \xrightarrow{g} p$. Now, if (T,Γ,T_0) admits invariants, then (T,Γ,T_0) has also the *interpolation property*, namely for $f \in T_0(m,n), g \in T_0(n,p), \alpha \in \Gamma(m,2m), \beta \in \Gamma(p,2p)$ the following implication is valid

$$(T \models \{\alpha\}f \cdot g\{\beta\}) \Rightarrow (\exists \delta \in \Gamma(n,2n).\ T \models \{\alpha\}f\{\delta\} \wedge T \models \{\delta\}g\{\beta\})$$

The completeness theorem may be stated as follows:

Theorem 7.6. *If (T,Γ,T_0) admits invariants, then*

$$Val(T,\phi,\Sigma \cup \Pi) \subseteq FH(T,\phi,\Sigma \cup \Pi)$$

where Γ (resp. T_0) is the image via ϕ of the Π-guard expressions (resp. program expressions over $\Sigma \cup \Pi$) in T.

PROOF: We show that

if $\{\alpha\}f\{\beta\}$ is valid in $(T,\phi,\Sigma \cup \Pi)$, then
$\{\alpha\}f\{\beta\}$ is deductible in $FH(T,\phi,\Sigma \cup \Pi)$

by structural induction on f, as follows:

(1) If f is an *atomic block or a connecting relation*, then this actually is an axiom;

(2) *Sum*: Let f be a sum $f = f' \oplus f''$, with $m \xrightarrow{f'} n$, $p \xrightarrow{f''} q$. Suppose that $(T,\phi,\Sigma \cup \Pi) \models \{[\alpha,\gamma]\}f' \oplus f''\{[\beta,\delta]\}$.

Write the above assertion as an equality. Then, by a left-side composition with $\mathsf{I}_m \oplus \bullet\!\!-_p$ and elimination of certain non-accessible parts (a valid fact in Căzănescu–Ungureanu theories), one gets an equality showing that $(T,\phi,\Sigma \cup \Pi) \models \{\alpha\}f'\{\beta\}$. Similarly, using $\bullet\!\!-_m \oplus \mathsf{I}_p$, one gets $(T,\phi,\Sigma \cup \Pi) \models \{\gamma\}f''\{\delta\}$. Next, by the inductive hypothesis, one gets $\{\alpha\}f'\{\beta\}, \{\gamma\}f''\{\delta\} \in FH(T,\phi,\Sigma \cup \Pi)$. Finally, using the Floyd–Hoare rule for sum one gets $\{[\alpha,\gamma]\}f' \oplus f''\{[\beta,\delta]\} \in FH(T,\phi,\Sigma \cup \Pi)$.

(3) *Composition:* This case easily follows applying the interpolation property, the inductive hypothesis, and the Floyd–Hoare rule for composition.

(4) *Feedback:* This case is similar, but now the existence of invariants and the Floyd–Hoare rule for feedback are used. □

♡ Notice that this is a *relative completeness theorem* in the following sense: the semantical properties of the atomic statements (assignments and tests) are lifted to the syntactic level as axioms for the Floyd–Hoare logic. ◇

Conclusions. We collect here the results on Floyd–Hoare logic correctness problems obtained so far.

Theorem 7.7. *$Val(T,\phi,\Sigma\cup\Pi) = FH(T,\phi,\Sigma\cup\Pi)$ iff (T,Γ,T_0) admits invariants, where Γ is the image via ϕ of the guard expressions over Π and T_0 is the image via ϕ of the program expressions over X.*

PROOF: For the right-to-left implication, one simply uses both the soundness and completeness theorems. For the opposite left-to-right implication, notice that if an assertion formula is valid, then it is derivable; from such derivations one picks up the required invariants. □

To conclude: Floyd–Hoare logic has been reduced to an equational logic in flowchart theories. Floyd–Hoare logic is sound when it is interpreted in a Căzănescu–Ungureanu theory T. If the semantic theory T admits invariants, then Floyd–Hoare logic is also complete with respect to T.

7.6 Duality: II. Control-Space

In this section we describe certain duality results between control- and space-based models. Roughly speaking, we show how the additive relational model AddRel for flowcharts (a control-flow model) and the multiplicative relational model MultRel used for logical networks (a data-flow like model) are related by certain (partially) dual transformations. For MultRel, two types of network algebra are used: one is based on an existential feedback, while the other one uses an iterative definition for feedback.

Bainbridge duality. Bainbridge duality uses the *iterative* additive feedback and the *existential* multiplicative feedback. The duality is partial, i.e., it preserves the network algebra operators, but fails to preserve identities.

Let us fix (recall) some notation: for both additive and multiplicative interfaces we use the same monoid $(\mathbb{N},+,0)$ having a,b,c,d as typical elements; aD denotes $D\oplus \overset{a}{\ldots} \oplus D$, where "$\oplus$" is disjoint union; AddRel$(D)(a,b)$ is the set of relations with source aD and target bD; $D^a = D\otimes \overset{a}{\ldots} \otimes D$ ("$\otimes$" is Cartesian product); MultRel$(D)(a,b)$ is the set of relations with source D^a and target D^b.

Notice that $2^{aD} \approx (2^D)^a$, i.e., a subset in aD may be identified with an a-tuple of subsets in D.

One may consider the following transformations $\mathsf{AddRel}(D) \xrightarrow{\bullet^\sharp} \mathsf{MultRel}(2^D) \xrightarrow{\bullet_\sharp} \mathsf{AddRel}(D)$:

(1) For $f \in \mathsf{AddRel}(D)(a,b)$ consider $f^\sharp \in \mathsf{MultRel}(2^D)(a,b)$ defined by

$$(\alpha,\beta) \in f^\sharp \quad \text{iff} \quad \alpha f \subseteq \beta$$

where αf denotes the f-image of α. (More precisely, this means the following: an element $\alpha \in (2^D)^a$ is identified with the corresponding subset in aD; to this subset one applies f, the resulting image being a subset in bD; this subset in bD is identified with the corresponding element in $(2^D)^b$, denoted simply αf.)

(2) For $\phi \in \mathsf{MultRel}(2^D)(a,b)$ consider $\phi_\sharp \in \mathsf{AddRel}(D)(a,b)$ defined by

$$(x,y) \in \phi_\sharp \quad \text{iff} \quad [x)\phi \subseteq [y)$$

where for $x \in aD$, $[x) = \{\alpha\colon \alpha \in (2^D)^a, \alpha \ni x\}$. (This definition uses the similar identification: $\alpha \in (2^D)^a$ is thought as a subset of aD; then $\alpha \ni x$ has a clear meaning. In brief: if x is in the i-th component of aD, then $\alpha \in [x)$ iff the i-th component of α contains x.)

Lemma 7.8. *Every $f \in \mathsf{AddRel}(D)(a,b)$ is closed, i.e., $(f^\sharp)_\sharp = f$.*

PROOF: By definition $(\{x\},\{x\}f) \in f^\sharp$ for all $x \in aD$. If $(x,y) \in (f^\sharp)_\sharp$, then $[x)f^\sharp \subseteq [y)$, hence $y \in \{x\}f^\sharp$. This implies $(x,y) \in f$, hence $(f^\sharp)_\sharp \subseteq f$. On the other hand one always has $f \subseteq (f^\sharp)_\sharp$, hence $(f^\sharp)_\sharp = f$. □

This shows that $\mathsf{AddRel}(D)$ may be identified with the family of closed subsets in $\mathsf{MultRel}(D)$, i.e., of those subsets ϕ satisfying $(\phi_\sharp)^\sharp = \phi$.

Theorem 7.9. *There is a (partial) duality between the additive $\mathsf{AddRel}(D)$ and the multiplicative $\mathsf{MultRel}(2^D)$ network algebra structures in the following sense:*

(a) transformation $^\sharp$ preserves the operations, i.e., for all appropriate f,g,h

$$(1)\ (f \oplus g)^\sharp = f^\sharp \otimes g^\sharp; \qquad (2)\ (h \uparrow^c_\oplus)^\sharp = h^\sharp \uparrow^c_\otimes; \qquad (3)\ (f \cdot g)^\sharp = f^\sharp \cdot g^\sharp$$

(b) for all appropriate f,g,h

$$(1)\ f \oplus g = (f^\sharp \otimes g^\sharp)_\sharp; \qquad (2)\ h \uparrow^c_\oplus = (h^\sharp \uparrow^c_\otimes)_\sharp; \qquad (3)\ f \cdot g = (f^\sharp \cdot g^\sharp)_\sharp$$

PROOF: By the above lemma all relations $f \oplus g$, $h \uparrow^c_\oplus$ and $f \cdot g$ are closed, hence (b) follows from (a) by a simple application of $_\sharp$. Therefore, the proof of (a) suffices.

In the following we write simply $\alpha \oplus \gamma$ for a subset in $(a+b)D$ corresponding to the subsets α in aD and γ in bD.

Let us start with (1). Suppose $f : a \to b$ and $g : c \to d$.

Notice that $(f \oplus g)^\sharp : (2^D)^{a+c} \to (2^D)^{b+d}$ is actually given by the relation

$$\begin{array}{ll} & ((\alpha,\gamma),(\beta,\delta)) \in (f \oplus g)^\sharp \\ \Leftrightarrow & \forall x \in aD \oplus cD, y \in bD \oplus dD.\ (x,y) \in f \oplus g \ \&\ x \in \alpha \oplus \gamma \to y \in \beta \oplus \delta \end{array}$$

But $(x, y) \in f \oplus g$ implies either $x \in aD,\ y \in bD$ and $(x, y) \in f$ or $x \in cD,\ y \in dD$ and $(x, y) \in g$. Hence the above statement is further equivalent to

$$\begin{array}{ll} \Leftrightarrow & \forall x \in aD, y \in bD.\ (x, y) \in f \ \&\ x \in \alpha \rightarrow y \in \beta \\ & \&\ \forall x \in cD, y \in dD.\ (x, y) \in g \ \&\ x \in \gamma \rightarrow y \in \delta \\ \Leftrightarrow & (\alpha, \beta) \in f^{\sharp} \ \&\ (\gamma, \delta) \in g^{\sharp} \\ \Leftrightarrow & ((\alpha, \gamma), (\beta, \delta)) \in f^{\sharp} \otimes g^{\sharp} \end{array}$$

The proof of (2) is slightly more complicated. First, by definition,

$$\begin{array}{ll} (i) & (\alpha, \beta) \in (h \uparrow^{c}_{\oplus})^{\sharp} \\ \Leftrightarrow & \forall x \in aD, y \in bD.\ (x, y) \in (h \uparrow^{c}_{\oplus})^{\sharp} \ \&\ x \in \alpha \rightarrow y \in \beta \end{array}$$

On the other hand,

$$\begin{array}{ll} (ii) & (\alpha, \beta) \in h^{\sharp} \uparrow^{c}_{\otimes} \\ \Leftrightarrow & \exists \gamma \in (2^{D})^{c}.\ ((\alpha, \gamma), (\beta, \gamma)) \in h^{\sharp} \\ \Leftrightarrow & \exists \gamma \in (2^{D})^{c}. \forall x \in aD \oplus cD, \forall y \in bD \oplus cD. \\ \Leftrightarrow & \qquad (x, y) \in h \ \&\ x \in \alpha \oplus \gamma \rightarrow y \in \beta \oplus \gamma \\ \Leftrightarrow & \exists \gamma \in (2^{D})^{c}. \forall x \in aD \oplus cD, \forall y \in bD \oplus cD \rightarrow \\ & \qquad \text{if } x \in aD,\ y \in bD,\ (x, y) \in h,\ x \in \alpha, \text{ then } y \in \beta \\ & \quad \text{and if } x \in aD,\ y \in cD,\ (x, y) \in h,\ x \in \alpha, \text{ then } y \in \gamma \\ & \quad \text{and if } x \in cD,\ y \in bD,\ (x, y) \in h,\ x \in \gamma, \text{ then } y \in \beta \\ & \quad \text{and if } x \in cD,\ y \in cD,\ (x, y) \in h,\ x \in \gamma, \text{ then } y \in \gamma \end{array}$$

We show that the statements (i) and (ii) are equivalent.

The implication $(ii) \rightarrow (i)$ easily follows by induction on the number of h computation steps used to connect x and y. If one starts with an $x \in \alpha \subseteq aD$, then by (ii) all the intermediary values of the computation in cD (if any) are in γ. Finally, one gets $y \in \beta$.

For the converse implication $(i) \rightarrow (ii)$ we take as the invariant $\gamma \in (2^{D})^{c}$ the subset of those $z \in cD$ such that:

(1) they are reached by h-computation paths from elements $x \in \alpha$; and
(2) *all* possible h-computation paths from such a z to an $y \in bD$ fulfil $y \in \beta$.

By a case analysis it is easy to see that γ fulfils (ii):

- Case $x \in aD,\ y \in bD,\ (x, y) \in h,\ x \in \alpha$: Notice that $(x, y) \in h \uparrow^{c}_{\oplus}$, hence, by (i), $y \in \beta$.
- Case $x \in aD,\ y \in cD,\ (x, y) \in h,\ x \in \alpha$: We show that $y \in \gamma$. First, y is reached from the element $x \in \alpha$, hence it fulfils condition (1). On the other hand, each element in bD reached from y is also reached from x. By (i) it is in β, hence y fulfils condition (2). Consequently, $y \in \gamma$.
- Case $x \in cD,\ y \in bD,\ (x, y) \in h,\ x \in \gamma$: Notice that $x \in \gamma$ shows that x is reached from an element in α, hence the same is true for y. By (i), $y \in \beta$.
- Finally, case $x \in cD,\ y \in cD,\ (x, y) \in h,\ x \in \gamma$: We show that $y \in \gamma$. First, $x \in \gamma$, hence x is reached from an element in α. Consequently, y is reached

from that element in α, too. Next, each element in bD which is reached from y is also reached from x. But $x \in \gamma$, hence by (2) that element of bD is in β. So, $y \in \gamma$, indeed.

(3) The proof for composition is similar. □

Corollary 7.10. *If one starts with a flowchart program, then takes its corresponding dual logical network and finally comes back, then one gets the same program. The converse may not be true. Actually, the flowchart programs may be identified with the particular logical networks ϕ which are "closed," i.e., $(\phi_\sharp)^\sharp = \phi$.*

However, this remarkable fact *fails* to be a fully network algebra relationship, in the sense that the above transformations do not preserve constants. For instance: $(\alpha, \beta) \in (\mathsf{I}_1)^\sharp$ iff $\alpha \subseteq \beta$, while in $\mathsf{MultRel}$ identity relates equal elements; $((\alpha, \beta), \gamma) \in (\succ\!\!\bullet_1)^\sharp$ iff $\alpha \cup \beta \subseteq \gamma$, while $\curlywedge_1$ relates only those combinations for which $\alpha = \beta = \gamma$; etc.

Second duality result. In this subsection we present another duality result of the same type. The network algebra structure on the additive model is the same, but for the multiplicative part an iterative feedback is used.

This new duality transformation well (e.g., it applies to the constants, as well) mainly because, as in the additive case, an iterative multiplicative feedback is used. It appears considering multiplicative relations as predicate transformers via weakest precondition technique.

Classes. The additional information of the data set M used in the multiplicative model $\mathsf{MultRel}(M)$ is that *everything is a class*; actually, $M = 2^D$, the collection of all classes/subsets in D. This is dual with the set-theoretic view used in $\mathsf{AddRel}(D)$, where everything is a set. Then, classes come with a free intersection (an arbitrary intersection of classes is a class), which allows for an iterative construction of the multiplicative feedback. Moreover, notice that with this class-theoretic view, one can easily give a meaning to the previously problematic equality test constant $\forall_a^k$: it simply gives the intersection of the corresponding input data, here classes. ◇

Let us start with a simple flowchart program:

$$\text{WHILE } x \leq 1000 \text{ DO } x := x * x$$

The corresponding function is easy to find: the output is the least power of the input of degree in $\{1, 2, 4, 8, \ldots\}$ which exceeds 1000. Let us describe in full detail how NA semantics leads to this result and then describe the duality machinery.

Suppose D denotes the set of states, here $\mathbb{N}$. The program (slightly modified, because of the use of the right feedback) may be represented by the NA expression

$$E = [\succ\!\!\bullet \cdot p \cdot (\mathsf{I} \oplus h)] \uparrow_{\oplus} : 1 \to 1$$

where $p = \boxed{x > 1000} : 1 \to 2$ and $h = \boxed{x := x * x} : 1 \to 1$. A simple graph-isomorphic transformation shows that $E = [(\mathsf{I} \oplus p \cdot (\mathsf{I} \oplus h))(\mathsf{X} \oplus \mathsf{I})(\mathsf{I} \oplus >\!\!\bullet)] \uparrow_{\oplus}$.
In terms of matrices what we have under feedback is $f = \begin{bmatrix} \emptyset & \mathsf{I} \\ p_1 & p_2 \cdot h \end{bmatrix}$, where $p_1 = \{(x, x) \colon x > 1000\}$ and $p_2 = \{(x, x) \colon x \leq 1000\}$.

Semantically, the term without feedback $f \in \mathsf{AddRel}(D)(2,2)$ is given by the following simple relations over D:

$$f_{11} = \emptyset_{D,D}, \ f_{12} = id_D,$$
$$f_{21} = \{(x, x) \colon x > 1000\}, \ f_{22} = \{(x, x * x) \colon x \leq 1000\}$$

Using the feedback definition we get the following evaluation for f:

$$\begin{array}{ll} & f \uparrow^1_{\oplus} \\ = & f_{11} \cup f_{12} \cdot (f_{22})^* \cdot f_{21} \\ = & f_{12} \cdot (id_D \cup f_{22} \cup f_{22} \cdot f_{22} \cup \ldots)^* \cdot f_{21} \\ = & (id_D \cup \{(x, x^2) \colon x \leq 1000\} \cup \{(x, x^4) \colon x^2 \leq 1000\} \cup \ldots) \cdot \{(x, x) \colon x > 1000\} \\ = & \{(x, x^{2^k}) \colon k = 0 \text{ or } (k > 0 \text{ and } x^{2^{k-1}} \leq 1000)\} \cdot \{(x, x) \colon x > 1000\} \\ = & \{(x, x^{2^k}) \colon x^{2^{k-1}} \leq 1000 < x^{2^k}\} \end{array}$$

We describe how in the dual model $\mathsf{MultRel}(2^D)(2,2)$ the weakest precondition machinery gets the same result. In the dual model MultRel the flow of information is reversed, from the outputs to the inputs. Notice that weakest preconditions are traditionally denoted as $wp(f, \alpha)$.

The dual of f is $f^{\vee} : 2 \leftarrow 2$ given by the relations

$$(f_{11})^{\vee} = (\emptyset_{D,D})^{\vee}, \ (f_{12})^{\vee} = (id_D)^{\vee},$$
$$(f_{21})^{\vee} = \{(x, x) \colon x > 1000\}^{\vee}, \ (f_{22})^{\vee} = \{(x, x * x) \colon x \leq 1000\}^{\vee}.$$

The final result is computed with the dual formula (we avoid some parentheses; hopefully, it will be clear to which arrow the dual operation applies)

$$\begin{array}{ll} & (f \uparrow^1_{\oplus})^{\vee} \\ = & f^{\vee} \uparrow^1_{\otimes} \\ = & f_{11}^{\vee} \cap f_{21}^{\vee} \cdot (f_{22}^{\vee})^* \cdot f_{12}^{\vee} \\ = & f_{11}^{\vee} \cap f_{21}^{\vee} \cdot (id_{2^D} \cap f_{22}^{\vee} \cap f_{22}^{\vee} \cdot f_{22}^{\vee} \cap \ldots) \cdot f_{12}^{\vee} \end{array}$$

Notice a few features of this duality machinery: the arrows change their direction; the multiplicative feedback is iterative, not existential; sequential composition reverses the terms.

Intuitively, the weakest precondition $wp(f, \alpha)$ to get states in α after f gives the largest set of initial states such that the execution from those states leads to states in α, only. Notice that *no* special requirement to assure termination is taken here; in particular, states where the execution does not terminate are included in these largest sets. Maybe the simplest question which produces $wp(f, \alpha)$ is: what is the largest set of states such that, starting with such a state, no state outside α is obtained?

Let us go in detail with the example.

The dual of an identity is just the identity. While this is trivial, it is an important fact. As we have seen, Bainbridge transformation does not preserve identities. But with *wp*-technique it is clear: what is the largest set of states such that applying the identity function you get no state outside α? Of course, α.

Empty relation(s) are more complex: notice that $(2^D)^0$ is a singleton, denoted $\{\bullet\}$. Each set $(2^D)^n$ has a maximal element, the whole state set nD, denoted Top_n. The duality acts on empty relations as follows:

—$\emptyset_{D,\emptyset} : D \to \emptyset$ is dualized in $(\emptyset_{D,\emptyset})^\vee : 2^D \leftarrow \{\bullet\}$,
that is the function which maps $\bullet$ in Top_1;
—$\emptyset_{\emptyset,D} : \emptyset \to D$ is dualized in $(\emptyset_{\emptyset,D})^\vee : \{\bullet\} \leftarrow 2^D$,
that is the function which maps all in $\bullet$;
—$\emptyset_{D,D} : D \to D$ is dualized in $\emptyset^\vee_{D,D} : 2^D \leftarrow 2^D$,
that is the function which maps all elements of 2^D in Top_1.

Similar questions and answers as for identity motivate these results. E.g., for the last one ask: which is the largest set such that doing the empty program (which breaks computation) you get no final state outside α? Of course, the full set.

We have seen that $f_{11}^\vee$ maps all subsets in Top_1. But Top_1 is neutral element for intersection, hence $f_{11}^\vee$ may be eliminated. Also identity $f_{12}^\vee$ (neutral element for composition) may be eliminated. One gets

$$f^\vee \uparrow^1_\otimes = f_{21}^\vee \cdot (id_{2^D} \cap f_{22}^\vee \cap f_{22}^\vee \cdot f_{22}^\vee \cap \ldots)$$

Now, take an output subset $\alpha \in 2^D$. The iterative formula gives the intersection of the sets of states such that by 0,1,2, or more computation paths, respectively they do not lead to states outside α. Hence, in the final set only those states are included which by arbitrary computations do not lead to states outside α. By reversing all relations and computing the result one gets the set of initial states from where each final path will lead to a state in the underlying output, only.

Notice that by this transformation the dual of a (even nondeterministic) flowchart program is a function. This shows the target may actually be shrunken to $\mathsf{MultPfn}^{-1}(2^D)$, the substructure of converses of partial functions in $\mathsf{MultRel}(2^D)$.

We collect the above observations in the following result, whose formal proof is not included here.

Theorem 7.11. *The transformation* $^\vee$: $\mathsf{AddRel}(D, \oplus, \cdot, \uparrow_\oplus, \mathsf{I}, \mathsf{X}, {}_k{>\!\bullet}, {\bullet\!<}_k) \to \mathsf{MultRel}(D, \otimes, \cdot, \uparrow_\otimes, \mathsf{I}, \mathsf{X}, \curlywedge_k, \curlyvee^k)$ *is a network algebra morphism. On the other hand,* $(x, y) \in f$ *iff* $x \in f^\vee(\{y\})$.

Exercises and Problems (sec. 7.6)

1. There is no injective network algebra mapping of $\mathsf{AddRel}(D)$ to $\mathsf{MultRel}(2^D)$. Such a mapping does exist if one uses multi-sets over D instead of subsets of D. (20)

7.7 Iteration and feedback in (co)algebraic theories

In this section we describe 4 iteration based systems of axioms I1–I4 for presenting BNAs over coalgebraic theories and prove they are equivalent. Then, we relate them with 2 feedback based systems of axioms F1–F2, showing that the underlying algebraic structure is a BNA over a coalgebraic theory, indeed.

Axioms for iteration. In this subsection we give a proof for the equivalence of the 4 axiomatic systems in Table 7.1; they were also presented in Chapter 1 (Section 1.7); read that part for a more detailed introduction to the problem to be solved here.

Table 7.1. Axiomatic systems for Căzănescu–Ungureanu theories (2nd posting)

I1.1	$(f \cdot (\mathsf{V}_a \star \mathsf{I}_b))^\dagger = f^{\dagger\dagger}$	I2.1	$f \cdot \langle f^\dagger, \mathsf{I}_b \rangle = f^\dagger$
I1.2	$(f \cdot (g \star \mathsf{I}_c))^\dagger = f \cdot \langle (g \cdot f)^\dagger, \mathsf{I}_c \rangle$	I2.2	$(f \cdot (\mathsf{V}_a \star \mathsf{I}_b))^\dagger = f^{\dagger\dagger}$
	for $f : a \to b \star c, g : b \to a$	I2.3	$g \cdot (f \cdot (g \star \mathsf{I}_c))^\dagger = (g \cdot f)^\dagger$
I1.3	$(f \cdot (\mathsf{I}_a \star g))^\dagger = f^\dagger \cdot g$		for $f : a \to b \star c, g : b \to a$
		I2.4	$(f \cdot (\mathsf{I}_a \star g))^\dagger = f^\dagger \cdot g$
I3.1	$(\mathsf{T}_a \star \mathsf{I}_a)^\dagger = \mathsf{I}_a$	I4.1	$(\mathsf{T}_a \star f)^\dagger = f$
I3.2	$\langle f, g \rangle^\dagger = \langle f^\dagger \cdot \langle h, \mathsf{I}_c \rangle, h \rangle$	I4.2	$\langle f, g \rangle^\dagger = \langle f^\dagger \cdot \langle h, \mathsf{I}_c \rangle, h \rangle$
	where $h = (g \cdot \langle f^\dagger, \mathsf{I}_{b \star c} \rangle)^\dagger$		where $h = (g \cdot \langle f^\dagger, \mathsf{I}_{b \star c} \rangle)^\dagger$
I3.3	$({}^a\mathsf{X}^b \cdot f \cdot ({}^b\mathsf{X}^a \star \mathsf{I}_c))^\dagger = {}^a\mathsf{X}^b \cdot f^\dagger$	I4.3	$({}^a\mathsf{X}^b \cdot f \cdot ({}^b\mathsf{X}^a \star \mathsf{I}_c))^\dagger = {}^a\mathsf{X}^b \cdot f^\dagger$
I3.4	$(f \cdot (\mathsf{I}_a \star g))^\dagger = f^\dagger \cdot g$	I4.4	$(f \star \mathsf{T}_c)^\dagger = f^\dagger \star \mathsf{T}_c$

Proposition 7.12. *Suppose T is an algebraic theory with a given iteration operation. Then the four axiomatic systems* I1, I2, I3 *and* I4 *are equivalent, providing different possible definitions for Căzănescu–Ungureanu theories.*

Proof: **(I1 ⇔ I2)** It is easy to see that I1 and I2 are equivalent. Indeed, (1) I1.2 for $g = \mathsf{I}_a$ gives I2.1; moreover, by I1.2 and I2.1 we get $g(f(g \star \mathsf{I}_c))^\dagger = gf\langle (gf)^\dagger, \mathsf{I}_c \rangle = (gf)^\dagger$, hence I1.2 also implies I2.3. (2) The converse implication "I2.1 & I2.3 implies I1.2" follows using I2.1 and I2.3 in $(f(g \star \mathsf{I}_c))^\dagger = f(g \star \mathsf{I}_c)\langle (f(g \star \mathsf{I}_c))^\dagger, \mathsf{I}_c \rangle = f\langle g(f(g \star \mathsf{I}_c))^\dagger, \mathsf{I}_c \rangle = f\langle (gf)^\dagger, \mathsf{I}_c \rangle$.

(I3 ⇔ I4) First, I3.1 and I3.4 imply I4.1 and I3.4 implies I4.4. Hence I3 implies I4. For the converse, first note that I4.1 implies I3.1. Then we prove that I3.4

follows from I4. Assume $f : a \to a \star b$ and $g : b \to c$. Then using I4.1 $((\top_{a\star b} \star g)\langle (f\star\top_c)^\dagger, \mathsf{I}_{b\star c}\rangle)^\dagger = (\top_b\star g)^\dagger = g$ and using I4.2 we get $\langle f\star\top_c, \top_{a\star b}\star g\rangle^\dagger = \langle (f\star \top_c)^\dagger\langle g, \mathsf{I}_c\rangle, g\rangle = \langle f^\dagger g, g\rangle$. Using I4.3 we get $\langle f\star\top_c, \top_{a\star b}\star g\rangle^\dagger = {}^a\mathsf{X}^b\langle \top_{b\star a}\star g, f\,{}^a\mathsf{X}^b\star \top_c\rangle^\dagger$. As $(\top_{b\star a}\star g)^\dagger = \top_a\star g$ and $(f\,{}^a\mathsf{X}^b\star\top_c)\langle \top_a\star g, \mathsf{I}_{a\star c}\rangle = f\langle \mathsf{I}_a\star\top_c, \top_a\star g\rangle = f(\mathsf{I}_a\star g)$ using I4.2 we get $\langle f^\dagger g, g\rangle = {}^a\mathsf{X}^b\langle (\top_a\star g)\langle (f(\mathsf{I}_a\star g))^\dagger, \mathsf{I}_c\rangle, (f(\mathsf{I}_a\star g))^\dagger\rangle = \langle (f(\mathsf{I}_a\star g))^\dagger, g\rangle$ hence I3.4 holds. Consequently I3 and I4 are equivalent.

(I1 ⇒ I3) In the next proof of I3 from I1 we also use I2. As I2.1 implies I3.1 and I2.3 implies I3.3 we only need to include a proof of I3.2:

$$
\begin{aligned}
\langle f,g\rangle^\dagger &= (\langle f,g\rangle(\mathsf{I}_a\star\top_{b\star a}\star\mathsf{I}_{b\star c})(\mathsf{V}_{a\star b}\star\mathsf{I}_c))^\dagger && \text{using I1.1}\\
&= (\langle f,g\rangle(\mathsf{I}_a\star\top_b\star\mathsf{I}_{b\star c})(\mathsf{I}_{a\star b}\star\top_a\star\mathsf{I}_{b\star c}))^{\dagger\dagger} && \text{using I1.3}\\
&= ((\langle f,g\rangle((\mathsf{I}_a\star\top_b)\star\mathsf{I}_{b\star c}))^\dagger(\top_a\star\mathsf{I}_{b\star c}))^\dagger && \text{using I1.2}\\
&= (\langle f,g\rangle\langle((\mathsf{I}_a\star\top_b)\langle f,g\rangle)^\dagger, \mathsf{I}_{b\star c}\rangle(\top_a\star\mathsf{I}_{b\star c}))^\dagger &&\\
&= (\langle f\langle f^\dagger, \mathsf{I}_{b\star c}\rangle, g\langle f^\dagger, \mathsf{I}_{b\star c}\rangle\rangle(\top_a\star\mathsf{I}_{b\star c}))^\dagger && \text{using I2.1}\\
&= (\langle f^\dagger, g\langle f^\dagger, \mathsf{I}_{b\star c}\rangle\rangle((\top_a\star\mathsf{I}_b)\star\mathsf{I}_c))^\dagger && \text{using I1.2}\\
&= \langle f^\dagger, g\langle f^\dagger, \mathsf{I}_{b\star c}\rangle\rangle\langle((\top_a\star\mathsf{I}_b)\langle f^\dagger, g\langle f^\dagger, \mathsf{I}_{b\star c}\rangle\rangle)^\dagger, \mathsf{I}_c\rangle &&\\
&= \langle f^\dagger, g\langle f^\dagger, \mathsf{I}_{b\star c}\rangle\rangle\langle(g\langle f^\dagger, \mathsf{I}_{b\star c}\rangle)^\dagger, \mathsf{I}_c\rangle && \text{using I2.1}\\
&= \langle f^\dagger\langle(g\langle f^\dagger, \mathsf{I}_{b\star c}\rangle)^\dagger, \mathsf{I}_c\rangle, (g\langle f^\dagger, \mathsf{I}_{b\star c}\rangle)^\dagger\rangle &&
\end{aligned}
$$

(I3 ⇒ I1) As I3 and I4 are equivalent we shall use axioms from I4, too. First, for every $g : b \to a$ and $v : a \to a\star b\star c$ we show that

(i) $v^\dagger\langle (gv^\dagger)^\dagger, \mathsf{I}_c\rangle = (v(\langle \mathsf{I}_a, g\rangle\star\mathsf{I}_c))^\dagger$ and (ii) $(gv^\dagger)^\dagger = g(v(\langle\mathsf{I}_a, g\rangle\star\mathsf{I}_c))^\dagger$.

As $((g\star\top_{b\star c})\langle v^\dagger, \mathsf{I}_{b\star c}\rangle)^\dagger = (gv^\dagger)^\dagger$ using I3.2 it follows that $\langle v, g\star\top_{b\star c}\rangle^\dagger = \langle v^\dagger\langle (gv^\dagger)^\dagger, \mathsf{I}_c\rangle, (gv^\dagger)^\dagger\rangle$. On the other hand, using I3.3 we get $\langle v, g\star\top_{b\star c}\rangle^\dagger = {}^a\mathsf{X}^b\langle \top_b\star g\star\top_c, v({}^a\mathsf{X}^b\star\mathsf{I}_c)\rangle^\dagger$. As $(\top_b\star g\star\top_c)^\dagger = g\star\top_c$ and $v({}^a\mathsf{X}^b\star\mathsf{I}_c)\langle g\star\top_c, \mathsf{I}_{a\star c}\rangle = v(\langle\mathsf{I}_a, g\rangle\star\mathsf{I}_c)$ using I3.2 we get $\langle v, g\star\top_{b\star c}\rangle^\dagger = {}^a\mathsf{X}^b\langle (g\star\top_c)\langle (v(\langle\mathsf{I}_a, g\rangle\star\mathsf{I}_c))^\dagger, \mathsf{I}_c\rangle, (v(\langle\mathsf{I}_a, g\rangle\star\mathsf{I}_c))^\dagger\rangle$, therefore $\langle v^\dagger\langle (gv^\dagger)^\dagger, \mathsf{I}_c\rangle, (gv^\dagger)^\dagger\rangle$
$= \langle (v(\langle\mathsf{I}_a, g\rangle\star\mathsf{I}_c))^\dagger, g(v(\langle\mathsf{I}_a, g\rangle\star\mathsf{I}_c))^\dagger\rangle$, hence we get (i) and (ii).

From (i) for $v := \top_a\star f$ where $f : a\to b\star c$ we get I1.2 and from (ii) for $v := f : a\to a\star a\star b$ and $g := \mathsf{I}_a$ we get I1.1. □

Table 7.2. Axiomatic systems for feedback for BNAs over coalgebraic theories (2nd posting)

*F*1.1	$\uparrow^a ({}^a\mathsf{X}^a) = \mathsf{I}_a$	*F*2.1	$\uparrow^a ({}^a\mathsf{X}^a\cdot(\mathsf{I}_a\star f)) = f$
*F*1.2	$\uparrow^b\uparrow^a f = \uparrow^{a\star b} f$	*F*2.2	$\uparrow^b\uparrow^a f = \uparrow^{a\star b} f$
*F*1.3	$\uparrow^{a\star b} g = \uparrow^{b\star a} f$ where $g = ({}^a\mathsf{X}^b\star\mathsf{I}_c)\cdot f\cdot({}^b\mathsf{X}^a\star\mathsf{I}_d)$	*F*2.3	$\uparrow^{a\star b} g = \uparrow^{b\star a} f$ where $g = ({}^a\mathsf{X}^b\star\mathsf{I}_c)\cdot f\cdot({}^b\mathsf{X}^a\star\mathsf{I}_d)$
*F*1.4	$(\uparrow^a f)\cdot g = \uparrow^a (f\cdot(\mathsf{I}_a\star g))$	*F*2.4	$\uparrow^a (f\star\top_d) = \uparrow^a f\star\top_d$
*F*1.5	$g\cdot(\uparrow^a f) = \uparrow^a ((\mathsf{I}_a\star g)\cdot f)$	*F*2.5	$\uparrow^a \langle f, g\rangle = g\cdot\langle\uparrow^a\langle f, \mathsf{I}_a\star\top_c\rangle, \mathsf{I}_c\rangle$ for $f : a\to a\star c, g : b\to a\star c$
*F*1.6	$\uparrow^a f\star g = \uparrow^a (f\star g)$		
*F*1.7	$\uparrow^a \mathsf{I}_a = \mathsf{I}_\epsilon$		

Iteration vs. feedback in algebraic theories. Let T be an algebraic theory and let $\mathbf{It}(T)$ (resp. $\mathbf{Fd}(T)$) be the set of all iteration operations (resp. feedback operations) defined on T. We define two functions

$$\alpha : \mathbf{Fd}(T) \to \mathbf{It}(T) \quad \text{and} \quad \beta : \mathbf{It}(T) \to \mathbf{Fd}(T)$$

as follows:

- $\alpha(\uparrow)$ is the iteration operation that maps an $f \in T(a, a \star b)$ into $\uparrow^a \langle f, \mathsf{l}_a \star \mathsf{T}_b \rangle$;
- $\beta(\dagger)$ is the feedback operation that for a, b and c objects in T maps $f = \langle f_1, f_2 \rangle \in T(a \star b, a \star c)$, with $f_1 : a \to a \star c$ and $f_2 : b \to a \star c$, into $f_2 \langle f_1^\dagger, \mathsf{l}_c \rangle$.

Let $\mathbf{Fd}_r(T)$ (resp. $\mathbf{Fd}_i(T)$) be the subset of all the feedback operations in $\mathbf{Fd}(T)$ that obey the axioms F1.4–F1.6 (resp. F2.5) and $\mathbf{It}_r(T)$ be the subset of all the iteration operations in $\mathbf{It}(T)$ that obey the axiom I3.4. Finally, let us consider the restrictions $\alpha_r : \mathbf{Fd}_r(T) \to \mathbf{It}_r(T), \beta_r : \mathbf{It}_r(T) \to \mathbf{Fd}_r(T), \alpha_i : \mathbf{Fd}_i(T) \to \mathbf{It}(T)$, and $\beta_i : \mathbf{It}(T) \to \mathbf{Fd}_i(T)$.

Theorem 7.13. *Suppose T is an algebraic theory. Then:*
a) The restrictions α_i and β_i are mutually inverse bijective functions; the same is true for α_r and β_r.
b) For $k \in [4]$, $\dagger$ satisfies I4.k *iff $\beta(\dagger)$ satisfies* F2.k.
c) For $k \in [4]$, $\dagger$ satisfies I3.k *iff $\beta(\dagger)$ satisfies* F1.k.
d) y is $\dagger$-functorial iff y is $\beta(\dagger)$-functorial.

PROOF: **a)** Note that $\uparrow := \beta(\dagger)$ satisfies F2.5; indeed, $g\langle \uparrow^a \langle f, \mathsf{l}_a \star \mathsf{T}_c \rangle, \mathsf{l}_c \rangle = g\langle (\mathsf{l}_a \star \mathsf{T}_c)\langle f^\dagger, \mathsf{l}_c \rangle, \mathsf{l}_c \rangle = g\langle f^\dagger, \mathsf{l}_c \rangle = \uparrow^a \langle f, g \rangle$. Consequently β_i is totally defined.

Obviously $\dagger = \alpha(\beta(\dagger))$. For the converse, note that $(\beta(\alpha(\uparrow)))^a$ maps $\langle f_1, f_2 \rangle \in T(a \star b, a \star c)$ (with $f_1 : a \to a \star c$ and $f_2 : b \to a \star c$) in $f_2 \langle \uparrow^a \langle f_1, \mathsf{l}_a \star \mathsf{T}_c \rangle, \mathsf{l}_c \rangle$. Hence $\uparrow = \beta(\alpha(\uparrow))$ for a $\uparrow \in \mathbf{Fd}_i(T)$.

For the second restriction, note that $\dagger$ satisfies I3.4 iff $\uparrow := \beta(\dagger)$ satisfies F1.4. Indeed, $\uparrow$ satisfies F1.4 iff for every $f = \langle f_1, f_2 \rangle : a \star b \to a \star c$ (with $f_1 : a \to a \star c$ and $f_2 : b \to a \star c$) and $g : c \to d$ the term $(\uparrow^a f)g = f_2 \langle f_1^\dagger, \mathsf{l}_c \rangle g = f_2 \langle f_1^\dagger g, g \rangle$ is equal to $\uparrow^a (f(\mathsf{l}_a \star g)) = \uparrow^a \langle f_1(\mathsf{l}_a \star g), f_2(\mathsf{l}_a \star g) \rangle = f_2(\mathsf{l}_a \star g)\langle (f_1(\mathsf{l}_a \star g))^\dagger, \mathsf{l}_d \rangle = f_2 \langle (f_1(\mathsf{l}_a \star g))^\dagger, g \rangle$. Consequently if an $\dagger$ satisfies I3.4, then the corresponding $\uparrow$ satisfies F1.4 and if an $\uparrow$ satisfies F1.4, then using $\mathsf{l}_a \star \mathsf{T}_c$ for f_2 above we conclude that the corresponding $\dagger$ satisfies I3.4. Hence we have got a bijective correspondence between $\mathbf{It}_r(T)$ and the subset of all the feedbacks in $\mathbf{Fd}(T)$ that satisfies F2.5 & F1.4.

The conclusion follows if we show that F2.5 & F1.4 is equivalent to F1.4-6. This equivalence may be proved as follows:

(1) F2.5 implies F1.5: if $f = \langle f_1, f_2 \rangle : a \star b \to a \star c$ (with $f_1 : a \to a \star c$ and $f_2 : b \to a \star c$) and $g : d \to b$, then by a double application of F2.5 we get $\uparrow^a ((\mathsf{l}_a \star g)f) = \uparrow^a \langle f_1, g f_2 \rangle = g f_2 \langle \uparrow^a \langle f_1, \mathsf{l}_a \star \mathsf{T}_c \rangle, \mathsf{l}_c \rangle = g \cdot \uparrow^a \langle f_1, f_2 \rangle = g \cdot \uparrow^a f$.

(2) F2.5 & F1.4 implies F1.6: if $f = \langle f_1, f_2\rangle : a \star b \to a \star c$ (with $f_1 : a \to a \star c$ and $f_2 : b \to a \star c$) and $g : d \to e$, then using in turn F2.5, F1.4 and again F2.5 we get $\uparrow^a (f \star g) = \uparrow^a \langle f_1 \star \mathsf{T}_e, f_2 \star g\rangle = (f_2 \star g)\langle \uparrow^a \langle f_1 \star \mathsf{T}_e, \mathsf{l}_a \star \mathsf{T}_{c\star e}\rangle, \mathsf{l}_{c\star e}\rangle = (f_2 \star g)\langle \uparrow^a \langle f_1, \mathsf{l}_a \star \mathsf{T}_c\rangle \star \mathsf{T}_e, \mathsf{l}_{c\star e}\rangle = (f_2 \star g)(\langle \uparrow^a \langle f_1, \mathsf{l}_a \star \mathsf{T}_c\rangle, \mathsf{l}_c\rangle \star \mathsf{l}_e) = f_2\langle \uparrow^a \langle f_1, \mathsf{l}_a \star \mathsf{T}_c\rangle, \mathsf{l}_c\rangle \star g = \uparrow^a f \star g$.

(3) F1.4–6 implies F2.5: If $f : a \to a \star c$ and $g : b \to a \star c$, then $\uparrow^a \langle f, g\rangle = \uparrow^a [(\mathsf{l}_a \star g)((\langle f, \mathsf{l}_a \star \mathsf{T}_c\rangle \star \mathsf{l}_c)(\mathsf{l}_a \star \mathsf{V}_c)] = g(\uparrow^a \langle f, \mathsf{l}_a \star \mathsf{T}_c\rangle \star \mathsf{l}_c)\mathsf{V}_c = g\langle \uparrow^a \langle f, \mathsf{l}_a \star \mathsf{T}_c\rangle, \mathsf{l}_c\rangle$.

b) Let $\uparrow$ and $\dagger$ be two corresponding operations, i.e. $\uparrow = \beta(\dagger)$. The equivalence for $k = 1$, say for $f : b \to c$, holds by $(\mathsf{T}_a \star f)^\dagger = \uparrow^a \langle \mathsf{T}_a \star f, \mathsf{l}_a \star \mathsf{T}_c\rangle = \uparrow^a ({}^a\mathsf{X}^a(\mathsf{l}_a \star f))$

For $k = 2$, note that if $f : a \to a \star b \star c, g : b \to a \star b \star c$ and $i : d \to a \star b \star c$, then $\uparrow^{a\star b} \langle f, g, i\rangle = i\langle \langle f, g\rangle^\dagger, \mathsf{l}_c\rangle$ and $\uparrow^b \uparrow^a \langle f, g, i\rangle = \uparrow^b (\langle g, i\rangle\langle f^\dagger, \mathsf{l}_{b\star c}\rangle) = \uparrow^b \langle g\langle f^\dagger, \mathsf{l}_{b\star c}\rangle, i\langle f^\dagger, \mathsf{l}_{b\star c}\rangle\rangle = i\langle f^\dagger, \mathsf{l}_{b\star c}\rangle\langle h, \mathsf{l}_c\rangle = i\langle\langle f^\dagger\langle h, \mathsf{l}_c\rangle, h\rangle, \mathsf{l}_c\rangle$ where $h = (g\langle f^\dagger, \mathsf{l}_{b\star c}\rangle)^\dagger$. Consequently $\uparrow$ satisfies F2.2 iff $\dagger$ satisfies I4.2.

For $k = 3$, note that if $f = \langle f_1, f_2\rangle : b\star a\star c \to b\star a\star d$ (with $f_1 : b\star a \to b\star a\star d$ and $f_2 : c \to b\star a\star d$), then $\uparrow^{a\star b} (({}^a\mathsf{X}^b \star \mathsf{l}_c) f ({}^b\mathsf{X}^a \star \mathsf{l}_d)) = \uparrow^{a\star b} \langle {}^a\mathsf{X}^b f_1({}^b\mathsf{X}^a \star \mathsf{l}_d), f_2({}^b\mathsf{X}^a \star \mathsf{l}_d)\rangle = f_2({}^b\mathsf{X}^a \star \mathsf{l}_d)\langle ({}^a\mathsf{X}^b f_1({}^b\mathsf{X}^a \star \mathsf{l}_d))^\dagger, \mathsf{l}_d\rangle = f_2\langle {}^b\mathsf{X}^a({}^a\mathsf{X}^b f_1({}^b\mathsf{X}^a \star \mathsf{l}_d))^\dagger, \mathsf{l}_d\rangle$ and $\uparrow^{b\star a} f = f_2\langle f_1^\dagger, \mathsf{l}_d\rangle$. Since ${}^a\mathsf{X}^b \cdot {}^b\mathsf{X}^a = \mathsf{l}_{a\star b}$ it follows that I4.3 is equivalent to F2.3.

For $k = 4$, note that the axioms F2.4 and I4.4 may be written as $\uparrow^a (f(\mathsf{l}_{a\star c} \star \mathsf{T}_d)) = (\uparrow^a f)(\mathsf{l}_c \star \mathsf{T}_d)$ and $(f(\mathsf{l}_{a\star b} \star \mathsf{T}_c))^\dagger = f^\dagger(\mathsf{l}_b \star \mathsf{T}_c)$, respectively. Now the equivalence of F2.4 and I4.4 directly follows from the above proof of the equivalence of F1.4 and I3.4.

c) The proof of c) is covered by the above proof of b).

d) Suppose that $y : a \to b$ is $\dagger$-functorial and $f = \langle f_1, f_2\rangle : a \star c \to a \star d$ (with $f_1 : a \to a \star d$ and $f_2 : c \to a \star d$) and $g = \langle g_1, g_2\rangle : b \star c \to b \star d$ (with $g_1 : b \to b \star d$ and $g_2 : c \to b \star d$) are such that $f(y \star \mathsf{l}_d) = (y \star \mathsf{l}_c)g$. Then $f_1(y \star \mathsf{l}_d) = yg_1$ and $f_2(y \star \mathsf{l}_d) = g_2$. By the $\dagger$-functoriality of y we get $f_1^\dagger = yg_1^\dagger$. Hence $\uparrow^a f = f_2\langle f_1^\dagger, \mathsf{l}_d\rangle = f_2\langle yg_1^\dagger, \mathsf{l}_d\rangle = f_2(y \star \mathsf{l}_d)\langle g_1^\dagger, \mathsf{l}_d\rangle = g_2\langle g_1^\dagger, \mathsf{l}_d\rangle = \uparrow^b g$.

Conversely, suppose that $y : a \to b$ is $\uparrow$-functorial and $f : a \to a \star c$ and $g : b \to b \star c$ are such that $f(y \star \mathsf{l}_c) = yg$. Then $\langle f, \mathsf{l}_a \star \mathsf{T}_c\rangle(y \star \mathsf{l}_c) = \langle f(y \star \mathsf{l}_c), y \star \mathsf{T}_c\rangle = \langle yg, y \star \mathsf{T}_c\rangle = (y \star \mathsf{l}_a)\langle g, y \star \mathsf{T}_c\rangle$. By the $\uparrow$-functoriality of y we get $\uparrow^a \langle f, \mathsf{l}_a \star \mathsf{T}_c\rangle = \uparrow^b \langle g, y \star \mathsf{T}_c\rangle$. As $\uparrow^a \langle f, \mathsf{l}_a \star \mathsf{T}_c\rangle = (\mathsf{l}_a \star \mathsf{T}_c)\langle f^\dagger, \mathsf{l}_c\rangle = f^\dagger$ and $\uparrow^b \langle g, y \star \mathsf{T}_c\rangle = (y \star \mathsf{T}_c)\langle g^\dagger, \mathsf{l}_c\rangle = yg^\dagger$ the result follows. □

Corollary 7.14. *For an algebraic theory with an iteration or a feedback operation, the axiomatic systems* F1, F2, I1, I2, I3 *and* I4 *are equivalent.*

Comments, problems, bibliographic remarks

Notes. The main result of this chapter is Theorem 7.7 on correctness of Floyd–Hoare logic. It is due to Bloom and Ésik (see, e.g., [BÉ93b]), but we have re-cast it in network algebra terms, i.e., by replacing iteration with feedback.

Structural programs were studied in the 1960s. The completeness of the structuring programming constructs was proved in [BJ66]. Elgot's extension was presented in [Elg76b]; the general representation is based on [Şte86b].

The idea of verifying programs by logical assertions was already stated by Turing in the 1950s, but it seems that this has no direct continuation. The principle of "logical assertions" for programs was really born with Floyd in the 1960s. It was "non-compositional", i.e., it was applied to the flowchart programs considered as indecomposable entities. A classical book is [Man74].

This method was later refined, formalized, and applied to structural programs by Hoare in the 1970s. The result is the current well known Floyd–Hoare logic. In the 1980s, Floyd–Hoare logic was extended to arbitrary flowcharts by Arbib–Manes and Bloom–Ésik; see [MA86] and [BÉ93b]. The outcome of the latter studies is that Floyd–Hoare logic is not a special logic: it is just the usual equational logic, but manifested in certain particular algebraic structures.

Boolean logic started with Boole, see, e.g., [Boo47]; a classical presentation may be found also in [Rud74]; its presentation within an algebraic theory is from [BÉ93b]. Căzănescu–Ungureanu theories were introduced in [CU82].

Bainbridge duality is presented in [Bai76]. The second duality between flowcharts and circuits presented in Section 7.6 seems to be known; see, e.g., [Möl98a]. The result comparing feedback and iteration based axiomatic systems is from [CŞ95].

The study of flowchart schemes was pioneered in [Yan58]; see also [Gre75, Kot78]. A similar approach to the one presented in this chapter is [Man92]. A relation algebra approach to program verification is presented in [Mad96]. Refinement calculus, presented, e.g., in [BvW98], also covers the program verification issues.

The exercises in Section 7.1 are from [Man74]; Exercise 7.6.1 is from [Sel98].

8. Automata

> Ideals are like the stars: we never reach them, but like the mariners of the sea, we chart our course by them.
>
> Carl Schurz

Finite state automata provide one of the oldest and best known models used in computer science. We take advantage here of certain recent results to give a short and elegant presentation of their algebraic theory.

Except for Section 8.11, providing an explicit connection with network algebra, this chapter may be read independently. From the network algebra point of view it requires to work in a Kleene theory ($d\delta$-flow). A large part of the algebraic theory of automata, including Kleene Theorem and the Normalization Theorem for regular expressions, may be done in a Conway theory (i.e., in a BNA over a matrix theory).

First, automata are represented by matrices and a simple linear algebra like calculus is used. Simulation via relations is introduced as a stepping stone for automata equivalence. Automata presentations are seen as normal form representations of regular expressions.

Next, the familiar power-set construction which reduces a nondeterministic automaton to a deterministic one is shown to produce an automaton similar to the original one. Minimization of deterministic automata is modeled with simulation via functions. Hence two automata are equivalent (i.e., they recognize the same language) if and only if they may be connected by a chain of simulations.

Then, we introduce two simple classes of algebras. One is the class of idempotent *Conway theories.* This setting suffices to prove Kleene Theorem and the Normalization Theorem for regular expressions. The second class is the stronger one provided by idempotent *Kleene theories.* Actually this is our main proposal for the present automata theory setting. The main result is that two regular expressions are equivalent iff they are equivalent within this axiomatic system. Therefore the axioms of idempotent Kleene theories give a correct (i.e., sound and complete) axiomatization for regular languages.

Finally, the connection with network algebra is clarified and three equivalent presentations of Conway theories are described using Kleene's star operation.

8.1 Finite automata

In this section we start with an informal presentation of finite automata, then we describe the formal representation of automata by matrices.

Informal presentation. To describe the informal model of finite state automata let us consider the example given in Fig. 8.1.

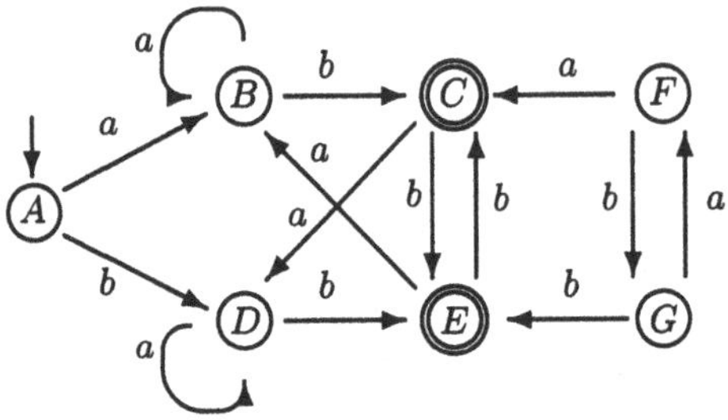

Fig. 8.1. A (complete, deterministic) finite state automaton

A *finite automaton* has a finite number of states, usually represented as circles. Some states may be distinguished as initial (respectively final) states. We use small incoming arrows to denote initial states and the double circle notation for final states. Note that a state may be simultaneously initial and final. For convenience automata states are labeled, but their labels have no special meaning/value. In the example in Fig. 8.1 there is a unique initial state A and two final states C and E.

Automata states are connected by labeled edges. The edge labels are from a given alphabet V. In our example $V = \{a, b\}$. A transition has precisely one input vertex and one output vertex. Note that a label in V may be attached to many edges. Actually the automaton in Fig. 8.1 has a special form: it is *complete* and *deterministic*. The former property says that in each state there is at least one outgoing arrow for each element in V. (The automaton is input enable.) The latter property says that in each state there is at most one outgoing arrow for each element in V. (The automaton's reaction to an input is unique.) The name *nondeterministic* finite automata is often used for the general class of finite automata to emphasize the fact that it is not required to be complete or deterministic.

Our finite automata lay in an *additive* world. This means that in a branching point, each time, *one* outgoing edge is chosen. The behaviour of automata is provided by the directed paths connecting initial states with final states, more precisely the set of sequences in V^* that label such paths. This set of sequences form the *language* recognized by the automaton. For the automaton in Fig. 8.1, *aab* leads from A to C, hence it is in the attached language, but *baba* leads from A to the non-final state B, hence it is not in the attached language. The languages recognized by finite automata are called *regular*.

Matricial presentations. A usual syntactical representation of automata is via systems of equations. Each automata state/nod gives rise to an equation. For instance, the equation corresponding to state C for the automaton in Fig. 8.1 is written

$$C = aD + bE + 1$$

(1 says that the state is final.) It looks like a linear algebra equation, but its meaning is completely different. One looks for the solution(s) of such systems in a standard semantical domain where variables are interpreted as languages (sets of words), sum as union, product as concatenation, 1 (respectively 0) as the language consisting of the empty word (respectively the empty language).

♡ A basic result says that a system of (guarded) equations has a unique solution and the language corresponding to an automaton is the union of the components of the solution corresponding to its initial states in the associated system. ◇

These systems of equations may also be represented by matrices. Their solutions may be obtained using a standard iterative process which is formalized using Kleene star operation. This leads to a sharp representation of automata in a linear algebra like setting as follows.

Definition 8.1. Suppose we have a semiring $(S, +, \cdot, 0, 1)$ and an operation "$*$" on (square) *matrices* over S. An *(abstract) matricial presentation* of a regular language is an extended matrix $M = \begin{matrix} & \uparrow & n \\ \rightarrow & A & B \\ n & C & D \end{matrix}$, (or $M = \begin{matrix} \rightarrow \\ n \end{matrix} \left[\begin{array}{c|c} A & B \\ \hline C & D \end{array}\right]$) over $(S, +, \cdot, 0, 1)$ of type $(1+n) \times (1+n)$, where:

$1, \ldots, n$ represent the states of the automaton;
A represents the direct input-output connection (if any);
$B = (b_j)_{j\in[n]}$ specifies the input states;
$C = (c_i)_{i\in[n]}$ specifies the final states;
$D = (d_{ij})_{(i,j)\in[n]\times[n]}$ specifies the transitions.

The *language* specified by such a matricial presentation is

$$L(M) = A + B \cdot D^* \cdot C$$

where "$+$" and "$\cdot$" are the usual operations on matrices induced by the operations in S.

The languages over an alphabet V may be organized as a semiring $S = (\mathcal{P}(V^*), \cup, \cdot, \emptyset, \lambda)$ where: $L_1 \cup L_2 = \{w\colon\ w \in L_1 \text{ or } w \in L_2\}$, $L_1 \cdot L_2 = \{w_1w_2\colon\ w_1 \in L_1 \text{ and } w_2 \in L_2\}$, $\emptyset$ is the empty set, and λ is the language consisting of the empty word.

A *concrete matricial presentation* is one where $(S,+,\cdot,0,1)$ is the semiring $S = (\mathcal{P}(V^*),\cup,\cdot,\emptyset,\lambda)$ and $D^* = Id \cup D \cup D^2 \cup D^3 \ldots$, where Id the appropriate identity matrix. The *presentations for nondeterministic finite automata* are concrete presentations with the following additional conditions: (1) $A = 0$; (2) $b_j = 1$ if j is initial state, otherwise 0; (3) $c_i = 1$ if i is final state, otherwise 0; and (4) $d_{ij} = \{a : i \xrightarrow{a} j\}$.

Let us note that:

(a) nondeterministic finite automata are concrete matricial presentations satisfying conditions: (1) D contains only zeros or sums of letters, (2) B and C are matrices with components 0 or 1, and (3) A is 0.

(b) if in the above setting one adds the restriction that in D in each line every letter a occurs exactly once, then the resulting presentations correspond to deterministic finite automata.

(c) if in (a) one allows moreover to have 1 in D, too, then one gets matricial presentations for (nondeterministic) automata with empty moves.

(d) finally, if one relaxes (a) allowing for finite sums of words (not letters) as elements in D and C, and B with a unique 1, otherwise 0, then one gets matricial presentations which model regular grammars.

In what follows we use the general notation $+,\cdot,0,1$ also for the operations on concrete presentations.

Proposition 8.2. *Let $\mathcal{A}$ be a nondeterministic finite automaton and* $M = \begin{matrix}\rightarrow \\ n\end{matrix}\left[\begin{array}{c|c} A & B \\ \hline C & D \end{array}\right]$ *the corresponding (concrete) matricial presentation. Then $L(M)$ coincides with the (usually defined) language $L(\mathcal{A})$ recognized by $\mathcal{A}$.*

PROOF: Let

$$D = D^1 = (d^1_{ij})_{ij}$$
$$D^k = D \cdot \overset{m}{\ldots} \cdot D = (d^k_{ij})_{ij} \text{ and}$$
$$D^* = (d^*_{ij})_{ij}.$$

Notice that:

- $d^1_{ij} = \sum_{a_1 : i \xrightarrow{a_1} j} a_1$ is the sum of the letters giving transitions from i to j;
- $d^2_{ij} = \sum_{a_1,a_2 : \exists r_1 .\ i \xrightarrow{a_1} r_1 \xrightarrow{a_2} j} a_1 a_2$ is the sum of the words of length 2 for which there are paths from i to j;

  ...
- $d^k_{ij} = \sum_{a_1,\ldots,a_k : \exists r_1,\ldots,r_{k-1} .\ i \xrightarrow{a_1} r_1 \xrightarrow{a_2} r_2 \ldots r_{k-1} \xrightarrow{a_k} j} a_1 \ldots a_k$ is the sum of the words of length k for which there are paths from i to j;

  ...

Hence $d^*_{ij} = \sum_{w : i \xrightarrow{w} j} w$ contains the words for which there are paths in the automaton from i la j (of arbitrary length). Multiplication by B and C selects from D^* the words for which there are paths from the initial states to final states, hence $L(\mathcal{A}) = BD^*C = A + BD^*C = L(M)$. □

Example 8.3. Let $\mathcal{A}$ be an automaton with the matricial presentation

$$M = \left[\begin{array}{c|ccc} 0 & 1 & 0 & 0 \\ \hline 1 & 0 & a & 0 \\ 0 & 0 & 0 & a \\ 1 & 0 & 0 & a \end{array}\right]$$

Then:

$$D^2 = \begin{bmatrix} 0 & 0 & a^2 \\ 0 & 0 & a^2 \\ 0 & 0 & a^2 \end{bmatrix}, D^3 = \begin{bmatrix} 0 & 0 & a^3 \\ 0 & 0 & a^3 \\ 0 & 0 & a^3 \end{bmatrix}, \ldots,$$

$$D^* = \begin{bmatrix} 1 & a & a^2 + a^3 + \ldots \\ 0 & 1 & a + a^2 + a^3 + \ldots \\ 0 & 0 & 1 + a + a^2 + a^3 + \ldots \end{bmatrix};$$

$$L(M) = A + BD^*C = \begin{bmatrix} 1 & 0 & 0 \end{bmatrix} \begin{bmatrix} 1 & a & a^{\geq 2} \\ 0 & 1 & a^{\geq 1} \\ 0 & 0 & a^{\geq 0} \end{bmatrix} \begin{bmatrix} 1 \\ 0 \\ 1 \end{bmatrix} = 1 + a^{\geq 2} = L(\mathcal{A})$$

EXERCISES AND PROBLEMS (SEC. 8.1)

1. What sequences are recognized by the automaton in Fig. 8.1? Give a matricial presentation of this automaton. (06)
2. Show that a system of guarded equations has a unique solution. Prove that the language recognized by an automaton is the union of the components of the solution corresponding to its initial states. (18)

8.2 Simulation

In this section we introduce simulation via (finite) relations as a basic tool to capture their equivalence.

More matricial presentations may specify the same regular language. So a basic question is: "How one may identify equivalent matricial presentations?" Our solution is to use simulation via relations as an import from the general network algebra principles presented in the previous part of the book. It is redefined here in this particular case.

Definition 8.4. Two matricial presentations $M_i = \begin{array}{c|cc} & \uparrow & n_i \\ \rightarrow & A_i & B_i \\ n_i & C_i & D_i \end{array}$, for $i = 1, 2$ are *similar* via a relation $\rho \subseteq \{1, \ldots, n_1\} \times \{1, \ldots, n_2\}$ (written $M_1 \rightarrow_\rho M_2$) if

$$M_1 \cdot (\mathsf{I}_1 \star \rho) = (\mathsf{I}_1 \star \rho) \cdot M_2$$

Notice that: $A \star B$ is $\begin{bmatrix} A & 0 \\ 0 & B \end{bmatrix}$; simulation means $\begin{bmatrix} A_1 & B_1\rho \\ C_1 & D_1\rho \end{bmatrix} = \begin{bmatrix} A_2 & B_2 \\ \rho C_2 & \rho D_2 \end{bmatrix}$, therefore its checking reduces to: $A_1 = A_2,\ B_1\rho = B_2,\ C_1 = \rho C_2,\ D_1\rho = \rho D_2$.

An important fact that has to be emphasized is that *this definition resembles the definition of the classical operator equivalence in linear algebra.*

Operator equivalence. While these two equivalences look quite similar, they are clearly very different: in linear algebra one uses inverseable matrices U, while in our case ρ may be an arbitrary 0-1 matrix. (One usually expresses the linear operator equivalence as $S = U^{-1}TU$; for inverseable ρ, the relation between the D-parts imposed by simulation may be written in the same way.) This simple observation may be corroborated to the basic fact, discovered in [JSV96], that, in vectorial space models, the feedback operation is a generalization of the classical trace. ◇

Theorem 8.5. *Simulation preserves the language (for concrete presentations).*

PROOF: Let $M_1 \rightarrow_\rho M_2$ be as above. Then

$$\begin{aligned} L(M_1) &= A_1 + B_1(1 + D_1 + D_1^2 + \ldots)C_1 \\ &= A_1 + B_1(1 + D_1 + D_1^2 + \ldots)\rho C_2 \\ &= A_1 + B_1\rho(1 + D_2 + D_2^2 + \ldots)C_2 \qquad \text{because } D_1^k\rho = \rho D_2^k \\ &= A_2 + B_2(1 + D_2 + D_2^2 + \ldots)C_1 \\ &= L(M_2) \end{aligned}$$

□

♡ This result may be stated in a more abstract setting, but it is not necessary for our purpose. To this end, one has to pick up the basic properties used in the proof (e.g., the existence of limits for increasing sums, the distributivity of product over such limits, etc.) and to lift them as axioms in an abstract setting. Then, the proof is similar. ◇

Notation 8.6. $\xrightarrow{inj}$, $\xrightarrow{sur}$, ... denote the particular types of simulation which use the specified classes of relations, i.e., injections, surjections, ..., respectively. Notice that simulation *is not* a symmetric relation! The opposite relation is denoted by $\longleftarrow$. Equivalent presentations are to be connected by chains of simulations in both directions.

8.3 From nondeterministic to deterministic automata

We present here the well-known power-set construction for passing from nondeterministic to deterministic automata. A technical result shows the construction produces an automaton similar (via relations) to the original one.

Deterministic automata have a few nice features. Among them we notice the easy process of minimization and the fact that minimal automata are unique up to an isomorphism – these will be shown soon in the forthcoming sections. These two properties are not shared by nondeterministic automata. Indeed, there exist non-isomorphic minimal nondeterministic automata and the minimization process for nondeterministic automata is still unclear. To conclude, it is desirable to have a procedure for transforming a nondeterministic automaton into a deterministic one.

♡ There are finitely many nondeterministic automata with fewer states than a given one and automata equivalence is decidable. Hence it is obvious that one may find a minimal equivalent automaton. The unclear problem with respect to the minimization of nondeterministic automata is whether there are some natural criteria to be tested for checking automata minimality. ◇

Actually, there exists a well-known method to eliminate the nondeterminism, the so-called "power-set construction." The states of the new (deterministic) automaton are the subsets of the states of the given automaton. By this procedure an automaton similar to the one given is produced.

Theorem 8.7. *Let $\mathcal{A}$ be a nondeterministic finite automaton. Then, there exists a deterministic finite automaton $D(\mathcal{A})$ and a relation ρ such that $D(\mathcal{A}) \to_\rho \mathcal{A}$.*

PROOF: The automaton $D(\mathcal{A})$ is $\begin{array}{ccc} & \uparrow & 2^n \\ \to & \overline{A} & \overline{B} \\ 2^n & \overline{C} & \overline{D} \end{array}$. Its states are elements in $[2^n]$. We use a more suggestive notation for them: such indices are identified with subsets $s \subseteq [n]$. The components of the matrix of $D(\mathcal{A})$ are defined as follows:

$\overline{A} = A$;
$\overline{b}_t = 1$ for $t = t_{in}$ where $t_{in} = \{j : b_j = 1\}$, otherwise 0;
$\overline{c}_s = \sum_{i:\, i \in s} c_i$; and
$a \in \overline{d}_{st}$ (or $s \xrightarrow{a} t$ in $\overline{D}$) iff $t = \{j : \exists i \in s \text{ with } i \xrightarrow{a} j \text{ in } D\}$.

A few verbal explanations may be useful. There is a unique initial state t_{in} which gathers all the initial states in $\mathcal{A}$. A state $\overline{c}_s$ is final if and only if it contains a final state of $\mathcal{A}$. For a state s in $D(\mathcal{A})$ and a letter a, there is a unique state t in $D(\mathcal{A})$ such that $s \xrightarrow{a} t$; this state t is obtained gathering all the states in $\mathcal{A}$ that are reached from states in s using a.

Take the relation "contain" $\rho \subseteq [2^n] \times [n]$ defined by

$$(s, j) \in \rho \quad \text{iff} \quad s \ni j$$

The proof is finished by showing that

$$D(\mathcal{A}) \to_\rho \mathcal{A}$$

(1) Clearly, $\overline{A} = A$;

(2) Then, $\overline{B}\rho = B$ is equivalent to $(\overline{B}\rho)_j = b_j, \forall j \in [n]$. We prove this equality by a double implication (the elements which occur so far are 0 or 1):

- Let $b_j = 1$. Since $j \in t_{in}$ we have $(\overline{B}\rho)_j = \sum_{s:\, s \ni j} \overline{b}_s \geq \overline{b}_{t_{in}} = 1$. Hence $(\overline{B}\rho)_j = 1$.
- Conversely, let $(\overline{B}\rho)_j = 1$. There exists s with $\overline{b}_s = 1$ and $\rho_{sj} = 1$. Hence $\overline{b}_s = 1$ and $j \in s$. As $\overline{b}_t$ is 1 only for $t = t_{in}$, we have $s = t_{in}$. Together with $j \in s$ it implies $b_j = 1$.

(3) Next, $\overline{C} = \rho C$: Clearly $(\rho C)_s = \sum_i \rho_{si} c_i = \sum_{i:\, i \in s} c_i$.

(4) Finally, $\overline{D}\rho = \rho D$ is equivalent to $(\overline{D}\rho)_{sj} = (\rho D)_{sj}$, $\forall s \subseteq [n], j \in [n]$. Using the definition of ρ this equality may be rewritten as $\sum_{t:\, t \ni j} \overline{d}_{st} = \sum_{i:\, i \in s} d_{ij}$, $\forall s \subseteq [n], j \in [n]$ We prove this latter equality by a double inclusion:

- Suppose that $a \in \sum_{t:\, t \ni j} \overline{d}_{st}$. Then $\exists t_0$ such that $t_0 \ni j$ and $s \xrightarrow{a} t_0$ in $\overline{D}$. From $s \xrightarrow{a} t_0$ in $\overline{D}$, by definition one gets $t_0 = \{k \colon \exists i \in s \text{ with } i \xrightarrow{a} k \text{ in } D\}$. Since $j \in t_0$, $\exists i_0 \in s$ with $i_0 \xrightarrow{a} j$ in D. Hence, $a \in \sum_{i:\, i \in s} d_{ij}$.
- Conversely, suppose $a \in \sum_{i:\, i \in s} d_{ij}$. Then $\exists i_0 \in s$ with $i_0 \xrightarrow{a} j$ in D. Let $t_0 = \{k \colon \exists i \in s \text{ with } i \xrightarrow{a} k \text{ in } D\}$. Clearly, $j \in t_0$. Then, by definition, $s \xrightarrow{a} t_0$ in $\overline{D}$ and $t_0 \ni j$. Hence $a \in \sum_{t:\, t \ni j} \overline{d}_{st}$. □

We already proved that simulation preserves automaton language. Hence the above construction is sound, i.e., the following corollary is obtained.

Corollary 8.8. *$L(\mathcal{A}) = L(D(\mathcal{A}))$, hence every nondeterministic finite automaton is equivalent to a deterministic finite automaton.* □

Example. A simple example illustrates the construction.

Let $\mathcal{A} = \left[\begin{array}{c|cc} 0 & 1 & 1 \\ \hline 0 & a & a+b \\ 1 & b & b \end{array}\right]$. Then: $D(\mathcal{A}) = \left[\begin{array}{c|cccc} 0 & 0 & 0 & 0 & 1 \\ \hline 0 & 0 & 0 & 0 & 0 \\ 0 & 0 & 0 & b & a \\ 1 & 0 & 0 & 0 & b \\ 1 & 0 & 0 & 0 & a+b \end{array}\right]$; $\rho = \begin{bmatrix} 0 & 0 \\ 1 & 0 \\ 0 & 1 \\ 1 & 1 \end{bmatrix}$;

(indices 1,2,3,4 denote the sets $\emptyset, \{1\}, \{2\}, \{1,2\}$, respcetively) A simple verification shows that:

$$D(\mathcal{A}) \begin{bmatrix} 1 & 0 \\ 0 & \rho \end{bmatrix} = \begin{bmatrix} 0 & 1 & 1 \\ 0 & 0 & 0 \\ 0 & a & a+b \\ 1 & b & b \\ 1 & a+b & a+b \end{bmatrix} = \begin{bmatrix} 1 & 0 \\ 0 & \rho \end{bmatrix} \mathcal{A}. \qquad \diamond$$

To minimize the effect of state explosion generated by this construction, one usually mixes the application of the above construction with the deletion of the non-accessible states. The combination is a sort of "on-the-fly" technique: the states are constructed step by step starting from the initial state; if a new state occurs, then its transitions are computed; these transitions eventually introduce new states and the procedure is repeated for them; finally, one gets the accessible part of the equivalent deterministic automaton.

8.4 Minimization: I. Accessibility

In the current and the forthcoming two sections we present an algebraic version of state minimization for finite automata. Here, we describe how reduction to the accessible part of an automaton is modeled by simulations via injective functions.

The main result achieved in the current and the forthcoming two sections is that the minimization process for deterministic automata is completely captured by simulation via functions. The present section starts with a description of the general minimization problem. Then we pass to the accessibility problem.

Generalities on automata minimization. In the minimization process one may consider two different problems: (1) reduction/minimization of the number of states, or (2) reduction/minimization of the number of edges. Here we study the minimization of the *state number* of *deterministic finite automata*.

The state number of deterministic finite automata may be reduced by the deletion of the states which are inaccessible from the initial states and by the identification of the states that have the same behaviour (i.e., those states which reach final states by the same set of words). We will show that these two minimization techniques are modeled using simulation via injective and surjective functions, respectively.

Let $\mathcal{A}$ be a deterministic automaton. Classically, it is represented as a 5-uple (S, V, s_0, δ, S_f), where S is the set of states, S_f is the set of final states, s_0 is the initial state, and δ is the transition function

$$\delta : S \times V \to S, \quad (\text{state, letter}) \xrightarrow{\delta} \text{state}$$

The transition function is extended to a function $\overline{\delta} : S \times V^* \to S$ on words by

$$\begin{aligned}&\overline{\delta}(s, \lambda) = s\\ &\overline{\delta}(s, wa) = \delta(\overline{\delta}(s, w), a), \quad w \in V^*, a \in V\end{aligned}$$

The language accepted by $\mathcal{A}$ may be written in terms of this function in the following way

$$L(\mathcal{A}) = \{w : \overline{\delta}(s_0, w) \in S_f\}$$

A state s is called *accessible* if there exists w with $s = \overline{\delta}(s_0, w)$. Two states s_1, s_2 have the same *(output) behaviour* if the language obtained starting from s_1 is equal to the language obtained starting from s_2; in other words, there are no words w with $\overline{\delta}(s_1, w) \in S_f$ and $\overline{\delta}(s_2, w) \notin S_f$ or vice-versa.

Accessibility. There is a simple algorithm which computes the set of accessible states. (For the second step in the forthcoming algorithm, notice that all S^k are subsets of a finite set, hence the increasing chain $S^0 \subset S^1 \subset \ldots$ should stabilize.)

ALGORITHM for detection of inaccessible states

1. Compute iteratively:

$$S^0 = \{s_0\}$$
$$\ldots$$
$$S^{k+1} = S^k \cup \{\delta(s,a) : s \in S^k, a \in V\}.$$
$$\ldots$$

2. Partial result: $S^0 \subset S^1 \subset S^2 \subset \ldots \subset S^{n_0} = S^{n_0+1} = \ldots$
3. Final result: S^{n_0} contains the accessible states and $S \setminus S^{n_0}$ the inaccessible states.

Example 8.9. Let us consider the automaton in Fig. 8.1. It has the following matricial presentation:

$$\mathcal{A} = \begin{array}{c} \rightarrow \\ A \\ B \\ C \\ D \\ E \\ F \\ G \end{array} \left[\begin{array}{c|ccccccc} 0 & 1 & 0 & 0 & 0 & 0 & 0 & 0 \\ \hline 0 & 0 & a & 0 & b & 0 & 0 & 0 \\ 0 & 0 & a & b & 0 & 0 & 0 & 0 \\ 1 & 0 & 0 & 0 & a & b & 0 & 0 \\ 0 & 0 & 0 & 0 & a & b & 0 & 0 \\ 1 & 0 & a & b & 0 & 0 & 0 & 0 \\ 0 & 0 & 0 & a & 0 & 0 & 0 & b \\ 0 & 0 & 0 & 0 & 0 & b & a & 0 \end{array} \right]$$

The algorithm produced the following sets:

$S^0 = \{A\}$;
$S^1 = \{A\} \cup \{B, D\} = \{A, B, D\}$;
$S^2 = \{A, B, D\} \cup \{B, D, C, E\} = \{A, B, C, D, E\}$;
$S^3 = \{A, B, C, D, E\} \cup \{B, D, C, E\} = \{A, B, C, D, E\} = S^2$

$\{A, B, C, D, E\}$ is the set of the accessible states and $\{F, G\}$ is the set of inaccessible states. Then, the accessible part of the automaton is

$$\mathcal{A}_{acc} = \begin{array}{c} \rightarrow \\ A \\ B \\ C \\ D \\ E \end{array} \left[\begin{array}{c|ccccc} 0 & 1 & 0 & 0 & 0 & 0 \\ \hline 0 & 0 & a & 0 & b & 0 \\ 0 & 0 & a & b & 0 & 0 \\ 1 & 0 & 0 & 0 & a & b \\ 0 & 0 & 0 & 0 & a & b \\ 1 & 0 & a & b & 0 & 0 \end{array} \right]$$

Theorem 8.10. *Let $\mathcal{A}$ be a deterministic finite automaton and $\mathcal{A}_{acc}$ its accessible part. There exists an injection ρ such that $\mathcal{A}_{acc} \rightarrow_\rho \mathcal{A}$.*

PROOF: Using certain permutations of the states of $\mathcal{A}$ we may suppose its matricial presentation is

$$\begin{array}{c} \rightarrow \\ n_1 \\ n_2 \end{array} \left[\begin{array}{c|cc} A & B_1 & B_2 \\ \hline C_1 & D_{11} & D_{12} \\ C_2 & D_{21} & D_{22} \end{array} \right]$$

where the accessible states are collected in n_1 and the inaccessible states are collected in n_2. Then $B_2 = 0$ (the states in n_2 are inaccessible, hence they are not initial) and $D_{12} = 0$ (if not so, $D_{12} \neq 0$ and there is a transition $s_1 \xrightarrow{a} s_2$ with s_1 in n_1 and s_2 in n_2, hence s_2 is accessible; but this is impossible since s_2 was supposed to be inaccessible). A simple computation shows that $\mathcal{A}_{acc} \rightarrow_{[\mathsf{I}_{n_1} \ \ 0_{n_1,n_2}]} \mathcal{A}$. □

A more general fact may be stated. For every $\mathcal{A}'$ we have:

$$\mathcal{A}' \rightarrow_\rho \mathcal{A} \text{ for an injection } \rho \text{ iff } \mathcal{A} \setminus \rho(\mathcal{A}') \text{ is non-accessible from } \rho(\mathcal{A}').$$

In other words, and read from right to left, simulation via injective functions allows one to delete a part from an automaton, provided that part is not accessible from the remaining one.

Again, taking into account the fact that simulation preserves the language we get the soundness of the algorithm.

Corollary 8.11. $L(\mathcal{A}) = L(\mathcal{A}_{acc})$, *hence every deterministic finite automaton is equivalent to an accessible deterministic finite automaton.*

8.5 Minimization: II. Reduction

In the present section we show that identification of the states with the same behaviour (the so-called "indistinguishable" states) is modeled by simulation via surjective functions.

The second possibility to minimize the number of states in a deterministic automaton is to identify those states which have "similar" behaviour. This intuitive idea is captured by the results presented in this subsection. (As before, the equivalences are over finite sets, hence the iterative process finishes is a finite number of steps.)

ALGORITHM for the detection of the indistinguishable states

1. Compute iteratively the equivalences:

$$s_1 \equiv_0 s_2 \text{ iff } (s_1 \in S_f \wedge s_2 \in S_f) \vee (s_1 \notin S_f \wedge s_2 \notin S_f), \quad \forall s_1, s_2 \in S$$

...

$$s_1 \equiv_{k+1} s_2 \text{ iff } s_1 \equiv_k s_2 \wedge \forall a \in V : \delta(s_1, a) \equiv_k \delta(s_2, a) \quad \forall s_1, s_2 \in S$$

...

(Notice that two states s_1, s_2 are: $\equiv_0$-equivalent iff they cannot be distinguished by 0-length transitions; $\equiv_1$-equivalent iff they cannot be distinguished by 1-length transitions; etc.)

2. Partial result: $\equiv_0 \supset \equiv_1 \supset \equiv_2 \supset \ \ldots \ \supset \equiv_{n_0} = \equiv_{n_0+1} = \ \ldots$

3. Final result: two states are considered $\equiv$-equivalent iff they are $\equiv_{n_0}$-equivalent.

Example 8.12. For the above automaton $\mathcal{A}_{acc}$, compute:

$$\begin{aligned}&\equiv_0: \{C,E\},\{A,B,D\}\\&\equiv_1: \{C,E\},\{A\},\{B,D\}\\&\equiv_2: \{C,E\},\{A\},\{B,D\}\end{aligned}$$

Hence we may identify C with E and B with D. This shows $\mathcal{A}_{acc}$ is reduced to $\mathcal{A}_{min}$, where

$$\mathcal{A}_{min} = \begin{array}{c} \to \\ A \\ B=D \\ C=E \end{array} \left[\begin{array}{c|cccc} 0 & 1 & 0 & 1 \\ \hline 0 & 0 & a+b & 0 \\ 0 & 0 & a & b \\ 1 & 0 & a & b \end{array}\right]$$

Theorem 8.13. *Let $\mathcal{A}$ be an automaton with n states, $\equiv$ the equivalence relation given by the above algorithm, and $\rho : [n] \to [m]$ a surjective function such that $\rho(i) = \rho(j)$ iff $i \equiv j$. Then, there exists an automaton $\mathcal{A}'$ with m states such that $\mathcal{A} \to_\rho \mathcal{A}'$.*

PROOF: By the definition

$$[s_1 \equiv_{k+1} s_2 \text{ iff } s_1 \equiv_k s_2 \wedge \forall a \in V : \delta(s_1,a) \equiv_k \delta(s_2,a)], \quad \forall s_1, s_2 \in S$$

There is an n_0 with $\equiv\, = \,\equiv_{n_0} \,=\, \equiv_{n_0+1}$. The above condition, written for $k = n_0$, implies:

$$[s_1 \equiv s_2 \text{ iff } s_1 \equiv s_2 \wedge \forall a \in V : \delta(s_1,a) \equiv \delta(s_2,a)], \quad \forall s_1, s_2 \in S$$

Hence,

$$(*) \left\{ \begin{array}{l} s_1 \equiv s_2 \Rightarrow \delta(s_1,a) \equiv \delta(s_2,a), \ \forall a \in V \\ s_1 \equiv s_2 \Rightarrow s_1 \equiv_0 s_2 \Rightarrow [s_1 \in S_f \text{ iff } s_2 \in S_f] \end{array} \right.$$

Let $M = \left[\begin{array}{c|c} A & B \\ \hline C & D \end{array}\right]$ be a matricial presentation of $\mathcal{A}$ and $\rho : [n] \to [m]$ a surjective function associated with $\equiv$. (I.e., ρ satisfies the relation $\rho(i) = \rho(j)$ iff $i \equiv j$. The elements of $[m]$ correspond to the equivalence classes of $\equiv$.)

The matrix $N = \left[\begin{array}{cc} A & B\rho \\ C & D\rho \end{array}\right]$ is obtained by adding all columns in M corresponding to $\equiv$-equivalent states. Condition (*) shows that: (1) the rows of N corresponding to $\equiv$-equivalent states are equal, hence there exist C' and D' such that $N = \left[\begin{array}{cc} A & B\rho \\ \rho C' & \rho D' \end{array}\right]$, and (2) D' is also deterministic. Take the automaton $\mathcal{A}' = \left[\begin{array}{c|c} A & B\rho \\ \hline C' & D' \end{array}\right]$. Then $\left[\begin{array}{cc} A & B\rho \\ C & D\rho \end{array}\right] = N = \left[\begin{array}{cc} A & B\rho \\ \rho C' & \rho D' \end{array}\right]$, hence $\mathcal{A} \to_\rho \mathcal{A}'$.

□

A more general fact may be stated in this case, too. For every automaton $\mathcal{A}'$

$\mathcal{A} \to_\rho \mathcal{A}'$ with ρ surjection iff $\mathcal{A}'$ may be obtained from $\mathcal{A}$ by identification of states with coherent output behaviour.

(By "coherent output behaviour" we mean the following: if two states are identified, then the states that are reached from them with the same letter are to be identified, too.)

Simulation preserves the language, hence the algorithm is sound. Formally,

Corollary 8.14. *The reduction algorithm based on the identification of equivalent states is correct, i.e., it produces an automaton $\mathcal{A}'$ with $L(\mathcal{A}') = L(\mathcal{A})$.* □

Let $L(\mathcal{A}; s)$ denotes the language accepted by an automaton $\mathcal{A}$ in which the initial state is changed to s. Then

Proposition 8.15. $s_1 \equiv s_2$ *iff* $L(\mathcal{A}; s_1) = L(\mathcal{A}; s_2)$

PROOF: For left-to-right implication, by the (*) property stated in the proof of Theorem 8.13 we have $s_1 \equiv s_2 \Rightarrow \forall w : \delta(s_1, w) \equiv \delta(s_2, w)$. As $\equiv\ \subseteq\ \equiv_0$ we have $\delta(s_1, w) \equiv_0 \delta(s_2, w)$, hence $w \in L(\mathcal{A}; s_1)$ iff $w \in L(\mathcal{A}; s_2)$.

Conversely, let $\sim$ be the equivalence

$$s_1 \sim s_2 \text{ iff } L(\mathcal{A}, s_1) = L(\mathcal{A}; s_2)$$

By induction it's easy to show that $\sim\ \subseteq\ \equiv_k, \forall k$. Indeed, for $k = 0$ one has

$$s_1 \sim s_2 \Rightarrow [\lambda \in L(\mathcal{A}; s_1) \text{ iff } \lambda \in L(\mathcal{A}; s_2)] \Rightarrow s_1 \equiv_0 s_2$$

If it is true for k, i.e., $\forall s_1, s_2 : s_1 \sim s_2 \Rightarrow s_1 \equiv_k s_2$, then

$$\begin{aligned} s_1 \sim s_2 \quad & \Rightarrow L(\mathcal{A}; \delta(s_1, a)) = L(\mathcal{A}; \delta(s_2, a)),\ \forall a \in V \\ & \Rightarrow \delta(s_1, a) \sim \delta(s_2, a),\ \forall a \in V \\ & \Rightarrow \delta(s_1, a) \equiv_k \delta(s_2, a),\ \forall a \in V (\text{by the inductive hypothesis}) \\ & \Rightarrow s_1 \equiv_{k+1} s_2 (\text{by } s_1 \equiv_k s_2) \end{aligned}$$

Both implications are proved, hence the proof is complete. □

8.6 Minimization: III. Deterministic automata

Finally, in this last section we conclude on deterministic automata minimization. It is shown that: (1) the minimal deterministic automaton is unique (up to an isomorphism); (2) an automaton may be connected with its minimal version by a chain of simulations via (injective and surjective) functions.

Minimization idea. Let L be an arbitrary language. One may define the following languages $L_w = \{v: wv \in L\}$. These languages may be obtained by taking the successive derivatives of the starting language. Another possibility to look on these languages is the following: if L is regular and we have a deterministic automaton which recognizes L, then L_w is the language recognized by the automaton starting from state $\overline{\delta}(s_0, w)$.

What can be said about L and the corresponding set of languages L_w, with $w \in V^*$? A few important facts are:

- The language L is regular iff the set of these L_w is finite.
- A deterministic automaton recognizing L passes through states associated with these languages L_w. (For nondeterministic automata the matter is more complicated: the languages L_w are split as sums. Actually, the minimization problem for nondeterministic automata is still open.)
- A deterministic automaton for L is minimal when exactly one state corresponds to each distinct L_w language.

Minimization via simulations. The constructions of the previous two sections for the deletion of *all* inaccessible vertices and for the identification of *all* the states which cannot be separated by paths to the final states have occurred as particular cases of simulations, i.e., via injective and surjective functions, respectively. So, we get the following result.

Corollary 8.16. *—The minimal deterministic finite automaton is unique, up to isomorphism.*
—A deterministic finite automaton is minimal iff it is accessible and reduced.
—Let $\mathcal{A}$ be an automaton and $\mathcal{A}_{min}$ the equivalent minimal automaton. Then, $\mathcal{A} \stackrel{inj}{\longleftarrow} \mathcal{A}_{acc} \stackrel{sur}{\longrightarrow} \mathcal{A}_{min}$.
—Two deterministic finite automata are equivalent iff via simulations $\stackrel{inj}{\longleftarrow}$ and $\stackrel{sur}{\longrightarrow}$ they may be reduced to the same automaton. □

Finally, we explicitly write a corollary which clarifies the important role played by simulation. Together with Kleene and Normalization Theorems it will be the main fact used for the Fundamental Theorem on the axiomatization of the algebra of regular events.

Corollary 8.17. *Two nondeterministic finite automata are equivalent (i.e., they recognize the same language) iff they may be connected by a chain of simulations.*

PROOF: Simulation preserves the associated language, hence the right-to-left implication is true.

For the left-to-right implication, suppose two equivalent nondeterministic finite automata $\mathcal{A}_1$ and $\mathcal{A}_2$ are given. Using simulations via relations they may be connected to the corresponding deterministic finite automata $D(\mathcal{A}_1)$ and $D(\mathcal{A}_2)$, respectively. Then $D(\mathcal{A}_1)$ and $D(\mathcal{A}_2)$ are equivalent, too. By simulations with surjective and injective functions we pass to the corresponding minimal deterministic finite automata $D(\mathcal{A}_1)_{min}$ and $D(\mathcal{A}_2)_{min}$, respectively. These minimal automata are again equivalent. From the uniqueness of minimal deterministic finite automaton (up to an isomorphism) it follows that

$$\mathcal{A}_1 \overset{rel}{\longleftarrow} D(\mathcal{A}_1) \overset{inj}{\longleftarrow} \cdot \overset{sur}{\longrightarrow} D(\mathcal{A}_1)_{min} = D(\mathcal{A}_2)_{min} \overset{sur}{\longleftarrow} \cdot \overset{inj}{\longrightarrow} D(\mathcal{A}_2) \overset{rel}{\longrightarrow} \mathcal{A}_2$$

This concludes our proof. □

8.7 Regular expressions and Kleene algebras

In this section we introduce regular expressions and two basic algebraic structures: (idempotent) Conway and Kleene theories.

Regular expressions. Starting with the fundamental study of McCulloch & Pitts in 1943 regarding the nervous activity, Kleene [Kle56] introduced *the algebra of regular events* as an abstract model for the relationship between an animal and its environment.

Intuitive meaning of "events". The underlying idea is rather simple. An animal can observe an event from its environment only by the analysis of the effect of the input stimuli to its nervous system. Then, an *event* in the environment is modeled by the class of the sequences of input stimuli that lead to the same final state. (One has to suppose that in the observation process the animal is "attentive"; this may be modeled by a requirement that it starts the observation from a fixed set of initial states.) ◇

The events are organized as an algebra

$$\emptyset, \{\lambda\}, E_1 \cup E_2, E_1 \cdot E_2, E^*$$

where:

— "$\cup$" is union;
— "$\cdot$" is catenation $E_1 \cdot E_2 = \{vw : v \in E_1, w \in E_2\}$; and
— "$*$" is nondeterministic repetition of a event zero, one, two, or more times: $E^* = \{\lambda\} \cup E \cup E \cdot E \cup \ldots$.

The *regular events* are those events that are obtained from $0, 1$, and atoms a by a finite number of applications of the regular operations (union, product, or star). Formally, they are defined as follows:

Definition 8.18. (regular expressions)

(1) 0, 1 and $a (\in V)$ are regular expressions;
(2) if E, F are regular expressions, then $E+F, E \cdot F$ and E^* are regular expressions;
(3) all regular expressions are obtained applying (1) and (2) by a finite number of times.

The *language (or event)* specified by a regular expression E, denoted $|E|$ $(\subseteq V^*)$, is inductively defined as follows:

$$|0| = \emptyset; \quad |1| = \{\lambda\}; \quad |a| = \{a\};$$
$$|E+F| = |E| \cup |F|; \quad |E \cdot F| = |E| \cdot |F|; \quad |E^*| = |E|^*$$

The relationship between regular events and regular expressions is given by the following fundamental theorem of Kleene: *A language is regular (e.g., it is recognized by a nondeterministic finite automaton) iff it may be specified by a regular expression.* In the coming sections we give an algebraic proof for this result.

Conway and Kleene algebras. In this subsection we describe the basic algebraic structures we are using in this chapter: Conway and Kleene algebras. They are defined as follows.

♡ For the readers who are familiar with the first part of this book we notice that Conway theories are BNAs over matrix theories and Kleene algebras are $d\delta$-flows. ◇

Definition 8.19. *Idempotent Conway/Kleene theories (regular algebras)*: Suppose a doubly-ranked family $\mathbf{M} = (\mathcal{M}(m,n)_{m,n}, +, \cdot, *, 0_{m,n}, \mathsf{I}_n)$ is given, where

$$0_{m,n} \in \mathcal{M}(m,n), \mathsf{I}_n \in \mathcal{M}(n,n);$$
$$+ : \mathcal{M}(m,n) \times \mathcal{M}(m,n) \to \mathcal{M}(m,n);$$
$$\cdot : \mathcal{M}(m,n) \times \mathcal{M}(n,p) \to \mathcal{M}(m,p);$$
$$* : \mathcal{M}(n,n) \to \mathcal{M}(n,n)$$

$(\mathcal{M}(m,n)_{m,n}, +, \cdot, 0_{m,n}, \mathsf{I}_n)$ is a *semiring of matrices* if the following axioms hold, whenever the terms are defined:

- with $+, 0$ it is a commutative monoid, i.e.,
  $(a+b)+c = a+(b+c); \quad a + 0_{m,n} = 0_{m,n} + a = a; \quad a+b = b+a$
- with $\cdot, \mathsf{I}_p$ it is a category, i.e.,
  $(a \cdot b) \cdot c = a \cdot (b \cdot c); \quad a \cdot \mathsf{I}_n = a = \mathsf{I}_m \cdot a$
- distributivity:
  $a \cdot (b+c) = (a \cdot b) + (a \cdot c); \quad (a+b) \cdot c = (a \cdot c) + (b \cdot c)$
- zero-laws:
  $0_{p,m} \cdot a = 0_{p,n}; \quad a \cdot 0_{n,p} = 0_{m,p}$

It is *idempotent* if $a + a = a$.

Axioms for repetition:

(I) $(\mathsf{I}_n)^* = \mathsf{I}_n$
(S) $(a+b)^* = (a^* \cdot b)^* \cdot a^*$
(P) $(a \cdot b)^* = \mathsf{I}_n + a \cdot (b \cdot a)^* \cdot b$
(Inv) $a \cdot \rho = \rho \cdot b \Rightarrow a^* \cdot \rho = \rho \cdot b^*$, where ρ is matrix over 0,1

All these axioms define *idempotent Kleene theories.* Without (Inv) one gets the axioms for *idempotent Conway theories.* Dropping idempotency, one gets the general notions.

A few simple consequences of the star axioms are:

(S1) $(a+b)^* = a^*(ba^*)^*$
(P1) $a^* = \mathsf{I}_n + aa^*$
(P2) $b(ab)^* = (ba)^*b$
(P3) $bb^* = b^*b$

Proposition 8.20. *The above laws are valid in the case of matrices of languages* $\mathcal{M}_{\mathcal{P}(V^*)}(m,n) = \{(a_{ij})_{i\in[m],j\in[n]} : a_{ij} \subseteq \mathcal{P}(V^*)\}$ *with the usual operations.*

PROOF: The soundness of (Inv) has already been proved. For (P) notice that: $(ab)^* = 1 + ab + abab + ababab + \ldots = 1 + a(1 + ba + baba + \ldots)b = 1 + a(ba)^*b$. For (S), notice that: $(a+b)^* = 1 + (a+b) + (a+b)(a+b) + (a+b)(a+b)(a+b) + \ldots$. Compute the product of sums and rearrange the terms according to the occurrences of b as $\{[(a\ldots a)b]\ldots[(a\ldots a)b]\}(a\ldots a)$, where $\ldots$ means zero, one, or more occurrences of the corresponding factors. Written in this way, the sum is equal to the sum resulting from the computation of $(a^*b)^*a^*$. □

EXERCISES AND PROBLEMS (SEC. 8.7)

1. Prove the following identities hold in an idempotent Conway theory: (S1) $(a+b)^* = a^*(ba^*)^*$; (P1) $a^* = \mathsf{I}_n + aa^*$; (P2) $b(ab)^* = (ba)^*b$; (P3) $bb^* = b^*b$. (06)
2. If S is an idempotent semiring, then the matrices over S forms an idempotent semiring. (06)
3. (derivatives) Let E be an event. Define

$$o[E] = \begin{cases} 1, & 1 \in E \\ 0, & 1 \notin E \end{cases} \qquad\qquad \partial_a[E] := \{w : aw \in E\}$$

The calculus is based on the following derivation rules:

$D1$	$o[0] = o[a] = 0$	$D5$	$\partial_a[0] = \partial_a[b] = 0, \partial_a[a] = 1 (a \neq b)$
$D2$	$o[E+F] = o[E] + o[F]$	$D6$	$\partial_a[E+F] = \partial_a[E] + \partial_a[F]$
$D3$	$o[E \cdot F] = o[E] \cdot o[F]$	$D7$	$\partial_a[E \cdot F] = \partial_a[E] \cdot F + o[E] \cdot \partial_a[F]$
$D4$	$o[E^*] = 1$	$D8$	$\partial_a[E^*] = \partial_a[E] \cdot E^*$

(a) Apply the derivation method to find an automaton associated with the

expression $E = a^*b + bb^*a$. (08)
(b) This calculus allows for a passing from regular expressions to automata: the automaton states are labeled with expressions; a state E is final iff $o[E] = 1$; the transitions are: $E \xrightarrow{a} F$ iff $F = \partial_a[E]$. Prove this method is correct, i.e., the result is an automaton which recognizes the language specified by the starting regular expression. (12)

4. If S is an idempotent semiring which obeys (I), (S) and (P), then the matrices over S also satisfies (I), (S) and (P). (Krob) (15)

8.8 Kleene Theorem: I. From automata to regular expressions

Here we give an algebraic version of the Kleene theorem that the languages recognized by automata may be specified with regular expressions. It turns out that within this algebraic framework the proof of the Kleene theorem is straightforward.

Let $\mathcal{A}$ be an automaton with the matricial presentation $\mathcal{A} = \left[\begin{array}{c|c} A & B \\ \hline C & D \end{array}\right]$. Then $L(\mathcal{A}) = A + BD^*C$, but ... this is not a regular expression. Indeed, the star operation acts on a matrix. So, the question is: can we express D^* in terms of the star of its components? The answer is positive and it is based on the following theorem.

Theorem 8.21. *(star of matrices in idempotent Conway theories)*
In an idempotent Conway theory one can prove that

$$\begin{bmatrix} a & b \\ c & d \end{bmatrix}^* = \begin{bmatrix} a^* + a^*bwca^* & a^*bw \\ wca^* & w \end{bmatrix}$$

where $w = (ca^*b + d)^*$.

PROOF: 1. First we prove that:

$$\begin{bmatrix} a & b \\ 0 & 0 \end{bmatrix}^* = \begin{bmatrix} a^* & a^*b \\ 0 & 1 \end{bmatrix}$$

Indeed,

$$\begin{aligned} \begin{bmatrix} a & b \\ 0 & 0 \end{bmatrix}^* &= \left(\begin{bmatrix} 1 \\ 0 \end{bmatrix} \begin{bmatrix} a & b \end{bmatrix} \right)^* \\ &=_P \begin{bmatrix} 1 & 0 \\ 0 & 1 \end{bmatrix} + \begin{bmatrix} 1 \\ 0 \end{bmatrix} \left(\begin{bmatrix} a & b \end{bmatrix} \begin{bmatrix} 1 \\ 0 \end{bmatrix} \right)^* \begin{bmatrix} a & b \end{bmatrix} \\ &= \begin{bmatrix} 1 + a^*a & a^*b \\ 0 & 1 \end{bmatrix} \\ &= \begin{bmatrix} a^* & a^*b \\ 0 & 1 \end{bmatrix} \end{aligned}$$

2. Similarly,

$$\begin{bmatrix} 0 & 0 \\ c & d \end{bmatrix}^* = \begin{bmatrix} 1 & 0 \\ d^*c & d^* \end{bmatrix}$$

3. Finally,

$$\begin{aligned}
\begin{bmatrix} a & b \\ c & d \end{bmatrix}^* &= \left(\begin{bmatrix} a & b \\ 0 & 0 \end{bmatrix} + \begin{bmatrix} 0 & 0 \\ c & d \end{bmatrix}\right)^* \\
&=_{(S1)} \begin{bmatrix} a & b \\ 0 & 0 \end{bmatrix}^* \left(\begin{bmatrix} 0 & 0 \\ c & d \end{bmatrix}\begin{bmatrix} a & b \\ 0 & 0 \end{bmatrix}^*\right)^* \\
&= \begin{bmatrix} a^* & a^*b \\ 0 & 1 \end{bmatrix}\begin{bmatrix} 0 & 0 \\ ca^* & ca^*b+d \end{bmatrix}^* \\
&= \begin{bmatrix} a^* & a^*b \\ 0 & 1 \end{bmatrix}\begin{bmatrix} 1 & 0 \\ wca^* & w \end{bmatrix} \\
&= \begin{bmatrix} a^*+a^*bwca^* & a^*bw \\ wca^* & w \end{bmatrix}
\end{aligned}$$

□

Remark 8.22. If π is a permutation, then $(\pi^{-1}A\pi)^* = \pi^{-1}A^*\pi$ (from (P2)), hence the extension of the star from elements to matrices does not depend on the order of the decomposition of the matrix in (block) components.

Theorem 8.23. *(Kleene Theorem, I: automata → regular expressions)*
(i) Every language accepted by an automaton may be specified with a regular expression.
(ii) There is a proof showing this result holds within an idempotent Conway theory.

PROOF: If we consider the matricial presentation of an automaton $\mathcal{A} = \left[\begin{array}{c|c} A & B \\ \hline C & D \end{array}\right]$, then $L(\mathcal{A}) = A + BD^*C$. Applying the above theorem, one may write D^* as a matrix of regular expressions. By substitution in the formula above we get a regular expression for $L(\mathcal{A})$. (All the rules we have applied so far are sound in the case of matrices of languages, hence the resulting regular expression represents the same language, indeed.) □

EXERCISES AND PROBLEMS (SEC. 8.8)

1. Show that Kleene Theorem I follows from the Normalization Theorem for flownomial expressions, provided a BNA over an idempotent matrix theory is used. (12)

8.9 Kleene Theorem: II. From regular expressions to automata

Here we give an algebraic Normalization Theorem for that part of the Kleene theorem saying that the languages specified with regular expressions are recognized by automata.

Usually, this part of the Kleene theorem is proved by some simple constructions on automata. Within the Conway theory framework the proof of this part of the Kleene theorem is slightly more involved.

Regular events form the least class of languages which contains $0, 1$, atoms and it is closed to the regular operations (union, product and star). In an informal way it may be shown that the languages recognized by automata have these properties, too. Hence they contain the regular languages. Below we give an algebraic proof for this result which is based on the *normalization of regular expressions*. This way, automata may be seen as regular expressions in normal form.

Theorem 8.24. *(Normalization Theorem for regular expressions)*
(i) For every regular expression E there exists a matricial presentation $N(E) = \left[\begin{array}{c|c} 0 & B \\ \hline C & D \end{array}\right]$ such that $|E| = BD^*C$.
(ii) There is a proof which shows this result holds in an idempotent Conway theory.

PROOF: For a matricial presentation $M = \left[\begin{array}{c|c} 0 & B \\ \hline C & D \end{array}\right]$ let us denote $L(M) = BD^*C$. The normalization function N (from regular expressions to matricial presentations) is inductively defined by:

– $N(0) = \left[\begin{array}{c|c} 0 & 1 \\ \hline 0 & 0 \end{array}\right]$; $N(1) = \left[\begin{array}{c|c} 0 & 1 \\ \hline 1 & 0 \end{array}\right]$; $N(a) = \left[\begin{array}{c|cc} 0 & 1 & 0 \\ \hline 0 & 0 & a \\ 1 & 0 & 0 \end{array}\right]$;

– if $N(E_i) = \left[\begin{array}{c|c} 0 & B_i \\ \hline C_i & D_i \end{array}\right]$, for $i = 1, 2$, then

$$N(E_1 + E_2) = \left[\begin{array}{c|cc} 0 & B_1 & B_2 \\ \hline C_1 & D_1 & 0 \\ C_2 & 0 & D_2 \end{array}\right]$$

– if $N(E_i) = \left[\begin{array}{c|c} A_i & B_i \\ \hline C_i & D_i \end{array}\right]$, for $i = 1, 2$, then

$$N(E_1 \cdot E_2) = \left[\begin{array}{c|cc} 0 & B_1 & B_1C_1B_2 \\ \hline C_1B_2C_2 & D_1 & D_1C_1B_2 \\ C_2 & 0 & D_2 \end{array}\right]$$

– if $N(E) = \left[\begin{array}{c|c} 0 & B \\ \hline C & D \end{array}\right]$, then

$$N\left(\left[\begin{array}{c|c} 0 & B \\ \hline C & D \end{array}\right]\right) = \left[\begin{array}{c|cc} 0 & B & 1 \\ \hline (CB)^*C & (CB)^*D & 0 \\ 1 & 0 & 0 \end{array}\right]$$

We show that the definitions are correct, namely $L(N(E)) = |E|$. The result follows by induction from the algebraic computation given below.

First, $L(N(f)) = f = |f|$, for $f \in \{0, 1\}$. Then, one can see that

$$\begin{aligned} L(N(a)) &= 0 + \begin{bmatrix} 1 & 0 \end{bmatrix} \begin{bmatrix} 0 & a \\ 0 & 0 \end{bmatrix}^* \begin{bmatrix} 0 \\ 1 \end{bmatrix} \\ &= 0 + \begin{bmatrix} 1 & 0 \end{bmatrix} \begin{bmatrix} 1 & a \\ 0 & 1 \end{bmatrix} \begin{bmatrix} 0 \\ 1 \end{bmatrix} \\ &= 0 + a = a = |a| \end{aligned}$$

The rule for sum follows by

$$\begin{aligned} L(N(E_1 + E_2)) &= 0 + \begin{bmatrix} B_1 & B_2 \end{bmatrix} \begin{bmatrix} D_1 & 0 \\ 0 & D_2 \end{bmatrix}^* \begin{bmatrix} C_1 \\ C_2 \end{bmatrix} \\ &= \begin{bmatrix} B_1 & B_2 \end{bmatrix} \begin{bmatrix} D_1^* & 0 \\ 0 & D_2^* \end{bmatrix} \begin{bmatrix} C_1 \\ C_2 \end{bmatrix} \\ &= B_1 D_1^* C_1 + B_2 D_2^* C_2 \\ &= L(N(E_1)) + L(N(E_2)) = |E_1| + |E_2| = |E_1 + E_2| \end{aligned}$$

The rule for product is satisfied as follows:

$$\begin{aligned} L(N(E_1 \cdot E_2)) &= 0 + \begin{bmatrix} B_1 & B_1 C_1 B_2 \end{bmatrix} \begin{bmatrix} D_1 & D_1 C_1 B_2 \\ 0 & D_2 \end{bmatrix}^* \begin{bmatrix} C_1 B_2 C_2 \\ C_2 \end{bmatrix} \\ &= \begin{bmatrix} B_1 & B_1 C_1 B_2 \end{bmatrix} \begin{bmatrix} D_1^* & D_1^* D_1 C_1 B_2 D_2^* \\ 0 & D_2^* \end{bmatrix} \begin{bmatrix} C_1 B_2 C_2 \\ C_2 \end{bmatrix} \\ &= B_1 D_1^* C_1 B_2 C_2 + B_1 D_1^* D_1 C_1 B_2 D_2^* C_2 + B_1 C_1 B_2 D_2^* C_2 \\ &= B_1 D_1^* C_1 B_2 C_2 + B_1 (D_1^* D_1 + 1) C_1 B_2 D_2^* C_2 \\ &= B_1 D_1^* C_1 B_2 (1 + D_2^*) C_2 \\ &= B_1 D_1^* C_1 B_2 D_2^* C_2 \\ &= L(N(E_1)) \cdot L(N(E_2)) = |E_1| \cdot |E_2| = |E_1 \cdot E_2| \end{aligned}$$

Finally, the rule for star follows by

$$\begin{aligned}
L(N(E^*)) &= [B\ \ 1] \cdot \begin{bmatrix} (CB)^*D & 0 \\ 0 & 0 \end{bmatrix}^* \begin{bmatrix} (CB)^*D \\ 1 \end{bmatrix} \\
&= [B\ \ 1] \cdot \begin{bmatrix} ((CB)^*D)^* & 0 \\ 0 & 1 \end{bmatrix} \begin{bmatrix} (CB)^*D \\ 1 \end{bmatrix} \\
&= B((CB)^*D)^*(CB)^*C + 1 \\
&= B(CB+D)^*C + 1 \\
&= B(D^*CB)^*D^*C + 1 \\
&= BD^*C(BD^*C)^* + 1 \\
&= (BD^*C)^* = |E|^* = |E^*| \qquad \square
\end{aligned}$$

Remark 8.25. There are some other natural candidates for the definition of normalization function. Another example is shown in Exercise 8.9.1.

EXERCISES AND PROBLEMS (SEC. 8.9)

1. Replace the normalization function $N(E^*)$ of Theorem 8.24 by $N1$ below and show that it still is a normalization function, i.e. $|N1(E)| = |E|$. For 0,1, and a the definition is the same. Next, $N1(E_1 + E_2) = \left[\begin{array}{c|cc} A_1 + A_2 & B_1 & B_2 \\ \hline C_1 & D_1 & 0 \\ C_2 & 0 & D_2 \end{array}\right]$;
$N1(E_1 \cdot E_2) = \left[\begin{array}{c|cc} A_1A_2 & B_1 & A_1B_2 \\ \hline C_1A_2 & D_1 & C_1B_2 \\ C_2 & 0 & D_2 \end{array}\right]$; $N1(E^*) = \left[\begin{array}{c|c} A^* & A^*B \\ \hline CA^* & CA^*B + D \end{array}\right]$.
Why was this form not used in Theorem 8.24? What sort of automata are produced by this normalization function? (14)
2. Find automata for $(y^2x + xy^*xy)^*(x^2 + yx^*)^*$. (09)

8.10 Axiomatization, regular expressions

This section collects all previous results: it is proved that Kleene theory axioms give a sound and complete axiomatization for regular languages.

Putting together all the above results, we get the following fundamental theorem for the axiomatization of the algebra of regular events. It shows two regular expressions specify the same language if and only if one may pass from one to the other using the axioms of idempotent Kleene theories.

Theorem 8.26. *(Fundamental Theorem) The idempotent Kleene theory axioms give a correct (i.e., sound and complete) axiomatization for the algebra of regular events/languages.*

PROOF: *Soundness part:* This part is clear, since the rules are sound in the case of matrices of languages.

Completeness part: For this part, let E_1, E_2 be regular expressions specifying the same language. Using the Normalization Theorem one find matricial presentations $\mathcal{A}_i = \left[\begin{array}{c|c} 0 & B_i \\ \hline C_i & D_i \end{array}\right]$, such that $|E_i| = B_i D_i^* C_i = L(\mathcal{A}_i)$, for $i = 1, 2$. Hence these matricial presentations specify equivalent nondeterministic finite automata. Then, by Corollary 8.17 these equivalent nondeterministic finite automata may be connected by a chain of simulations; this transformation is modeled by the invariance rule (Inv). Therefore, one may pass from E_1 to E_2 using idempotent Kleene theory axioms, only. □

EXERCISES AND PROBLEMS (SEC. 8.10)

1. No finite equational axiomatization may be found for regular expressions, even for one-letter alphabets. (15)
2. The *classical axioms* (i.e., axioms of Conway theories and "$C14n.\ a^* = (1 + a + \ldots + a^{n-1})(a^n)^*,\ (n \geq 2)$") are complete for languages over a one-letter alphabet. (15)
3. $C14.n$ implies $C14.n$ for matrices, provided the other classical axioms are valid; $C14.p$ is independent of $C14.n$, whenever n is not divisible by p. (12)
4. The classical axioms, supplemented with $C^+1.\ ab = ba$ and $C^+2.\ a^*b^* = (ab)^*(a^* + b^*)$ are complete for commutative languages. (15)
5. Replacing (Inv) by the equations:

   (I_G) $E_{11} + \ldots + E_{1n} = (x_1 + \ldots + x_n)^*$, where $G = (\{1, 2, \ldots, n\}, \circ, 1)$ is an arbitrary finite group and $E_{11}, \ldots, E_{1n}$ are the components of the matrix $[(a_{ij})_{i,j \in [n]}]^*$, where $a_{ij} = x_k$ iff $i \circ k = j$

   we still have a complete set of axioms for regular events, but now in a pure equational setting. (Krob) (40)
6. Regular expressions may be interpreted as expressions over real numbers, with $a^* = 1/(1 - a)$ and the usual meaning for the remaining operations. Show that if two unambiguous regular expressions are equal when they are interpreted as languages, then they are equal when interpreted as arithmetic expressions. (Unambiguous regular expressions (UREs) are defined by: (1) 0,1, and $a \in A$ are UREs; (2) $E_1 + E_2$ is URE if $|E_1| \cap |E_2| = \emptyset$ and E_1, E_2 are UREs; (3) $E_1 \cdot E_2$ is URE if each $w \in |E_1| \cap |E_2|$ is uniquely decomposed as $w = w_1 w_2$ with $w_i \in |E_i|$ and E_1, E_2 are UREs; (3) E^* is URE if $E \neq 1$ and each nonempty $w \in |E^*|$ is uniquely decomposable into $w = w_1 \ldots w_k$ with $w_i \in |E|$ and E is URE.) (16)
7. Show that the Gauss–Jordan elimination method for determining the inverse of a real matrix may be used to solve the all-paths problem on directed graphs. (12)

8.11 Repetition, iteration, and feedback in matrix theories

In this section we prove 3 axiomatic systems for Conway theories are equivalen to corresponding iteratiobased axiomatic systems, provided a matrix theory structure is used.

Conway theories. For convenience, the 3 systems of axioms for Conway theories from Chapter 1 are displayed in Table 8.1, again. More on the technical problem to be solved in this section may be found in Chapter 1, Section 1.7.

Table 8.1. Axiomatic systems for Conway theories (2nd posting)

R1.1	$(f+g)^* = (f^* \cdot g)^* \cdot f^*$	R2.1	$f^* = \mathsf{I}_a + f \cdot f^*$
R1.2	$(f \cdot g)^* = \mathsf{I}_a + f \cdot (g \cdot f)^* \cdot g$	R2.2	$(f+g)^* = (f^* \cdot g)^* \cdot f^*$
		R2.3	$(f \cdot g)^* \cdot f = f \cdot (g \cdot f)^*$
R3.1	$(0^a_a)^* = \mathsf{I}_a$		
R3.2	$\begin{bmatrix} f & g \\ h & i \end{bmatrix}^* = \begin{bmatrix} x & f^* \cdot g \cdot w \\ w \cdot h \cdot f^* & w \end{bmatrix}$ where $w = (h \cdot f^* \cdot g + i)^*$ and $x = f^* + f^* \cdot g \cdot w \cdot h \cdot f^*$		
R3.3	$({}^a\mathsf{X}^b \cdot f \cdot {}^b\mathsf{X}^a)^* = {}^a\mathsf{X}^b \cdot f^* \cdot {}^b\mathsf{X}^a$		

Repetition, iteration, and feedback. Let T be a matrix theory and $\mathbf{Rp}(T)$ be the set of all the repetition operations defined on T. We use the translations

$$\sigma : \mathbf{It}(T) \to \mathbf{Rp}(T) \quad \text{and} \quad \tau : \mathbf{Rp}(T) \to \mathbf{It}(T)$$

defined by

- $\dagger\sigma$ is the repetition operation that for an object a in T maps $f \in T(a,a)$ into $[f, \mathsf{I}_a]^\dagger$; and
- $*\tau$ is the iteration operation that for a and b objects in T maps $f = [f_1, f_2] \in T(a, a \star b)$, with $f_1 : a \to a$ and $f_2 : a \to b$, into $f_1^* \cdot f_2$.

Finally, let us consider the restrictions $\sigma_r : \mathbf{It}_r(T) \to \mathbf{Rp}(T)$ and $\tau_r : \mathbf{Rp}(T) \to \mathbf{It}_r(T)$ induced by σ and τ.

Theorem 8.27. *Suppose T is a matrix theory. Then:*
a) The restrictions σ_r and τ_r are mutually inverse bijective functions.
b) For $k \in [3]$, $*$ satisfies R3.k *iff $*\tau$ satisfies* I3.k.
c) For $k \in [3]$, $*$ satisfies R2.k *iff $*\tau$ satisfies* I2.k.
d) For $k \in [2]$, $*$ satisfies R1.k *iff $*\tau$ satisfies* I1.k.
e) y is $*$-functorial iff y is $*\tau$-functorial.

PROOF: **a)** Note that $\dagger := \tau(*)$ satisfies I3.4. Indeed, if $f = [f_1, f_2] : a \to a \star b$ (with $f_1 : a \to a$ and $f_2 : a \to b$) and $g : b \to c$, then $(f(\mathsf{I}_a \star g))^\dagger = [f_1, f_2 g]^\dagger = f_1^* f_2 g = f^\dagger g$. Consequently τ_r is totally defined. Obviously $* = \sigma(\tau(*))$. For the converse, note that $\tau(\sigma(\dagger))$ maps $f = [f_1, f_2] = [f_1, \mathsf{I}_a](\mathsf{I}_a \star f_2) \in T(a, a\star b)$ (with $f_1 : a \to a$ and $f_2 : a \to b$) in $[f_1, \mathsf{I}_a]^\dagger f_2$. Hence $\dagger = \tau(\sigma(\dagger))$, for $\dagger \in \mathbf{It}_r(T)$.

b) Let $\dagger$ and $*$ be corresponding operations, i.e., $\dagger = \tau(*)$. The equivalence in the case $k = 1$ holds by $(\top_a \star \mathsf{I}_a)^\dagger = [0_a^a, \mathsf{I}_a]^\dagger = (0_a^a)^* \mathsf{I}_a = (0_a^a)^*$.

For $k = 2$ note that if $f = [f_1, f_2, f_3] : a \to a \star b \star c$ (with $f_1 : a \to a, f_2 : a \to b$ and $f_3 : a \to c$) and $g = [g_1, g_2, g_3] : b \to a \star b \star c$ (with $g_1 : b \to a, g_2 : b \to b$ and $g_3 : b \to c$), then $\langle f, g\rangle^\dagger = \begin{bmatrix} f_1 & f_2 & f_3 \\ g_1 & g_2 & g_3 \end{bmatrix}^\dagger =$ $\begin{bmatrix} f_1 & f_2 \\ g_1 & g_2 \end{bmatrix}^* \begin{bmatrix} f_3 \\ g_3 \end{bmatrix}$ and $h := (g\langle f^\dagger, \mathsf{I}_{b\star c}\rangle)^\dagger = ([g_1 \;\; g_2 \;\; g_3] \begin{bmatrix} f_1^* f_2 & f_1^* f_3 \\ \mathsf{I}_b & 0 \\ 0 & \mathsf{I}_c \end{bmatrix})^\dagger =$ $[g_1 f_1^* f_2 + g_2 \;\;\; g_1 f_1^* f_3 + g_3]^\dagger = w(g_1 f_1^* f_3 + g_3)$ where $w = (g_1 f_1^* f_2 + g_2)^*$. Hence $\langle f^\dagger\langle h, \mathsf{I}_c\rangle, h\rangle = \begin{bmatrix} f_1^* f_2 h + f_1^* f_3 \\ h \end{bmatrix} = \begin{bmatrix} f_1^* f_2 w g_1 f_1^* + f_1^* & f_1^* f_2 w \\ w g_1 f_1^* & w \end{bmatrix} \begin{bmatrix} f_3 \\ g_3 \end{bmatrix}$. Consequently, if $*$ satisfies R3.2 then $\dagger$ satisfies I3.2.

Conversely, if $\dagger$ satisfies I3.2, then from the above computation it follows that $\begin{bmatrix} f_1 & f_2 \\ g_1 & g_2 \end{bmatrix}^* \begin{bmatrix} f_3 \\ g_3 \end{bmatrix} = \begin{bmatrix} f_1^* f_2 w g_1 f_1^* + f_1^* & f_1^* f_2 w \\ w g_1 f_1^* & w \end{bmatrix} \begin{bmatrix} f_3 \\ g_3 \end{bmatrix}$. Applying this first for $f_3 = \mathsf{I}_a, g_3 = 0_a^b$ and then for $f_3 = 0_b^a, g_3 = \mathsf{I}_b$ we get R3.2.

For $k = 3$, note that if $f = [f_1, f_2] : b \star a \to b \star a \star c$ (with $f_1 : b \star a \to b \star a$ and $f_2 : b\star a \to c$), then $({}^a\mathsf{X}^b f({}^b\mathsf{X}^a \star \mathsf{I}_c))^\dagger = [{}^a\mathsf{X}^b f_1 \cdot {}^b\mathsf{X}^a, {}^a\mathsf{X}^b f_2]^\dagger = ({}^a\mathsf{X}^b f_1 \cdot {}^b\mathsf{X}^a)^* \cdot {}^a\mathsf{X}^b f_2$ and ${}^a\mathsf{X}^b f^\dagger = {}^a\mathsf{X}^b f_1^* f_2$. Since ${}^b\mathsf{X}^a \cdot {}^a\mathsf{X}^b = \mathsf{I}_{a\star b}$, it follows that R3.3 and I3.3 are equivalent.

c) Let $\dagger$ and $*$ be corresponding operations. For $k = 1$, note that if $f = [f_1, f_2] : a \to a \star b$ (with $f_1 : a \to a$ and $f_2 : a \to b$), then $f^\dagger = f_1^* f_2$ and $f\langle f^\dagger, \mathsf{I}_b\rangle = f_1 f_1^* f_2 + f_2 = (f_1 f_1^* + \mathsf{I}_a) f_2$. Hence R2.1 and I2.1 are equivalent.

For $k = 2$, note that if $f = [f_1, f_2, f_3] : a \to a \star a \star b$ (with $f_1 : a \to a, f_2 : a \to a$ and $f_3 : a \to b$), then $f^{\dagger\dagger} = [f_1^* f_2, f_1^* f_3]^\dagger = (f_1^* f_2)^* f_1^* f_3$ and $(f(\vee_a \star \mathsf{I}_b))^\dagger = [f_1 + f_2, f_3]^\dagger = (f_1 + f_2)^* f_3$. Hence R2.2 and I2.2 are equivalent.

For $k = 3$, suppose that $f = [f_1, f_2] : a \to b \star c$ (with $f_1 : a \to b$ and $f_2 : a \to c$) and $g : b \to a$. Then $g(f(g \star \mathsf{I}_c))^\dagger = g[f_1 g, f_2]^\dagger = g(f_1 g)^* f_2$ and $(gf)^\dagger = [gf_1, gf_2]^\dagger = (gf_1)^* g f_2$ show that R2.3 and I2.3 are equivalent.

d) The case $k = 1$ is covered by c). For $k = 2$, note that if $f = [f_1, f_2] : a \to b\star c$ (with $f_1 : a \to b$ and $f_2 : a \to c$) and $g : b \to a$, then $(f(g \star \mathsf{I}_c))^\dagger = [f_1 g, f_2]^\dagger = (f_1 g)^* f_2$ and $f\langle (gf)^\dagger, \mathsf{I}_c\rangle = [f_1, f_2]\langle (gf_1)^* g f_2, \mathsf{I}_c\rangle = f_1 (gf_1)^* g f_2 + f_2 = (\mathsf{I}_a + f_1 (gf_1)^* g) f_2$. Hence R1.2 and I1.2 are equivalent.

e) Suppose that $y : a \to b$ is $*$-functorial and $f = [f_1, f_2] : a \to a \star c$ (with $f_1 : a \to a$ and $f_2 : a \to c$) and $g = [g_1, g_2] : b \to b \star c$ (with $g_1 : b \to b$ and $g_2 : b \to c$) are such that $f(y \star \mathsf{I}_c) = yg$. Then $f_1 y = y g_1$ and $f_2 = y g_2$. By the

$*$-functoriality of y we get $f_1^* y = y g_1^*$. Consequently, $yg^\dagger = yg_1^* g_2 = f_1^* y g_2 = f_1^* f_2 = f^\dagger$. Conversely, suppose that $y : a \to b$ is $\dagger$-functorial and $f : a \to a$ and $g : b \to b$ are such that $fy = yg$ Then $[f, y](y \star \mathsf{I}_b) = [fy, y] = [yg, y] = y[g, \mathsf{I}_b]$ hence $[f, y]^\dagger = y[g, \mathsf{I}_b]^\dagger$ showing that $f^* y = yg^*$. □

Corollary 8.28. *a) The restrictions* $\alpha_r \cdot \sigma_r$ *and* $\tau_r \cdot \beta_r$ *are mutually inverse bijective functions.*
b) For $k \in [3]$, $*$ *satisfies* R3.k *iff* $*\tau\beta$ *satisfies* F1.k.
c) y *is* $*$*-functorial iff* y *is* $*\tau\beta$*-functorial.*

Corollary 8.29. *For a matrix theory enriched with a repetition, an iteration, or a feedback operation, the axiomatic systems* F1, F2, I1, I2, I3, I4, R1, R2 *and* R3 *are equivalent.*

Comments, problems, bibliographic remarks

Notes. Finite state automata are one of the oldest, best known, and widest used tools in computer science. With respect to its algebraic counterpart we may distinguish three stages of evolution.

(1) The roots of the algebraic theory of finite automata lay in the seminal paper by Kleene [Kle56] on the representation of events in neural nets and automata published in the early 1950s.

(2) In the 1960s its development was very intensive - see, e.g., [CEW58], [Brz64], [Red64], or [Sal66]. A few books recording the results are [Gin68], [Con71], [Eil74] and [KS85].

(3) A last stage may be located in the 1980s, when one may see a reawaken interest on this topics; an outstanding new result is Krob's solution [Kro91] for two deep conjectures of Conway; see Exercise 8.10.5; Exercises 8.10.6,7 are from [Tar81]. See also [Şte87b], [BÉ93a], [AFI98], or [Pre99] to mention just a few more recent related papers.

The main axiomatization result presented in the current chapter (Theorem 8.26) is interesting, nontrivial, but weaker than Krob's Theorem. We have used simulation via relations as an import from the general network algebra principles presented in the previous part of the book. The key fact that the power-set construction produces a similar deterministic automaton is based on [Koz91]. Kleene theorem(s) are well-known; we included them accompanied with simple algebraic proofs. The minimization of deterministic automata is also well-known; we re-cast them to show that actually, along the way, similar automata are obtained.

The results on the translation between repetition based structures and analogous ones using feedback or iteration are from [CŞ95]. The example on minimization of nondeterministic automata presented in Appendix E is from [Bra84].

9. Process algebra

> Between two evils choose neither; between two goods, choose both.
>
> Tryon Eduards

Process algebra provides an algebraic description of communicating processes. In this chapter we present a particular algebraic approach to processes, known as ACP (Algebra of Communicating Processes). It is particularly more suited for our aims since it has an arbitrary sequential composition. (The other standard process algebra versions CSP and CCS use prefix composition with atomic elements, only.)

The chapter starts with an introduction on the main problems risen by parallel process modeling, including the motivation for interleaving semantics, action atomicity, and the fairness problems. Then, we describe the transition systems modulo bisimulation and the associated algebraic counterpart, i.e., the BPA (Basic Process Algebra) with recursion.

Parallel operators are introduced thereafter, both without and with communications. This way, the basic ACP setting is obtained. The main result included in the present chapter is the soundness and completeness theorem for ACP with respect to transition systems modulo bisimulation.

A brief account of the abstraction operator and a case study (the verification of the correctness for an Alternating Bit Protocol) complete this chapter.

9.1 An overview on parallel processes

We start here with a brief presentation of the problems risen by modeling parallel processes, together with the main solutions proposed by the process algebra approach. Along the way, the problems of interleaving, program interference, and fairness are discussed.

An example. To start the discussion about parallel processes let us consider a simple program aiming at computing the binomial coefficients $\left(\begin{smallmatrix} k \\ n \end{smallmatrix}\right) = [n \cdot (n-1) \cdot \ldots \cdot (n-k+1)]/[1 \cdot 2 \cdot \ldots \cdot k]$. The program is given in Table 9.1.

First, a few words about the syntax. The program is written in a pseudo-code. The statements are considered to be transitions and are labeled by t_0–t_3, r_0–r_4. The states are labeled by l_0–l_2, m_0–m_3. We used "$\rightarrow$" on the right-hand-side for the GOTO statement; on the left-hand-side "$\rightarrow$" indicates the starting states.

Table 9.1. Program for binomial coefficients

in k, n: integer where $0 \leq k \leq n$				
local y_1, y_2: integer where $y_1 = n, y_2 = 1$				
out b: integer where $b = 1$				
$P1 ::$	$\rightarrow$	l_0	$t_0 : (y_1 > (n-k))?$	$\rightarrow l_1$
			$t_3 : (y_1 \leq (n-k))?$	$\rightarrow exit$
		l_1	$t_1 : b = b \cdot y_1$	$\rightarrow l_2$
		l_2	$t_2 : y_1 = y_1 - 1$	$\rightarrow l_0$
$\|\|$				
$P2 ::$	$\rightarrow$	m_0	$r_0 : (y_2 \leq k)?$	$\rightarrow m_1$
			$r_4 : (y_2 > k)?$	$\rightarrow exit$
		m_1	$r_1 : ((y_1 + y_2) \leq n)?$	$\rightarrow m_2$
		m_2	$r_2 : b = b \ div \ y_2$	$\rightarrow m3$
		m_3	$r_3 : y_2 = y_2 + 1$	$\rightarrow m_0$

The program has two subprocesses $P1$ and $P2$ which run in parallel. The memory locations are n, k, y_1, y_2, b and they are shared by both subprocesses. One process ($P1$) is charged to do the multiplications and the other ($P2$) to do the divisions.

Interleaving. A technique which is very often used to analyze parallel processes is that known as the "interleaving" method. By this technique a system of parallel sequential processes is reduced to a nondeterministic sequential process. For the parallel processes P_1 and P_2 of the above example, one may image a unique sequential program Q which simulate them by a random switch from one process to the other. The behaviour of Q may be specified by the *traces* (i.e., sequences of execution) followed in the program $P1||P2$.

For instance, the sequence $t_0t_1t_2t_0t_1t_2r_0r_1r_2r_3r_0r_1r_2r_3r_0t_0t_1t_2t_3r_1r_2r_3r_4$ is a possible trace for the execution of Q for $n = 5, k = 3$, but $t_0t_1t_2r_0r_1r_2r_3r_0r_1\ldots$ is not an appropriate trace.

The main quality of this method is that it is *exhaustive*, i.e., all combinations are taken into account. Hence, this method is good for checking program correctness in those areas where this is a vital problem, e.g., for safety critical systems. Along with this quality, the main defect of the method comes: this is the *state explosion* problem. Indeed, a naive application of the technique leads to a large number of combinations which have to be analyzed. This number is very large, even for relatively small programs. Many sophisticated variants were developed to apply the method for larger and larger systems, but the problem is still unsolved and it narrows the area of applications of this method.

Notice that the processes $P1$ and $P2$ used in Table 9.1 are not independent. They communicate via some shared variables. A critical example is in the statement with label r_1. Indeed, while P_2 basically uses the local variable y_2, in the statement with label r_1 the variable y_1 of the first process is used to force the number of division steps to be at most equal to the number of multiplication steps in any prefix of computation. This way, it is assured that the division process gives an exact result over integer numbers and finally the correct result is obtained.

Program interference. At a first glance the interleaving technique looks to be a good solution to verify the correctness of parallel programs. However, the program interference via shared variables is quite complicated and cannot be solved so simply. One further step to be taken into account is the necessity to consider "atomic" statements.

Consider for instance the following program: $t_1 : y = y + 1 \parallel t_2 : y = y - 1$. What result may be obtained? The answer is: it depends.

Indeed, if t_1 and t_2 are atomic statements, then the possible traces are $t_1 t_2$ or $t_2 t_1$, and, in both cases, the result is the expected one: 1 (the default initial value of a variable is supposed to be 0). But what happens if one consider a lower level of granularity? Then, for instance, the assignment statement $y = y + 1$ may be split into three simpler statements, say, $read(y); compute(y + 1); write(y)$. (Their meaning should be clear: $read(y)$ loads the value from the memory cell of y into an appropriate register; $compute(y + 1)$ computes the requested value into the register, and finally, $write(y)$ stores the resulting value into the memory cell corresponding to y.) If one abstracts from the computation process, which is irrelevant here, then he concludes that the set of possible traces includes: $r_1(1)r_2(1)w_1(2)w_2(0)$, $r_1(1)r_2(1)w_2(0)w_1(2)$, and $r_1(1)w_1(2)r_2(2)w_2(1)$, leading to 0,2, and 1 as the corresponding results. Actually, the set of all possible results is $\{0, 1, 2\}$.

The conclusion of this simple example is important: in order to get the desired result one has to consider that the assignment statement is atomic, hence its result is not affected by a possible interference with other processes.

A similar conclusion may be inferred for test statements. For instance, let us consider the program *when* $y = y$ *do* S. It apparently loops forever. However, it may terminate, provided the test statement is not atomic. Indeed, in such a case another program may interfere between the moments when the values for the left and the right occurrences of y are loaded. If the value of y is changed, then the test fails and the program terminates. Consequently, to behave in a natural way the test statement should be considered as an atomic action, too.

All the above remarks show that in order to control the communication between processes one needs "atomic" actions; once they are activated they run till the very end as simple sequential processes, without any interference.

This "atomicity" assumption appears as a very restrictive hypothesis. While theoretically one may say that this hypothesis is not restrictive (one may choose any level of granularity he wants), practically there may be problems. Indeed, a low level of granularity, combined with the state explosion induced by the interleaving method, may lead to an equivalent sequential process which is too large to be completely analyzed by any practical means.

Another problem is generated by process refinement: as we have seen in the above examples, refinement of (atomic) actions is not compatible with the interleaving technique.

Fairness. With the action atomicity and interleaving hypotheses one may simulate the behaviour of a system of parallel processes $P_1||\dots||P_n$ with a (nondeterministic) sequential process, say P. However, not all possible traces of P describe *real runs* of the system $P_1||\dots||P_n$! What is meant here is the emphasis on "fair" activation. The traces of P which actually activate a unique process P_i are compatible with the interleaving semantics, but not with a run of the system $P_1||\dots||P_n$ in a truly parallel environment. In this latter case each process evolves independently and will do its job (if possible).

As an example, consider the program

$$\begin{array}{llll} P1 :: \ \rightarrow & l_0 & t_0 : (x = T)? \rightarrow l_1 & || \ P2 :: \ \rightarrow \ m_0 \ \ r_0 : x = F \rightarrow exit \\ & & t_2 : (x = F)? \rightarrow exit & \\ & l_1 & t_1 : y = y + 1 \rightarrow l_0 & \end{array}$$

The trace $(t_0t_1)^\infty = t_0t_1t_0t_1\dots$ is possible in the interleaving model, but not in a truly parallel model.

The fairness property is not easy to model. A few possibilities which have been considered in the literature so far are described below. Roughly speaking, they are:

(1) *Impartiality* (or unconditional fairness): each process is activated an infinite number of times, regardless of whether it is able to execute an appropriate action or not;
(2) *Justice* (or weak fairness): a process which, except for a finite number of times, is able to execute one of its actions, is to be activated an infinite number of times.
(3) *Fairness* (or strong fairness): a process which is able to execute a given statement an infinite number of times, is to be activated an infinite number of times.

In process algebra terms, the fairness is related to the rules used for abstraction (the rules for the τ constant) in a cyclic setting.

Ergodic theory. The problems related to the fairness property are inherently difficult. At the 1994 Marktoberdorf summer school, asked on the role of fairness, E. Dijkstra, one of the fathers of the theory of distributed systems, was very "fair": he suggested that fairness should be simply ignored! Probably this is the natural attitude of a computer scientist. For a mathematician the matter may be somehow related to statistical mechanics, ergodic theory, to those transformations which mix the space in a very good way. For instance, this is the case for the "baker transformation": in this transformation, one takes a square rectangle (of dough), and deform it (by throwing on the table) to make it 2 times shorter and 2 times larger; then, cut the right half part and put it on top over the left part to get a full square rectangle again; if you start with a small marked area (a piece of yeast in a dough of flour), then after a number of such transformations each area in the rectangle will contain marked parts (yeast). ◊

Much more natural may be to have an algebraic framework to represent and verify the parallel processes in a truly concurrent way, but unfortunately such a calculus is not available at present.

As we noticed above, the truly concurrent programs may not be reduced to simple nondeterministic sequential programs with interleaving only. However, it was argued that to some extent: *Concurrency = Interleaving + Fairness.* Anyway, fairness will not be a central topic for this chapter/book, so we stop the comments here.

9.2 Transition systems

Transition systems and structural operational semantics are briefly presented here.

Origin. The idea may be traced back at least to the application of this classical concept to automata theory. The process of taking the "derivative" of regular expressions has been extensively studied starting with the 1960s, see e.g., [Brz64], [Con71]. In the present extended form, called "Structural Operational Semantics," it was introduced by G. Plotkin in the early 1970s.

Derivatives of regular expressions, recalled. We briefly recall how derivatives of regular expressions may be defined. Let E be an event; mathematically it is represented as the set of sequences of actions that lead to the same final state which correspond to that event. Then define the derivative and the termination property as

$$\partial_a[E] = \{w : aw \in E\} \text{ and } o[E] = \begin{cases} 1, & 1 \in E \\ 0, & 1 \notin E \end{cases}$$

The derivation (mixed with the termination) rules are:

$D1$	$o[0] = o[a] = 0 (a \in V)$	$D5$	$\partial_a[0] = \partial_a[b] = 0, \partial_a[a] = 1 (a \neq b, a, b \in V)$
$D2$	$o[E+F] = o[E] + o[F]$	$D6$	$\partial_a[E+F] = \partial_a[E] + \partial_a[F]$
$D3$	$o[E \cdot F] = o[E] \cdot o[F]$	$D7$	$\partial_a[E \cdot F] = \partial_a[E] \cdot F + o[E] \cdot \partial_a[F]$
$D4$	$o[E^*] = 1$	$D8$	$\partial_a[E^*] = \partial_a[E] \cdot E^*$

This mechanism allows for a simple passing from regular expressions to automata:
—the automata states are labeled by expressions $\boxed{E}$;
—a state $\boxed{E}$ is final iff $o[E] = 1$;
—the automata transitions are: $\boxed{E} \xrightarrow{a} \boxed{F}$ iff $F = \partial_a[E]$. ◇

Transition systems. One may adapt the above method to give meaning to expressions describing parallel processes. However, the world where the result is considered is not the familiar one of finite automata, but a new one: the world of *transition systems.* This departs from the world of finite automata mainly in the following respect: *the equivalence relation on transition systems is* not *the one given by the input–output sequences*; the new equivalence takes into account certain information on the intermediary states of the processes, as well. (The need for considering these intermediary states comes from the necessity to model process interaction. In order to communicate, to cooperate the processes have to know something of the intermediary results of the partners.)

The basic equivalence relation used in this context is *bisimulation*: roughly speaking, two processes are bisimilar if they have the same traces *and* they exhibit the same branching structure. Before giving the formal definition let us fix a notation. We use $\surd$ to specify that a process terminates after the execution of an action.

Definition 9.1. Let T and T' be two transition systems. A relation $R \subseteq nod(T) \times nod(T')$ between the vertices of the systems is called a *(strong) bisimulation* from T to T', denoted $R : T \leftrightarrows T'$, if it obeys the following conditions:

- If $(r, s) \in R$ and $r \xrightarrow{a} r'$ in T, then there exists s' and $s \xrightarrow{a} s'$ in T', such that $(r', s') \in R$.
- If $(r, s) \in R$ and $r \xrightarrow{a} \surd$ in T, then $s \xrightarrow{a} \surd$ in T'.
- The above two conditions with the roles of T and T' interchanged.

Usually, the transition systems have a distinguished initial vertex $root(T)$, called *root*. In such a case the definition of bisimulation has an additional condition:

- $(root(T), root(T')) \in R$.

Structural operational semantics - SOS. Transition systems give an operational (context-sensitive) semantics for concurrent processes. We present a few examples of derivations using the derivation rules for BPA_δ. These rules are presented in Table 9.2. Certain derivations for $a \cdot b + a \cdot c \xrightarrow{a} b$ and for $a \cdot (b + c) \xrightarrow{a} b + c$ are described below.

$$lplus_1 \frac{secv_0 \dfrac{atom\ \ a \xrightarrow{a} \surd}{a \cdot b \xrightarrow{a} b}}{a \cdot b + a \cdot c \xrightarrow{a} b} \qquad\qquad secv_0 \frac{atom\ \ a \xrightarrow{a} \surd}{a \cdot (b + c) \xrightarrow{a} b + c}$$

The SOS rules for ACP with recursion (the main axiomatic system we shall use in this chapter) are collected in Table 9.2. They will be described in detail in the forthcoming sections.

Notation 9.2. Let $T(p)$ denote the transition system associated by the SOS rules with p. Notice that $T(p)$ has labels on nodes, too. The node labels in $T(p)$ are BPA_δ-terms and the root is labeled by p.

Table 9.2. The structural operational semantic rules for ACP

SOS rules for BPA_δ

$$atom : a \xrightarrow{a} \surd$$

$$secv_0 : \frac{p \xrightarrow{a} \surd}{p \cdot q \xrightarrow{a} q} \qquad secv_1 : \frac{p \xrightarrow{a} p'}{p \cdot q \xrightarrow{a} p' \cdot q}$$

$$lplus_0 : \frac{p \xrightarrow{a} \surd}{p + q \xrightarrow{a} \surd} \qquad lplus_1 : \frac{p \xrightarrow{a} p'}{p + q \xrightarrow{a} p'}$$

$$rplus_0 : \frac{p \xrightarrow{a} \surd}{q + p \xrightarrow{a} \surd} \qquad rplus_1 : \frac{p \xrightarrow{a} p'}{q + p \xrightarrow{a} p'}$$

SOS for $||$ and $⌊⌊$

$$\frac{x \xrightarrow{a} x'}{x||y \xrightarrow{a} x'||y} \qquad \frac{x \xrightarrow{a} \surd}{x||y \xrightarrow{a} y} \qquad \frac{x \xrightarrow{a} x'}{y||x \xrightarrow{a} y||x'} \qquad \frac{x \xrightarrow{a} \surd}{y||x \xrightarrow{a} y}$$

$$\frac{x \xrightarrow{a} x'}{x ⌊⌊ y \xrightarrow{a} x' ⌊⌊ y} \qquad \frac{x \xrightarrow{a} \surd}{x ⌊⌊ y \xrightarrow{a} y}$$

SOS for recursion

$$\frac{\langle t_X|E\rangle \xrightarrow{a} p}{\langle X|E\rangle \xrightarrow{a} p} \qquad \frac{\langle t_X|E\rangle \xrightarrow{a} \surd}{\langle X|E\rangle \xrightarrow{a} \surd}$$

SOS rules for ρ_f

$$\frac{x \xrightarrow{a} x' \quad (f(a) \neq \delta)}{\rho_f(x) \xrightarrow{a} \rho_f(x')} \qquad \frac{x \xrightarrow{a} \surd \quad (f(a) \neq \delta)}{\rho_f(x) \xrightarrow{a} \surd}$$

SOS rules for $|$

$$\frac{x \xrightarrow{a} x' \;\&\; y \xrightarrow{b} y' \;\&\; \gamma(a,b) = c}{x||y \xrightarrow{c} x'||y'} \qquad \frac{x \xrightarrow{a} \surd \;\&\; y \xrightarrow{b} y' \;\&\; \gamma(a,b) = c}{x||y \xrightarrow{c} y'}$$

$$\frac{x \xrightarrow{a} x' \;\&\; y \xrightarrow{b} y' \;\&\; \gamma(a,b) = c}{x|y \xrightarrow{c} x'|y'} \qquad \frac{x \xrightarrow{a} \surd \;\&\; y \xrightarrow{b} y' \;\&\; \gamma(a,b) = c}{x|y \xrightarrow{c} y'}$$

$$\frac{x \xrightarrow{a} x' \;\&\; y \xrightarrow{b} \surd \;\&\; \gamma(a,b) = c}{x||y \xrightarrow{c} x'} \qquad \frac{x \xrightarrow{a} \surd \;\&\; y \xrightarrow{b} \surd \;\&\; \gamma(a,b) = c}{x||y \xrightarrow{c} \surd}$$

$$\frac{x \xrightarrow{a} x' \;\&\; y \xrightarrow{b} \surd \;\&\; \gamma(a,b) = c}{x|y \xrightarrow{c} x'} \qquad \frac{x \xrightarrow{a} \surd \;\&\; y \xrightarrow{b} \surd \;\&\; \gamma(a,b) = c}{x|y \xrightarrow{c} \surd}$$

9.3 Nondeterministic sequential processes; BPA plus recursion

In this section nondeterministic sequential processes are re-cast into a setting to be used for parallel processes. Specifically, nondeterministic sequential processes modulo bisimulation are studied, first without recursion, then in the case where recursion is present.

BPA_δ is an algebraic theory for the specification of sequential nondeterministic processes. We have to start by specifying the signature. The basic operations are: *nondeterministic choice* "+" (binary operation); *sequential composition* "·" (binary operation); and *deadlock* "δ" (constant). Hence the syntax for BPA_δ-terms is given by the grammar

$$p ::= a \;\big|\; p + p \;\big|\; p \cdot p \;\big|\; \delta$$

where $a \in A$ (A represents the set of atomic actions; A_δ denotes $A \cup \{\delta\}$). The SOS rules are given in Table 9.2.

Operations. A comment about the choice of operations may be useful here. A form of sequential composition is necessary, for sure. The use of nondeterministic choice may require more motivation; while internally a process may very well be deterministic, it cannot control the other processes it interacts with; hence, a form of (external) nondeterminism may appear to be justified. Finally, the deadlock constant is included here as a form of describing the impossibility of a process to perform any further action; checking the deadlock-free behaviour of a system is an important current practical issue.
◇

BPA_δ is the specification consisting of this signature and the axioms A1–A7 in Table 9.3. The axiomatic system without δ and the axioms A6–A7 is denoted by BPA.

Table 9.3. Axioms for BPA_δ

Axioms for BPA_δ	
$x + y = y + x$	A1
$(x + y) + z = x + (y + z)$	A2
$x + x = x$	A3
$(x + y) \cdot z = (x \cdot z) + (y \cdot z)$	A4
$(x \cdot y) \cdot z = x \cdot (y \cdot z)$	A5
$x + \delta = x$	A6
$\delta \cdot x = \delta$	A7

- A *basic term* is a term p of the following form

  $\sum_{i\in[k]} a_i \cdot p_i + \sum_{j\in[l]} b_j \quad (a_i, b_j \in A)$
  (No repeated equal summands are present in the sum. When $k = l = 0$ the sum is by definition δ.)
- A basic term p as above is in a *prefix normal form* if all its sub-terms are basic terms.

Remark 9.3. The terms in prefix normal form may be identified with *synchronization trees.* A synchronization tree is an edge-labeled tree with labels from A_δ. Two more conditions are imposed on these trees: (1) a δ edge has no continuations and no brothers and (2) no two edges with the same label $a \in A$ and starting from the same node have equal continuation subtrees.

One may easily introduce BPA operations on these synchronization trees: the δ tree has only one edge with this label; sequential composition corresponds to appending copies of the second tree to all non-δ leaves of the first; sum is modeled by glutting the roots of the trees (with certain further simplifications to get a synchronization tree).

The main interest in the synchronization trees (or prefix normal forms) is that they provide representatives for the classes of transition systems modulo bisimulation.

Soundness and completeness for BPA$_\delta$-terms modulo bisimulation. The BPA$_\delta$ rules are of interest because they are sound and complete with respect to the interpretation of terms as transition systems modulo bisimulation. (As usual, here and in the rest of the chapter, " $\vdash$ " denotes equational deduction.)

Proposition 9.4. *For each BPA$_\delta$-term p there exists a basic term p′ such that BPA$_\delta$ $\vdash p = p'$.*

PROOF: We prove the result by structural induction. It is obvious for $p = a$ or $p = p' + p''$. For composition, it is enough to prove that the composite of two basic terms $z \cdot z'$ may be reduced to a basic term by means of the BPA$_\delta$-axioms. This is proved by structural induction on z: (1) the case $z = a$ is trivial; (2) if $z = a \cdot z_1$, then use A5; (3) if $z = z_1 + z_2$, then use A4 and the inductive hypothesis. □

A different possibility is to use the following rewriting system, modulo A1–A2

$$\begin{array}{rcl} x+x & \rightarrow & x \\ (x+y)z & \rightarrow & xz+yz \\ (xy)z & \rightarrow & x(yz) \\ x+\delta & \rightarrow & x \\ \delta x & \rightarrow & \delta \end{array}$$

which is confluent and terminating. This rewriting system will be reconsidered in Section 9.6. Properly extended, it will give a stepping stone for proving that in ACP parallel processes may be reduced to sequential processes.

Confluent and terminating rewriting systems. The *confluence* property refers to the situation when two rules may be applied to the same term; the system is confluent if in such a situation there are further applications of the rewriting rules which brought the resulting terms to the same term. *Termination* property is clear: no infinite chain of rewriting rules may be applied to a term. (To have a terminating rewriting system for ACP we had to throw away the axioms A1 and A2, hence to use rewriting modulo such axioms.) If both properties are valid, then each term may be brought to a (unique) normal form, independent of the way the rules are applied. ◇

Theorem 9.5. *(BPA$_\delta$ and bisimulation) BPA$_\delta$ $\vdash p = q$ iff $T(p) \leftrightarrows T(q)$. In words, two BPA$_\delta$-terms are equivalent within BPA$_\delta$ iff their corresponding transition systems are bisimilar.*

PROOF: 1. Soundness "⇒": The first observation is that $\leftrightarrows$ is a congruence relation on the trees associated with BPA$_\delta$-terms, hence the equational-logic rules are valid. Consequently, it is enough to prove that the terms of an equation are bisimilar. By the SOS rules, the transition system $T(p)$ associated with a term p is labeled on nodes with BPA$_\delta$-terms. Then, a bisimulation relation is to be specified by a relation between such terms. For A1 the relation $Id \cup \{(p+q, q+p)\} : T(p+q) \leftrightarrows T(q+p)$ gives the required bisimulation (the semantics of "+" is commutative). It is equally easy for the other axioms: $Id \cup \{(p+q)r, pr+qr)\}$ does the job for A4; $Id \cup \{(p+\delta, p)\}$ for A6.

2. Completeness "$\Leftarrow$": Let p, q be BPA$_\delta$-terms with $T(p) \leftrightarrows T(q)$. By Proposition 9.4 there are basic terms p_b, q_b such that BPA$_\delta \vdash p = p_b$ and BPA$_\delta \vdash q = q_b$. Using the already proved Soundness part "$\Rightarrow$" one gets $p \leftrightarrows p_b$ and $q \leftrightarrows q_b$, hence $p_b \leftrightarrows q_b$. This reduced the proof to the implication "$\Leftarrow$" for basic terms. Let $R : T(p_b) \leftrightarrows T(q_b)$ be a bisimulation. (1) If p_b has a summand ap', then $p_b \xrightarrow{a} p'$ in $T(p_b)$. As $(p_b, q_b) \in R$ and R is bisimulation, there exists q' and $q_b \xrightarrow{a} q'$ in $T(q_b)$ such that $(p', q') \in R$. This shows aq' is a summand in q_b and $(p', q') \in R$. (2) In a similar way, if p_b has a summand a, then $p_b \xrightarrow{a} \surd$ in $T(p_b)$. Hence there exists $q_b \xrightarrow{a} \surd$ in $T(q_b)$. This shows a is a summand in q_b. (3) By induction, using (1) and (2), it follows that the summands in p_b are summands in q_b. But the sum is idempotent, hence BPA$_\delta \vdash p_b + q_b = q_b$. (4) Finally, in a similar way one gets BPA$_\delta \vdash p_b + q_b = p_b$, hence BPA$_\delta \vdash p_b = q_b$ and the theorem is proved. □

One can add a constant ϵ to simulate the identity (*skip*). The resulting system is BPA$_{\delta,\epsilon}$ consisting of the BPA$_\delta$ axioms and (A8) $\epsilon \cdot x = x$ & (A9) $x \cdot \epsilon = x$. However, ϵ does not fit nicely with the (interleaving based) parallel operators to be introduced in the next section, so it will not be used in the main axiomatic system ACP.

Recursion. The basic model is BPA$_\delta$ with recursion. A process is defined by *recursion* if it is the solution of a finite or infinite system of equations

$$\begin{aligned} X_1 &= t_{X_1} \\ &\dots \\ X_k &= t_{X_k} \\ &\dots \end{aligned}$$

where for each variable X the corresponding term t_X is finite over the variables $X_1, \dots, X_k, \dots$. Of course, with recursion certain infinite trees or transition systems may be finitely specified.

X occurring in an equation $X = Xa$ is undefined. Indeed, the equation has infinitely many solutions in the algebra of infinite trees. (Which ones?) An equation $X = t_X$ is *guarded* (against this undefinedness) if each occurrence of a variable Y in t_X is in a sub-term of the form $a \cdot t$, for $a \in A$. If E is a system of guarded equations, then it has a unique solution in the tree model. The component of the solution corresponding to X will be denoted by $\langle X|E\rangle$. This notation is further extended to terms, namely $\langle t_X|E\rangle$ denotes the term obtained from t_X by substituting $\langle Y|E\rangle$ for each variable Y in t_X. The SOS rules for recursion are given in Table 9.2.

A few more definitions are useful here: (1) A process is *BPA_δ-definable* if it may be obtained from atomic actions and δ by a finite number of applications of sum, composition, and recursion. (2) It is *finitely (definable)*, if it is obtained as above, but with finite recursions, only. (3) The "unique solution for systems of guarded equations" property, in brief the *Unique-rule*, assures that each system of guarded equations has a unique solution.

As an example we take the system E given by the equations

$$\begin{cases} X &= aX + bY \\ Y &= d(X + bY) + cY \end{cases}$$

Then $\langle Y|E\rangle = d(\langle X|E\rangle + b\langle Y|E\rangle) + c\langle Y|E\rangle$. By the SOS rules, $\boxed{\langle Y|E\rangle} \xrightarrow{d} \boxed{\langle X|E\rangle + b\langle Y|E\rangle}$, $\boxed{\langle Y|E\rangle} \xrightarrow{c} \boxed{\langle Y|E\rangle}$, etc.

Example 9.6. (specification of a stack over $\{0,1\}$)
A natural specification with infinitely many linear equations is

$$(*)\begin{cases} S(\lambda) = push(0) \cdot S(0) + push(1) \cdot S(1) \\ S(d \frown \sigma) = push(0) \cdot S(0 \frown d \frown \sigma) + push(1) \cdot S(1 \frown d \frown \sigma) + pop(d) \cdot S(\sigma) \end{cases}$$

Without the restriction to have linear equations only, a finite specification may be given, as well. E.g.,

$$\begin{aligned} S &= T \cdot S \\ T &= push(0) \cdot T_0 + push(1) \cdot T_1 \\ T_0 &= pop(0) + T \cdot T_0 \\ T_1 &= pop(1) + T \cdot T_1 \end{aligned}$$

Then S (and some derived terms) satisfy the same system $(*)$. By applying the Unique-rule, it foolows that these two specifications are equivalent.

EXERCISES AND PROBLEMS (SEC. 9.3)

1. Describe in detail the BPA structure of the synchronization trees; prove that a BPA model is obtained. (07)
2. Use the SOS rules to show that the specifications in Example 9.6 lead to bisimilar transition systems. (09)

9.4 Coloured traces

Coloured traces are traces of actions that carry with them certain information (colour) of the nodes, too. It turn out that, with an appropriate definition of graphs colouring, coloured traces plays a similar role for transition graphs modulo bisimulation as the usual traces are playing for IO equivalence.

We start with a few remarks regarding usual traces. Then we pass to coloured traces.

Definition 9.7. A *(concrete) trace* in a graph f is a sequence $(a_1, a_2, \ldots, a_k)$, such that there is a sequence of transitions $r_0 \xrightarrow{a_1} r_1 \xrightarrow{a_2} r_2 \ldots \xrightarrow{a_k} r_k$ in f starting from the root $r_0 = root(f)$.

Coloured traces contain additional information about the intermediary states. Let (f, C) be a coloured graph, namely each vertex r in f has a corresponding colour $C(r)$. A *coloured (concrete) trace* is a sequence $(c_0, a_1, c_1, \ldots, a_k, c_k)$, where $r_0 \xrightarrow{a_1} r_1 \xrightarrow{a_2} r_2 \ldots \xrightarrow{a_k} r_k$ is a sequence of transitions in f with $r_0 = root(f)$ and $c_i = C(r_i)$, for $i = 0, \ldots, k$. A *consistent colouring* of a graph f is a colouring satisfying the following two conditions: (1) two vertices have the same colour *only if* they have the same coloured traces (that is, the sets of coloured traces generated starting from these vertices are equal) and (2) if two vertices have the same colour, then they are simultaneously final or not final.

Notations: $f \equiv_{cctr} g$ if there exist two consistent colourings C_f for f and C_g for g such that (f, C_f) and (g, C_g) have the same set of coloured (concrete) traces.

Theorem 9.8. *$f \leftrightarroweq g$ iff $f \equiv_{cctr} g$ (That is, two transition systems are bisimilar iff they have the same coloured traces for appropriate consistent colouring.)*

PROOF: "Left to right implication": suppose R is a bisimulation from f to g. Let $\overline{R}$ denote the equivalence relation generated by R on the disjoint union set $nod(f) \sqcup nod(g)$. Take $C = (nod(f) \sqcup nod(g)) \,/\, \overline{R}$ as the desired set of colours. Finally, consider the coloured graphs (f, C_f) and (g, C_g) obtained using this colour set; that is, the colour of a node is its equivalence class.

C_f is a consistent colouring for f. Indeed: (1) let s_0 be a node in f and $(c_0, a_1, c_1, \ldots, a_k, c_k)$ a coloured trace in (f, C_f) starting from s_0. Hence there is a path $s_0 \xrightarrow{a_1} s_1 \xrightarrow{a_2} s_2 \ldots s_{k-1} \xrightarrow{a_k} s_k$ in f such that $c_i = C_f(s_i)$, for $i = 0, \ldots, k$. We first observe that any node t_0 in g which is R-related with s_0 has this coloured trace $(c_0, a_1, c_1, \ldots, a_k, c_k)$. (This is a straightforward consequence of the definition of bisimulation.) Thereafter, we notice that this property is propagated to all nodes in f or g which are connected by a chain of R-related nodes. Hence two nodes with the same colour in (f, C_f) or (g, C_g) have the same coloured traces. (2) Similarly, two nodes s and t in f with $C_f(s) = C_f(t)$ are simultaneously final or not final.

A similar argument shows that (g, C_g) is a consistent colouring for g. Finally, as the roots in f and g are R-related, they have the same colour, hence the same set of coloured traces. This shows $f \equiv_{cctr} g$.

"Right to left implication": if $f \equiv_{cctr} g$, then there exist certain consistent colourings C_f and C_g such that (f, C_f) and (g, C_g) have the same sets of coloured traces. The desired relation is roughly the one which relates nodes with the same colour. However, it is slightly more complicated: it is necessary to restrict ourselves to the accessible vertices, only. (What happens with the non-accessible vertices does not matter.) Formally, one takes the relation $R = \{(s,t): s$ is an accessible node in f, t is an accessible node in g, and $C_f(s) = C_g(t)\}$ and shows that R is a bisimulation, indeed. □

Minimal normal forms. The set of bisimulations between two graphs is closed under union, hence there exists a maximal bisimulation. In particular, there exists a maximal auto-bisimulation on a graph (between the graph and itself) which gives the minimal graph that is bisimilarly equivalent to the given graph. The *canonic colouring* of a graph is the colouring associated to the maximal auto-bisimulation as in Theorem 9.8. The *(minimal) normal form* $nf(f)$ of a graph f is the graph obtained from its accessible part by the identification of the nodes with the same canonic colouring.

Theorem 9.9. *(a)* $f \leftrightarrows nf(f)$; *(b)* $f \leftrightarrows g$ *iff* $nf(f) \approx nf(g)$. *(This shows the minimal normal form is unique up to an isomorphism, denoted* $\approx$*.)*

PROOF: By definition f and $nf(f)$ have the same sets of coloured traces. By Theorem 9.8 this proves (a).

Now (b) is reduced to $nf(f) \leftrightarrows nf(g) \quad \Leftrightarrow \quad nf(f) \approx nf(g)$. For this latter equivalence, the implication "$\Leftarrow$" is obvious. For the converse implication "$\Rightarrow$", suppose $nf(f) \leftrightarrows_R nf(g)$. Then R is a bijective function. Indeed: (1) The graph $nf(g)$ is accessible, hence R is surjective. (Each node t in g may be reached by a path from the root. By bisimulation there exists a corresponding path in f from the root to a node s such that $(s,t) \in R$.); (2) If $(s, r_1), (s, r_2) \in R$, then r_1 and r_2 have the same sets of coloured traces, hence they belong to the maximal auto-bisimulation on g. But this is identity. Hence $r_1 = r_2$ showing the relation R is injective. (3) In a similar way it follows that R^{-1} is surjective and injective.

All the above show R is a bijective function. Moreover, since $nf(g) \leftrightarrows_R nf(h)$, it follows that $(s \xrightarrow{a} s') \Leftrightarrow (R(s) \xrightarrow{a} R(s'))$ and (s is final) $\Leftrightarrow$ ($R(s)$ is final), hence $nf(g) \approx nf(h)$. □

9.5 Communicating processes; ACP

In this section parallel operators are introduced. The first variant PA is without communication. Then, parallel operators with communication are considered; making use of certain renamings (encapsulation), the basic ACP setting is finally introduced.

Parallel processes without communication. The syntax of PA is

$$p ::= p + p \,\big|\, p \cdot p \,\big|\, p \| p \,\big|\, p \,\lfloor\!\lfloor\, p \,\big|\, a \quad (a \in A)$$

This is an extension of the BPA signature. The new operators used here are $\|$ (*merge*) and $\lfloor\!\lfloor$ (*left merge*). The SOS rules for the new operators $\|$ and $\lfloor\!\lfloor$ are given in Table 9.2.

The intuitive meaning of the operator $⌊\!⌊$ is: "take a first action of the left process, then continue with the remaining part of the left process in parallel with the right process."

Left merge. The role of the left merge $⌊\!⌊$ is to produce a finite system of axioms for the reduction of parallel processes to nondeterministic sequential processes. Bergstra and Klop conjectured that no finite system of axioms does exist without auxiliary operators as $⌊\!⌊$; this conjecture was affirmatively proved by F. Moller. ◇

PA is the axiomatic system obtained by enriching BPA with $||$ and $⌊\!⌊$ and satisfying axioms M1–M4 from Table 9.4.

A few key properties of PA are listed here. (1) The *elimination property* holds, i.e., all parallel processes may be reduced within PA to simple sequential processes. (2) The *preservation property* is valid, i.e., the structure of BPA is not changed, namely no more BPA-terms are identified in PA. (3) As a corollary of these results it follows that a process is PA-definable (i.e., using systems of guarded equations with PA-terms) iff it is BPA-definable.

♡ For the sake of brevity these results are not proved here. More general results, proved in the presence of communication, will be given in the next subsections. ◇

Example 9.10. (the specification of a bag over $\{0,1\}$) A linear (infinite) specification is (in the last 3 equations, $m, n \geq 1$):

$$\begin{array}{rcl} B_{0,0} & = & in(0)\cdot B_{1,0} + in(1)\cdot B_{0,1} \\ B_{m,0} & = & in(0)\cdot B_{m+1,0} + in(1)\cdot B_{m,1} + out(0)\cdot B_{m-1,0} \\ B_{0,n} & = & in(0)\cdot B_{1,n} + in(1)\cdot B_{0,n+1} + out(1)\cdot B_{0,n-1} \\ B_{m,n} & = & in(0)\cdot B_{m+1,n} + in(1)\cdot B_{m,n+1} + out(0)\cdot B_{m-1,n} + out(1)\cdot B_{m,n-1} \end{array}$$

$B_{m,n}$ represents a bag with m of 0 and n of 1. Within PA one can find a simple finite specification:

$$X = in(0)\cdot(X||out(0)) + in(1)\cdot(X||out(1))$$

Then X and the derived expressions $X_{m,n} = X||out(0)^m||out(1)^n$ satisfy the same equations as $B_{m,n}$ of the first specification. By the Unique-rule, the specifications are equivalent.

Renaming. The *renaming* operator is defined using a renaming function $f : A \to A \cup Const$. ($Const$ is the set of constants, including δ, ϵ, maybe τ – to be introduced later). It is denoted by ρ_f. The SOS rules are given in Table 9.2. *Encapsulation* ∂_H is a special renaming, namely the actions in H are renamed into δ and the remaining ones are left unchanged. The natural axioms D1–D4 in Table 9.4 are to be used.

Processes with communication. The new ingredient here is a (partially defined) *communication function*

Table 9.4. ACP-axioms

A1	$x + y = y + x$	CF1	$a\|b = \gamma(a,b)$ $\gamma(a,b)$ defined
A2	$(x+y)+z = x+(y+z)$	CF2	$a\|b = \delta$ otherwise
A3	$x + x = x$	CM1	$x\|\|y = x \mathbin{\|\!\!\_} y + y \mathbin{\|\!\!\_} x + x\|y$
A4	$(x+y)\cdot z = (x\cdot z)+(y\cdot z)$	CM2	$a \mathbin{\|\!\!\_} x = a \cdot x$
A5	$(x\cdot y)\cdot z = x\cdot(y\cdot z)$	CM3	$a\cdot x \mathbin{\|\!\!\_} y = a\cdot(x \mathbin{\|\!\!\_} y)$
A6	$x + \delta = x$	CM4	$(x+y) \mathbin{\|\!\!\_} z = x \mathbin{\|\!\!\_} z + y \mathbin{\|\!\!\_} z$
A7	$\delta \cdot x = \delta$	CM5	$a\cdot x \mid b = (a\|b)\cdot x$
D1	$\partial_H(a) = a$ if $a \notin H$	CM6	$a \mid b\cdot x = (a\|b)\cdot x$
D2	$\partial_H(a) = \delta$ if $a \in H$	CM7	$a\cdot x \mid b\cdot y = (a\|b)\cdot(x\|\|y)$
D3	$\partial_H(x+y) = \partial_H(x)+\partial_H(y)$	CM8	$(x+y) \mid z = x\|z + y\|z$
D4	$\partial_H(x\cdot y) = \partial_H(x)\cdot\partial_H(y)$	CM9	$x \mid (y+z) = x\|y + x\|z$

$$\gamma : A \times \ldots \times A \rightarrow A$$

If $\gamma(a_1, \ldots, a_n)$ is defined and equal to c, then the intuition is that the atomic action c consists of the cooperation of n atomic actions $a_1, \ldots, a_n$. Hence, c specifies the simultaneous execution of a tuple of actions from different processes.

For many purposes it is more convenient to use *binary* communication functions which are (1) commutative, i.e., $\gamma(a,b) = \gamma(b,a)$ and (2) associative, i.e., $\gamma(\gamma(a,b),c) = \gamma(a,\gamma(b,c))$.

The syntax for ACP is given by the following grammar:

$$p ::= p' + p'' \,\big|\, p' \cdot p'' \,\big|\, p'\|p'' \,\big|\, p' \mathbin{\|\!\!\_} p'' \,\big|\, p'|p'' \,\big|\, \delta_H(p') \,\big|\, a(\in A) \,\big|\, \delta$$

The SOS rules for the new parallel operator | are given in Table 9.2. Except for δ_H, a new operator here is the *communication merge* |. Its intuitive meaning is firstly to execute a communication action and then to continue with the remaining processes in parallel.

ACP is the axiomatic system obtained from BPA_δ enriched with: (1) parallel operators $\|$, $\mathbin{\|\!\!\_}$ and | satisfying the additional axioms CF1–CF2 and CM1–CM9 from Table 9.4 and (2) the encapsulation operator ∂_H satisfying the axioms D1–D4 in Table 9.4. (For convenience, all the axioms for ACP, a basic axiomatic system for process algebra, have been collected in Table 9.4. Notice that in this table $a \in A_\delta$.)

9.6 Soundness and completeness of ACP

We prove that the ACP axioms are sound and complete for ACP-processes modulo bisimulation.

Proof plan. These theorem is based on the following theorem for elimination and conservative extension. Then, by the elimination part of the theorem below the completeness is achieved: the axioms are complete for reducing all terms to terms without parallel operators; then the completeness of the BPA_δ system may be used. The soundness part follows from the conservative-extension part: the former equivalence on BPA_δ terms is not changed, hence all newly introduced rules are sound. ◇

Theorem 9.11. *(elimination) For every ACP-term t, there exists a BPA_δ-term s such that $ACP \vdash t = s$.*
(conservative extension) For any closed BPA_δ-terms t, s (i.e., without variables), $ACP \vdash t = s$ iff $BPA_\delta \vdash t = s$.

PROOF: We use the rewriting system R in Table 9.5, modulo A1–A2.

Table 9.5. Rewriting system for ACP

R1	$x + x$	$\to$	x
R2	$(x+y) \cdot z$	$\to$	$(x \cdot z) + (y \cdot z)$
R3	$(x \cdot y) \cdot z$	$\to$	$x \cdot (y \cdot z)$
R4	$x + \delta$	$\to$	x
R5	$\delta \cdot x$	$\to$	δ
R6	$a \mid b$	$\to$	$\gamma(a,b)$ if $\gamma(a,b)$ is defined
R7	$a \mid b$	$\to$	δ otherwise
R8	$x \parallel y$	$\to$	$x \mathbin{\lfloor\!\lfloor} y + y \mathbin{\lfloor\!\lfloor} x + x \mid y$
R9	$a \mathbin{\lfloor\!\lfloor} x$	$\to$	$a \cdot x$
R10	$a \cdot x \mathbin{\lfloor\!\lfloor} y$	$\to$	$a \cdot (x \mathbin{\lfloor\!\lfloor} y)$
R11	$(x+y) \mathbin{\lfloor\!\lfloor} z$	$\to$	$x \mathbin{\lfloor\!\lfloor} z + y \mathbin{\lfloor\!\lfloor} z$
R12	$a \cdot x \mid b$	$\to$	$(a \mid b) \cdot x$
R13	$a \mid b \cdot x$	$\to$	$(a \mid b) \cdot x$
R14	$a \cdot x \mid b \cdot y$	$\to$	$(a \mid b) \cdot (x \parallel y)$
R15	$(x+y) \mid z$	$\to$	$x \mid z + y \mid z$
R16	$x \mid (y+z)$	$\to$	$x \mid y + x \mid z$
R17	$\partial_H(a)$	$\to$	a if $a \notin H$
R18	$\partial_H(a)$	$\to$	δ if $a \in H$
R19	$\partial_H(x+y)$	$\to$	$\partial_H(x) + \partial_H(y)$
R20	$\partial_H(x \cdot y)$	$\to$	$\partial_H(x) \cdot \partial_H(y)$

The first task is to prove that the system is confluent and terminating. (These terms were explained in Section 9.3.)

We begin with the termination property, which is slightly more difficult. As usual, in this case we try to pick up a function with values in an ordered set such that no infinite strictly decreasing chains exist and the value of the function strictly decreases at each application of a rule in R.

Our terminating function is based on the function fr which returns the length of the frontier of the trees representing process algebra terms. Then,

if x is a tree and $\alpha \in \{\parallel, \mathbin{\lfloor\!\lfloor}, \mid, \cdot, +\}$, then we denote by
$h_\alpha(x)$ the maximum of the frontier length for subtrees in x with root α.

(If there are no such subtrees the value is 0.) Then, let $\mathbb{N}_{\frac{1}{2}} = \{0, \frac{1}{2}, 1, 1\frac{1}{2}, 2, \ldots\}$ and $P = (\mathbb{N}_{\frac{1}{2}})^5$, considered as an ordered set $(P, \prec)$ using the usual lexicographical order: $(a_1, a_2, a_3, a_4, a_5) \prec (a'_1, a'_2, a'_3, a'_4, a'_5)$ iff there exists $k \leq 5$ such that $a_i = a'_i$, for $i \leq k$ and $a_k < a'_k$). Now our terminating function f is defined as follows:

$$f = (f_1, f_2, f_3, f_4, f_5) : \text{ACP-terms} \to P$$

where

$$f_1(x) = \begin{cases} max\{h_{||}(x), h_{\lfloor\!\lfloor}(x), h_{|}(x)\} & \text{if} \quad h_{||}(x) < max\{h_{||}(x), h_{\lfloor\!\lfloor}(x), h_{|}(x)\} \\ h_{||}(x) + \frac{1}{2} & \text{if} \quad h_{||}(x) = max\{h_{||}(x), h_{\lfloor\!\lfloor}(x), h_{|}(x)\} \end{cases}$$

$f_2(x) = h_{\cdot}(x)$ (h for the composition sign)

$f_3(x) = h_{+}(x)$

$f_4(x) = h_{\partial_H}(x)$

$f_5(x)$ is inductively defined by: $f_5(a) = 1 (a \in A_\delta)$; $f_5(x \cdot y) = 2^{f_5(x)} f_5(y)$; $f_5(x \;\alpha\; y) = max\{f_5(x), f_5(y)\}$, for $\alpha \in \{||, \lfloor\!\lfloor, |, +\}$; and $f_5(\partial_H(x)) = f_5(x)$

Table 9.6. Termination property

R1	$x + x$	(=,=,>,?,?)	x
R2	$(x + y) \cdot z$	(=,>,?,?,?)	$(x \cdot z) + (y \cdot z)$
R3	$(x \cdot y) \cdot z$	(=,=,=,=,>)	$x \cdot (y \cdot z)$
R4	$x + \delta$	(=,=,>,?,?)	x
R5	$\delta \cdot x$	(≥,>,?,?,?)	δ
R6	$a \mid b$	(>,?,?,?,?)	$\gamma(a, b)$
R7	$a \mid b$	(>,?,?,?,?)	δ
R8	$x \| y$	(>,?,?,?,?)	$x \lfloor\!\lfloor y + y \lfloor\!\lfloor x + x \mid y$
R9	$a \lfloor\!\lfloor x$	(>,?,?,?,?)	$a \cdot x$
R10	$a \cdot x \lfloor\!\lfloor y$	(>,?,?,?,?)	$a \cdot (x \lfloor\!\lfloor y)$
R11	$(x + y) \lfloor\!\lfloor z$	(>,?,?,?,?)	$x \lfloor\!\lfloor z + y \lfloor\!\lfloor z$
R12	$a \cdot x \mid b$	(>,?,?,?,?)	$(a \mid b) \cdot x$
R13	$a \mid b \cdot x$	(>,?,?,?,?)	$(a \mid b) \cdot x$
R14	$a \cdot x \mid b \cdot y$	(>,?,?,?,?)	$(a \mid b) \cdot (x \| y)$
R15	$(x + y) \mid z$	(>,?,?,?,?)	$x \mid z + y \mid z$
R16	$x \mid (y + z)$	(>,?,?,?,?)	$x \mid y + x \mid z$
R17	$\partial_H(a)$	(=,=,=,>,?)	a
R18	$\partial_H(a)$	(=,=,=,>,?)	δ
R19	$\partial_H(x + y)$	(=,=,=,>,?)	$\partial_H(x) + \partial_H(y)$
R20	$\partial_H(x \cdot y)$	(=,=,=,>,?)	$\partial_H(x) \cdot \partial_H(y)$

Some simple computations show that the value of f strictly decreases at each application of a rule in R. The collected results are presented in Table 9.6. A detailed comment on $R1$ will clarify the meaning of the items in this table. The possible occurrences of parallel or composition operators in the left term $x + x$ actually are in x, hence the first two weighting functions f_1, f_2 remain unchanged, i.e., $f_1(x+x) = f_1(x)$ and $f_2(x+x) = f_2(x)$. The 3rd value strictly decreases, $f_3(x + x) = fr(x + x) > fr(x) \geq f_3(x)$. Hence $f(x + x) \succ f(x)$. The situation here is written simply as $x + x \xrightarrow{(=,=,>,?,?)} x$, i.e., on the first 2 components the values are equal, the 3rd value is strictly larger on the left term, and the relationship between the 4th and 5th values do not matter.

For the other rules we make a few comments just when some special situations occur. For $R3$ all the weighting functions f_1, f_2, f_3, f_4 remain unchanged, hence we had to introduce the auxiliary function f_5. Then $f_5((x \cdot y) \cdot z) = 2^{2^{f_5(x)} f_5(y)} f_5(z) > 2^{f_5(x)} 2^{f_5(y)} f_5(z) = f_5(x \cdot (y \cdot z))$. In $R5$ the left-hand-side term may contain parallel operators, hence $f_1(\delta \cdot x) \geq f_1(\delta)$. In $R6$ the right-hand-side has frontier 1 ($\gamma(a, b)$ is an atomic action). A problematic rule is $R8$. Function h returns the same value for the parallel operators for both left and right-side terms. To make the left-hand-side term of greater weight we had to slightly increase the weight of $||$.

For the confluence property we check for the critical pairs, i.e., for that pair of rules that may overlap. For R the critical pairs are $(R1, R2)$, $(R1, R4)$, $(R1, R11)$, $(R1, R15)$, $(R1, R16)$, $(R1, R19)$, $(R3, R20)$, $(R4, R11)$, $(R4, R15)$, $(R4, R19)$, $(R5, R10)$, $(R5, R12)$, $(R5, R13)$, $(R5, R14)$, $(R5, R20)$, and $(R15, R16)$. The confluence property is easy to check. We exemplify this with two examples.

$(R1, R2)$: Let $x \cdot z \xleftarrow{R1} \boxed{(x + x) \cdot z} \xrightarrow{R2} x \cdot z + x \cdot z$. Then $x \cdot z + x \cdot z \xrightarrow{R1} x \cdot z$ and that is all.

$(R3, R20)$: Let $\partial_H(x \cdot (y \cdot z)) \xleftarrow{R3} \boxed{\partial_H((x \cdot y) \cdot z)} \xrightarrow{R20} \partial_H(x \cdot y) \cdot \partial_H(z)$. By a repeated application of $R20$ we have $\partial_H(x \cdot (y \cdot z)) \xrightarrow{*} \partial_H(x) \cdot (\partial_H(y) \cdot \partial_H(z))$. For the other term, $\partial_H(x \cdot y) \cdot \partial_H(z) \xrightarrow{R20} (\partial_H(x) \cdot \partial_H(y)) \cdot \partial_H(z) \xrightarrow{R3} \partial_H(x) \cdot (\partial_H(y) \cdot \partial_H(z))$.

Now the first part of the theorem (the elimination property) is almost proved. It is easy to see that a normal form, with respect to the rewriting system R, is a basic term, actually in prefix normal form. Indeed, with $R1$–$R5$ each BPA-term may be rewritten as a basic term; then, if a term has parallel or ∂_H operators, then take such an occurrence q having only BPA-sub-terms; rewrite the BPA-sub-terms as basic terms and apply to q the corresponding rules in R; finally, a term without parallelism is obtained. (Actually, R is tailored on this proof: each basic term is either a sum $x + y$, or a simple action a, or a term $a \cdot x$, so we have rewriting rules for these situations in R.)

The proof of the second part (the conservative extension property) is based on the observation that BPA-terms can be rewritten only with $R1$–$R5$ and these rule are valid in BPA. □

Complementary results in ACP. A few more results on ACP are inserted here.

Theorem 9.12. *(standard concurrency properties) For closed ACP-terms x, y, z (i.e., without variables) the following identities are valid:*

1. $x|y = y|x$
2. $x||y = y||x$
3. $(x|y)|z = x|(y|z)$
4. $x ⌊⌊ (y||z) = (x ⌊⌊ y) ⌊⌊ z$
5. $x|(y ⌊⌊ z) = (x|y) ⌊⌊ z$
6. $x||(y||z) = (x||y)||z$

PROOF: (sketch) 1 and 2 are obvious. For 3-6 uses a simultaneous induction on the number of atomic symbols that occur in all the terms x, y, z. □

Theorem 9.13. *(expansion theorem for handshaking communication, i.e., $a|b|c = \delta, \forall a, b, c \in A$) If $x_1, \ldots, x_n$ are closed ACP-terms then*

$$x_1||\ldots||x_n = \sum_{i\in[n]} x_i ⌊⌊ (||_{j\in[n]\setminus\{i\}} x_j) + \sum_{i,j\in[n],i\neq j} (x_i|x_j) ⌊⌊ (||_{k\in[n]\setminus\{i,j\}} x_k)$$

PROOF: The proof is by induction on n. The basic case $n = 2$ is covered by axiom CM1. The inductive step is easy, too: $x_1||\ldots||x_{n+1} = E$, where

$$E = \underbrace{(x_1||\ldots||x_n) ⌊⌊ x_{n+1}}_{T1} + \underbrace{x_{n+1} ⌊⌊ (x_1||\ldots||x_n)}_{T2} + \underbrace{(x_1||\ldots||x_n)|x_{n+1}}_{T3}$$

Notice that

$$\begin{aligned} T1 &=_{IPH} [(\textstyle\sum_{i\in[n]} x_i ⌊⌊ (||_{k\in[n]\setminus\{i\}} x_k)) \\ &\quad + \textstyle\sum_{i,j\in[n],i\neq j} (x_i|x_j) ⌊⌊ (||_{k\in[n]\setminus\{i,j\}} x_k)] ⌊⌊ x_{n+1} \\ &=_4 (\textstyle\sum_{i\in[n]} x_i ⌊⌊ (||_{k\in[n+1]\setminus\{i\}} x_k)) \\ &\quad + \textstyle\sum_{i,j\in[n],i\neq j} (x_i|x_j) ⌊⌊ (||_{k\in[n+1]\setminus\{i,j\}} x_k) \end{aligned}$$

$$\begin{aligned} T3 &=_{IPH} [(\textstyle\sum_{i\in[n]} x_i ⌊⌊ (||_{k\in[n]\setminus\{i\}} x_k)) \\ &\quad + \textstyle\sum_{i,j\in[n],i\neq j} (x_i|x_j) ⌊⌊ (||_{k\in[n]\setminus\{i,j\}} x_k)] \mid x_{n+1} \\ &=_5 (\textstyle\sum_{i\in[n]} (x_i|x_{n+1}) ⌊⌊ (||_{k\in[n]\setminus\{i\}} x_k)) \\ &\quad + \textstyle\sum_{i,j\in[n],i\neq j} ((x_i|x_j)|x_{n+1}) ⌊⌊ (||_{k\in[n]\setminus\{i,j\}} x_k) \end{aligned}$$

{ by the handshaking hypothesis the last term is equal to δ }

$$\begin{aligned} &= \textstyle\sum_{i\in[n]} (x_i|x_{n+1}) ⌊⌊ (||_{k\in[n]\setminus\{i\}} x_k) \\ &= \textstyle\sum_{i\in[n],j=n+1} (x_i|x_j) ⌊⌊ (||_{k\in[n+1]\setminus\{i,j\}} x_k) \end{aligned}$$

By adding $T3$ to the expressions that have been obtained so far for $T1$ and $T2$ one gets the required identity for $n + 1$. □

Recursion in ACP. As we shall see, recursion in ACP is more powerful than in PA.

Theorem 9.14. *(infinite specifications) A process is ACP-definable iff it is BPA-definable.*
(finite specifications) With respect to finite specification ACP is more expressive than PA.

PROOF: (sketch) The first observation is a straightforward consequence of the Elimination Theorem 9.11.

For the second statement, consider the following finite specification over ACP:

$$\begin{array}{rcl} Q & = & \partial_H(dcY\|z) \\ X & = & cXc+d \\ Y & = & dXY \\ Z & = & dXcZ \end{array}$$

where $H = \{c, d\}$ and $\gamma(c, c) = a$, $\gamma(d, d) = b$, otherwise undefined.

It is not very difficult to see that this specification is equivalent in ACP with the following one, which actually lays in BPA:

$$\begin{array}{rcll} P & = & P_1 & \\ P_n & = & ba^n ba^{n+1} P_{n+1}, & n \geq 1 \end{array}$$

In fact, the process P is of the following form $ba(ba^2)^2(ba^3)^2 \ldots$. It is a bit more complicated to show that this process P is not PA-definable. A proof may be found in [BK94]. □

Example 9.15. (buffer of capacity 2 as a composite of 2 buffers of capacity 1) The specification of a buffer of capacity one $B1^{ij}$ which receives a datum $d \in D$ from channel i and sends it on channel j is

$$B1^{ij} = \sum_{d \in D} r_i(d) \cdot s_j(d) \cdot B1^{ij}$$

A buffer of capacity two $B2^{ij}$ which receives data from channel i and sends them on channel j may be specified by

$$\begin{array}{rcl} B2^{ij} & = & \sum_{d \in D} r_i(d) \cdot B_d \\ B_d & = & s_j(d) \cdot B2^{ij} + \sum_{e \in D} r_i(e) \cdot s_j(d) \cdot B_e \end{array}$$

Finally, the composite of $B1^{ik}$ and $B1^{kj}$ is

$$X = \partial_H(B1^{ik} \| B1^{kj})$$

where $H = \{r_k(d), s_k(d) \colon d \in D\}$ and $\gamma(s_k(d), r_k(d)) = c_k(d)$ for $d \in D$, otherwise undefined.

Representing communication networks. This example is typical with respect to the way process algebra handle such communications on channels. Function γ simply says that (trying to) send $s_k(d)$ and (trying to) receive $r_k(d)$ of the same data d using the same channel k is a successful attempt, the result being denoted as a successful communication $c_k(d)$. To avoid the case where only one of the two actions is made, a special set H is considered where all such individual attempts for communications are put; thereafter, all these actions are renamed in δ. This explain the form of the above buffer system X. ◇

It is not difficult to show within ACP that X and the derived term $X_d = \partial_H(B1^{ik}||s_j(d)B1^{kj})$ satisfy the system of equations

$$\begin{array}{rcl} X & = & \sum_{d\in D} r_i(d)\cdot c_k(d)\cdot X_d \\ X_d & = & s_j(d)\cdot X + \sum_{e\in D} r_i(e)\cdot s_j(d)\cdot c_k(e)\cdot X_e \end{array}$$

The specification satisfied by processes (X, X_d) is almost the same as for processes $(B2^{ij}, B_d)$: one moreover has to "make abstraction" of the fact that in the former system one can see an internal action of type $c_k(a)$.

Abstraction, to be briefly treated in the next section, is a renaming in a special action τ. This new constant τ has the intuitive meaning of a silent action - "the machine is working." Technically speaking, there are some different competitive calculi to handle abstraction, but all of them use the rule $(*)$ $\;x\tau = x$. Renaming $c_k(a)$ in τ and then using this unproblematic rule $(*)$ lead to the original specification for a buffer of capacity two. Hence by the Unique-rule the equality between processes X and $B2^{ij}$ is proved. See the next section for more on the abstraction mechanism.

9.7 Abstraction

A brief account of the abstraction is given here. Actually, the abstraction is a mechanism which makes certain actions to be invisible for the observer.

The role of the abstraction is to hide some actions, to make them silent, invisible for the observer. It is a simple renaming operation, but into a constant τ with specific properties.

The first observation is that τ cannot be ϵ. (Constant ϵ, read "skip," satisfies $x\cdot\epsilon = x$ & $\epsilon\cdot x = x$.) Indeed, if τ is ϵ and $b\to\tau$, then $a + b\delta$ is renamed into $a+\delta$, equal to a. Then, a term with deadlock is equivalent with one without deadlock. Such a result is highly undesirable due to the effect the deadlock has in the context of parallel processes.

The solution is to introduce a new constant τ, a ***silent action***. Then abstraction is just renaming into τ. The abstraction operator is τ_I, for $I\subseteq A$; it renames into τ the actions in I.

The main problem is to find the rules satisfied by τ. This is not a simple task, even for acyclic processes. For cyclic processes it is much more difficult since the rules of τ are related to the fairness properties. We gives a short account here for one proposal, i.e., the ***branching bisimulation***; see, e.g., [vG90] and [BW90a].

The key property of a branching bisimulation $R: g \leftrightarrows_b h$ is that an action $s \xrightarrow{a} s'$ is simulated by a chain of actions $t \xrightarrow{\tau^*} t_1 \xrightarrow{a} t'$ *and the intermediary node t_1 is related to s*. Some other proposed equivalences are: the ***delay*** equivalence, which does not require the relation of t_1 and s; and the *weak* equivalence, which is like the delay equivalence, but it also uses τ^* actions in between a and t'.

We list a few results. (1) The following (acyclic) laws for τ

$$(B1)\ x \cdot \tau = x \quad \& \quad (B2)\ x \cdot (\tau \cdot (y + z) + y) = x \cdot (y + z)$$

are sound and complete for finite acyclic graphs modulo branching bisimulation. (2) Branching bisimulation also has a natural characterization in terms of coloured traces. An ***abstract coloured trace*** is a coloured trace where all sequences $(c, \tau, c, \ldots, \tau, c)$ are replaced by c. That means, τ between two nodes with the same colour may be eliminated, but it is to be preserved when it occurs in between nodes with different colours. The result is: *two graphs are branching bisimilar iff they have the same abstract coloured traces with respect to consistent colourings.*

EXERCISES AND PROBLEMS (SEC. 9.7)

1. Take the processes P and P' defined by: (1) $P = aQ + c$, $Q = bP + d$ and (2) $P' = \tau_{\{i\}}\partial_H((a(b + dk_2) + ck_2)R \parallel S)$, where $R = k_1(a(b + dk_2) + ck_2)R + k_3$, $S = k_1S + k_3k_2$, $H = \{k_1, k_2, k_3\}$, and $\gamma(k_1, k_1) = \gamma(k_2, k_3) = \gamma(k_3, k_2) = i$, othervise undefined. Show that P and P' are rooted branching bisimilar and their equality may be formally derived in ACP+(B1–B2).(12)

9.8 A case study: Alternating Bit Protocol

We show here how ACP may be used to verify a simple distributed protocol, i.e., the Alternating Bit Protocol. Because of the lack of abstraction results (our previous section was rather short) the verification is partial. For a full verification certain fairness properties, captured by rules of τ in the cyclic setting, are to be used.

Informal presentation. The requirement is to solve the following problem (presented in informal terms):

> Find a protocol for safe transmission of data via unsafe channels (that is, channels which may lose data).

A possible solution is to use a transmission protocol with alternating bits for acknowledgment. It is known as the Alternating Bit Protocol – ABP. In still informal, but more detailed terms, the protocol consists of the following steps:

(1) The sender receives a datum from its external interface, appends its current control bit and sends the composed message to the sender. It waits for an acknowledgment from the receiver and repeats (re)transmission of the underlying

datum till it receives as acknowledgment the current bit. Then it changes its current bit and repeats the above behaviour.

(2) The communication channels randomly choose to pass the received datum or an erroneous element and repeat this behaviour.

(3) When the receiver gets a composed message with the complementary bit (with respect to its current bit) it passes the datum to its external output channels, changes its own bit, and sends this (current) bit as an acknowledgment to the sender. Then it repeats to send this confirmation as long as it receives a composed message with the current bit or an erroneous message.

Formal specification

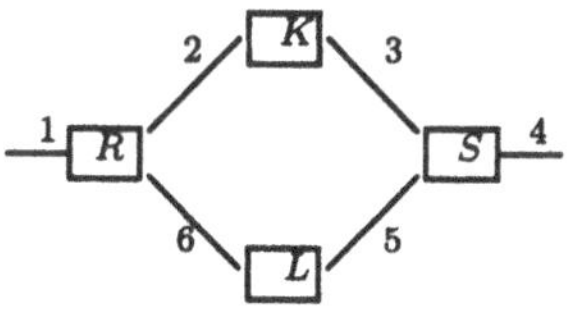

This "Alternating Bit Protocol" ABP may be specified in ACP as follows. Suppose D is the set of data to be transmitted and consider the set of frames $F = \{d0, d1 : d \in D\}$. Denote by i an atomic action which models an internal choice between a correct and an erroneous transmission on a channel and $H = \{s_k(x), r_k(x) : x \in F \cup \{0, 1, \perp\},\ k \in \{2, 3, 5, 6\}\}$.

Then:

$$ABP = \partial_H(K||L||S||R)$$

where

$$\begin{aligned}
K &= \textstyle\sum_{x \in F} r_2(x) \cdot (i \cdot s_3(x) + i \cdot s_3(\perp)) \cdot K \\
L &= \textstyle\sum_{n \in \{0,1\}} r_5(n) \cdot (i \cdot s_6(n) + i \cdot s_6(\perp)) \cdot L \\
S &= S0 \cdot S1 \cdot S, \quad \text{where for } n \in \{0,1\} \text{ and } d \in D \\
&\qquad Sn = \textstyle\sum_{d \in D} r_1(d) \cdot Sn_d \\
&\qquad Sn_d = s_2(dn) \cdot Tn_d \\
&\qquad Tn_d = [r_6(1-n) + r_6(\perp)] \cdot Sn_d + r_6(n) \\
R &= R1 \cdot R0 \cdot R, \quad \text{where for } n \in \{0,1\} \\
&\qquad Rn = [\textstyle\sum_{d \in D} r_3(dn) + r_3(\perp)] \cdot s_5(n) \cdot Rn \\
&\qquad\qquad + \textstyle\sum_{d \in D} r_3(d(1-n)) \cdot s_4(d) \cdot s_5(1-n)
\end{aligned}$$

Partial verification in ACP. We consider certain relevant states of the system and denote them as follows:

$$\begin{aligned}
X &= \partial_H(S||K||L||R) \\
X1_d &= \partial_H(S0_d \cdot S1 \cdot S||K||L||R) \\
X2_d &= \partial_H(T0_d \cdot S1 \cdot S||K||L||s_5(0) \cdot R0 \cdot R) \\
Y &= \partial_H(S1 \cdot S||K||L||R0 \cdot R) \\
Y1_d &= \partial_H(S1_d \cdot S||K||L||R0 \cdot R) \\
Y2_d &= \partial_H(T1_d \cdot S||K||L||s_5(1) \cdot R)
\end{aligned}$$

The following identities relating these states are valid:

$$\begin{array}{rl}
X= & \sum_{d\in D}\underline{r_1(d)}\cdot X1_d \\
X1_d= & c_2(d0)\cdot[i\cdot c_3(d0)\cdot \underline{s_4(d)}\cdot X2_d + i\cdot c_3(\bot)\cdot c_5(1)\cdot(i\cdot c_6(\bot)+i\cdot c_6(1))\cdot X1_d] \\
X2_d= & i\cdot c_6(0)\cdot Y + c_5(0)\cdot[i\cdot c_6(\bot)\cdot c_2(d0)\cdot(i\cdot c_3(d0)+i\cdot c_3(\bot))]\cdot X2_d \\
Y= & \sum_{e\in D}\underline{r_1(e)}\cdot Y1_e \\
Y1_e= & c_2(e1)\cdot[i\cdot c_3(e1)\cdot \underline{s_4(e)}\cdot Y2_e + i\cdot c_3(\bot)\cdot c_5(0)\cdot(i\cdot c_6(\bot)+i\cdot c_6(0))\cdot Y1_e] \\
Y2_e= & i\cdot c_6(1)\cdot X + c_5(1)\cdot[i\cdot c_6(\bot)\cdot c_2(e1)\cdot(i\cdot c_3(e1)+i\cdot c_3(\bot))]\cdot Y2_e
\end{array}$$

From these identities it follows that if one abstracts from all the internal actions, then only the underlined actions remain, hence with respect to these visible actions the protocol behaves like a buffer of capacity 1. The complete proof needs sophisticated rules for abstraction (including some sorts of fairness rules) to eliminate the cyclic, internal actions and is not given here. See e.g., [BW90a] for a complete verification.

As an example of the way process algebra is used we put its machinery to work and prove the validity of the above-mentioned identities. It is easy to see that

$$\begin{array}{rcl}
X & = & \partial_H(S||K||L||R) \\
 & = & \sum_{d\in D} r_1(d)\partial_H(S0_d\cdot S1\cdot S||K||L||R) \\
 & = & \sum_{d\in D} r_1(d)\cdot X1_d
\end{array}$$

Let $d \in D$. Then the identity for $X1_d$ follows using the Expansion Theorem as follows (in parentheses there are short hints to explain the transformation):

$$\begin{array}{rl}
 & X1_d \\
= & \partial_H(\underline{S0_d\cdot S1\cdot S}||\underline{K}||L||R) \\
= & \{\!\!\{ \text{ due to } \partial_H \text{ only the communication between the underlined terms remains } \}\!\!\} \\
 & c_2(d0)\cdot[i\cdot\partial_H(T0_d\cdot S1\cdot S||\underline{s_3(d0)\cdot K}||L||\underline{R} \\
 & \;+i\cdot\partial_H(T0_d\cdot S1\cdot S||\underline{s_3(\bot)\cdot K}||L||\underline{R})] \\
= & \{\!\!\{ \text{ idem } \}\!\!\} \\
 & c_2(d0)\cdot[i\cdot c_3(d0)\cdot\partial_H(T0_d\cdot S1\cdot S||K||\underline{L}||\underline{s_4(d)\cdot s_5(0)\cdot R0\cdot R}) \\
 & \;+i\cdot c_3(\bot)\cdot\partial_H(T0_d\cdot S1\cdot S||K||\underline{L}||\underline{s_5(1)\cdot R1\cdot R0\cdot R})] \\
= & \{\!\!\{ \text{ idem } \}\!\!\} \\
 & c_2(d0)\cdot[i\cdot c_3(d0)\cdot s_4(d)\cdot\underline{\partial_H(T0_d\cdot S1\cdot S||K||L||s_5(0)\cdot R0\cdot R)} \\
 & \;+i\cdot c_3(\bot)\cdot c_5(1)\cdot\partial_H(\underline{T0_d\cdot S1\cdot S}||K||\underline{(i\cdot s_6(1)+i\cdot s_6(\bot))\cdot L}||R)] \\
= & \{\!\!\{ \text{ idem for the last two terms; the first underlined term is } X2_d \;\}\!\!\} \\
 & c_2(d0)\cdot[i\cdot c_3(d0)\cdot s_4(d)\cdot X2_d+i\cdot c_3(\bot)\cdot c_5(1)\cdot(i\cdot c_6(1) \\
 & \;\cdot\underline{\partial_H(S0_d\cdot S1\cdot S||K||L||R)}+i\cdot c_6(\bot)\cdot\underline{\partial_H(S0_d\cdot S1\cdot S||K||L||R)}))] \\
= & \{\!\!\{ \text{ the underlined terms are equal to } X1_d \text{ and use distributivity } \}\!\!\} \\
 & c_2(d0)\cdot[i\cdot c_3(d0)\cdot s_4(d)\cdot X2_d \\
 & \;+i\cdot c_3(\bot)\cdot c_5(1)\cdot(i\cdot c_6(1)+i\cdot c_6(\bot))\cdot X1_d))]
\end{array}$$

For $X2_d$ the computation looks as follows:

$$
\begin{array}{rl}
& X2_d \\
= & \partial_H(T0_d \cdot S1 \cdot S||K||\underline{L}||\underline{s_5(0)} \cdot R0 \cdot R) \\
= & \{\!\![\text{ due to } \partial_H \text{ only the communication between the underlined terms remains }]\!\!\} \\
& c_5(0) \cdot \partial_H(\underline{T0_d \cdot S1 \cdot S}||K||\underline{(i \cdot s_6(0) + i \cdot s_6(\bot))} \cdot L||R0 \cdot R) \\
= & \{\!\![\text{ idem }]\!\!\} \\
& c_5(0) \cdot [i \cdot c_6(0) \cdot \partial_H(S1 \cdot S||K||L||R0 \cdot R) \\
& + i \cdot c_6(\bot) \cdot \partial_H(\overline{\underline{S0_d \cdot S1 \cdot S}||\underline{K}||L||R0 \cdot R})] \\
= & \{\!\![\text{ idem for the last 2 terms; the first underlined term is } Y\ ]\!\!\} \\
& c_5(0) \cdot [i \cdot c_6(0) \cdot Y + i \cdot c_6(\bot) \cdot c_2(d0) \\
& \cdot \partial_H(T0_d \cdot S1 \cdot S||\underline{(i \cdot s_3(d0) + i \cdot s_3(\bot))} \cdot K||L||\underline{R0 \cdot R})] \\
= & \{\!\![\text{ idem }]\!\!\} \\
& c_5(0) \cdot [i \cdot c_6(0) \cdot Y + i \cdot c_6(\bot) \cdot c_2(d0) \\
& \cdot [i \cdot c_3(d0) \cdot \underline{\partial_H(T0_d \cdot S1 \cdot S||K||L||s_5(0) \cdot R0 \cdot R)} \\
& + i \cdot c_3(\bot) \cdot \underline{\partial_H(T0_d \cdot S1 \cdot S||K||L||s_5(0) \cdot R0 \cdot R)}]] \\
= & \{\!\![\text{ the underlined terms are equal to } X1_d \text{ and use distributivity }]\!\!\} \\
& c_5(0) \cdot [i \cdot c_6(0) \cdot Y + i \cdot c_6(\bot) \cdot c_2(d0) \cdot (i \cdot c_3(d0) + i \cdot c_3(\bot)) \cdot X2_d]
\end{array}
$$

Finally, the last two equations are simple variations of the above ones using $(Y, 1, e)$ instead of $(X, 0, d)$.

Comments, problems, bibliographic remarks

Problems. A few problems related to the complex relationship between process algebra and network algebra are inserted here.

Process algebra as a time network algebra. As in [BBŞ95] one may simply add feedback to the process algebra operators, but the result is not really a network algebra: it lacks well-behaving identities. Another possibility, which will be described in some detail in the forthcoming chapter, is to build up a network algebra on top of process algebra, using an underlying data flow model; however, the two formalisms are mainly separated in this combination, therefore the result is also not fully satisfying. Other difficulties are generated by the fairness problems, the lack of a simple relational model for process semantics, the large number of different syntactic models, etc.

To overcome (some of) these problems, we propose to recast process algebra in a full network algebra framework. The key idea is to consider processes as job transformers; moreover, the associated time network algebra is to be used to specify and study processes. Protocol verification would be reduced to the usual verification of programs (the "Floyd–Hoare logic" of the previous chapter), but in the orthogonal time network algebra model.

Notes. Process algebra was introduced in the mid 1970s. There is a huge literature on this approach to concurrent processes, including the well-known books [Hen88] and [Mil89].

Our approach here uses the ACP model introduced by Bergstra and Klop [BK84, BK85]; in book form, it is presented in [BW90a]; for a time version see [BB92].

The main result is the correctness (soundness and completeness) of ACP with respect to the bisimulation equivalence; we have included a detailed proof for this result.

The bisimulation relation was introduced by Park in [Par81]; its introduction has been subsequently based on Milner's observation equivalence defined in [Mil80]. Branching bisimulation was introduced by van Glabbeek–Weijland; see, e.g., [vG90] and [BW90a].

A large number of papers are dedicated to decidability issues. A seminal paper is [BBK93], where the decidability of bisimulation equivalence for processes generating context-free languages is proved.

The interaction of Kleene-like iteration and process algebra is studied in [BBP94]; the interplay between the feedback operator and process algebra is described in [BBŞ95].

Samples of protocol correctness verification may be found in many papers, including [vGV93, BP96, Şor97]. A denotational verification of the Alternating Bit Protocol is presented in [Möl94] by means of streams manipulations. Fairness is a central topic in [Par83, Par85]. Transition systems (and communicating automata) are also studied [Arn94]. Exercise 9.7.1 is from [BBP94].

10. Data-flow networks

To copy others is necessary, but to copy oneself is pathetic.

Pablo Picasso

Data-flow networks are a model of concurrent computation. They consist of a collection of concurrent processes which communicate by sending data over FIFO channels.

The relational models associated with data-flow networks are multiplicative network algebra models based on streams of data; the temporal order on streams is used here to support an iterative definition of the multiplicative feedback.

We start with a brief presentation of the synchronous data-flow networks. This model is rather simple, but it has some shortcomings: one has to mix the one-time delay model for cells with the zero-time delay model for wires.

Passing to the asynchronous case, first we deal with deterministic networks. We show the axioms of dual Elgot theories ($d\alpha$-flows) give a correct (i.e., sound and complete) axiomatization for the relational input–output semantics.

Next, we pass to nondeterministic networks. The relational input–output semantics of nondeterministic networks is not compositional; this is known as a "time anomaly." Certain fully abstract semantics (i.e., maximal compositional semantics finer than the input–output one) are briefly presented, e.g., the trace semantics or the semantics based on time intervals. In more detail the *oracle based semantics* is presented; it is also compositional, but not fully abstract; mathematically, it consists of sets of stream processing functions and it may be seen as a simple extension of the model used for deterministic data-flow networks. We show that the BNA axioms are valid in this model. We also identify certain additional axioms satisfied by the branching constants in this model.

Finally, a general process algebra-based model for data-flow networks is presented. This way network algebra models (BNAs) are built up on top of process algebra. The model is further specialized to the case of asynchronous networks and applied to study the branching structure of the networks in this process algebra based setting.

10.1 Data-flow networks; general presentation

A brief presentation of the data-flow model is given here. It includes some examples, the dichotomy between synchronous vs. asynchronous networks, the time anomaly, and the dichotomy between multiplicative vs. additive branching connections (that is, copy/equality test vs. split/merge interpretation).

In the model of data-flow networks, as well as in the case of Petri nets (to be studied in the next chapter), in order to get a true concurrency model one has to enter into the structure of the states. The hypotheses that are used here are: (1) The global state is divided into pieces, called *channels*. (2) The channels are considered as (unbounded) FIFO queues; consequently, they are modeled as data streams; such a data stream is a finite or infinite sequence that records the history of the data communicating along a specific channel. (3) The network cells act by consuming data from their input channels and producing data to the output channels.

Data-flow networks. A network can be any labeled directed hyper-graph that represents some kind of flow between the components of a system. Data-flow networks are networks concerning flow of data.

Example 10.1. An example of a data-flow network is given in the figure below.

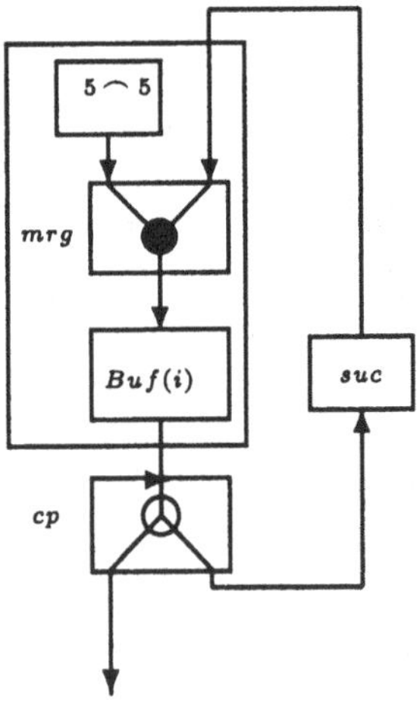

Fig. 10.1. Asynchronous, nondeterministic data-flow network

Explanation for Fig. 10.1.
- $5 \frown 5$ is a cell that generates the sequence 5 followed by 5, then stop;
- *mrg* is a particular merge cell that (nondeterministically) merges $5 \frown 5$ and the first data of the right channel (if any), then stop;

- *cp* is a copy cell that copies the input datum to both output channels, then repeat this behaviour;
- *suc* is a cell that eats a datum, produces its successor, then repeat this behaviour;
- $Buf(1)$ is a cell that allow to pass one datum, then another one, then stop;
- finally, $Buf(2)$ eats one datum, then another one, then produces at the output the eaten data (in the same order), then stop. ◇

Mismatch. The following behaviour is possible in the network using $Buf(1)$, but not in the network using $Buf(2)$: $5 \frown 5$ generates one 5 which is passed by mrg to $Buf(1)$, then copied to the output and also passed to *suc*; then *suc* generates a 6 that is passed by mrg to $Buf(1)$ cell and copied to output and also passed to *suc*; to conclude, the network with $Buf(1)$ is able to output the sequence $5 \frown 6$. On the other hand, the instance of the network with $Buf(2)$ has no such ability. ◇

This is a typical example of a *nondeterministic asynchronous* data-flow network. The asynchrony refers to the lack of a global clock mechanism forcing the cells to fire simultaneously; instead of this, the cells fire when there are data on their input channels. The nondeterminism is manifested in the mrg cell.

There are various data-flow models. The main difference between them is in the interpretation of the identity connections, called *wires*. Except for this different meaning of the wires, in data-flow networks one can also find distinct interpretations for the branching constants.

Network representation. As in the case of general networks, data-flow networks can be built up from their components and (possibly branching) connections using juxtaposition (or parallel composition $\otimes$), sequential composition $(\cdot)$ and feedback $(\uparrow)$. The connections needed are at least the identity (I) and transposition (X) connections, but branching connections may also be needed for specific classes of networks – e.g., the binary ramification $(\wedge)$ and identification $(\vee)$ connections and their nullary counterparts $(\perp$ and $\top)$. ◇

The cells are interpreted as processes that consume data at their input ports, compute new data, deliver the new data at their output ports, and then start over again. The wires are interpreted as queues of some kind. The classical data-flow networks are classified according to the type of wire as follows: firstly, queues that deliver data with a negligible delay and never contain more than one datum (minimal stream delayers), and secondly, unbounded, delaying queues (stream delayers).

A stream is a sequence of data consumed or produced by a component of a data-flow network on a specific channel. A flow of data is a transformation of a tuple of data streams into a tuple of data streams. Then, a wire behaves as an identity flow. If the wire is a stream delayer, then data pass through it with a time delay. If the wire is a minimal stream delayer, then data enter and leave it with a negligible delay – i.e., within the same time slice, provided time is divided into time slices.

Synchronous case. In synchronous data-flow networks, the wires are *minimal stream delayers*. Such a data-flow network use a global clock. On ticks of the clock, cells can start up the consumption of exactly one datum from each of their input ports and the production of exactly one datum at each of their output ports. A cell that started up with that completes the production of data before the next tick, and it completes the consumption of data as soon as a new datum has been delivered at all input ports. On the first tick following the completion of both, the cell concerned starts up again. In order to start the computation in a synchronous data-flow network, every cell is supposed to have, for each of its output ports, an initial datum available to be delivered at the initial clock tick.

Minimal stream delayers. The underlying idea of synchronous data-flow is that computation takes a good amount of time, whereas storage and transport of data takes a negligible amount of time. Phrased differently, data always pass through a wire between two consecutive ticks of the global clock. So minimal stream delayers fit in exactly with this kind of data-flow network. ◇

The semantics of synchronous data-flow networks turns out to be rather simple; however, there are some complications generated by the fact that all proper cells have one-time delay, while the wires have no delay.

Asynchronous case. In asynchronous data-flow networks, the wires are *stream delayers*. The underlying idea of asynchronous data-flow is that computation as well as storage and transport of data takes a good deal of time, which is sometimes more realistic for large systems. In such cases, it is more natural to have computation driven by the arrival of the data needed – instead of clock ticks. Therefore, there is no global clock in an asynchronous data-flow network.

Stream delayers. Cells may independently consume data from their input ports, compute new data, and deliver the new data at their output ports. There may be data produced by cells but not yet consumed by other cells, hence these networks need wires that are able to buffer an arbitrary amount of data. So, stream delayers fit in exactly with this kind of data-flow network. ◇

However, the semantics of asynchronous data-flow networks turns out to be rather problematic. The main semantic problem is a *time anomaly*, also known as the Brock–Ackermann anomaly. There are networks that have the same input–output behaviour, but if they are placed in the same context, then networks with different input–output behaviours may be produced.

Time anomaly, Example 10.1. Such an example is provided by the networks N_i encapsulated in the big rectangle of Example 10.1. Then N_1 (using $Buf(1)$) and N_2 (using $Buf(2)$) have the same behaviour, but placed in the context described in Example 10.1, the resulting networks have different behaviours. The feedback is responsible for this: timing differences in producing data may become important when both feedback and nondeterminism are present. ◇

Gilles Kahn suggested a mathematical model for asynchronously communicating agents that could be used as a model for *deterministic* data-flow networks. Due to the deterministic assumption the input–output relation specified by such a process actually is a (continuous) function. The Khan theorem asserts that the function specified by a deterministic network may be obtained from the functions specified by its components in a compositional way using the least fixed-point construction.

Such a result cannot be easily extended to *nondeterministic* data-flow networks. The above explained time anomaly has shown a mismatch between the operational meaning of the networks and their relational input–output behaviour. In other words, the input–output behaviour of the components no longer suffices to compute the whole network behaviour. This situation was solved by adding information to the input–output behaviour, e.g., using traces, oracles, time intervals, etc.

Branching constants. As one probably expects, data-flow networks also need branching connections.

What to do in a branching point. The branching structure of data-flow networks is more complex than the branching structure of flowcharts. Indeed, in the case of (sequential) flowcharts, there is a flow of control which is always at one point in the flowchart concerned; consequently, the interpretation of the branching connections is rather obvious. However, in the case of data-flow networks, there is a flow of data which may be spread in the network; hence, the interpretation of the branching connections is not immediately clear. ◇

In this chapter, two kinds of interpretation for branching connections are considered. They are the multiplicative *copy/equality test* interpretation and the additive *split/merge* interpretation.

Two interpretations. The first kind of interpretation fits in with the idea of permanent flows of data which naturally go in all directions at branchings. Synchronous data-flow reflects this idea most closely. For example, the ramification (copy) constant *cp* in Example 10.1 is based on such an interpretation. The second kind of interpretation fits in with the idea of intermittent flows of data which go in one direction at branchings. Such a situation may be present in an asynchronous data-flow. An example is provided by the identification (merge) constant *mrg* in Example 10.1. ◇

In order to distinguish between the branching constants with these different interpretations, different symbols for $\wedge^a$ and $\vee_a$ are used: $\overset{\circ}{\wedge}{}^a$ and $\underset{\circ}{\vee}{}_a$ for the copy/equality test interpretation, $\overset{\bullet}{\wedge}{}^a$ and $\underset{\bullet}{\vee}{}_a$ for the split/merge interpretation. Likewise, different symbols for the nullary counterparts $\perp^a$ and $\top_a$ are used: $\underset{\circ}{\mid}{}^a$ and $\overset{\circ}{\mid}{}_a$ versus $\underset{\bullet}{\mid}{}^a$ and $\overset{\bullet}{\mid}{}_a$. Of these last constants, $\underset{\circ}{\mid}{}^a$ and $\underset{\bullet}{\mid}{}^a$ are called *sink* and *dummy sink*, respectively; and $\overset{\circ}{\mid}{}_a$ and $\overset{\bullet}{\mid}{}_a$ are called *source* and *dummy source*, respectively.

10.2 Synchronous networks

This section contains a brief presentation of synchronous networks. Both the operational and relational semantics are presented in a generic example.

Global clock; synchronous networks. In this case there is a global clock. All the cells react at the clock tick. In between two time ticks a cell consumes one datum from each of its incoming channels, uses these data in the appropriate computation, and then produces one datum for each of its outgoing channels. All the cells act as above on each time slice. In order to start the computation, it is supposed that each cell has a default tuple of values to be used for its outgoing channels.

Warming up. There is a clear analogy here with many usual physical devices; one has to wait for "warming up" the network; at this period the output is irrelevant; after that, the device works properly: depending on its construction and functionality it reacts to the input stimuli and, with an appropriate delay, produces the corresponding outputs. ◊

Example: Fibonacci sequence. As an example, let us consider the network

$$E = [(\mathsf{l}_1 \otimes del) \cdot add \cdot \mathcal{R}_3^1] \uparrow^2 : 0 \to 1$$

The cells $\mathsf{l}_1 : 1 \to 1$, $\mathcal{R}_3^1 : 1 \to 3$, $del : 1 \to 1$, $add : 2 \to 1$ have the following meaning (a stream x is written as $x(0) \frown x(1) \frown \ldots$, the sequence of its components separated by "$\frown$"):

$-\mathsf{l}_1(x) = x;$
$-\mathcal{R}_3^1(x) = (x, x, x);$
$-del(x(0) \frown x(1) \frown \ldots) = 1 \frown x(0) \frown x(1) \frown \ldots;$
$-add(x(0) \frown x(1) \frown \ldots, y(0) \frown y(1) \frown \ldots) = 1 \frown x(0) + y(0) \frown x(1) + y(1) \frown \ldots.$

The result of this network, which has no input channels, is the Fibonacci sequence at the output channel (the recursive relation for defining this sequence is $x(n+2) = x(n+1) + x(n)$, where $x(0) = 1$ and $x(1) = 2$):

$$||E||() = 1 \frown 2 \frown 3 \frown 5 \ldots$$

Explanation for Fig. 10.2.
- As one may see from the figure, *add* is a cell with two input channels and one output channel; it adds the corresponding input data, i.e., $x(0)+y(0)$, $x(1)+y(1)$, etc. However, the output is not instantaneous, namely a one-time step delay is in between the arrival of the input data and the corresponding resulting output; the default output value for the *add* cell is 1, to be used as the output at the first time slice; it is indicated in the bottom-right part of the cell.
- The *del* (delay) cell is like an identity, but it delays the input with one time step; its

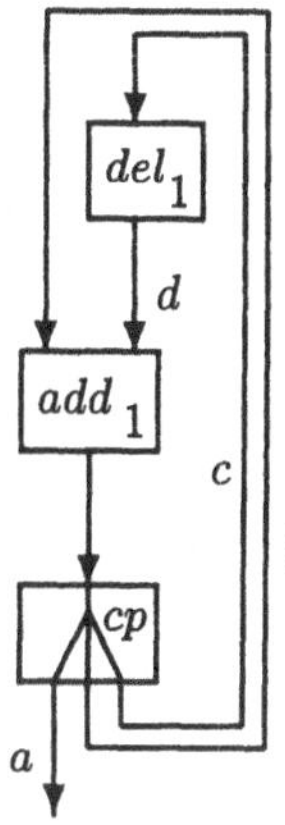

Fig. 10.2. Synchronous data-flow network

default output value for the first time slice is 1, too.
- Finally, the network uses certain wires (I_1 and $\wedge^1_3$) which copy the input to the corresponding output(s) without time delay. ◇

The result may be obtained either monolithic or modular:

(monolithic). This case is based on the operational semantic. One simply uses one (meta)variable for each channel; then he completes the initial values for all variables (using the fact that the wires have no time delay); after this, the values of the variables for the next time slice are computed using the cell meanings; and so on.

For the given example, let a be the channel that connects the output of *add* to the network output; b the one that connects the output of *add* to the 1st input of *add*; c the one that connects the output of *add* to the input of *del*; and d the channel that connects the output of *del* to the 2nd input of *add*. Then,

time	0	1	2	3	4	5	6	...
a	1	2	3	5	8	13	21	
b	1	2	3	5	8	13	21	
c	1	2	3	5	8	13	21	
d	1	1	2	3	5	8	13	

The recursive definition is very simple here: $a(n+1) = b(n+1) = c(n+1) = a(n) + d(n)$; $d(n+1) = c(n)$.

(modular). In the second case, starting with the semantic of the cells, step-by-step one gets the semantics of larger and larger parts of the network expression. This procedure is based on the relational semantics. Except for feedback, the meaning of the network algebra operators and constants are the standard ones used in the multiplicative interpretation MultRel. The feedback operation is defined using the default value on the feedback loop. (Such a value exists if the input and the output channels to be connected by feedback are not connected

by a simple wire, hence in between them there is at least one proper cell. Notice that this feedback operation is partially defined. E.g., it is not defined for the copy constant; that is, $\wedge_2^1 \uparrow^1$ is undefined.)

For the given example, notice that:

$$(\mathsf{I}_1 \otimes \mathit{del})(x(0) \frown x(1) \frown \ldots, y(0) \frown y(1) \frown \ldots)$$
$$= (x(0) \frown x(1) \frown \ldots, 1 \frown y(0) \frown y(1) \frown \ldots)$$
$$[(\mathsf{I}_1 \otimes \mathit{del}) \cdot \mathit{add}](x(0) \frown x(1) \frown \ldots, y(0) \frown y(1) \frown \ldots)$$
$$= 1 \frown x(0) + 1 \frown x(1) + y(0) \frown \ldots$$
$$[(\mathsf{I}_1 \otimes \mathit{del}) \cdot \mathit{add} \cdot \wedge_3^1](x(0) \frown x(1) \frown \ldots, y(0) \frown y(1) \frown \ldots)$$
$$= (1 \frown x(0) + 1 \frown x(1) + y(0) \frown \ldots, 1 \frown x(0) + 1 \frown x(1) + y(0) \frown \ldots,$$
$$1 \frown x(0) + 1 \frown x(1) + y(0) \frown \ldots)$$
$$\{[(\mathsf{I}_1 \otimes \mathit{del}) \cdot \mathit{add} \cdot \wedge_3^1] \uparrow^1\}(x(0) \frown x(1) \frown \ldots)$$
$$= (1 \frown x(0) + 1 \frown x(1) + 1 \frown \ldots, 1 \frown x(0) + 1 \frown x(1) + 1 \frown \ldots)$$
$$\{\{[(\mathsf{I}_1 \otimes \mathit{del}) \cdot \mathit{add} \cdot \wedge_3^1] \uparrow^1\} \uparrow^1\}$$
$$= 1 \frown 2 \frown 3 \frown \ldots$$

Notice that the feedback is computed identifying the streams of the corresponding channels: for the 1st feedback, $y(0) \frown y(1) \frown y(2) \frown \ldots = 1 \frown x(0) + 1 \frown x(1) + y(0) \frown \ldots$; for the 2nd, $x(0) \frown x(1) \frown x(2) \frown \ldots = 1 \frown x(0) + 1 \frown x(1) + 1 \frown \ldots$.

EXERCISES AND PROBLEMS (SEC. 10.2)

1. Describe in detail the relational model corresponding to synchronous data-flow networks. Check the validity of the (B)NA axioms in this model. (12)
2. Show that with copy/equality test interpretation, the forward demonic axioms in Appendix C give a (sound and complete) axiomatization for the branching structure of synchronous data-flow networks. (10)

10.3 Asynchronous networks

In this section asynchronous networks and their stream processing function based semantics are described.

Deterministic networks. An example of asynchronous network was already presented in the beginning of this chapter (Example 10.1). Actually, due to the use of the nondeterministic *mrg* component, the result is a nondeterministic network. One may get a deterministic network if the *mrg* cell is replaced by a deterministic merge component.

Deterministic merge. Such a deterministic merge may be $mrg1 = 1 \frown 2 \frown 1$; this takes a first datum from its 1st (left) input channel, then one from its 2nd (right) channel, finally one from its 1st channel, then stops; this means we have a merge which has a fixed merge politics; for instance, if 2 data are present at the 1st input channel of *mrgl* and no data at its 2nd input channel, then after its delivering of the first datum from the 1st input channel, the *mrgl* enter into a deadlock (it waits forever to receive a datum from its 2nd input channel), while there are still data on its 1st channel. ◇

Example 10.2. We use the data-flow networks described in Figure 10.3.(a)–(c) to illustrate the forthcoming algebraic techniques.

Figure explanation. These data-flow networks may be represented by flownomial expressions as well. For instance, the data-flow network shown in (a) may be represented by the following expression:

$$\wedge^1 \cdot [(\mathsf{I}_1 \otimes f \cdot {}^1\mathsf{X}^1) \cdot (f \otimes \mathsf{I}_1) \cdot (\wedge^1 \cdot (f' \uparrow^1 \otimes \mathsf{I}_1) \otimes {}^1\mathsf{X}^1)] \uparrow^1$$

As we will see below, all the networks in Figure 10.3 compute the same stream processing function, provided the ramification constant $\wedge$ is interpreted as the copy constant and the cells are considered to be deterministic components. Moreover, we show their equality may be proved using the *da*-flow axioms. In (d) we have drawn the common infinite tree obtained by unfolding the networks in (a)–(c). In that picture f_1 (resp. f_2) represents the cell obtained from f by restriction to the first (resp. second) output only. ◇

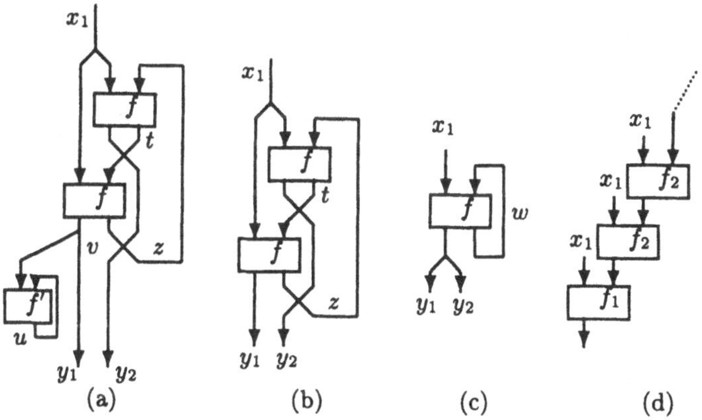

Fig. 10.3. Network equivalence

Stream processing functions. In this subsection we construct a semantic model SPF(M) for the interpretation of deterministic data flow networks. It is based on stream processing functions.

Certain multiplicative (logical) networks have already been used in Chapter 7 related to Floyd–Hoare logics. Different from the situation there, a *temporal order* on data is used here to support an iterative definition of the multiplicative feedback.

A *stream* represents a communication history of a channel. A stream of messages over a given message set D is a finite or infinite sequence of messages. We define the set of streams M as

$$M = D^\omega \quad \text{where } D^\omega =_{def} D^* \cup D^\infty$$

By $x \frown y$ we denote the result of concatenating two streams x and y. We assume that $x \frown y = x$, if x is infinite. By $\langle\,\rangle$ we denote the empty stream. If a stream x is a *prefix* of a stream y, we write $x \sqsubseteq y$. The relation $\sqsubseteq$ is called *prefix order*. It is formally specified as follows: $x \sqsubseteq y \quad =_{def} \quad \exists z \in M.\, x \frown z = y$. The behaviour of deterministic interactive systems with m input channels and n output channels is modeled by functions $f : M^m \to M^n$ called (m,n)*-ary stream processing functions*. A stream processing function is called *prefix monotonic*, if for all tuples of streams $x, y \in M^m$ we have $x \sqsubseteq y \quad \Rightarrow \quad f(x) \sqsubseteq f(y)$. This particular ordering is extended to tuples and functions point-wise in a straightforward way. By $\bigsqcup S$ we denote the least upper bound of a set S, if it exists. A set S is called *directed*, if for any pair of elements x and y in S there exists an upper bound in S. The set of streams is complete in the sense that for every directed set of streams there exists a least upper bound. A stream processing function f is called *continuous*, if f is monotonic and for every directed set $S \subseteq M$ we have: $f(\bigsqcup S) = \bigsqcup\{f(x) : x \in S\}$.

SPF*-model.* In the following we define the model in the many sorted case. It is a specialization of the general $\mathsf{MultRel}$ model to the case when we have a stream structure on data. This new ingredient is the key feature which help to define a feedback operation and to have a BNA structure on SPF.

Let S be a set of sorts. Let $D = \{D_s\}_{s \in S}$ be an S-sorted set of messages and $M_s := D_s^* \cup D_s^\infty$ be the set of streams over D_s representing communication histories of channels of type s. Concatenation in $S^\otimes$ is denoted by $\otimes$; an element $a \in S^\otimes$ is written as $a = a_1 \otimes \ldots \otimes a_{|a|}$, where a_i is the i-th letter of a. Finally, M^a denotes the product $M_{a_1} \times \ldots \times M_{a_{|a|}}$.

Given an S-sorted set D of messages and the corresponding sets of streams M_s for $s \in S$, we define the set of stream processing functions with input sorts $a \in S^\otimes$ and output sorts $b \in S^\otimes$ by

$$\mathsf{SPF}(M)(a,b) = \{f : M^a \to M^b : f \text{ is prefix continuous }\}.$$

The $d\beta$-symocat structure on $\mathsf{SPF}(M)$ is standard:
- Parallel composition is $(f \otimes g)(x,y) = (f(x), g(y))$
- Sequential composition is $(f \cdot g)(x) = g(f(x))$
- Identity is $\mathsf{I}_a(x) = x$, for $x \in M^a$
- Transposition is ${}^a\mathsf{X}^b(x,y) = (y,x)$, for $x \in M^a, y \in M^b$
- Copy is $\wedge^a(x) = (x,x)$, for $x \in M^a$
- Sink is $\text{!}^a(x) = (\,)$, for $x \in M^a$; $(\,)$ is the empty tuple
- Dummy source is $\text{?}_a(\,) = (\langle\,\rangle_a)$, i.e., the a-tuple of empty streams
- (The constant $\vee$ is left *uninterpreted* in this deterministic case.)

Feedback: For $f \in \mathsf{SPF}(M)(a \otimes c, b \otimes c)$ the feedback $f \uparrow^c_\otimes \in \mathsf{SPF}(M)(a,b)$ is defined as follows: for streams $x \in M^a$, we specify

$$f \uparrow^c_\otimes (x) = \bigsqcup_{k \geq 1} y_k$$

where the streams $y_k \in M^b$ and $z_k \in M^c$ are inductively defined by

$$(y_1, z_1) = f(x, \langle\,\rangle), \quad \text{(recall that } \langle\,\rangle \text{ denotes the empty stream) and}$$
$$(y_{k+1}, z_{k+1}) = f(x, z_k), \quad \text{for } k \geq 1.$$

(f is continuous, hence this definition is well formed.)

Notice that the above definition of the feedback produces the minimal fixed-point solution. In particular, $\wedge^a \uparrow^a_\otimes = \top_a$; that is, the constant T is to be interpreted as a dummy source $\top$. ◇

Nondeterministic networks. Algebraic models for nondeterministic data-flow cannot be easily obtained as extensions of those for deterministic data-flow, mainly due to the unsoundness of the fixed-point equation. This fixed-point equation gives a fundamental technique for defining the semantics of cyclic deterministic networks. It corresponds to defining the semantics of a cyclic process by unfolding the loop.

Duplication of nondeterministic cells. Formally, for $f : a \otimes b \to b$ this fixed-point equation may be written as $(f \cdot \wedge^b) \uparrow^b_\otimes = \wedge^a \cdot (\mathsf{l}_a \otimes (f \cdot \wedge^b) \uparrow^b_\otimes) \cdot f$. The left-hand-side of the equation contains an occurrence of f, while the right hand side has two such occurrences. Due to this reason the corresponding networks are not echivalent in the nondeterministic case. Indeed, if f is nondeterministic, then different behaviours may be selected for the two occurrences of f in the right-hand-side term, while a unique choice for f exists for the left-hand-side term; hence, the resulting terms may be different.

By a similar argument, the strong commutation axiom Sb in Table 6.3 is valid for deterministic networks, but not for nondeterministic ones. ◇

With its top graph isomorphism equivalence the calculus of flownomials overcomes this problem. A simple extension of the SPF model may be introduced to handle nondeterministic networks; this model uses sets of stream processing functions and it will be studied in a forthcoming section.

10.4 Axiomatization: asynchronous, deterministic case

In this section we present two axiomatization results. The first one shows that SPF is a BNA, hence the relational semantic is compositional for deterministic networks. The second one shows that the axioms for dual Elgot theories are sound and complete for deterministic networks with respect to the input–output semantics.

Graph isomorphism properties. First we check what graph isomorphism axioms are valid in the SPF model. One can easily see that the feedback operation produces the dummy-source constant $\top$ when it is applied to the copy constant $\wedge$; hence this constant has to be included in the signature.

Theorem 10.3. *The model* $(\mathsf{SPF}(M), \otimes, \cdot, \uparrow_{\otimes}, \mathsf{I}_a, {}^a\mathsf{X}^b, \curlywedge, \flat, \dagger)$ *is a* $d\beta$*-symocat with feedback.* □

Proof options. The proof is rather straightforward; it is given in detail in Appendix F. There are basically two options: (1) either one defines the feedback for atomic sorts and extends it to arbitrary sorts by the identities used in BNA rules R4–R5. Then, these rules are obviously valid, but the check of the validity of the remaining axioms is more complicated. (2) The other option, which we followed in Appendix F, is to use the iterative definition for feedback on arbitrary sorts; then the check of the BNA rule R5 is more complex (it says a simultaneous feedback may be replaced with repeated unary feedbacks), but the proof of the validity of the remaining axioms is smoother. ◇

Input–output behaviour. The role of this subsection is to present a correct and complete axiomatization for deterministic networks with respect to the input–output behaviour.

Dual Elgot theory. One resulting algebraic structure that is of central interest for the study of deterministic data-flow networks is the dual Elgot theory (or $d\alpha$-flow algebra). We recall its definition here. A *dual Elgot theory* is defined by: (1) the graph isomorphism axioms with the branching constants $\perp, \wedge, \top$; (2) the critical axioms Sa–Sb; and (3) the enzymatic axiom for converses of functions, i.e., for terms written with $\star, \cdot, \mathsf{I}, \mathsf{X}, \perp, \wedge$. The axioms are presented in Table 6.3 (p. 192); for the present context, the general juxtaposition $\star$ used in Table 6.3 has to be replaced by the current $\otimes$ operation used in SPF; also the branching constants $\wedge, \perp, \top$ have slightly different notation in the current context, i.e., they are $\curlywedge, \flat, \dagger$, respectively. ◇

It is natural to take into account some equivalences on networks which are coarser than graph-isomorphism. Here we consider the equivalence which identifies the networks that compute the same input–output function.

Plan for soundness and completeness proof. We use the following plan for the proof of the soundness and completeness theorem.
(1) The key result we are starting with is the axiomatization of the unfolding equivalence for abstract networks given in Chapter 5. It is used here in the dual version, namely as consisting of the dual Elgot theory axioms.
(2) Next we prove the soundness part. To this end, we check the validity of the dual Elgot theory axioms in SPF. Then, by a general result, the dual Elgot theory axioms imply the fix-point identity. This and the continuity property show the network unfolding procedure is sound, i.e., it preserves the SPF semantics.
(3) For the completeness part, we show that if two networks have the same SPF semantics, then they are unfolding equivalent.
(4) Combined with the soundness part, this latter result shows the input–output equivalence on deterministic networks coincides with the unfolding equivalence, hence the dual Elgot theory axioms gives a correct axiomatization for such networks modulo SPF-equivalence. ◇

We start with the following result which follows from Theorem 5.17:

Fact 10.4. *The axioms of dual Elgot theories are sound and complete for deterministic data-flow networks modulo unfolding equivalence.* □

Next we show the unfolding procedure is sound with respect to the SPF semantics. We say two networks F and G are *input-output (or* SPF*-) equivalent*, and write $F \equiv_{IO} G$, iff they have the same SPF-semantics, i.e., for any interpretation of the cells as stream processing functions the networks compute the same stream processing function.

Notice that the following results are valid:

Fact 10.5. *(1) The strong axioms Sa–Sb of Table 6.3 hold in* SPF*.*
(2) $ENZ_{F_n^{-1}}$ *holds in* SPF*.* □

From (1), the graph-isomorphism axioms, and the dual version of Proposition 4.4 it follows that SPF is an algebraic theory. This shows each multiple-output function in SPF is equal to a tuple of one-output functions. Hence we may suppose each cell in a network and the network itself have exactly one output.

A network as above (with one output and such that each cell has also one output) may be unfolded towards inputs into a tree. The unfolding of a multiple output network is by definition equal to the tuple of the unfoldings corresponding to each output.

Unfolding. The unfolding procedure is the standard one. It starts with a copy of the output network cell which gives the 1st level of the tree; at the 2nd level we put copies of the network cells whose outputs are inputs for the cells of the 1st level; then copies of the network cells whose outputs are inputs for the cells of the 2nd level; and so on. Moreover, a proper variable is used for each input channel and it is used whenever it is necessary as a terminal vertex in the tree.

An example of such an unfolding process is given in Figure 10.3(d), namely it represents the first component of the unfolding of the networks in (a)–(c) of the same figure. We want to point out that this kind of unfolding may lead to infinite trees. E.g., the tree in Figure 10.3(d) is infinite on the top. ◇

We say two deterministic data-flow networks F and G are *unfolding equivalent*, and write $F \equiv_{unfold} G$, iff both networks F and G unfold into the same tuple of trees.

The soundness of the unfolding procedure with respect to the SPF semantics is easy to prove. The fix-point equation, i.e., $(f \cdot \wedge^b) \uparrow^b_\otimes = \wedge^a \cdot (\mathsf{I}_a \otimes (f \cdot \wedge^b) \uparrow^b_\otimes) \cdot f$, for $f : a \otimes b \to b$ is valid in SPF. Hence, if one unfold a loop once, then an SPF-equivalent network is obtained. Iterating this result one gets the soundness of an arbitrary, but finite, number of loop unfoldings. Finally, the continuity assumption on stream processing functions implies the correctness of the infinite unfolding process. So we have got the following result

Fact 10.6. *If* $F' \equiv_{unfold} F''$*, then* $F' \equiv_{IO} F''$*.* □

Next, we show the converse: if two deterministic data-flow networks compute the same input–output function for all functional interpretations of the atomic cells, then they unfold into the same tuple of trees. Instead of this, we prove the equivalent negated statement: if F' and F'' are different trees, then there exists a functional interpretation of the atoms such that F' and F'' compute different functions.

Let D be a domain of data consisting of partial Σ-terms over X ("partial" means that terms $\sigma(x_1, \ldots, x_n)$ with some undefined arguments are allowed; such undefined elements are denoted by "?"), where

- X is an infinite set of variables and
- Σ is a signature containing a symbol σ_f for each atom f which occurs either in F' or F''; it inherits the corresponding arity.

The cells act on streams of such data according to the following interpretation (let m denotes $|a|$):

- Case $m \geq 1$: A cell $f : a \to s$ ($|a| = m, |s| = 1$) acts by:

$$\begin{aligned}
&f(v_1, \ldots, v_m) = \sigma_f(?, \ldots, ?) \frown g(v_1, \ldots, v_m) \\
&g(t_1 \frown w_1, \ldots, t_m \frown w_m) = \sigma_f(t_1, \ldots, t_m) \frown g(w_1, \ldots, w_m) \\
&\text{where } t_1, \ldots, t_m \in D \text{ and } v_1, \ldots, v_m, w_1, \ldots, w_m \in D^\omega
\end{aligned}$$

  (This definition also works for finite streams. According to the definition, the components of the output stream are defined up to the maximal length of the input streams; thereafter the output is empty.)
- Case $m = 0$: A cell $f : \epsilon \to s$ ($|s| = 1$) produces the output $(\sigma_f)^\infty$, or formally

$$f(\,) = \sigma_f(\,) \frown f(\,)$$

Take a distinguished variable x_i (of sort a_i) for each input i and consider as input the tuple of streams $((x_1)^\infty, \ldots, (x_m)^\infty)$. The output $|F|((x_1)^\infty, \ldots, (x_m)^\infty)$ produced by a tree $F : a \to s$ ($|a| = m, |s| = 1$) is a stream of terms $t_1 \frown t_2 \frown \ldots$, where t_i is the partial approximation of F up to level i.

Since F' and F'' are different, there is a level i such that they are different up to level i, hence $|F'|((x_1)^\infty, \ldots, (x_m)^\infty) \neq |F''|((x_1)^\infty, \ldots, (x_m)^\infty)$. The claim is proved, hence we have got the following result

Fact 10.7. *If $F' \equiv_{IO} F''$ for all functional interpretations of the atoms, then $F' \equiv_{unfold} F''$.* □

Proof illustration. A simple example illustrates the proof. Under the displayed interpretation, the output computed by the tree-network in Figure 10.3(d) is

$$\begin{aligned}
&\sigma_{f_1}(?, ?) \frown \\
&\sigma_{f_1}(x_1, \sigma_{f_2}(?, ?)) \frown \\
&\sigma_{f_1}(x_1, \sigma_{f_2}(x_1, \sigma_{f_2}(?, ?))) \frown \\
&\sigma_{f_1}(x_1, \sigma_{f_2}(x_1, \sigma_{f_2}(x_1, \sigma_{f_2}(?, ?)))) \frown \ldots
\end{aligned}$$

One may see that the first output data gives the approximation of the tree up to level 1, the second up to level 2, and so on. ◊

The above two facts show that unfolding equivalence coincides with input–output equivalence, with respect to the functional interpretation of the atoms. Now the main result of this subsection may be stated.

Theorem 10.8. *(input–output behaviour, deterministic networks) The dual Elgot theory axioms give a sound and complete axiomatization for deterministic data-flow networks with respect to their* SPF *semantics.* □

♡ An interesting thing here is the fact that dual Elgot theory axioms suffice to get a sound and complete axiomatization for the IO-behaviour; as we have seen, to model the IO-behaviour of flowcharts, the stronger Park axioms were needed. ◊

10.5 Time anomaly for nondeterministic networks

We describe the time anomaly using both Brock–Ackermann's and Russell's examples. Then, a brief account of the proposed solutions is included.

As we have seen in the previous section, an abstract model based on stream transformers can be given for deterministic asynchronous data-flow networks and this model is compositional. Unfortunately, as we briefly explained in the first section, this model is not compositional in the nondeterministic case. Some networks that are equivalent (i.e., they realize the same relation between their input and output streams) cannot be substituted for each other in a larger network because equivalence will get lost. This deviation, known also as the Brock–Ackermann, or merge anomaly, is a time anomaly.

Feedback, as a source of the anomaly. This anomaly is related to the feedback operation. Consider an arbitrary deterministic data-flow network with a feedback loop. If the network gets data faster from its feedback loop, the additional data do not change any prefix of the streams being produced because the network is deterministic. So only the relation between the input and output streams matters. However, in the nondeterministic case, the timing differences in producing the data that are fed back become important. The Brock–Ackermann example relies on such timing differences to show that the feedback of certain networks with the same relation between its input and output streams are different. ◊

One may try to solve this anomaly in two ways: either (1) weaken the abstract model, or (2) strengthen the operational model.

On the lines of (1), several models have been proposed. The general approach of these proposals can be described as follows: add more information to the model, but keep unchanged the operational interpretation of wires as unbounded queues.

In this way the simple stream transformer model is sacrificed and other models emerge. A few of them are:

Trace model: This model is based on a global time information. All the local streams are shuffled into a unique stream. (All the input and output data have now distinct time stamps. With this additional information N_1 and N_2 in Example 10.1 are distinguished: the trace $out_1(5) \frown in_1(6) \frown out_1(6)$ is possible in N_2, but not in N_1.)

The model based on time intervals: In this case, the streams are divided into streams of finite sequences; the data in a sequence are supposed to be used in a unique time cycle. More or less, this way the model is reduced to a sort of synchronous model, where the time anomaly is not present.

Oracle based model: In this model nondeterministic behaviour is reduced to deterministic ones up to certain oracles. Then the compositionality of the stream transformer model for deterministic data-flow networks is used.

From these models the first two are *fully abstract.* This means, they are compositional and in between these models and the relational input–output model there are no other compositional models. The oracle based model is not fully abstract.

♡ Fully abstractness may be an important property to have. However, we prefer to study in more detail the oracle based model since it lies in the true-concurrency world and this semantics may be presented as a simple and natural extension of the natural SPF semantics of deterministic data-flow networks. ◇

Below, the time anomaly is illustrated by means of two examples: the original Brock–Ackermann example and an example due to Russell.

⦶ Network behaviour formalization: to have a succinct presentation of these examples, we use process algebra terms to describe the behaviour of cells; more on process algebra formalism may be read from Chapter 9. A few explanations may help here: the cells have the input channels denoted by $1, 2, \ldots$; a term $\sum_{d\in D} r_1(d)$ describes the nondeterministic choice of waiting for reading a datum d on channel 1; $s_1(d)$ describes the sending of datum d on the first output channel. Finally, network algebra terms are used to describe the network architecture. ⦶

Example 10.9. (Brock–Ackermann example) The Brock–Ackermann's example is depicted in Figure 10.4. The atomic cells used in this example are described by the following process algebra terms:

$$\begin{array}{lcl} \mathsf{SUC} & = & (\sum_{d\in D} r_1(d)\cdot s_1(d+1))\cdot \mathsf{SUC} \\ \mathsf{DUP} & = & (\sum_{d\in D} r_1(d)\cdot s_1(d)\cdot s_1(d))\cdot \mathsf{DUP} \\ \mathsf{2BUF} & = & (\sum_{d\in D} r_1(d)\cdot (\sum_{e\in D} r_1(e)\cdot s_1(d)\cdot s_1(e))\cdot \mathsf{2BUF} \end{array}$$

Moreover, the constants $\vee_1$ and $\wedge^1$ are used, i.e., the merge and copy constants, respectively. They may be defined by the processes

$$\begin{array}{lcl} \mathsf{MRG} & = & (\sum_{d\in D}(r_1(d)+r_2(d))\cdot s_1(d))\cdot \mathsf{MRG} \\ \mathsf{CP} & = & (\sum_{d\in D} r_1(d)\cdot (s_1(d)\|s_2(d))\cdot \mathsf{CP} \end{array}$$

The following networks are built from these atomic cells:

$$\begin{array}{lcl} f & = & (\mathsf{DUP}\otimes \mathsf{SUC}\cdot \mathsf{DUP})\cdot \curlyvee_1 \cdot \mathsf{2BUF}\cdot \curlywedge^1 \\ f' & = & (\mathsf{DUP}\otimes \mathsf{SUC}\cdot \mathsf{DUP})\cdot \curlyvee_1 \cdot \mathsf{I}_1 \cdot \curlywedge^1 \end{array}$$

It is easy to see that the networks f and f' realize the same relation between their input and output streams, written $f \equiv_{\mathsf{history}} f'$. However, f and f' cannot always be substituted for each other in a larger network. Consider, for instance, the networks $f \uparrow^1$ and $f' \uparrow^1$. The stream $1\frown 2\frown 2\frown 3\frown \ldots$ is contained in $[f' \uparrow^1](1)$ but not in $[f \uparrow^1](1)$ because 2BUF must have consumed both 1's yielded by the duplication of the input before the feedback loop can contribute to the output. So, $f \uparrow^1 \not\equiv_{\mathsf{history}} f' \uparrow^1$.

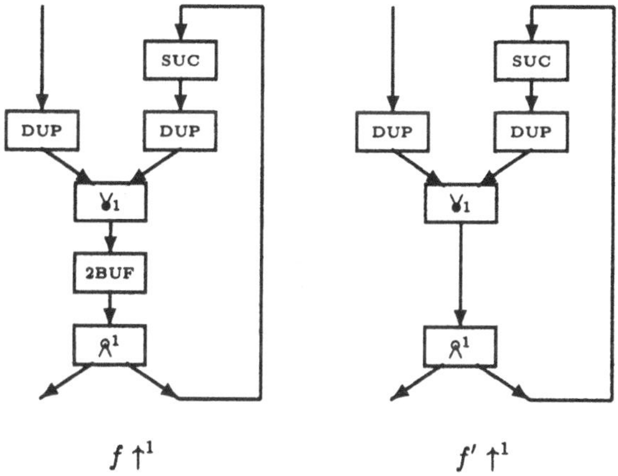

Fig. 10.4. Brock–Ackermann example. Two data-flow networks with the same relational input–output semantics (the ones obtained dropping the feedback), producing different input–output semantics when they are placed as parts of a larger network (the full networks in the figure).

Example 10.10. (Russell's example) In the previous example, the atomic cells were all deterministic. The merge constant introduces nondeterminism in that example. Russell's example shows that the time anomaly may also occur by nondeterministic atomic cells. Consider the following atomic cells:

$$\begin{array}{lcl} G & = & (\sum_{d\in D} r_1(d)\cdot s_1(0)\cdot s_1(1) + s_1(0)\cdot (\sum_{d\in D} r_1(d)\cdot s_1(0)))\cdot G \\ G' & = & (\sum_{d\in D} r_1(d)\cdot s_1(0)\cdot s_1(1)) + s_1(0)\cdot (\sum_{d\in D} r_1(d)\cdot s_1(0)) \\ & & + s_1(0)\cdot (\sum_{d\in D} r_1(d)\cdot s_1(1)))\cdot G' \end{array}$$

It is easy to see that the atomic cells G and G' realize the same relation between their input and output streams, i.e., $G \equiv_{\text{history}} G'$. Furthermore, notice that the possible output streams are independent of the input streams, provided the input streams have at least one datum. However, the stream $0 \frown 1 \frown \ldots$ is contained in $[(G' \cdot ᖇ^1) \uparrow^1]()$ but not in $[(G \cdot ᖇ^1) \uparrow^1]()$, because G must already have produced one datum before it can contribute a 0 followed by a 1 to the output. So $(G \cdot ᖇ^1) \uparrow^1 \not\equiv_{\text{history}} (G' \cdot ᖇ^1) \uparrow^1$.

EXERCISES AND PROBLEMS (SEC. 10.5)

1. How do the model based on time intervals and the oracle based model solve the time anomaly described in Example 10.1? (08)

10.6 Axiomatization: asynchronous, nondeterministic case

In this section we present some axioms for nondeterministic data-flow networks under the oracle-based semantics. It is shown that the BNA axioms are valid, hence the model is compositional. Certain additional axioms for branching constants under the split/merge interpretation are described.

Deterministic data-flow nets have more or less a canonical denotational semantics, which was used in the previous chapter. To find such a semantics for nondeterministic networks is less obvious. The semantics of nondeterministic data-flow networks may be reduced to the semantics of deterministic networks using oracles. Such an oracle fixes a priori the behaviour of the network regarding the nondeterministic points. Given a fixed oracle, a nondeterministic network becomes deterministic and it computes a stream processing function. Varying the oracle we obtain the semantics of a nondeterministic network as a set of stream processing functions.

Sets of stream processing functions. Formally, we construct the model $\mathcal{P}\mathsf{SPF}(M)$ for the interpretation of nondeterministic data-flow networks as follows. First, for sorts $a, b \in S^*$ define

$$\mathcal{P}\mathsf{SPF}(M)(a,b) := \{F \mid F \subseteq \mathsf{SPF}(a,b)\}$$

Next, the operations $\otimes, \cdot, \uparrow_\otimes$ are defined in an element-wise manner by

$$\begin{aligned} F \otimes G &= \{f \otimes g : f \in F, g \in G\} \\ F \cdot G &= \{f \cdot g : f \in F, g \in G\} \\ F \uparrow^a_\otimes &= \{f \uparrow^a_\otimes : f \in F\} \end{aligned}$$

Finally, each constant $c \in \{\mathsf{I}, \mathsf{X}, ᖇ, \flat, \mathsf{?}\}$ of SPF induces a corresponding constant $\{c\}$ of $\mathcal{P}\mathsf{SPF}$.

Additional nondeterministic branching constants:

— (BLOCK) SPLIT: $\wedge^a = \{ \wedge^a_\phi \mid \phi : \omega \to \{1,2\}\}$, for $a \in S$, where for an oracle ϕ, $\wedge^a_\phi(x)$ is (y,z), with y and z obtained by splitting x according to ϕ. That is, if $\phi(i) = 1$ then the i-th input is delivered on output channel 1, otherwise on output channel 2. This definition is extended to arbitrary words $a \in S^{\otimes}$ using the identities in A18–A19. (Notice that we have *independent oracles* for each input channel in a and *not* a unique oracle for all the inputs in a. Axiom A19 would fail with a definition that uses the latter version.)

— (BLOCK) MERGE: $\vee_a = \{ \vee^\phi_a \mid \phi : \omega \to \{1,2\}\}$, for $a \in S$, where for an oracle ϕ, $\vee^\phi_a(x,y) = z$ with z obtained from x and y according to ϕ. This means, the 1st output element in z is taken from the $\phi(1)$-th input channel, the 2nd from the $\phi(2)$-th input channel, and so on. If the oracle requires a data from an input channel which is empty, then the merge cell is blocked and no more output data are delivered. With A14–A15 this definition is extended to arbitrary $a \in S^{\otimes}$.

— (BLOCK RICH) SOURCE: $\lozenge_a = \{g_x \mid x \in M_a\}$, for $a \in S^*$, where for $x \in M_a$, $g_x : \epsilon \to a$ is the function given by $g_x(\,) = x$.

By split, merge, and source we have introduced three nondeterministic constants for data-flow nodes.

Graph–isomorphism properties. It is easy to see that the oracle-based semantics leads to a compositional model. Indeed,

Theorem 10.11. *($\mathcal{P}\mathsf{SPF}, \otimes, \cdot, \uparrow_{\otimes}, \mathsf{I}, \mathsf{X}$) is a BNA model.*

PROOF: The proof follows directly from the corresponding result in the deterministic case, i.e., Theorem 10.3. The key point is the observation that all the BNA axioms are identities with both the left-hand-side and the right-hand-side terms containing *at most one occurrence* of a variable and each variable that occurs in one part of an identity *occurs in the other part*, as well. Hence the validity of the proof of a BNA axiom in $\mathcal{P}\mathsf{SPF}$ may be checked on elements and its validity follows from the validity of the corresponding axiom in SPF. □

Extending nonlinear identities from SPF *to* $\mathcal{P}$SPF. To see better the point, let us consider a *nonlinear* axiom which is valid in the deterministic case, e.g., $f \cdot \wedge^b = \wedge^a(f \otimes f)$. When we lift this equation to the $\mathcal{P}\mathsf{SPF}$ model, we have to check the identity $F \cdot \wedge^b = \wedge^a \cdot (F \otimes F)$, where F is a set of (stream processing) functions. The left-to-right inclusion holds, but not the right-to-left one; indeed, provided there are two different functions f_1, f_2 in F, one can consider the function $\wedge^a(f_1 \otimes f_2) \in \wedge^a(F \otimes F)$, for which no corresponding element in $F \cdot \wedge^b$ may be found (for each input data each function in the latter term gets equal data for both output channels; on the other hand, the former term with f_1 and f_2 different, may produce different data on the output channels for the same input data). ◇

To these basic axioms for graph isomorphism we can add equations for the branching constants interpreted as split/merge cells.

Theorem 10.12. *(Graph isomorphism with various constants)* $(\mathcal{P}\mathsf{SPF}, \otimes, \cdot, \uparrow_{\otimes}, \mathsf{I}, \mathsf{X}, \wedge, \bot, \vee, \top)$ *is a BNA and obeys the additional axioms in Table 10.1, which parallel the standard network algebra axioms for branching constants in Table 6.3. Notice that only one inclusion holds for 5 axioms, i.e., "$\supset$" for A1b*, A1d*, A3d* and R10* and "$\subset$" for A4*.*

Proof: The proof is given in Appendix F. □

For convenience, the resulting axioms are collected in Table 10.1. Since axioms A1a, A2a and A1c, A2b are valid, the oracle-based semantics of the nondeterminism is associative and commutative. Hence we may equivalently use the extended branching constants $\underset{\phi}{\wedge}{}_k^a : a \to ka$ and $\overset{\phi}{\vee}{}_a^k : ka \to a$ for $k \geq 1$, where $\phi : \omega \to \{1, \ldots, k\}$ is a k-oracle. On the other hand, axioms A1b and A1d do not hold, hence we have a non fair split/merge and therefore a non-angelic calculus of relations modeling the split and merge branching structure of nondeterministic networks.

Exercises and Problems (sec. 10.6)

1. Find a complete axiomatization for the split/merge branching constants with respect to the PSPF model. (**)

10.7 Fully abstract models

The role of this section is to present a few results related to fully abstract models for nondeterministic data-flow networks.

In the trace model all input and output actions of the network are ordered into a stream. We distinguish them by writing $in_i(x)$ for the eating of x at the network input channel i and $out_j(y)$ for the producing of y at network output channel j.

Theorem 10.13. *The trace model is fully abstract with respect to the input–output relational semantics.*

Proof: (sketch) Let $[\![\ ]\!]_{Tr}$ denote the trace semantics on data-flow networks; similarly, $[\![\ ]\!]_H$ and $[\![\ ]\!]_O$ are used for the relational input–output (or "history")

Table 10.1. Axioms for split/merge branching constants in $\mathcal{P}$SPF (wrt oracle-based semantics)

III′. Axioms for the additional constants $\top, \bot, \vee, \wedge$, without feedback		IV′. Feedback on split/merge branching constants	
A1a	$\wedge^a \cdot (\wedge^a \otimes \mathsf{I}_a) = \wedge^a \cdot (\mathsf{I}_a \otimes \wedge^a)$	R8	$\vee_a \uparrow^a_\otimes = \bot^a$
A1b*	$\wedge^a \cdot (\bot^a \otimes \mathsf{I}_a) \supset \mathsf{I}_a$	R9	$\wedge^a \uparrow^a_\otimes = \top_a$
A1c	$(\vee_a \otimes \mathsf{I}_a) \cdot \vee_a = (\mathsf{I}_a \otimes \vee_a) \cdot \vee_a$	R10*	$[(\mathsf{I}_a \otimes \wedge^a) \cdot ({}^a\mathsf{X}^a \otimes \mathsf{I}_a)$
A1d*	$(\top_a \otimes \mathsf{I}_a) \cdot \vee_a \supset \mathsf{I}_a$		$\cdot (\mathsf{I}_a \otimes \vee_a)] \uparrow^a_\otimes \supset \mathsf{I}_a$
A2a	$\wedge^a \cdot {}^a\mathsf{X}^a = \wedge^a$		
A2b	${}^a\mathsf{X}^a \cdot \vee_a = \vee_a$		
A3a	$\top_a \cdot \bot^a = \mathsf{I}_\epsilon$		
A3b	$\top_a \cdot \wedge^a = \top_a \otimes \top_a$		
A3c	$\vee_a \cdot \bot^a = \bot^a \otimes \bot^a$		
A3d*	$\vee_a \cdot \wedge^a \subset (\wedge^a \otimes \wedge^a)$		
	$\cdot (\mathsf{I}_a \otimes {}^a\mathsf{X}^a \otimes \mathsf{I}_a) \cdot (\vee_a \otimes \vee_a)$		
A4*	$\wedge^a \cdot \vee_a \supset \mathsf{I}_a$		
A5a	$\bot^{a\otimes b} = \bot^a \otimes \bot^b$		
A5b	$\wedge^{a\otimes b} = (\wedge^a \otimes \wedge^b) \cdot (\mathsf{I}_a \otimes {}^a\mathsf{X}^b \otimes \mathsf{I}_b)$		
A5c	$\top_{a\otimes b} = \top_a \otimes \top_b$		
A5d	$\vee_{a\otimes b} = (\mathsf{I}_a \otimes {}^b\mathsf{X}^a \otimes \mathsf{I}_b) \cdot (\vee_a \otimes \vee_b)$		
[A5e	$\bot^\epsilon = \mathsf{I}_\epsilon$		
A5f	$\wedge^\epsilon = \mathsf{I}_\epsilon$		
A5g	$\top_\epsilon = \mathsf{I}_\epsilon$		
A5h	$\vee_\epsilon = \mathsf{I}_\epsilon$]		

semantics, and for the operational semantics, respectively. The proof is based on the following 3 facts:

1. $[\![N_1]\!]_{Tr} = [\![N_2]\!]_{Tr}$ implies $[\![N_1]\!]_O = [\![N_2]\!]_O$;
2. $[\![N_1]\!]_{Tr} = [\![N_2]\!]_{Tr}$ implies $\forall C.[\![C(N_1)]\!]_{Tr} = [\![C(N_2)]\!]_{Tr}$;
3. $[\![N_1]\!]_{Tr} \neq [\![N_2]\!]_{Tr}$ implies $\exists C.[\![C(N_1)]\!]_H \neq [\![C(N_2)]\!]_H$

The proof of the first two facts is fairly easy and it is omitted. For the last fact, one has to provide a context C such that the trace-differences of two networks $N_1, N_2 : p \to q$ may be lifted as input–output differences in the larger networks $C(N_i), i \in [2]$. The context C is using special cells $MergeMark : 1+q \to 1$ and $CopyRelese : 1 \to 1+p$ as follows:

$$C(N) = (MergeMark \cdot CopyRelease \cdot (\mathsf{I}_1 \otimes N)) \uparrow^q$$

We suppose the set of data D allows us to codify the pairs $\langle in_j, d\rangle$ and $\langle out_j, d\rangle$.

♡ This is not a restrictive condition: it is enough to have at least 2 elements in D; then, each additional information in_j or out_j is codified by a sequence of data in D of a fixed length k; this sequence is to be transmitted before the relevant data d; this way, instead of a pair, a sequence of length $k+1$ is processed. ◇

The key cells have the following behaviour:

MergeMark: this cell merges the data from its 1st input channel with the ones from the following q input channels; in addition: (1) each datum d from the

$1+j$-th channel is transformed into the pair $\langle out_j, d\rangle$; (2) the data of type $\langle in_j, d\rangle$ from the 1st channel are passed unchanged; and (3) the merge is fair, i.e., no channel is neglected infinitely often.

CopyRelease: this cell copy the input data at the 1st output channel; in addition, if the current incoming datum is of type $\langle in_j, d\rangle$ then it also transmits the datum d on the $1+j$-th output channel.

We supposed $[\![N_1]\!]_{Tr} \neq [\![N_2]\!]_{Tr}$, hence there exists a trace $t \in [\![N_1]\!]_{Tr} \setminus [\![N_2]\!]_{Tr}$ or conversely. Notice that t is a finite or infinite sequence of events $\langle in_j, d\rangle$ or $\langle out_j, d\rangle$. Let $in(t) = t|_{in}$ be its restriction to the in events. Then a careful analysis shows that

$$(in(t), t) \in [\![N_i]\!]_H \quad \text{iff} \quad t \in [\![N_i]\!]_{Tr}$$

Hence trace differences between N_1 and N_2 are seen as input–output differences between the networks $C(N_1)$ and $C(N_2)$. □

A similar result holds for the model with time intervals:

Theorem 10.14. *The model with time intervals is fully abstract with respect to the input–output relational semantics.* □

These theorems show that the involved models (the trace model and the model with time intervals) add the smallest amount of information to the input–output relational model in order to transform it in a compositional model.

EXERCISES AND PROBLEMS (SEC. 10.7)

1. Prove that the trace model is compositional. Do the same for the model with time intervals. (10)
2. Fill in the detail for the proof of Theorem 10.13. (11)
3. Describe in details the model with time intervals. Prove that it is fully abstract with respect to the input–output semantics. (16)
4. Show that the oracle based semantics is not fully abstract. (*Hint*: use the networks $P = \wedge^1 \cdot \vee_1$ and $P' = \wedge^1 \cdot (Buf(2) \otimes P) \cdot \vee_1$ with streams over $\{0,1\}$.) (09)

10.8 Network algebra on top of process algebra

In this section we sketch the construction of a (data-flow based) network algebra model on top on process algebra. Process algebra is used to describe the cell behaviour and network algebra to describe the network architecture. Finally, everything is translated in the process algebra formalism which is used to reason about network properties.

General process algebra model of BNA. A general process algebra model for BNA is described here. Network algebra can be regarded as being built on top of process algebra as follows.

Let D be a fixed, but arbitrary, set of data. The involved processes use standard actions $r_i(d)$, $s_i(d)$ and $c_i(d)$ for $d \in D$ and nothing more. They stand for read, send, and communicate, respectively, datum d at port i. Communication is defined on these actions such that $r_i(d)|s_i(d) = c_i(d)$ (for all $i \in \mathbb{N}$ and $d \in D$); in all other cases, it yields δ.

We write $H(i)$, where $i \in \mathbb{N}$, for the set $\{r_i(d) : d \in D\} \cup \{s_i(d) : d \in D\}$ and $I(i)$ for $\{c_i(d) : d \in D\}$. In addition, we write $H(i,j)$ for $H(i) \cup H(j)$, $H(i+[k])$ for $H(i+1) \cup \ldots \cup H(i+k)$ and $H(i+[k], j+[l])$ for $H(i+[k]) \cup H(j+[l])$. The abbreviations $I(i,j)$, $I(i+[k])$ and $I(i+[k], j+[l])$ are used analogously.

$in(i/j)$ denotes the renaming function defined by

$$\begin{array}{lcll} in(i/j)(r_i(d)) & = & r_j(d) & \text{for } d \in D \\ in(i/j)(a) & = & a & \text{for } a \notin \{r_i(d) \mid d \in D\} \end{array}$$

So $in(i/j)$ renames port i into j in read actions. $out(i/j)$ is defined analogously, but renames send actions. We write $in(i+[k]/j+[k])$ for $in(i+1/j+1) \cdot \ldots \cdot in(i+k/j+k)$ and $in([k]/j+[k])$ for $in(0+[k]/j+[k])$. The abbreviations $out(i+[k]/j+[k])$ and $out([k]/j+[k])$ are used analogously.

With these notations a *network* $f \in \mathsf{Proc}(D)(m,n)$ is a triple $f = (m,n,P)$, where P is a process with actions in $\{r_i(d) : i \in [m], d \in D\} \cup \{s_i(d) : i \in [n], d \in D\}$.

A *wire* is a network $\mathsf{I} = (1,1,w_1^1)$, where w_1^1 satisfies the following 3 conditions: for all networks $f = (m,n,P)$, if w_v^u denotes $\rho_{in(1/u)}(\rho_{out(1/v)}(w_1^1))$, then
(P1) $\tau_{I(u,v)}(\partial_{H(v,u)}(w_u^v \| w_v^u)) ||| P = P$
(P2) $\tau_{I(u,v)}(\partial_{H(u,v)}(\rho_{in(i/u)}(P) ||| w_u^v ||| w_v^j)) = \rho_{in(i/j)}(P)$ for all $i,j \in [m]$
(P3) $\tau_{I(u,v)}(\partial_{H(u,v)}(\rho_{out(i/u)}(P) ||| w_v^u ||| w_j^v)) = \rho_{out(i/j)}(P)$ for all $i,j \in [n]$.
($|||$ denotes the communication-free merge between the corresponding processes.)

♡ These, somehow obscure properties correspond to the following network algebra axioms over arbitrary wires, translated into the process algebra terms: (P1) to $\mathsf{I}_\epsilon \star f = f = f \star \mathsf{I}_\epsilon$, (P2) to $\mathsf{I}_m \cdot f = f$ and (P3) to $f \cdot \mathsf{I}_n = f$. ◊

The operations and constants of BNA are defined on $\mathsf{Proc}(D)$ as is shown in Table 10.2. (Note that the standard representation of networks is used: the input ports are numbered $1,\ldots,k$ and the output ones $1,\ldots,l$.)

Hiding communication ports. The definitions of sequential composition and feedback illustrate clearly the differences between the mechanisms for using ports in network algebra and process algebra. In network algebra the ports that become internal after composition are hidden. In process algebra based models these ports are still visible; a special operator must be used to hide them. ◊

Theorem 10.15. $(\mathsf{Proc}(D), \otimes, \cdot, \uparrow, \mathsf{I}, \mathsf{X})$ *is a model of BNA.*

Table 10.2. Operations in $\mathsf{Proc}(D)$

$$
\begin{array}{l}
(m,n,P) \star (p,q,R) = (m+p, n+q, R), \text{ where} \\
\qquad\qquad R = P|||\rho_{in([n]/m+[n])}(\rho_{out([q]/n+[q])}(Q)) \\
(m,n,P) \cdot (n,p,R) = (m,p,R), \text{ where, for } u = max(m,p), v = u+n \\
\qquad\qquad R = \tau_{I(u+[n],v+[n])}\partial_{H(u+[n],v+[n])} \\
\qquad\qquad\qquad ((\rho_{out([n]/u+[n])}(P)|||\rho_{in([n],v+[n])}(Q))||w_{v+1}^{u+1}||\dots||w_{v+n}^{u+n}) \\
(m+p, n+p, P)\uparrow^p = (m,n,Q), \text{ where, for } u = max(m,n), v = u+p \\
\qquad\qquad Q = \tau_{I(u+[p],v+[p])}\partial_{H(u+[p],v+[p])} \\
\qquad\qquad\qquad ((\rho_{in(m+[p]/v+[p])}(\rho_{out(n+[p],u+[p])}(P))||w_{v+1}^{u+1}||\dots||w_{v+p}^{u+p})) \\
\mathsf{I}_n = (n,n,P), \text{ where} \\
\qquad\qquad P = w_1^1|||\dots|||w_n^n, \ \text{ if } n > 0 \\
\qquad\qquad P = \tau_{I(1,2)}(\partial_{H(1,2)}(w_2^1||w_1^2)), \ \text{ otherwise} \\
{}^m\mathsf{X}^n = (m+n, n+m, P), \text{ where} \\
\qquad\qquad P = w_{n+1}^1|||\dots|||w_{n+m}^m|||w_1^{m+1}|||\dots|||w_n^{m+n}, \ \text{ if } m+n > 0 \\
\qquad\qquad P = \tau_{I(1,2)}(\partial_{H(1,2)}(w_2^1||w_1^2)), \ \text{ otherwise}
\end{array}
$$

PROOF: The proof of this theorem is left as an exercise; except for some simple process algebra properties, the proof used the crucial properties P1–P3 which are supposed to be satisfied in view of the wire definition. □

Process algebra model for asynchronous data-flow networks. In this subsection we exemplify the above construction with the particular case of asynchronous data-flow networks. This appears using stream delayers as wires, i.e., unbounded FIFO queues.

In the asynchronous case, the identity constant, called *stream delayer*, is the wire $\mathsf{I}_1 = (1, 1, sd(\varepsilon))$, where sd is defined by

$$sd(\sigma) = (\Sigma_{d \in D} r_1(d) \cdot sd(\sigma \frown d)) \lhd |\sigma| > 0 \rhd s_1(hd(\sigma)) \cdot sd(tl(\sigma))$$

(hd denotes the head of the queue and tl its tail; ε is the empty queue; moreover the conditional notation $x \lhd b \rhd y$ is used, where $x \lhd T \rhd y = x$ and $x \lhd F \rhd y = y$.) The constants I_n, for $n \neq 1$, and ${}^m\mathsf{X}^n$ are defined by the equations occurring as BNA axioms B6 and B8, respectively. The restriction of $\mathsf{Proc}(D)$ to the processes that can be built under this actualization is denoted by $\mathsf{AProc}(D)$.

The definition of sd simply expresses that it behaves as a queue. The definition of the process P_f, the process associated with a cell working under function f, is the natural one (it is not formally given; see the forthcoming comments for a verbal explanation). Finally, all the cells and the branching network algebra constants are pre- and post-fixed with identities I_n.

Cell behaviour. A definition of P_f may be given to express the following: P_f waits until one datum is offered at each of the input ports $1, \dots, m$. When data is available at all input ports, P_f proceeds with producing data at the output ports $1, \dots, n$. The datum produced at the i-th output port is the i-component of the value of the function

f for the consumed input tuple. When data is delivered at all output ports, P_f proceeds with repeating itself. ◇

The following result shows that a network algebra model over process algebra may be introduced using *sd*.

Lemma 10.16. *The wire* $\mathsf{I}_1 = (1, 1, sd)$ *gives an identity flow of data, i.e., for all* $f = (m, n, P)$ *in* $\mathsf{AProc}(D)$, $\mathsf{I}_m \cdot f = f = f \cdot \mathsf{I}_n$.

PROOF: For I_1, it is well known that $\mathsf{I}_1 \cdot \mathsf{I}_1 = \mathsf{I}_1$ (see, e.g., [vGV93]). So, these equations hold for I_1. Due to the pre- and postfixing with identities in the definitions of the other constants and the atomic cells, it follows immediately that these equations hold for all processes. □

For $\mathsf{AProc}(D)$, the BNA operations and constants are defined as in $\mathsf{Proc}(D)$, but using *sd* as wire. From the general results, one gets the following corollary.

Corollary 10.17. $(\mathsf{AProc}(D), \otimes, \cdot, \uparrow, \mathsf{I}, \mathsf{X})$ *is a model of BNA.* □

Now, the additional constants for asynchronous data-flow networks may be defined. With respect to these branching constants, both split/merge and copy/equality test interpretations are used. For $n \neq 1$, these constants are defined by the equations occurring as axioms of type A5. Notice that ↓ (dummy sink) and ♀ (rich source) are not used here.

$$
\begin{array}{lcll}
\curlywedge^1 & = & \mathsf{I}_1 \cdot (1, 2, \mathit{split}^1) \cdot \mathsf{I}_2 & \mathit{split}^1 = (\Sigma_{x \in D} r_1(x) \cdot (s_1(x) + s_2(x))) \cdot \mathit{split}^1 \\
\multimap^1 & = & \mathsf{I}_1 \cdot (1, 0, \mathit{sink}^1) & \mathit{sink}^1 = (\Sigma_{x \in D} r_1(x) \cdot \tau) \cdot \mathit{sink}^1 \\
\curlyvee_1 & = & \mathsf{I}_2 \cdot (2, 1, \mathit{merge}_1) \cdot \mathsf{I}_1 & \mathit{merge}_1 = (\Sigma_{x \in D} (r_1(x) + r_2(x)) \cdot s_1(x)) \cdot \mathit{merge}_1 \\
\bar{\bullet}_1 & = & (0, 1, \mathit{source}_1) \cdot \mathsf{I}_1 & \mathit{source}_1 = \delta \\
\curlywedge^1_{\circ} & = & \mathsf{I}_1 \cdot (1, 2, \mathit{acopy}^1) \cdot \mathsf{I}_2 & \mathit{acopy}^1 = (\Sigma_{x \in D} r_1(x) \cdot (s_1(x) \| s_2(x))) \cdot \mathit{acopy}^1 \\
\curlyvee_{\circ 1} & = & \mathsf{I}_2 \cdot (2, 1, \mathit{aeq}_1) \cdot \mathsf{I}_1 & \mathit{aeq}_1 = [(\Sigma_{x_1 \in D} r_1(x_1) \| \Sigma_{x_2 \in D} r_2(x_2)) \\
 & & & \qquad \cdot\, s_1(x_1) \langle x_1 = x_2 \rangle s_1(err)] \cdot \mathit{aeq}_1
\end{array}
$$

Certain axioms for the split/merge constants are given in Table 10.3. These axioms agree with the standard ones of the algebra of flownomials (Table 6.3) with four exceptions: A1a, A1b, A4 and R10 – they all concern the split constant.

Theorem 10.18. $(\mathsf{AProc}(D), \otimes, \cdot, \uparrow, \mathsf{I}, \mathsf{X})$ *is a model of BNA. The constants* $\curlywedge, \multimap, \curlyvee, \bar{\bullet}$ *satisfy the additional axioms for asynchronous data-flow networks in Table 10.3. The constants* $\curlywedge_{\circ}, \multimap, \curlyvee_{\circ}, \bar{\bullet}$ *satisfy the additional forward-demonic axioms for synchronous data-flow networks of Appendix C.*

PROOF: Using the above lemma and Theorem 10.15, the first part follows easily. The second and third parts have also simple proofs. □

Axioms for branching constants. It may be interesting to look at the axioms satisfied by the split/merge branching constants in the oracle-based semantics and in the current process algebra based semantics. As a consequence of bisimulation, the split constant has poor properties in the latter approach, but the merge constant behaves well. On the other hand, the former oracle-based semantics is more symmetric with respect to split and merge constants.

The third part of this theorem expresses the preservation of the algebraic structure of the synchronous branching constants by passing to the asynchronous case. That is, the asynchronous data-flow networks built up using copy and equality test constants satisfy the same axioms as in the synchronous case. ◇

Table 10.3. Axioms for split/merge branching constants in bisimulation-based asynchronous data-flow networks

III″. Axioms for the additional constants $\top, \perp, \vee, \wedge$, without feedback		IV″. Feedback on split-merge branching constants	
A1a*	(*)	R8	$\wedge^a \uparrow^a = \top_a$
A1b*	(*)	R9	$\vee_a \uparrow^a = \perp^a$
A1c	$(\vee_a \otimes \mathsf{I}_a) \cdot \vee_a = (\mathsf{I}_a \otimes \vee_a) \cdot \vee_a$	R10*	(*)
A1d	$(\top_a \otimes \mathsf{I}_a) \cdot \vee_a = \mathsf{I}_a$		
A2a	$\wedge^a \cdot {}^aX^a = \wedge^a$		
A2b	${}^aX^a \cdot \vee_a = \vee_a$		
A3a	$\top_a \cdot \perp^a = \mathsf{I}_\epsilon$		
A3b	$\top_a \cdot \wedge^a = \top_a \otimes \top_a$		
A3c	$\vee_a \cdot \perp^a = \perp^a \otimes \perp^a$		
A3d	$\vee_a \cdot \wedge^a = (\wedge^a \otimes \wedge^a) \cdot (\mathsf{I}_a \otimes {}^aX^a \otimes \mathsf{I}_a) \cdot (\vee_a \otimes \vee_a)$		
A4*	(*)		
A5a	$\perp^{a\otimes b} = \perp^a \otimes \perp^b$		
A5b	$\wedge^{a\otimes b} = (\wedge^a \otimes \wedge^b) \cdot (\mathsf{I}_a \otimes {}^aX^b \otimes \mathsf{I}_b)$		
A5c	$\top_{a\otimes b} = \top_a \otimes \top_b$		
A5d	$\vee_{a\otimes b} = (\mathsf{I}_a \otimes {}^bX^a \otimes \mathsf{I}_b) \cdot (\vee_a \otimes \vee_b)$		
[A5e	$\perp^\epsilon = \mathsf{I}_\epsilon$		
A5f	$\wedge^\epsilon = \mathsf{I}_\epsilon$		
A5g	$\top_\epsilon = \mathsf{I}_\epsilon$		
A5h	$\vee_\epsilon = \mathsf{I}_\epsilon$		

Other models for wires. Our models of synchronous and asynchronous data-flow networks allow for a finite amount of nondeterminism: for each tuple of input data, a cell may nondeterministically produce a tuple of output data from a finite set. The wires are fully deterministic.

Minimal stream delayer. Synchronous data-flow networks are based on this type of wire. However, for the description of this setting a more sophisticated process algebra version is needed, namely the timed process algebra. The interested reader may find the details in [BŞ94a].

Guess-and-borrow queues. Different from a synchronous data-flow network, an asynchronous network may delay the use of its resources. In case asynchronism is exploited fully, it should also be possible to use the resources in advance.

In the Brock–Ackermann example (10.9), $1 \frown 2 \frown 2 \frown 3 \frown \ldots$ is in $[f' \uparrow^1](1)$, but not in $[f \uparrow^1](1)$, provided the data-flow network wires are treated as unbounded queues. With a more powerful kind of identity connections, viz. guess-and-borrow queues, this anomaly disappears.

A *guess-and-borrow queue* may deliver any number of arbitrary data to a cell while it is empty. However, if this turns out not to be in agreement with the actual data subsequently received, the cell has to drop the computation based on the wrong data.

In the Brock–Ackermann example, the critical stream $1 \frown 2 \frown 2 \frown 3 \frown \ldots$ is in $f' \uparrow^1$ if the identity connections are interpreted as (unbounded) guess-and-borrow queues. One only needs such a queue before the 2BUF cell, which now may borrow the necessary tokens in order to annihilate the differences between the 2BUF cell and an identity connection. More generally, one may see that $f \uparrow^1$ and $f' \uparrow^1$ compute the same input–output relation on streams with this more powerful operational interpretation for the identity connections. A similar argument works in the case of Russell's example, as well.

Bags. With the constants $\overset{\bullet}{\wedge}, \underset{\circ}{|}, \underset{\bullet}{\vee}, \overset{\bullet}{|}$, the axiom R10 does not hold. It is interesting that the left-hand-side of this equation describes a bag. Although a bag is not a queue, it behaves as an identity flow; writing I'_1 for the left-hand-side, i.e., $((\mathsf{I}_1 \otimes \overset{\bullet}{\wedge}{}^1) \cdot ({}^1\mathsf{X}^1 \otimes \mathsf{I}_1) \cdot (\mathsf{I}_1 \otimes \underset{\bullet}{\vee}{}_1)) \uparrow^1$, we have $\mathsf{I}'_1 \cdot \mathsf{I}'_1 = \mathsf{I}'_1$. By the general result on network algebra construction on top of process algebra, it appears that a process algebra model of BNA with wires that behave as bags may be constructed.

Exercises and Problems (sec. 10.8)

1. Prove Theorem 10.15. (10)
2. Show that in the synchronous case with minimal stream delayers as identity connections and the copy/equality test interpretation of the branching connections the forward demonic axioms in Appendix C give a sound and complete set of axioms. (12)
3. Study the relationship between the branching constants from both kinds: the copy/equality test and split/merge interpretations. (- -)

Comments, problems, bibliographic remarks

Problems A few, maybe technical, open problems are inserted here.

Communication is time, as a source of anomaly. If one looks carefully at the given time anomaly examples, then one can see that all these examples use *non-deterministic* components and *communication between piped processes.* Roughly speaking, the latter means that certain independent processes generated by data may interfere in time in some components.

Interference in space is normal in data-flow network, all cells with more input channels potentially having such an ability. Communication in time for piped processes is also natural via cell memory (a first process may change the internal state of a cell according to its will; the next process may detect and use this information). However, the situation is different in the case of $Buf(2)$ component in Example 10.1; in this case the process corresponding to the first datum actually *waits* for the process corresponding to the next datum; this is a sort of synchronization between the corresponding piped processes. In Russell's example, a similar situation is in the first term of the sum when a datum generates two data piped in time, each participating in a independent process; this may be seen as a fork of a process in two piped processes. In the presence of feedback, such a communication may lead to a (sometime impossible) communication of a process with itself. In the Brock–Ackermann example, a process entering in the $Buf(2)$ cell stops and waits for itself to do something more, a clearly impossible behaviour.

The comments above may suggest the following conjecture. There is no time anomaly for nondeterministic networks if there is no communication of piped processes, except for the one provided by cell memory. In other words, no time anomaly does exist when the process terms representing the cell behaviours are linear, i.e., each read action is followed by a unique write action.

Split/merge axiomatization. Find a (sound and complete) axiomatization for networks consisting of split and merge nodes only under oracle equivalence. This problem may be related to the difficult problem of comparing the power of sorting networks; see, e.g., [Knu73].

Notes. The idea of data-flow had mainly two sources. (1) Single assignment languages are based on the concept of a set of (non-recursive) declarations. The order of the evaluation of the declarations is then only determined by their data dependencies. These dependencies can be shown in an acyclic graph called their data-flow graph. Influenced by these ideas and by the concept of Petri-nets and their firing rules, Jack Dennis suggested data-flow graphs and gave firing rule semantics for them. (2) Quite independently, versions of data-flow graphs can be found in many software engineering methods and also for the description of switching circuits.

The fixed-point semantics for deterministic data-flow is based on [Kah74]; see also [Bro83]; Theorem 10.8 is from [BŞ96].

The time anomaly (sometimes called "merge anomaly") for nondeterministic networks was noticed in [Kel78, BA81, Rus89]. Methods to avoid this anomaly are presented in [Bro88]. Enriched BNA models for nondeterministic data-flow are presented in [Sta92, BŞ94a, BŞ96]; a different approach is presented in [Par87, Par93]; see [CŞ90b] for a different presentation for BNA which is related to this latter axiomatization.

The full abstractness for trace semantics is shown in [Jon94]; for interval-based model, it is shown in [Kok87]; the lack of such a result for oracle-based semantics is noticed in [Bro93, BŞ94a]. The construction of network algebra models on top of process algebra is from [BŞ94a, BMŞ97].

For circuits, one may have a look in [Klo87]; in particular, identities similar to the BNA axioms are inserted on p. 27; see also [Möl98b] for a NA-based circuit design technique. Data-flow computation is also described in the book [Böh84]. Demonic behaviour on related models is studied in [Jos92, dNS95]; the axiomatization of the relations with demonic operators presented in Appendix C is from [BŞ98]. Other results on data-flow networks may be found in [KP86, SN85, Sta87, TZ92, BWM94].

11. Petri nets

> It does not take much strength to do things, but it requires great strength to decide what to do.
>
> Herbert Hubbard

In this chapter we make a tour of Petri nets, one of the oldest and most interesting concurrency model. Our aim here is rather modest. Indeed, the chapter is mainly devoted to the presentation of concurrent regular expressions; this is an extension of regular expressions with parallel operators. The choice is motivated by the strong relationship between classical regular expressions and network algebra, documented in Chapters 6 and 8.

Concurrent regular expressions are used as a tool for representing the behaviour of Petri nets. The main result is the equivalence of Petri nets and concurrent regular expressions with respect to their associated languages. To this end, a useful decomposition of Petri nets in units is introduced, as well as certain intermediary expressions corresponding to these units.

The chapter includes some intriguing open problems. In particular, notice that the study here is less algebraic and more effort is necessary to present the included results in an elegant algebraic way, as was the case with classical regular expressions and automata.

11.1 Introducing the model

The basic notions are introduced here. An example and a few comments on the type of parallelism modeled within Petri nets are included, as well.

Petri nets vs. other concurrency models. Process algebra, e.g., the ACP version briefly presented in Chapter 9, is based on *global states*. As a direct consequence of this assumption, no true concurrency model may be easily developed for process algebra; indeed, the standard models are interleaving based. If one is allowed to enter into the state structure, then true concurrency models may be naturally developed. E.g., in such a case the processes that do not interfere may be modeled as really taking place in parallel, for instance on different processors.

Petri nets provide one of the oldest models for parallel and distributed computation. They use the *multiset* model for states; multisets are sets with repetition, i.e., each element has an associated natural number indicating how many times the element is repeated in that set. Roughly speaking, there is a set P for the types of the elements that enter into the state structure. A multiset over P is formally defined as a function

$$M : P \to \mathbb{N} \cup \{\infty\}$$

which specify how many items of a given type are present in state M.

Syntactically, a Petri net is similar to a (process algebra like) transition system built up over such states. But now, one can specify the types of the elements that are used in a given transition. The result is now a more detailed specification as an *hyper-graph*, namely each transition may have many inputs and many outputs.

If one compares the process algebra and Petri net models, then the following relative advantages may be observed.

(1) Process algebra is a *structured theory* which allows for a modular specification of processes; that is, all processes occur using a number of specific operations, starting from certain atomic elements. Such a theory is still a desiderata for Petri nets, even if certain proposals have occurred in the literature, e.g., the Box calculus, models based on symmetric strict monoidal categories, or the model with concurrent regular expressions; this latter model will be studied in detail in the forthcoming sections.

(2) On the other hand, Petri nets give a more realistic representation of parallel processes; due to the possibility to enter into the state structure, they are able to model the processes in a *truly concurrent* way. Petri nets are much closer to automata/machines, while process algebra is more related to grammars/languages; consequently, a lot of features related to parallel computers are naturally modeled by Petri nets.

Petri nets and data-flow networks. As we have already pointed out, Petri nets may be considered as a particular instance of the data-flow model. For historical reasons, it may be more accurate to read this in the opposite direction: the data-flow model is a generalization of the Petri net model; later extensions of the Petri nets, including coloured Petri nets, are even closer related to the data-flow model. ◇

Basic definitions. A *net* is a triple (P, T, F) where: (1) P is the set of places; (2) T is the set of transitions; (3) $F \subseteq P \times T \cup T \times P$ is the relation of incidence; in addition the following axioms are valid:

A1. $P \cap T = \emptyset$; and
A2. $F \neq \emptyset$ & $(\forall x \in P \cup T, \exists y \in P \cup T.xFy \ \vee yFx)$

A useful notation is: ${}^\bullet x = \{y : yFx\}$, $x^\bullet = \{y : xFy\}$.

Sometimes the following auxiliary hypothesis is used, too:

A3. $\forall p_1, p_2 \in P : ({}^\bullet p_1 = {}^\bullet p_2)$ & $(p_1^\bullet = p_2^\bullet) \Rightarrow (p_1 = p_2)$

It allows us to minimize the number of places; the price to be paid is the increasing of the edge weights; as we will see below, to be equivalent with arbitrary Petri nets, ordinary Petri nets have to avoid such a condition.

A *Petri net* is a 5-uple $N = (P, T, F, W, M_0)$ such that: (1) (P, T, F) is a finite net (hence P, T are finite sets); (2) $M_0 : P \to \mathbb{N}$ specify the initial state/mark; and (3) $W : F \to \mathbb{N} \setminus \{0\}$ specifies the edge weights.

Place meaning. A place is a sort of deposit which contains resources of a unique type. The term "marking" is often used as a substitute for "state" in this area of Petri nets. The marking specifies how many resources of a given kind are available in the current state. Then, the weight of an edge specifies how many resources from the place are used (respectively, produced) during the firing of the transition to which the edge is connected. ◇

An *ordinary Petri net* is a Petri net where all edges have weight 1.

Next we define transition firing. Before this, let us make a few conventions. Let $P = \{p_1, \ldots, p_n\}$. The elements in P are ordered, say as $p_1 \prec \ldots \prec p_n$. Now a marking may be identified with a vector $M = (m_1, \ldots, m_n)$, where $m_i = M(p_i), i \in [n]$. Some more notations are:

(1) $F(x,y) = \begin{cases} n & \text{iff } xFy \wedge W(x,y) = n \\ 0 & \text{iff } \neg xFy \end{cases}$
(2) $^\bullet F(t) = (F(p_1,t), \ldots, F(p_n,t))$ – the resources needed for the firing of t;
(3) $F^\bullet(t) = (F(t,p_1), \ldots, F(t,p_n))$ – the resources produced by the firing of t;
(4) "$>$", "$+$" and "$-$" on vectors are obtained by applying the corresponding operations component-wise.

A transition t may be *fired* in the state specified by the marking M if

$$M \geq F^\bullet(t)$$

As a result of the firing of the transition the net goes in the state

$$M' = M - {}^\bullet F(t) + F^\bullet(t)$$

Some additional notation:
(1) $M[t\rangle M'$ is used to denote the fact that t may be fired in state M, and, as the result of this action, the net goes to state M';
(2) $M[t_1 \ldots t_k\rangle M'$ iff there exist states $M_1, \ldots, M_k$ such that
$$M[t_1\rangle M_1 \ldots M_{k-1}[t_k\rangle M';$$
(3) $M[\,\rangle M'$ iff $\exists w \in T^*.\ M[w\rangle M'$;
(4) $R(N, M) = \{M' \colon M[\,\rangle M'\}$ is the set of states that can be reached in N from marking M;
(5) $R(N)$ is the set of states that can be reached in N from its initial marking;
(6) if a set Φ of final places is specified, then the language associated with a Petri net N is $L(N) = \{w \colon M_0[w\rangle M \text{ and } M(p) = 0 \text{ if } p \notin \Phi\}$.

Example 11.1. An example of a Petri net is given in the figure below. The above definitions are illustrated in the figure as follows:
– a, b, c, d are transitions; p, q, r are places;

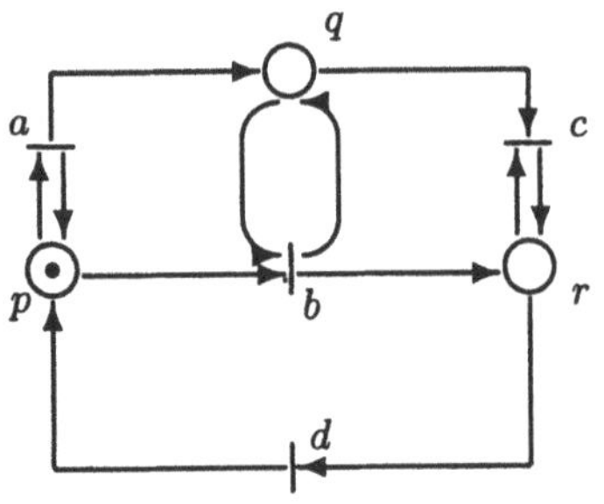

Fig. 11.1. A Petri net

- the initial configuration contains one token (point) in place p, the other places having no tokens;
- to fire, a requires a token in place p; during its firing the token is consumed and new tokens are produced in places p and q; we write this as $a : p \to p \otimes q$;
- similarly, $b : p \otimes q \to q \otimes r$, $c : r \otimes q \to r$ and $d : r \to p$.

The typical behaviour of this net is as follows: while a token is in place p transition a may fire an arbitrary number of times producing an amount of tokens in place q; then b may fire (provided there is at least one token in q) and the net goes in a configuration where there is a token in place r; while this token is in r transition c may fire several times, but no more than the number of the tokens in q; then d may take place and the net goes into a configuration where it can repeat this behaviour. (This may be seen as a model for a producer–consumer system: a produces a number of items that are consumed by c.)

Remark 11.2. The definition of the language recognition we have given so far is via *final places*, not using a finite set of final configurations, as is sometimes used. In other words, any configuration with no points in the non-final places is terminal; so an infinite number of "classic" terminal configurations may occur. The robustness of this accepting criterion is emphasized by the fundamental theorem we shall prove later in this chapter.

Parallelism in space and time. A useful distinction may be drawn between two forms of parallelism. The first one is that exhibited in spatially parallel processes; these are the usual parallel processes, i.e., the ones that may run in the same time on different processors. The second type of parallelism is that manifested in temporally parallel processes; these processes, sometime called piped processes, may be handled by an agent at different time periods.

In Petri nets these two different types of parallelism are identified under the usual operational semantics. Indeed, the tokens that come in a place are identified, no matter how they arrived there. Sometimes the "collective token philosophy" phrase is used to describe this situation.

EXERCISES AND PROBLEMS (SEC. 11.1)

1. Describe the behaviour of the Petri net in Example 11.1, but now starting with 3 tokens in place p. (07)

11.2 Concurrent regular expressions (CRegExp)

In this section concurrent regular expressions CRegExp are introduced. They extend the usual regular expressions by parallel operators (shuffle, iterated shuffle, and synchronization). The standard language interpretation of these expressions is described.

Signature. The symbols used to describe concurrent regular expressions are

$$+, \cdot, {}^*, 0, 1, ||, {}^\alpha, [], \sigma$$

Of these, 0,1 are constants (zero-ary operations), the next ones *, $^\alpha$, σ are unary operators, and finally, $+, \cdot, ||, []$ are binary operators.

Language semantics. The semantics for concurrent regular expressions we are describing here is based on interleaving. Consequently, the final aim is to reduce the description of a system of concurrent processes to that of a unique nondeterministic sequential process. In addition, the behaviour of a single agent is described using the trace semantics, hence what we actually get as the semantics for a concurrent regular expression is a set of strings, i.e., a language.

We briefly describe the intuitive meaning and the semantics of the operators used to build up concurrent regular expressions. The resulting languages are considered together with their corresponding alphabets; e.g., we will distinguish between the empty set 0 over the alphabet Σ_1 and 0 over the alphabet Σ_2. The notation $sign(E)$ is used for the alphabet associated with the concurrent regular expression E.

Nondeterministic choice: "+" expresses the agent possibility to choose between two sets of behaviours; in the language semantics it corresponds to set union, namely

$$|E_1 + E_2| = |E_1| \cup |E_2|; sign(E_1 + E_2) = sign(E_1) \cup sign(E_2)$$

Sequential composition: "·" expresses the possibility to sequentialize the actions of an agent; it is modeled by concatenation of the words in the corresponding languages;

$$|E_1 \cdot E_2| = |E_1| \cdot |E_2| =_{\mathsf{def}} \{x_1x_2 : x_1 \in |E_1|, x_2 \in |E_2|\};$$
$$sign(E_1 \cdot E_2) = sign(E_1) \cup sign(E_2)$$

Iterate sequential composition (Star): "$\_^*$" describes the possibility of an agent to repeat an action for an indefinite number of times

$$|E^*| = \{\epsilon\} \cup |E| \cup |E|^2 \cup |E|^3 \ldots ; sign(E^*) = sign(E)$$

where $A^k = A \cdot \overset{k}{\ldots} \cdot A$.

Parallel composition: "$||$" models the parallel execution of two processes without any interaction between them; more precisely, the result of this operation is the set of the random interleavings of the atomic actions of these processes. In terms of languages the inductive definition is as follows:

On strings, $||$ is inductively defined as follows.

$$s \,||\, \epsilon = \epsilon \,||\, s = s$$

$$as \,||\, bt = a(s \,||\, bt) \cup b(as \,||\, t)$$

for a, b letters and s, t strings. This is naturally extended to the set of strings by

$$A \,||\, B = \bigcup_{s \in A, t \in B} s \,||\, t$$

With these preliminaries we can now define

$$|E_1 \cdot E_2| = |E_1| \;||\; |E_2|; sign(E_1 + E_2) = sign(E_1) \cup sign(E_2)$$

Iterate parallel composition (Alpha): "$\_^\alpha$" describes the possibility of an agent to replicate itself in an indefinite number of copies which are run in parallel without any interaction between these copies. It is formally defined in a similar way as the star operator, but now parallel composition is used instead of the sequential composition. Denote

$$A^{(k)} = A \,||\, \overset{k}{\ldots} \,||\, A$$

Then the operation is defined by

$$|E^\alpha| = \{\epsilon\} \cup |E|^{(1)} \cup |E|^{(2)} \cup |E|^{(3)} \ldots ; sign(E^\alpha) = sign(E)$$

Parallel composition with synchronization: "$[\,]$" is defined similarly with the usual parallel composition, but now the interleaving of the atomic actions of the processes are randomly done only for the proper actions of the processes; the actions in the common alphabet of the processes are synchronized.

$$\begin{aligned} |E_1 [\,] E_2| = \{w :\; & w \in (sign(E_1) \cup sign(E_2))^*, w|_{sign(E_1)} \in |E_1|, \\ & w|_{sign(E_2)} \in |E_2|\}; \\ sign(E_1 \cdot E_2) = {} & sign(E_1) \cup sign(E_2) \end{aligned}$$

Here $w|_A$ denotes the restriction of the word w to the alphabet A, i.e., all the letters in w that do not belong to A are deleted.

Renaming: If $\sigma : \Sigma_1 \to \Sigma_2 \cup \{\epsilon\}$ is a renaming function which renames a letter into a letter or the empty string, then

$$|\sigma(E)| = \sigma(|E|); sign(\sigma(E)) = sign(E)$$

Concurrent regular expressions – CRegExp. Now we can syntactically define *concurrent regular expressions.* Due to some technical problems, an intermediary class of expressions, the class of *unary* ones, is also introduced.

Definition 11.3. (1) *Regular expressions* RegExp(V) *over the alphabet* V are built up from $0, 1$ and letters $a \in V$ by applying the operations of sum, sequential composition and star, i.e., $+, \cdot, ^*$.

(2) *Unitary expressions* UnitExp(V) *over the alphabet* V are obtained starting with regular expressions in RegExp(V) and applying the operations of parallel composition and iterate parallel composition, i.e., $||, ^\alpha$.

(3) *Concurrent regular expressions* CRegExp(V) *over the alphabet* V are obtained starting with unitary expressions in UnitExp(V) and applying the operations of parallel composition, synchronized parallel composition and renaming, i.e., $||, [\,], \sigma$.

Remark 11.4. (1) Let us note that there are two levels which are not mixed here. One basic level is the additive, sequential one (generated by the operators $+, \cdot, ^*$) and it is used to describe the behaviour of the individual agents. Another level is a multiplicative, parallel one (generated by the operators $||, ^\alpha, [\,], \sigma$) built up over the previous additive level. The application of the additive operators *over* the multiplicative ones is forbidden; e.g., an expression such as $(E_1 \,||\, E_2) + E_2^\alpha + (E_1 \,[\,]\, E_3)$ is not relevant here.

(2) Due to some technical reasons, it is useful to separate the application of the synchronization operator from the other parallel operators and to use it in a further stage of the construction. The separate study of the resulting unitary expressions, built up using only parallel composition and iterate parallel composition, proves to be useful here.

(3) Notice that in the definition of the iterative parallel composition (alpha) one uses the parallel composition and *not* the parallel composition with synchronization. In some sense, the latter may be more consonant to the sequential composition operation used for star. (To be nontrivial, the iterative parallel composition with synchronization one has to consider here should use a subset of actions that have to be synchronized in all copies of the agent; the remaining actions are renamed to be proper for each instance of the replicated agent and are freely interleaved.)

The restrictions used in the definition of CRegExp are motivated by the strong relationship between the class of languages specified by such expressions and the languages recognized by Petri nets.

Exercises and Problems (sec. 11.2)

1. Show that the following identities hold, provided the language interpretation is used:

$$
\begin{array}{ll}
A \,||\, B = B \,||\, A & (A^\alpha)^\alpha = A^\alpha \\
A \,||\, (B \,||\, C) = (A \,||\, B) \,||\, C & (A \,||\, B^\alpha)^\alpha = A^\alpha \,||\, B^\alpha \\
A \,||\, \{\epsilon\} = A & A \,[]\, A = A \\
A \,||\, 0 = 0 & A \,[]\, B = B \,[]\, A \\
(A + B) \,||\, C = A \,||\, C + B \,||\, C & A \,[]\, (B \,[]\, C) = (A \,[]\, B) \,[]\, C \\
(\Sigma_{i \in I} A_i) \,||\, (\Sigma_{j \in J} B_j) = \Sigma_{i \in I, j \in J} A_i \,||\, B_j & A \,||\, 0_{sign(A)} = 0_{sign(A)} \\
(A + B)^\alpha = A^\alpha \,||\, B^\alpha & A \,||\, sign(A)^* = A \\
(A^*)^\alpha = A^\alpha & (A + B) \,[]\, C = A \,[]\, C + B \,[]\, C
\end{array}
$$

(10)

2. Show that RegExp $\prec$ UnitExp $\prec$ CRegExp, where "$\prec$" means less powerful with respect to the class of specified languages. (*Hint*: $(ab)^\alpha$ is unitary, but not regular; $\{a^n b^n c^n : n \geq 0\}$ is concurrent-regular, but not unitary.) (14)

11.3 Decomposed Petri nets

In this section we introduce decomposed Petri nets as a key step towards a representation of Petri net languages by CRegExp. A Petri net is seen as a collection of parallel synchronized units (a unit is a kind of finite automata).

The representation of Petri net languages by concurrent regular expressions will be obtained using some "decomposed Petri nets." These decomposed Petri nets may be seen as systems of nondeterministic finite automata with synchronization. However, an extension is necessary in order to allow for an infinite number of copies of an automaton acting in parallel.

ⓘ Such a situation occurs in the case where we have a Petri net transition without, say, input places; in that case the transition can independently fire at any time. ⓘ

Definition 11.5. A *decomposed Petri net* is a pair (T, U) where: (1) T is a finite set specifying the alphabet for transitions; (2) U is a finite set of units $(U_1, U_2, \ldots, U_k)$, where each unit is a 5-uple $U_i = (P_i, \Sigma_i, \delta_i, M_i, \Phi_i)$ consisting of:

P_i - a finite set of "places";
Σ_i - a finite set of labels for transitions;
$\delta_i \subseteq (P_i \times \Sigma_i) \times P_i$ - the transition relation
M_i - an initial configuration, where $M_i : P_i \to \mathbb{N} \cup \{\infty\}$;
$\Phi_i \subseteq P_i$ - the set of finite places.

♡ Let us emphasize that the units use ordinary transitions with one input and one output. ◊

A *running* in a decomposed Petri net is defined similarly as in the case of usual Petri nets, with one (important!) difference: all the transitions in different units that share the same label are to be fired simultaneously.

ⓘ More precisely, a transition with label a may fire in the decomposed Petri net if each unit U_i that has a transition with such a label can fire that transition (i.e., it has at least a point in its incoming place); after the firing of the transition a point is moved from the input place to the output place of the transition a in each unit U_i that contains a. (Adding or deleting a point to/from a place with mark ∞ does not modify that mark.) ⊕

Before showing how Petri nets are simulated by decomposed Petri nets, let us prove a simple lemma.

Lemma 11.6. *Each Petri net is equivalent with an* ordinary *Petri net, i.e., with a Petri net where the weight of each edge is 1.*

PROOF: Indeed, let p be a place and k a number such that all incoming or outgoing edges of p have a weight less than or equal to k. Such a place may be replaced with a group of k places connected in a circular way (as a "ring") by ϵ-transitions. Then each edge connected to p, say having weight n, $n \leq k$ is replaced with a group of n edges connected to different places in the ring. This transformation is applied to all the places in the net. The result is an ordinary Petri net that is equivalent to the original one. □

♡ The usefulness of this transformation comes from the fact that for an ordinary Petri net the section of a transition corresponding to a unit is unique. (See below for the construction of the units corresponding to a Petri net.) Consequently, there are no edges with the same label in the same unit that have to be synchronized, hence the threads corresponding to different initial points in a unit run independently. ◇

Theorem 11.7. *Every Petri net N may be decomposed, i.e., there exists a decomposed Petri net N' such that $L(N) = L(N')$.*

PROOF: Using the above lemma we may now restrict ourselves to ordinary Petri nets. For such a net $N = (P, T, F, W, M_0, \Phi)$ an equivalent decomposed Petri net is constructed as follows. Each place of the given Petri net N becomes a place in a unit of the decomposed Petri net via an *assignment function*, namely

$$f : P \to \{1, 2, \ldots, k\} \text{ such that}$$
$$\forall t \in T, p_1, p_2 \in P : [(\{p_1, p_2\} \subseteq {}^\bullet t \text{ or } \{p_1, p_2\} \subseteq t^\bullet) \Rightarrow f(p_1) \neq f(p_2)]$$

That is, two different inputs (resp. outputs) of a transition have to be placed in different units.

The associated decomposed Petri net $D(N, f) = (T, \{U_1, \ldots, U_k\})$ is defined as follows: a unit $U_i = (P_i, \Sigma_i, \delta_i, M_i, \Phi_i)$ consists of the following ingredients:

- $P_i = \{p \in P : f(p) = i\} \cup \{sp_i\}$ (sp_i is a special place of the i-th unit U_i which contains an infinite number of points)
- $\Sigma_i = \{t \in T : \exists p \in P_i, p \in {}^\bullet t \cup t^\bullet\}$ (a transition in N induces a corresponding transition in each unit containing at least one of its input or output places)

- $\delta_i = \{(p_j, t, p_k) : t \in \Sigma_i, p_j \in {}^\bullet t \text{ and } p_k \in t^\bullet\}$
  $\cup\{(p_j, t, sp_i) : t \in \Sigma_i, p_j \in {}^\bullet t \text{ and } t^\bullet \cap P_i = \emptyset\}$
  $\cup\{(sp_i, t, p_k) : t \in \Sigma_i, {}^\bullet t \cap P_i = \emptyset \text{ and } p_k \in t^\bullet\}$
- $M_i(p) = M_0(p), \forall p \in P_i$ and $M_i(sp_i) = \infty$
- $\Phi_i = (P_i \cap \Phi) \cup \{sp_i\}$

Then $L(N) = L(D(N, f))$. Indeed: (1) the initial configurations of both nets N and $D(N, f)$ are the same, except for the special places sp_i; (2) the transitions that can be fired starting with equal configurations, except for sp_i places, are the same in N and $D(N, f)$; (3) the final configurations, except for sp_i places, are the same; and (4) the special sp_i places create no problems in firing the transitions as they hold an infinite number of tokens. Consequently, each terminating sequence of actions in a net is possible in the other net and vice-versa. Hence $L(N) = L(D(N, f))$. □

Example 11.8. Let us consider $N = (P, T, F, W, M_0, \Phi)$, the Petri net drawn in Example 11.1, where the set of final places is $\Phi = \{r\}$. Because of the restriction that no two input (or output) places of a transition can be put into the same unit, it follows that p and q are to be placed in separate units, and also r and q. To minimize the number of units, we put p and r in the same unit. Hence the choice to follow.

The assignment function is $f(p) = f(r) = 1$ and $f(q) = 2$. Now:

$a : p \to p \otimes q$ gives two transitions $a_{p,p} : p \to p$ and $a_{*,q} : * \to q$;
$b : p \otimes q \to r \otimes q$ produces $b_{p,r} : p \to r$ and $b_{q,q} : q \to q$;
$c : r \otimes q \to r$ is decomposed into $c_{r,r} : r \to r$ and $c_{q,*} : q \to *$; and finally,
$d : r \to r$ gives nothing special; for uniformity, we just rewrite it as $d_{r,r} : r \to r$.

This leads to a decomposition in two units (the subscripts of the transitions are used to help the description only):
- the 1st unit has: places p, r, and the special sp_1 place (this place plays no rôle here); transitions $a_{p,p} : p \to p$, $b_{p,r} : p \to r$, $c_{r,r} : r \to r$, $d_{r,p} : r \to p$; an initial token in p and an ∞ number of tokens in sp_1; r and sp_1 are final places;
- the 2nd unit has: a place q and the special place sp_2; the transitions are $a_{*,q} : sp_2 \to q$, $b_{q,q} : q \to q$, $c_{q,*} : q \to sp_2$; there are initially an ∞ number of tokens in sp_2; sp_2 is final.

11.4 From Petri net languages to CRegExp

In this section we prove that the languages accepted by Petri nets may be specified by concurrent regular expressions.

We recall that our accepting criterion for Petri net languages is using *final places* and not certain *finite sets of final markings*, as is sometimes used.

Accepting criterion. The current criterion is more relaxed. Indeed, an infinite set of final markings may occur, i.e., all markings with no points in the non-final places are considered to be final. The justification of this enlarged criterion is supported by the close relationship with the languages specified by concurrent regular expressions. ◇

From decomposed Petri net to CRegExp. The passing from decomposed Petri nets to CRegExp requires some auxiliary transformations.

Lemma 11.9. *A unit with several places marked with ∞ is equivalent to one with a single place marked with ∞.*

PROOF: By the identification of the places marked with ∞ the language is preserved. □

Lemma 11.10. *A unit is equivalent to another one having at most two components of the following type:*
- one with a single initial point (of type I) and
- one with an initial place marked ∞ and without other initial points (of type II).

PROOF: By the above lemma we may suppose there is a unique place p_∞ marked as ∞. Let C be the connected component which contains this place. If C has initial points in certain places, different from p_∞, as they evolve in an independent way (recall that each transition in a unit has a unique input and a unique output, and no pair of transitions in a unit are synchronized) one can decompose C in two components: a first component $C1$ that has initial points in places different from p_∞, and a second one $C2$ with all its initial points (infinitely many) in p_∞. The components $C1$ and the remaining part of U not included in C may be combined into a unique component with a unique initial point recognizing the same language. (The language is the interleaving of a finite number of regular languages, hence it is also regular.) □

Lemma 11.11. *The language recognized by a unit of type II may be represented by an expression of the type* RegExp$^\alpha$.

PROOF: Let U be a unit with the initial configuration M satisfying $M(p_\infty) = \infty$, otherwise 0. If A is the finite automaton with the initial state p_∞ and the same transitions and finite states as U, then $L(U) = L(A)^\alpha$. The easy proof is sketched below.

For $L(U) \subseteq L(A)^\alpha$, let $s \in L(U)$. In the process of the recognition of s the unit U uses a finite number of points, say n, from its initial infinite place. As these points evolve independently, they trace some strings $s_1, \ldots, s_n$ such that $s \in s_1 \,||\, \ldots \,||\, s_n$. It is obvious that each $s_i \in L(A)$, hence $s \in L(A)^\alpha$.

For the converse inclusion $L(A)^\alpha \subseteq L(U)$, let $s \in L(A)^\alpha$, hence $s \in s_1 \,||\, \ldots \,||\, s_n$, for certain strings $s_1, \ldots, s_n$ in $L(A)$. Then s may be simulated in U with n threads generated by points from p_∞, each point tracing one string s_i. □

Theorem 11.12. *The language recognized by a decomposed Petri net may be represented by a concurrent regular expression.*

PROOF: With Lemma 11.10 each unit may be decomposed in at most two components, one of type I and the other of type II. Therefore the language recognized by a unit is the interleaving of a regular language with a language represented by an expression of type RegExp^α. By the synchronized parallel composition of these languages one gets the language recognized by the decomposed Petri net. Finally, using the renaming operator one may replace the transitions with the corresponding labels to get a concurrent regular expression corresponding to the given labeled decomposed Petri net. □

Remark 11.13. In addition to the theorem, notice that the passing is algorithmic. I.e., there is an algorithm which, given a Petri net, produces an equivalent concurrent regular expression.

EXERCISES AND PROBLEMS (SEC. 11.4)

1. Show that the language associated with the Petri net in Example 11.1 (with r the unique final place) is described by $(a+bc^*d)^* \,[\,]\, (ab^*c)^\alpha$. (07)
2. Describe a CRegExp-expression for the case when the initial configuration for the above net has 3 tokens in place p. (08)
3. Find various CRegExp-representations for the language associated with the Petri net in Example 11.1 using different assignment functions for placing the places into units. (10)

11.5 From CRegExp to Petri net languages

We prove that the languages specified by CRegExp operations are accepted by Petri nets.

Lemma 11.14. *Let A, B be regular expressions. Then: (1) $A^\alpha \,||\, B^\alpha = (A+B)^\alpha$; (2) $(A \,||\, B^\alpha)^\alpha = A^\alpha \,||\, B^\alpha$*

PROOF: (1) If $s \in A^\alpha \,||\, B^\alpha$, then $s \in a_1 \,||\, \ldots \,||\, a_m \,||\, b_1 \,||\, \ldots \,||\, b_n$, for some $a_i \in A$, $b_j \in B$. Hence $s \in (A+B)^\alpha$.
Conversely, if $s \in (A+B)^\alpha$, then $s \in c_1 \,||\, \ldots \,||\, c_k$ with $c_i \in A+B$. Using the commutativity of "||" the factors of the parallel composite may be rearranged by the occurrences from A and then from B. One gets $s \in A^\alpha \,||\, B^\alpha$.

(2) If $s \in (A \,||\, B^\alpha)^\alpha$, then $s \in s_1 \,||\, \ldots \,||\, s_m$ with $m \geq 0$ and for each i one has $s_i \in a_i \,||\, b_1^i \,||\, \ldots \,||\, b_{n_i}^i$, with $a_i \in A$ and $b_j^i \in B$. Rearranging the factors of the parallel composite by the occurrences from A and then from B one gets

$s \in A^\alpha \,||\, B^\alpha$.
Conversely, if $s \in A^\alpha \,||\, B^\alpha$, then $s \in a_1 \,||\, \ldots \,||\, a_m \,||\, b_1 \,||\, \ldots \,||\, b_n$, for some $a_i \in A$, $b_j \in B$; then, $s \in (a_1 \,||\, \epsilon) \,||\, \ldots \,||\,(a_{m-1} \,||\, \epsilon) \,||\,(a_m \,||\, b_1 \,||\, \ldots \,||\, b_n) \in (A \,||\, B^\alpha)^\alpha$.

□

Lemma 11.15. *A unitary expression is equivalent to a "normal form" one of the following type* RegExp || RegExp$^\alpha$.

PROOF: The proof is by induction on the number of the occurrences of || and $^\alpha$. The initial step is easy. The inductive step follows by the observation that

$$\begin{array}{lcll} U_1 \,||\, U_2 & = & (A_1 \,||\, B_1^\alpha) \,||\,(A_2 \,||\, B_2^\alpha) & \text{Hypothesis} \\ & = & A_1 \,||\, A_2 \,||\, B_1^\alpha \,||\, B_2^\alpha & \\ & = & (A_1 \,||\, A_2) \,||\,(B_1 + B_2)^\alpha & \text{Lemma 11.14} \\ & = & A \,||\, B^\alpha & \end{array}$$

$$\begin{array}{lcll} U^\alpha & = & (A \,||\, B^\alpha)^\alpha & \\ & = & (A + B)^\alpha & \text{Lemma 11.14} \end{array}$$

where A, A_1, A_2, B, B_1, B_2 are regular expressions. □

Lemma 11.16. *The class of languages recognized by Petri nets contains the class of regular expressions and it is closed to* ||, [ ], σ.

PROOF: The fact that regular languages are recognized by Petri nets is trivial. In addition we may restrict ourselves to Petri nets with a unique initial point. (Nothing is lost: with ϵ transitions one may generate any initial configuration that is wanted.) Then, the closure to the remaining operations is easy, too.

Indeed, for parallelism ||, let L_1, L_2 be recognized by Petri nets N_1, N_2 as above. The language represented by $L = L_1 \,||\, L_2$ is recognized by the Petri net N consisting of the disjoint union of N_1 and N_2, together with a new initial place; from this place there is an ϵ transition which puts one point in the initial places of N_1 and N_2.

For synchronization [ ], let L_i, N_i be as above. We repeat the above construction, but now the transitions with labels in $sign(L_1) \cap sign(L_2)$ which occur in both nets are synchronized (glued together). (More precisely, for each pair of transitions t_1 in N_1 and t_2 in N_2 that share the same label a a new transition, denoted $t_1 \otimes t_2$, is introduced which collects the connections of t_1 and t_2 and has label a; finally, the old transitions t_1 and t_2 are removed.)

For renaming σ is obvious: simply rename by σ the labels of the transitions in the given Petri net. □

Theorem 11.17. *The languages represented by concurrent regular expressions are recognized by Petri nets.*

PROOF: By the above lemmas it is enough to show that there are Petri nets equivalent to expressions of type $E_1 \,||\, E_2^{\alpha}$. Take the finite automata that recognize the languages represented by E_1 and E_2 and combine them in a synchronized way, throwing away the place marked by ∞. This way one gets a Petri net equivalent to the given expression. □

Remark 11.18. Notice again that the passing above is algorithmic. I.e., there is an algorithm which, given a concurrent regular expression, produces an equivalent Petri net.

11.6 Equivalence of CRegExp and Petri net languages

From the main theorems on the previous 3 sections, one gets the following fundamental result.

Theorem 11.19. *(equivalence of* CRegExp *and Petri nets)*
The class of languages specified by concurrent regular expressions coincides with the class of languages recognized by Petri nets. □

Comments, problems, bibliographic remarks

Problems The presentation of this chapter naturally leads to a few interesting open questions, which are collected here.

Algebraic rules for CRegExp. From our algebraic point of view, the most important question is to find sound and complete algebraic rules for passing from one expression to an equivalent one. In particular, this has to be used to find an algebraic proof for Theorem 11.19, similar to the one on automata that has been described in Chapter 8.

Other versions of CRegExp. The importance of the restrictions used in the definition of CRegExp needs to be clarified. So:

1. Study the concurrent regular expressions CRegExp1 obtained by throwing away the restrictions used in the definition of CRegExp, hence using any expression defined using the operations in the CRegExp signature.
2. Study the concurrent regular expressions CRegExp2 obtained using iterate parallel composition with synchronization instead of the iterate parallel composition used by CRegExp.
3. Is the new class of expressions CRegExp1 (resp. CRegExp2) strictly more expressive than standard CRegExp ones?

4. If CRegExp1 (respectively, CRegExp2) is strictly more expressive than CRegExp, is there a natural model of parallel computation which corresponds to these expressions?

Notes. The model of Petri nets was introduced by C.A. Petri in the 1960s, [Pet62a, Pet62b] and it is quite popular now. An introduction to the basic facts related to Petri nets is given in [Rei85]; see also [Kot84].

Our presentation is based on [GR92]; the fundamental theorem (Theorem 11.19) is from this paper. Petri nets are represented by regular expressions, enriched with parallel operators. The central position placed by regular expressions in the network algebra approach was singled out in Chapter 8, hence we have a straightforward way to link network algebra and Petri nets.

Petri nets are also studied in a monoidal setting in [MM90]; another calculus for Petri nets is the box calculus presented in [KB99]. Results relating linear logic and Petri nets are presented in [Kan95]. Of interest is also [Old91] or [DE95]. The category of Petri nets creates many difficult design problems, which have been tackled in [Win87, WN94].

Part IV

Towards an algebraic theory for software components

12. Mixed Network Algebra

> Justice and power must be brought together, so that whatever is just may be powerful, and whatever is powerful may be just.
>
> Blaise Pascal

We have seen that network algebra with its flexibility is able to formalize a lot of apparently unrelated models used in computer science. The present chapter is dedicated to the search for a *unique* model for the full software field, if possible. The basic idea is to use mixtures of network algebras with their different Cantorian or Cartesian interpretations. The resulting mixed network algebra model (MixNA) looks very promising, but many technical problems are still open.

The present chapter starts with a study of the acyclic part of the mixed network algebra model. Algebraic presentations for mixed relations are given. These relations are used to model transformations of the interface types in parallel programs. The involved algebras are enriched sysecats (symmetric semiringal categories). A particular algebra covered by this class of algebras is the *discat* (distributive category) structure; it may play the role the algebraic theory is playing for the usual unmixed model. We also present the *plans*, a concrete construction of the free discat algebra.

Next, a few results on the cyclic part of the model are inserted. On a generic example, we show how the mixed network algebra model may be used to parallelize programs, a basic technique to get more efficient code. While the axiomatization for the full MixNA model is still an open question, along the way we identify a few useful algebraic rules to be used in this context. Next, we pass to the fully mixed model, which covers both parallel and object-oriented settings. In this model, a space-time duality thesis is described, accompanied with the presentation of a few facts which support the space-time duality machinery. (This thesis is to be used to keep the overall complexity of the mixed model at a reasonable complexity size.)

12.1 Why mixed network algebra models?

Let us start with the classical example of a One-Agent System whose behaviour is specified by a flowchart scheme. Syntactically a flowchart is a diagram of blocks/statements which are connected by arrows. Such diagrams represent the control flow. There is *one* agent which is moving along the arrows according to the specification. When the agent reaches a block the corresponding statement is executed. There are two basic types of statements: assignments (they update the memory states) and tests (they give information on the next block to be executed). In such a diagram "the action" is always localized at one spot. See Chapter 7 for more on this model.

A diagram such as above may be considered as a Multi-Agent System, as well. For instance, one may consider each block to be an agent, i.e., to be active. In order to get the same result as in the above interpretation it is enough to require that each block *reacts* precisely when the controlling agent in the One-Agent interpretation reaches that block. We have a multi-agent system, but due to the control restriction at most one agent is active at any time.

Starting with the above observation, one may consider the extremal situation when there is no control at all. In such a case each block is always active and we get a *purely reactive* system. One may use the same syntactic diagrams as in the flowchart case, but now with a different "reactive" meaning. Petri nets and data-flow networks are classical examples of reactive systems. In such diagrams the action is everywhere.

This last view is very close to the classical model of *dynamical systems* used in mechanics. The behaviour of a multi-agent computational system is different from a classical dynamical system essentially due to its control features.

A good model of distributed computation systems has to obey the following condition: *the behaviour of a multi-agent system should be described by diagrams where the action is distributed somewhere.* That is, the model should not require that the action is always localized at one spot or that it is always everywhere.

♡ The need for parallelism is widely recognized in the field of sequential computation. On the opposite side it is also recognized that reactive systems have to use certain control features. Example of such control mechanisms are: inhibitory arcs in Petri nets, merge or split nodes in data-flow networks, automata over differential equations in hybrid systems, membrane encapsulation in chemical abstract machines, tube encapsulation and splicing restrictions in DNA computation, etc. ◇

12.2 Mixing control, space, and time

As we already pointed out in Chapter 1, three radically different models for network algebra may be considered. They are related to control, space, and time. Here, we briefly describe how to mix them.

Mixed behavioural–spatial setting. Here we describe a mixed formalism which combine the control- and data-flow based network algebras. Mathematically, this means to combine additive and multiplicative network algebras.

If we collect the signatures of the AddNA and MultNA algebras, then we get the following set of operations and constants: $\oplus, \cdot_\oplus, \uparrow_\oplus, \mathsf{I}^\oplus, \mathsf{X}, {\bullet\!\!<}_k, {}_k{>\!\!\bullet}$ and $\otimes, \cdot_\otimes, \uparrow_\otimes, \mathsf{I}^\otimes, \mathsf{X}, \curlywedge_k, \curlyvee^k$. The first hypothesis is that both network algebras share the same categorical structure. Therefore, $\cdot_\oplus = \cdot_\otimes$ (denoted $\cdot$) and $\mathsf{I}^\oplus = \mathsf{I}^\otimes$

(denoted I). In addition MixNA will have some new constants $\delta_;$ and $\rho_;$, which are mutually inverse; they link the involved NA structures. Consequently, the MixNA signature is given by the following operations and constants:

$$\oplus \quad \otimes \quad \cdot \quad \uparrow_\oplus \quad \uparrow_\otimes \quad \mathsf{I} \quad \delta_; \quad \rho_; \quad \mathsf{X} \quad \mathsf{X} \quad \prec_k \quad {}_k\succ \quad R_k \quad V^k$$

This mixed network algebra setting is denoted by $\mathsf{Mix}_{\mathrm{BS}}\mathsf{NA}$, provided the mixed control-data-flow interpretation is considered. It will be used to describe parallel programs.

Let S be a set of atomic sorts. The set $S^{\oplus,\otimes}$ of arbitrary sorts is obtained with the rules

$$a = a \oplus a \mid a \otimes a \mid 0 \mid 1 \mid s(\in S)$$

Such a sort may be used to describe the network communication interface. For the semantics, suppose we have given a set D_s for each atomic sort s modeling the data that are sent along channels of type s. We extend the notation D_a to arbitrary sorts $a \in S^{\oplus,\otimes}$ by the rules ($\sqcup$ denotes disjoint union on sets): $D_0 = \emptyset$, $D_{a\oplus b} = D_a \sqcup D_b$, $D_1 = \{(\,)\}$, $D_{a\otimes b} = D_a \times D_b$.

The basic semantic model is $\mathsf{MixRel}_S(D)$:

$$\mathsf{MixRel}_S(D)(a,b) = \{f : f \subseteq D_a \times D_b\}$$

where $a, b \in S^{\oplus,\otimes}$.

Besides the additive and multiplicative network algebra operators this mixed setting uses certain distributivity constants

- *Distributivity:* $\delta_{a;b_1,\ldots,b_n} : a \otimes (b_1 \oplus \ldots \oplus b_n) \to a \otimes b_1 \oplus \ldots \oplus a \otimes b_n$ (In the relational model it denotes the natural isomorphism between the corresponding sets.)

These constants may be used to state the following distributivity axiom

$(f \otimes (g_1 \oplus \ldots \oplus g_n)) \cdot \delta_{a';b'_1,\ldots,b'_n} = \delta_{a;b_1,\ldots,b_n} \cdot (f \otimes g_1 \oplus \ldots \oplus f \otimes g_n)$,
where $f : a \to a'$ and $g_j : b_j \to b'_j$.

Mixed behavioural-temporal setting. The mixed setting of the *behavioural* and *timed* aspects of a systems (denoted by $\mathsf{Mix}_{\mathrm{BT}}\mathsf{NA}$) is fully orthogonal to the $\mathsf{Mix}_{\mathrm{BS}}\mathsf{NA}$ model. (Instead of "$\otimes$", the multiplicative operators are denoted by "$\frown$" in this context.) Besides the behavioural and the temporal network algebra operators this mixed behavioural–temporal setting contains the distributivity constants $\mathsf{d}_;$

$$\mathsf{d}_{a;b_1,\ldots,b_k} : a \frown (b_1 \oplus \ldots \oplus b_k) \to a \frown b_1 \oplus \ldots \oplus a \frown b_k$$

These constants allow us to express the following interesting distributivity law: if $f : a \to a'$ and $g_j : b_j \to b'_j$, for $j \in [n]$, then

$$(f \frown (g_1 \oplus \ldots \oplus g_n)) \cdot \mathsf{d}_{a';b'_1,\ldots,b'_n} = \mathsf{d}_{a;b_1,\ldots,b_n} \cdot ((f \oplus g_1) \frown \ldots \frown (f \oplus g_n))$$

Mixed control-space-time setting. The most expressive (but also most complicated) setting is $\mathsf{Mix_{BST}NA}$ which mixes the behavioural, spatial, and temporal aspects of a system. At the current stage of research the picture is not completely clear to the author. Its study requires much effort, with the hope of providing a sharp, clean and natural model for concurrent, real-time, object-oriented programming.

12.3 Acyclic models

This long section is dedicated to several results related to the acyclic part of the mixed network algebra model. We introduce sysecats (symetric semiringal categories), relate them with discats (distributive categories), axiomatize mixed relations, introduce mixalgebras, and describe plans as freely generated discats.

12.3.1 Sysecats

A sysecat (symmetric semiringal category) is a mixtures of two symocats. This and its enriched versions are presented here.

Let us start with the MixNA setting given by the signature $\oplus, \otimes, \cdot, \uparrow_\oplus, \uparrow_\otimes, \mathsf{I}, \delta_;, \rho_;, \chi, \mathsf{X}, \bullet\!\!<_k, {}_k\!>\!\!\bullet, \mathcal{R}_k, \mathcal{V}^k$. The associated relational model $\mathsf{MixRel}(D)$ has been described in the previous section. The new MixNA constants $\delta_;$ and $\rho_;$ model the isomorphisms corresponding to the distribution of Cartesian product over disjoint sum.

Distributivity constants in $\mathsf{MixRel}(D)$. They have the following meaning:
$\delta_{a_1,\ldots,a_m;b_1,\ldots,b_n} : (a_1 \oplus \ldots \oplus a_m) \otimes (b_1 \oplus \ldots \oplus b_n) \to (a_1 \otimes b_1) \oplus \ldots \oplus (a_1 \otimes b_n) \oplus \ldots \oplus (a_m \otimes b_1) \oplus \ldots \oplus (a_m \otimes b_n)$ is defined by
$\delta_{a_1,\ldots,a_m;b_1,\ldots,b_n} = \{((x,y),z):$ if $x = i.x_i$ and $y = j.y_j$, then $z = (i-1)n + j.(x_i, y_j)$, where $i \in [m], j \in [n], x_i \in D_{a_i}, y_j \in D_{b_j}\}$ ◇

The algebraic structures used for the feedback-free fragment of the MixNA model are "enriched sysecats" (shorthand for symmetric semiringal categories enriched with additive and multiplicative branching constants). They are defined below.

A *sysecat* $(M, \oplus, \otimes, \cdot, \mathsf{I}, \delta_;, \rho_;, \chi, \mathsf{X})$ consists of: (1) a category $(M, \cdot, \mathsf{I})$, with (2) two symmetric strict monoidal structures $(\oplus, \chi)$ and $(\otimes, \mathsf{X})$ (hence both $(M, \oplus, \cdot, \mathsf{I}, \chi)$ and $(M, \otimes, \cdot, \mathsf{I}, \mathsf{X})$ are symocats) and (3) such that the distributivity/reduction constants fulfil the (coherence) rules in Table 12.1.

If C is a subset of morphisms in M, then we say $(M, \oplus, \otimes, \cdot, \mathsf{I}, \delta_;, \rho_;, \chi, \mathsf{X})$ is a *C-weak sysecat* if the strong distributivity axioms D10–D11 are required for morphisms f in C, only (and arbitrary g, h in M).

Various enriched symmetric semiringal categories are obtained introducing additive or multiplicative branching constants: ${}_a\!\bullet\!\!<_k, k \geq 0$ (additive ramification);

Table 12.1. Axioms for sysecats

($\delta_{a,b;c}$ as a short notation for ${}^{a\oplus b}\mathsf{X}^{c} \cdot \delta_{c;a,b} \cdot ({}^{c}\mathsf{X}^{a} \oplus {}^{b}\mathsf{X}^{a})$)

D1	$\delta_{a;0,d} = \mathsf{I}_{a\otimes d}$
D2	$\delta_{a;b\oplus c,d} \cdot (\delta_{a;b,c} \oplus \mathsf{I}_{a\otimes d}) = \delta_{a;b,c\oplus d} \cdot (\mathsf{I}_{a\otimes b} \oplus \delta_{a;c,d})$
D3	$\delta_{a;b,c} \cdot {}^{a\otimes c}_{a\otimes b}\mathsf{X} = (\mathsf{I}_a \otimes {}^{c}_{b}\mathsf{X}) \cdot \delta_{a;c,b}$
D4	$\delta_{1;b,c} = \mathsf{I}_{b\oplus c}$
D5	$\delta_{a'\otimes a'';b,c} = (\mathsf{I}_{a'} \otimes \delta_{a'';b,c}) \cdot \delta_{a';a''\otimes b,a''\otimes c}$
D6	$\delta_{a'\oplus a'';b,c} \cdot (\delta_{a',a'';b} \oplus \delta_{a',a'';c}) =$ $\delta_{a',a'';b\oplus c} \cdot (\delta_{a';b,c} \oplus \delta_{a'';b,c}) \cdot (\mathsf{I}_{a'\otimes b} \oplus {}^{a''\otimes b}_{a'\otimes c}\mathsf{X} \oplus \mathsf{I}_{a''\otimes c})$
D7	$\delta_{1;} = \mathsf{I}_0$
D8	$\delta_{a'\otimes a'';} = (\mathsf{I}_{a'} \otimes \delta_{a'';}) \cdot \delta_{a';}$
D9	$\delta_{a'\oplus a'';} = \delta_{a',a'';0} \cdot (\delta_{a;} \oplus \delta_{a'';})$
D10	$(f \otimes (g \oplus h)) \cdot \delta_{a';b',c'} = \delta_{a;b,c} \cdot ((f \otimes g) \oplus (f \otimes h))$
D11	$(f \otimes \mathsf{I}_0) \cdot \delta_{a';} = \delta_{a;}$

${}_{k}{>\!\!\bullet}_{a}, k \geq 0$ (additive identification); $\curlywedge^{a}_{k}, k \geq 0$ (multiplicative ramification); $\curlyvee^{k}_{a}, k \geq 0$ (multiplicative identification). The restrictions a ($k = 1$), b ($k \leq 1$), c ($c \geq 1$), and d (arbitrary k) may be freely used for each type of branching constant ${}_{a}{\bullet\!\!<}_{k}$, ${}_{k}{>\!\!\bullet}_{a}$, $\curlywedge^{a}_{k}$, or $\curlyvee^{k}_{a}$. (For notational reasons, the previous distinction using Latin and Greek letters for such restrictions is dropped here.)

12.3.2 Mixed relations

Mixed relations describe the connections between complex interfaces used in parallel programs, where both sum and product are present. Their presentation as enriched sysecats is described here.

Finite relations were studied in detail in the first parts of the book; see, e.g., Chapter 3. In order to obtain a full model for parallel programs one needs to mix control and reactive parts, hence a richer theory of finite relations is needed. The relations corresponding to a particular choice (x_1, x_2, x_3, x_4) for the branching constants ${}_{a}{\bullet\!\!<}_{k}$, ${}_{k}{>\!\!\bullet}_{a}$, $\curlywedge^{a}_{k}$ and $\curlyvee^{k}_{a}$ with $x_i \in \{a, b, c, d\}$, are called $x_1x_2x_3x_4$-relations. We present axiomatizations as enriched (weak) sysecats for the resulting classes of interface-changing relations.

Axiomatization. The point of view taken in Chapter 3 was to get axiomatizations in a "linear" setting, i.e., to avoid the use of the strong axioms of type Sa–Sd (on p. 192); these axioms allow us to make copies of or to delete arbitrary cells.

Remark 12.1. A first observation is that the additive strong axioms follow from the distributivity axiom of the semiringal categories.

Proof. Indeed, using $a \stackrel{{}_{a}{\bullet\!<}_{k}}{\longrightarrow} a\oplus \overset{k}{\ldots} \oplus a = 1 \otimes a \stackrel{{}_{1}{\bullet\!<}_{k}\otimes \mathsf{I}_a}{\longrightarrow} (1\oplus \overset{k}{\ldots} \oplus 1) \otimes a \stackrel{\delta_{1,\overset{k}{\ldots},1;a}}{\longrightarrow} (1 \otimes a)\oplus \overset{k}{\ldots} \oplus(1 \otimes a)$ we get

$$f \cdot {}_b{\bullet\!<}_k = f \cdot ({}_1{\bullet\!<}_k \otimes \mathsf{I}_b) \cdot \delta_{1,\overset{k}{\dots},1;b} = (\mathsf{I}_1 \otimes f) \cdot ({}_1{\bullet\!<}_k \otimes \mathsf{I}_b) \cdot \delta_{1,\overset{k}{\dots},1;b}$$
$$= ({}_1{\bullet\!<}_k \otimes \mathsf{I}_a) \cdot (\mathsf{I}_k \otimes f) \cdot \delta_{1,\overset{k}{\dots},1;b} = ({}_1{\bullet\!<}_k \otimes \mathsf{I}_a) \cdot ((\mathsf{I}_1 \oplus \overset{k}{\dots} \oplus \mathsf{I}_1) \otimes f) \cdot \delta_{1,\overset{k}{\dots},1;b}$$
$$= ({}_1{\bullet\!<}_k \otimes \mathsf{I}_a) \cdot \delta_{1,\overset{k}{\dots},1;a} \cdot (\mathsf{I}_1 \otimes f \oplus \overset{k}{\dots} \oplus \mathsf{I}_1 \otimes f) = {}_a{\bullet\!<}_k \cdot (f \oplus \overset{k}{\dots} \oplus f) \qquad \diamond$$

A similar fact holds for $>\!\bullet$, too.

For this reason, we have to weaken the semiringal category structure by requiring that the distributivity axioms hold for particular classes of morphisms only. An $x_1x_2x_3x_4$*-weak sysecat* is one where the distributivity axioms D10–D11 are required for morphisms which are represented by $x_1x_2x_3x_4$-terms.

Let us start with a *dddd*-weak sysecat

$$\overline{M} = (M, \oplus, \otimes, \cdot, \mathsf{I}, \delta_{;}, \rho_{;}, \mathsf{X}, \mathsf{X}, {\bullet\!<}_k, {}_k{>\!\bullet}, \mathsf{R}_k, \mathsf{V}^k)$$

We are also considering the following groups of axioms:

M1 $(M, \oplus, \cdot, \mathsf{I}, \mathsf{X}, {\bullet\!<}_k, {}_k{>\!\bullet})$ fulfils the angelic additive $d\delta$-symocat axioms in Chapter 3;

M2 $(M, \otimes, \cdot, \mathsf{I}, \mathsf{X}, \mathsf{R}_k, \mathsf{V}^k)$ fulfils the forward-demonic multiplicative $d\delta$-symocat axioms in Chapter 3 / Appendix C;

M3 the scalar–vectorial axioms of type A5 for the additive and multiplicative branching constants hold.

Theorem 12.2. *The dddd-weak sysecat axioms gives an axiomatization for the corresponding mixed relations. From this general result, particular axiomatizations for classes of $x_1x_2x_3x_4$-relations are obtained.* □

A sketch of the proof may be found in Appendix G.

Example 12.3. As an example, consider a relation $f : a \to b$, where $a = (x \otimes y \otimes z) \oplus (z \otimes y)$ and $b = (x \oplus (y \otimes z)) \otimes y$, for certain atomic sorts x, y, z. The additive normal form of b is $(x \otimes y) \oplus (y \otimes z \otimes y)$, hence f may be given as a relation $\{1.1, 1.2, 1.3, 2.1, 2.2\} \times \{1.1, 1.2, 2.1, 2.2, 2.3\}$. For instance, suppose f is specified by the pairs
$\{(1.1, 1.1), (1, 2, 1.2), (1.2, 2.1), (1.2, 2.3), (1.3, 2, 2), (2.1, 2.2), (2.2, 2.1), (2.2, 2.3)\}$
It is easy to see that f preserves the sorts. E.g., the sort corresponding to 2.1 in the domain is z, the same with that of 2.2 in the codomain. An additive mixed normal forms for f is
$f = \mathsf{I}_{(x\otimes y\otimes z)\oplus(z\otimes y)} \cdot ({}_{x\otimes y\otimes z}{\bullet\!<}_2 \oplus \mathsf{I}_{z\otimes y}) \cdot [(\mathsf{I}_{x\otimes y} \otimes \mathsf{R}_2^z) \oplus (\mathsf{R}_0^x \otimes \mathsf{R}_2^y \otimes \mathsf{I}_z) \cdot (\mathsf{I}_y \otimes {}^y\mathsf{X}^z) \oplus {}^z\mathsf{X}^y \cdot (\mathsf{I}_{y\otimes z} \otimes \mathsf{V}_y^0)] \cdot (\mathsf{I}_{x\otimes y} \oplus {}_2{>\!\bullet}_{y\otimes z\otimes y}) \cdot \rho_{x,(y\otimes z);y}$

12.3.3 Distributive categories (discats)

Here we study the relationship between discats (distributive categories) and enriched sysecats.

For the network algebra purpose we needed certain very weak monoidal structures which still contain finite relations. In the previous subsection enriched sysecats (symmetric semiringal categories enriched with additive and multiplicative branching constants) have been introduced as abstract models for classes of finite mixed relations. These structures are closely related to certain classical structures used in the literature.

A category with coproducts and products is called *bicartesian*. Such a category is called *distributive* if moreover products distribute over coproducts.

Semiringal categories are more general than distributive categories as they require the mixing of "$\oplus$" and "$\otimes$" at the monoidal level. As it was already noticed in Proposition 12.1, the distributivity condition implies the strong commutation axiom for the additive identification branching constants (if present), hence a semiringal category with such constants has ordinary coproducts. Consequently, in the presence of multiplicative ramification and additive identification branching constants, semiringal categories are more general than distributive categories, just because they do not require general ordinary products.

Proposition 12.4. *A bicartesian category is a sysecat and obeys the strong axioms for ${}_k{>\!\!\bullet}$ and ${\curlywedge}_k$ whenever the following are isomorphisms:*

$$\rho_{a;b,c} := [\mathsf{I}_a \otimes (\mathsf{I}_b \oplus {}_0{>\!\!\bullet}_c) \oplus \mathsf{I}_a \otimes ({}_0{>\!\!\bullet}_b \oplus \mathsf{I}_c)] \cdot {}_2{>\!\!\bullet}_{a\otimes(b\oplus c)}$$
$$\rho_{a;} := {}_0{>\!\!\bullet}_{a\otimes 0}$$ □

A sketch of the proof may be found in Appendix H.

As we have seen in Remark 12.1, in a sysecat with natural additive identification constants ${}_k{>\!\!\bullet}$ the distributivity condition implies the strong axiom for ${}_k{>\!\!\bullet}$, hence such a category has coproducts. With the above result, this shows that

Proposition 12.5. *Distributive categories coincides with sysecats with branching constants ${\curlywedge}_k$ and ${}_k{>\!\!\bullet}$ which are ${\curlywedge}_k$-strong.* □

12.3.4 Mixalgebras

This subsection is devoted to the introduction of mixalgebras. They occur as a common generalization of both algebras and coalgebras. The subsection includes examples and certain procedures to decompose morphisms in discats.

Introductory remarks. A mixalgebra is an algebraic structure based on (mix)functions

$$(*) \quad f_A : A_s \to A_{t^1} \oplus \ldots \oplus A_{t^n}$$

with $s, t^1, \ldots, t^n$ products of atomic sorts. The restriction to functions with $n = 1$ (no sum) produces classical algebras, while the dual restriction to functions with all $s, t^1, \ldots, t^n$ atoms (no product) produces coalgebras.

Atomic operations. A comment on the type of morphisms used in this signature may be useful. The above (*) gives the most general form of the indecomposable elements in a distributive category. E.g., if one considers functions of the following simpler type

$$(**) \quad f_A : A_{s_1} \otimes \ldots \otimes A_{s_m} \to A_{t_1} \oplus \ldots \oplus A_{t_n} \text{ with } s_i,\ t_j \text{ atoms}$$

then the reduction of discat morphisms to such terms is partial. The point is that in a term as in (*) the partial functions corresponding to different outputs of a certain t^k have synchronized tests, i.e., in all interpretations they are defined on the same subset on input data. One cannot simulate such a situation with functions as in (**), as they may be arbitrary interpreted. This shows a signature with functions of type (**) is strictly less expressive than one using functions of type (*). ◇

The proper setting to speak about (co)algebras is that of a (co)algebraic theory: one needs tuples of terms when dealing with term substitution. In a similar way, an appropriate setting to handle mixalgebra substitution is one using tuples and cotuples of ordinary terms. A natural such a setting consists of categories with coproducts and products such that products distribute over coproducts, i.e., the setting of discat structures.

Further, distributivity requires the splitting of ordinary coalgebraic terms in "paths." A path is modeled as an extended conditional term and it is called "elementary plan." Plans are tuples and cotuples of elementary plans with certain completeness and consistency conditions. They play the role the ordinary terms are playing for (co)algebras.

Free (co)algebras, revised. In the next subsection we will present a construction for the free mixalgebra generated by a mixalgebra signature. Specialized to the case when there are no sums the construction gets the usual construction with terms for the free many-sorted algebras, whilest in the case when there are no products, the construction gets the free coalgebras, but presented in a slightly different way. We give some hints here for getting the intuition on plans from the well-known construction of free algebras and coalgebras.

The construction of the free algebra is well-known. It may be represented as an algebra of terms or trees.

Coalgebras are dual to algebras. Free coalgebras may be represented by terms (or trees), too. However, the interpretation is different, since the trees are interpreted in an additive setting; they correspond to acyclic programs (with global states) using **case** statements. Suppose A is a set of variables for leaves; the elements in A may be thoughts of as representing possible worlds. Then, a tree may be interpreted as a particular collection of paths towards some such output worlds A,B,... in A.

Naming variables. A comment on naming variables is necessary. Two ways of naming variables may be used. The first, more traditional way, is to use letters; we use lowercase SMALL CAPS letters for object variables (obvars), e.g., A,B,... and lowercase *italic* letters for property variables (propvars), e.g., $x, y, z, \ldots$. The second way provides a standard default naming procedure using numbers as it used in the network algebra approach. To simplify the understanding, we will not use this second possibility here. ◇

Free coalgebra may be described in the following equivalent way. To each operation symbol $f : t \to t_1 \oplus \ldots \oplus t_n$ we associate a generalized test symbol $f^c : t \to n$ and n unary operation symbols $f^s_j : t \to t_j,\ j \in [n]$. The tree corresponding to a term, say, $f\langle \text{A}, g\langle f\langle \text{B}, \text{A}\rangle, h\rangle\rangle$, where $f, g : t \to 2t,\ h : t \to 0$ may be represented as:

$$\begin{array}{lcl}\{\text{A}.f^s_1(x) & \Leftarrow & f^c(x) = 1,\\ \ \text{B}.f^s_1(g^s_1(f^s_2(x))) & \Leftarrow & f^c(x) = 2 \ \&\ g^c(f^s_2(x)) = 1 \ \&\ f^c(g^s_1(f^s_2(x))) = 1,\\ \ \text{A}.f^s_2(g^s_1(f^s_2(x))) & \Leftarrow & f^c(x) = 2 \ \&\ g^c(f^s_2(x)) = 1 \ \&\ f^c(g^s_1(f^s_2(x))) = 2,\\ & \Leftarrow & f^c(x) = 2 \ \&\ g^c(f^s_2(x)) = 2 \ \&\ h(g^s_2(f^s_2(x)))\}\end{array}$$

Each leaf has a corresponding path to root. The body of an implication gives a label of a possible world and a term for the resulting state in that world; sometimes, it may be absent. The condition gives the guard of that path; sometimes, it may be absent, too. Of course, some additional restrictions on these collections of conditional terms are required to represents trees and only trees.

Examples of mixalgebras. We present a few simple, but maybe typical, examples.

Mix-functions. As a first example, we describe MixFn, the model with mix-functions. Actually, it gives a description for the initial discat. In the one-sorted case, a normal form sort (i.e., written as a sum of products of atomic sorts) may be represented as a set α, where $\alpha = \{i.j : i \in [k], j \in [m_i]\}$. Take another sort $\beta = \{i.j : i \in [l], j \in [n_i]\}$. The arrows between such sorts are mix-functions, i.e., relations $\phi \subseteq \alpha \times \beta$ such that $\exists f : k \to l$ and $f^i : n_{f(i)} \to m_i,\ i \in [k]$ such that: $(i.j, i'.j') \in \phi$ iff $f(i) = i'$ and $f^i(j') = j$. Then, MixFn (with appropriate operations) is the initial distributive category. In the many-sorted case, the construction is similar, but the mix-functions should preserve the sorts. (Actually mix-functions are particular instances of "plans," namely when the mixalgebra signature is empty; the plans will be defined in the next subsection.) ◇

Arithmetic mixalgebra. As a second example we describe an arithmetic mixalgebra. The signature is:
$\Sigma_0 = \{\mathbf{s} : n \to n,\ \mathbf{z} : 1 \to n,\ \mathbf{k} : n \to 0,\ \mathbf{a} : n^2 \to n,\ \mathbf{p} : n \to 2n,\ \mathbf{d} : n^2 \to 2n\}$
A particular Σ_0-algebra over natural numbers is:
- $\mathbf{s}$ is the successor function $\mathbf{s}(x)\langle \text{A}\rangle = \{\text{A}.x + 1 \Leftarrow\}$;
- $\mathbf{z}$ is constant zero $\mathbf{z}()\langle \text{A}\rangle = \{\text{A}.0 \Leftarrow\}$;
- $\mathbf{k}$ is the kill constant $\mathbf{k}(x)\langle\rangle = \{\Leftarrow k(x)\}$;
- $\mathbf{a}$ is the addition function $\mathbf{a}(x, y)\langle \text{A}\rangle = \{\text{A}.x + y \Leftarrow\}$;
- $\mathbf{p}$ is the predecessor operation $\mathbf{p}(x)\langle \text{A}, \text{B}\rangle = \{\text{A}.x - 1 \Leftarrow x \geq 1,\ \text{B}.0 \Leftarrow x = 0\}$;
- $\mathbf{d}$ is the difference operation $\mathbf{d}(x, y)\langle \text{A}, \text{B}\rangle = \{\text{A}.x - y \Leftarrow x \geq y,\ \text{B}.y - x \Leftarrow x < y\}$.

In the particular algebra we have described, the interpretation of a mixalgebra operation requires two sorts of variables: input propvars for the incoming values to be used and output propvars where the results have to be put. It is supposed that it is a unique input obvar whose name is omitted, and a unique output propvar whose name is omitted, too. To be fully described, an operation as above, say p, has to be written as $\mathrm{p}(\mathrm{C}.x)\langle \mathrm{A}.y, \mathrm{B}.y\rangle = \{\mathrm{A}.y := \mathrm{C}.x - 1 \Leftarrow \mathrm{C}.x \geq 1,\ \mathrm{B}.y := 0 \Leftarrow \mathrm{C}.x = 0\}$ ◇

Boolean logic. Boolean logic may be incorporated into such a framework in various ways. One way was already presented in some detail in Chapter 7. In the current stronger setting one has some other natural possibilities. E.g., for two predicates $p, q : s \to 2$ one may define the "and" operator by

$$p \,\&\, q = \wedge_2^s (p \otimes q)\delta_{1,1;1,1}(\mathsf{I}_1 \oplus {}_3\!\succ\!\!\bullet_1)$$

Such a definition is more related to a linear logic definition and it works in the cases there are side effects. The "sequential definition" $p \,\&\, q = p \cdot (q \oplus \mathsf{I}_s) \cdot (\mathsf{I}_s \oplus {}_2\!\succ\!\!\bullet_s)$, for $p, q : s \to s$ does not work properly in the case the starting predicates have side effects. In such a case q is tested in a possible different state which may result after the application of p. ◇

Discat decompositions. A few particular decompositions of discat arrows are presented here.

Mixalgebra vs. relational structures. We first compare mixalgebras with relational structures, cf. [Coh81]. Obviously, a relational structure is a mixalgebra. We describe certain conditions which assure they are equivalent.

In the one-sorted case, given a (single output) mixalgebra operation $f : A^{i(f)} \to o(f)A$ one may define

— a generalized $o(f)$-valued $i(f)$-ary test $f^c = f \cdot (o(f)\wedge_0^A) : A^{i(f)} \to o(f)$ and
— a usual $i(f)$-ary operation $f^s = f \cdot {}_{o(f)}\!\succ\!\!\bullet_A : A^{i(f)} \to A$

where A^k (resp. kA) is the product (resp. sum) of k copies of A.

Proposition 12.6. *In a distributive category setting, given f^c and f^s as above, the original mixalgebra operation f may be recovered by the formula*

$$f = \wedge_2^{A^{i(f)}}(f^c \otimes \mathsf{I}_{A^{i(f)}})\delta_{1,\overset{o(f)}{\dots},1;A^{i(f)}}(o(f)f^s)$$

A proof may be found in Appendix I. ◇

Path-decomposition. Another useful decomposition is in terms of paths. To a mixalgebra operation $f : A_{s_1} \otimes \ldots \otimes A_{s_m} \to A_{t_1} \oplus \ldots \oplus A_{t_n}$, with $s_i \in S$ and $t_i \in S \cup \{1\}$ one may associate some path-operations

$$f_j : A_{s_1} \otimes \ldots \otimes A_{s_m} \to A_{t_j} \oplus 1$$

removing the values from the output summands different from j, and identifying all the resulting summands that are equal to 1.

Proposition 12.7. *(1) If a distributive category has a morphism $k : 1 \to 0$, then a morphism $f : A \to B \oplus C$ may be described using the associated path operations f_j as follows*

$$f = \wedge_2^A (f_1 \otimes f_2)\delta_{B,1;1,C}(\mathsf{I}_B \oplus \wedge_0^{B\otimes C} \cdot k \oplus k \oplus \mathsf{I}_C)$$

(2) A similar decomposition exists when the mixalgebra signature is such that there is a ground term $c_a : 1 \to a$, for each sort a, e.g.,

$$f = \wedge_2^A (f_1 \otimes f_2)\delta_{B,1;1,C}(\mathsf{I}_B \oplus (\wedge_0^{B\otimes C} \cdot c_C \oplus c_C \oplus \mathsf{I}_C) \cdot {}_3\!\succ\!\!\bullet_C)$$

Appendix I may be consulted for a proof. ◇

12.3.5 Plans

In this subsection we formally define the plans, describe their discat structure, and state their freeness property.

An example. Suppose we start with a mixalgebra signature which contains operations f, g of sorts $f : \mathbf{a} \otimes \mathbf{b} \to \mathbf{e} \oplus \mathbf{f}$, $g : \mathbf{b} \otimes \mathbf{b} \otimes \mathbf{d} \to \mathbf{a} \oplus \mathbf{g}$. Take an interpretation which associates the set A to sort a, B to b, and so on. Then consider the following term written in the corresponding distributive category:

$$
\begin{array}{l}
A{\otimes}B{\otimes}C{\otimes}B{\otimes}D \\
\qquad \downarrow \mathsf{I}_A \otimes \wedge_2^B \otimes \wedge_0^C \otimes [(\wedge_2^B \otimes \mathsf{I}_D) \cdot (\mathsf{I}_B \otimes {}^B\mathsf{X}^D)] \\
A{\otimes}B{\otimes}B{\otimes}B{\otimes}D{\otimes}B \\
\qquad \downarrow f \otimes g \otimes \mathsf{I}_B \\
(E{\oplus}F){\otimes}(A{\oplus}G){\otimes}B \\
\qquad \downarrow \mathsf{I}_{E\oplus F} \otimes [\delta_{A,G;B} \cdot (\mathsf{I}_{A\otimes B} \oplus (\mathsf{I}_G \otimes \wedge_0^B))] \\
(E{\oplus}F){\otimes}(A{\otimes}B{\oplus}G) \\
\qquad \downarrow \mathsf{I}_{E\oplus F} \otimes (f \oplus \mathsf{I}_G) \\
(E{\oplus}F){\otimes}(E{\oplus}F{\oplus}G) \\
\qquad \downarrow \delta_{E,F;E,F,G} \\
(E{\otimes}E){\oplus}(E{\otimes}F){\oplus}(E{\otimes}G){\oplus}(F{\otimes}E){\oplus}(F{\otimes}F){\oplus}(F{\otimes}G)
\end{array}
$$

This morphism may be graphically represented as in Fig.12.1.

The occurrence of "plans". The free algebra/algebraic theory is described using terms over some variables. Then the algebraic theory rules allow us to model the composition (substitution) making a copy of the term to be substituted for each occurrence of the corresponding variable. Similarly for coalgebras and coalgebraic theories.

The main problem when one tries to find a construction for the free mixalgebra or distributive category is to model the distributivity of products over coproducts. The distributivity rule requires us to think of the additive terms (or trees) as consisting of collections of paths corresponding to the leaf variables. A path consists of a guard which has to be satisfied in order to follow that path and a tuple of terms for the final leaf (if any). Then, the distributivity rule applied to a product of additive terms is easy to model at the path level: take all combinations of paths (one from each factor of the product) and consider the corresponding conjunction of guards and tuple of leaf terms.

Technically, the problem is to describe conditions which are to be satisfied by such collections of paths in order to model precisely the terms in the free distributive category. To this end certain completeness, consistency, demonicity, and relevancy conditions are necessary. The resulting collections of paths are called *plans*. ◇

The plan corresponding to the above term is given below.

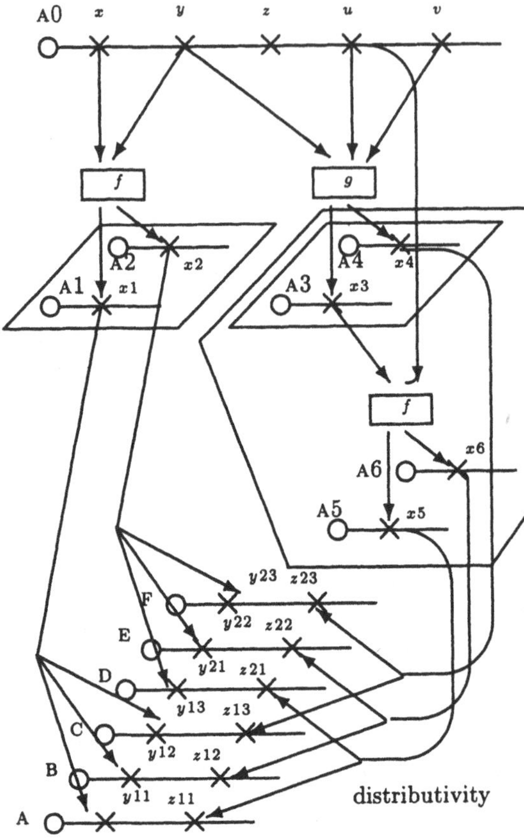

Fig. 12.1. Terms in distributive categories

$$
\begin{array}{lcl}
\{\mathrm{A}.(f_1^s(x,y),f_1^s(g_1^s(y,u,v),u)) & \Leftarrow & f^c(x,y)=1\ \&\ g^c(y,u,v)=1\ \&\ f^c(g_1^s(y,u,v),u))=1,\\
\mathrm{B}.(f_1^s(x,y),f_2^s(g_1^s(y,u,v),u)) & \Leftarrow & f^c(x,y)=1\ \&\ g^c(y,u,v)=1\ \&\ f^c(g_1^s(y,u,v),u))=2,\\
\mathrm{C}.(f_1^s(x,y),g_2^s(y,u,v)) & \Leftarrow & f^c(x,y)=1\ \&\ g^c(y,u,v)=2,\\
\mathrm{D}.(f_2^s(x,y),f_1^s(g_1^s(y,u,v),u)) & \Leftarrow & f^c(x,y)=2\ \&\ g^c(y,u,v)=1\ \&\ f^c(g_1^s(y,u,v),u))=1,\\
\mathrm{E}.(f_2^s(x,y),f_2^s(g_1^s(y,u,v),u)) & \Leftarrow & f^c(x,y)=2\ \&\ g^c(y,u,v)=1\ \&\ f^c(g_1^s(y,u,v),u))=2;\\
\mathrm{F}.(f_2^s(x,y),g_2^s(y,u,v)) & \Leftarrow & f^c(x,y)=2\ \&\ g^c(y,u,v)=2\}
\end{array}
$$

Definition of plans. In the construction of the plans we start with a pair of sets $(\mathcal{O},\mathcal{P})$. The elements of $\mathcal{O}$ are object variables, briefly called obvars. The elements in $\mathcal{P}$ are property variables, briefly called propvars. Each occurrence of a variable should be fully specified as A.x, which may be read, e.g., as propvar x applied to obvar A. Notations $\mathcal{O}.x$ and A.$\mathcal{P}$ for the corresponding sets of occurrences are used, as well. We also use a short notation x for such an occurrence when the object variable is clear. Similarly for object variables.

Basic plans. Our basic objects are called plans. Roughly speaking, a plan is a collection of "extended conditional terms" fulfilling certain restrictions. An extended conditional term by itself is called an elementary plan.

To simplify the description we use simplified signatures Σ of type (**). The construction may be naturally lifted to signatures of type (*). We define two derived signatures Σ^c, Σ^s as follows: for each operation symbol $f : s_1 \otimes \ldots \otimes s_m \to t_1 \oplus \ldots \oplus t_n$, the signature Σ^c contains a (generalized) test symbol $f^c : s_1 \otimes \ldots \otimes s_m \to n$ and Σ^s contains an operation symbol $f_i^s : s_1 \otimes \ldots \otimes s_m \to t_i$ for each $i \in [n]$ with $t_i \in S$. For a usual algebra signature Φ, the set of Φ-terms over a set of variables X is denoted by $T_\Phi(X)$.

An *elementary Σ-plan* is an extended conditional term (that is similar to a usual conditional term, but sometimes the left-hand-side term may be absent)

$$\text{B}.y := t \quad \Leftarrow \quad C$$

where

- $\text{B}.y := t$ may be present or not; if it is present, then $\text{B} \in \mathcal{O}$ and $t \in T_{\Sigma^s}(\text{A}.\mathcal{P})$; and
- C is a (possibly empty) conjunction of conditions $f^c(t_1, \ldots, t_m) = j$, with t_i terms in $T_{\Sigma^s}(\text{A}.\mathcal{P})$ and appropriate j (not exceeding the coarity of f). (If f has arity 0, then one may write simply $f^c = j$, while in the case f has coarity 0, the condition is written $f^c(t_1, \ldots, t_m)$.)

An elementary Σ-plan should be

- *complete*, i.e., C includes $f^c(t_1, \ldots, t_m) = j$ for each subterm $f_j^s(t_1, \ldots, t_m)$ occurring in t or in another condition in C and
- *consistent*, i.e., no contradictory conditions $f^c(t_1, \ldots, t_m) = j'$ and $f^c(t_1, \ldots, t_m) = j''$ with $j' \neq j''$ occur in C.

This definition is further extended to include the inconsistent plans

$$\text{B}.y := dd \quad \Leftarrow \quad ff$$

where ff denotes the "false" boolean logic value, and dd the "dark" (undefined) term. (It is used to give a meaning to the distributive category term ${}_0\!\succ\!\!\bullet_a$.)

Notice that plans without left-hand-side term are allowed. E.g., the plans corresponding to operations in Σ of coarity 0 are of such a form.

A *Σ-plan* is a collection P of elementary plans which is:

- *complete*, i.e., for each condition $f^c(t_1, \ldots, t_m) = j$ of an elementary plan any other "complementary" condition $f^c(t_1, \ldots, t_m) = j'$ occurs in an elementary plan in P;

- *consistent*, i.e., if two elementary plans have different obvars A, B, then their guards contain at least one condition $f^c(t_1, \ldots, t_m)$ which separates A and B; moreover, there are no different elementary plans with the same condition;
- *demonic*, i.e., if a guard is false, then the term is "dark" and this darkness is propagated to further (multiplicative) terms it interacts with;
- *relevant*, i.e., for a given condition $f^c(t_1, \ldots, t_m) = i$ there is a different term (i.e., with different obvars, or defining different tuple of terms) for a complementary condition $f^c(t_1, \ldots, t_m) = j$, with $j \neq i$.

A few examples of plans were given before.

General plans. The construction is further extended to fit the setting of distributive categories. Notice that till now each plan consists of elementary plans with reference to the same input obvar A and output propvar y.

First, we extend the plans to the case of tuples of output terms

$$\text{B}.y_1 := t_1, \ldots, y_n := t_n \quad \Leftarrow \quad (C_1, \ldots, C_n)$$

for distinct variables $y_1, \ldots, y_n$ and $t_i \in T_{\Sigma^\bullet}(\text{A}.\mathcal{P})$.

Then, we extend the plans to the case of cotuples of input obvars. Such a plan is a union of one-input-obvar plans all with the same output interface. More precisely,

$$\{P_1; \ldots; P_k\}$$

for plans P_i with (different) one-obvar input interfaces, but the same (general) output interface. This way we get the general notion of plan.

Discat structure on plans. We describe how a discat structure may be introduced on plans; the corresponding category is denoted by $\mathsf{Plan}(\Sigma)$.

Composition: $\alpha \xrightarrow{P} \beta \xrightarrow{Q} \gamma$ is modeled by substitution in Q of the variables A.x in β with the corresponding terms of the elementary plans in P written with reference to the interface α. Moreover, the guards are collected, namely the resulting guard is the conjunction of the current guard of Q (written with reference to interface α using the current substitution) and of the guard of elementary plan in P used for substitution.

Moreover: (1) if a condition in an elementary plan in P has a ff guard, then the resulting guard of the composite is ff, as well and the resulting term is *dd* (undefined), i.e., "darkness is propagating." (2) Then, one takes the union of the conditions which define equal terms and use some logical simplifications. That is, if *tt* denotes the "true" boolean logic value and the output interface of f has k summands, then $(f(t_1, \ldots, t_m) = 1 \vee \ldots \vee f(t_1, \ldots, t_m) = k) \equiv tt$. Notice that if a condition becomes irrelevant, then it may be eliminated in this way.

Cotupling: for $P^i : \alpha^i \to \beta$, the plan $\langle P^1, P^2 \rangle : \alpha^1 \oplus \alpha^2 \to \beta$ is the union $\{P_1; P_2\}$ of the corresponding plans.

Tupling: Let $P^i : \alpha \to \beta^i$. Notice that the normal form associated with $\beta^1 \otimes \beta^2$ consists of pairs of obvars in β^1 and β^2 and for each such pair the propvars are obtained taking the union of the corresponding propvars in β^1 and β^2. Then, the tuple of the plans is obtained taking all the pairs of elementary plans in P^1 and P^2 and for each such pair taking the elementary plan obtained by the tuple of leaf terms and the conjunction of the path guards.

However, the construction requires some further simplifications. All the inconsistent elementary plans obtained so far are removed and the resulting leaf terms are considered undefined, i.e., *dd*.

Freeness. The main interest in the distributive category of plans lays in the following result: Plan(Σ) *is the free distributive category generated by* Σ.

The proof of this result requires a detailed analysis of the structure of plans. It is not presented here. A sketch, including the main points of the proof, is given in Appendix J.

♡ The problem of generating free distributive categories seems to be studied by various researchers. Some results are mentioned in Cockett's paper [Coc93],p.279, concerning works by Bob Walters, Shu-Hao Sun, and Steve Schanuel, but no proper references are inserted. Since distributive categories are equationally defined the free structure may be trivially described using classes of terms. As far as we know no other concrete construction is available at present. ◇

EXERCISES AND PROBLEMS (SEC. 12.3)

1. Study the arithmetic mixalgebra described in this subsection (with $2 = 1 \oplus 1$ interpreted as the boolean set: $tt = 1$ and $ff = 2$). Give mixalgebra terms to describe usual arithmetic tests or operations, for instance, "$x \neq 0$", "$x = 0$", "$x = y$", $|x - y|$. (- -)
2. Get experience in using the plans and their operations. Try to prove the freeness theorem. (- -)
3. Study mixalgebra congruences. (*Hint*: in algebras, simulation via functions capture the notion of congruence, while in coalgebras, it captures bisimulation. Then, simulation via mix-functions is a good candidate for the notion of mixalgebra congruence to generalize both bisimulation and congruence relations.) (**)
4. Study mixalgebra equations. (*Hint*: plans, the elements in the free mixalgebras, may be used to specify mixalgebraic equations. How may one interpret the validity of such an equation in a particular mixalgebra? A general question is to lift classical results on (universal) algebras to mixalgebras.) (**)
5. Study infinite plans. (*Hint*: one goal is to extend the model with plans to the full MixNA setting, i.e., to use it to give meaning to the feedback operators, as well. The model seems to be well-suited for this purpose. To this end

one may easily define infinite plans and use them to give meaning to the feedback operators via unfolding.) (**)

12.4 Compilers, code generation

In this section we describe, on a generic example, how the MixNA setting may be used to parallelize programs, to get more efficient program code.

Parallel programs. We start with a sequential program which computes the binomial coefficients. It may be obtained from the flownomial expression

$$BIN1 = i_1 \cdot i_2 \cdot i_3 \cdot i_4 \cdot ({}_2\!\!>\!\!\bullet_w \cdot t_1 \cdot (\mathsf{I}_w \oplus a_1 \cdot a_2)) \uparrow_{\oplus}^{w} \cdot ({}_2\!\!>\!\!\bullet_w \cdot t_2 \cdot (\mathsf{I}_w \oplus b_1 \cdot b_2)) \uparrow_{\oplus}^{w} \cdot e$$

Interpretation. Indeed, use the following interpretation I, where the sort w, the assignment cells $i_1, i_2, i_3, i_4, a_1, a_2, b_1, b_2, e$ and the test cells t_1, t_2 are:

$$
\begin{array}{rl}
D_w = Z^6 & \\
i_1 : (n,k,z_1,z_2,y_1,y_2) \mapsto & (n,k,1,z_2,y_1,y_2) \\
i_2 : (n,k,z_1,z_2,y_1,y_2) \mapsto & (n,k,z_1,z_2,n,y_2) \\
i_3 : (n,k,z_1,z_2,y_1,y_2) \mapsto & (n,k,z_1,1,y_1,y_2) \\
i_4 : (n,k,z_1,z_2,y_1,y_2) \mapsto & (n,k,z_1,z_2,y_1,1) \\
t_1 : (n,k,z_1,z_2,y_1,y_2) \mapsto & \left\{ \begin{array}{lll} 1.(n,k,z_1,z_2,y_1,y_2) & if & y_1 \le n-k \\ 2.(n,k,z_1,z_2,y_1,y_2) & if & y_1 > n-k \end{array} \right. \\
a_1 : (n,k,z_1,z_2,y_1,y_2) \mapsto & (n,k,z_1 * y_1,z_2,y_1,y_2) \\
a_2 : (n,k,z_1,z_2,y_1,y_2) \mapsto & (n,k,z_1,z_2,y_1-1,y_2) \\
t_2 : (n,k,z_1,z_2,y_1,y_2) \mapsto & \left\{ \begin{array}{lll} 1.(n,k,z_1,z_2,y_1,y_2)) & if & y_2 > k \\ 2.(n,k,z_1,z_2,y_1,y_2)) & if & y_2 \le k \end{array} \right. \\
b_1 : (n,k,z_1,z_2,y_1,y_2) \mapsto & (n,k,z_1,z_2 * y_2,y_1,y_2) \\
b_2 : (n,k,z_1,z_2,y_1,y_2) \mapsto & (n,k,z_1,z_2,y_1,y_2+1) \\
e : (n,k,z_1,z_2,y_1,y_2) \mapsto & (n,k,z_1/z_2,z_2,y_1,y_2)
\end{array}
$$

The function specified by $BIN1$ in AddRel(Z^6) under the above interpretation computes the binomial coefficients, i.e., it maps $(n,k,..,..,..,..)$ to $(..,..,\binom{k}{n},..,..,..)$. $\diamond$

The first step is to get a more detailed specification at the syntactic level. In this new specification the assignment and test statements are related to the variables they are actually using. First of all we introduce notations for a few basic atoms.

$$
\begin{array}{ll}
one: \;\to s & one(\,) = 1 \\
suc : s \to s & suc(x) = x+1 \\
pred : s \to s & pred(x) = x-1 \\
add : s \otimes s \to s & add(x,y) = x+y \\
dif : s \otimes s \to s & dif(x,y) = x-y \\
mult : s \otimes s \to s & mult(x,y) = x * y \\
div : s \otimes s \to s & div(x,y) = x/y \\
leq : s \otimes s \to 1 \oplus 1 & leq(x,y) = \left\{ \begin{array}{lll} 1.(\,) & if & x \le y \\ 2.(\,) & if & x > y \end{array} \right.
\end{array}
$$

A description at the memory-cell level for the functions involved in the BIN1 specification is given below. To ease the understanding we indicate the correspondence between the name-free sorts below and the variable names we were using before. The correspondence between the name-free string of sorts and the corresponding tuple of names is

$$s \otimes s \otimes s \otimes s \otimes s \otimes s \quad \longrightarrow \quad (n, k, z_1, z_2, y_1, y_2)$$

The new specification is obtained from the above one using the following substitutions:

$$\begin{array}{lcl}
i_1 & = & \mathsf{I}_{s^2} \otimes (\bot^{s} \otimes one) \otimes \mathsf{I}_{s^3} \\
i_2 & = & (\curlywedge^{s} \otimes \mathsf{I}_{s^3} \otimes \bot^{s} \otimes \mathsf{I}_{s}) \cdot ({}^{s}\mathsf{X}^{s^4} \otimes \mathsf{I}_{s}) \\
i_3 & = & \mathsf{I}_{s^3} \otimes (\bot^{s} \otimes one) \otimes \mathsf{I}_{s^2} \\
i_4 & = & \mathsf{I}_{s^5} \otimes (\bot^{s} \otimes one) \\
t_1 & = & (\curlywedge^{s^2} \otimes \mathsf{I}_{s^2} \otimes \curlywedge^{s} \otimes \mathsf{I}_{s}) \cdot ({}^{s^6}\mathsf{X}^{s} \otimes \mathsf{I}_{s^2}) \cdot ((\mathsf{I}_{s} \otimes dif) \cdot leq \otimes \mathsf{I}_{s^6}) \cdot \delta_{1,1;s^6} \\
a_1 & = & (\mathsf{I}_{s^4} \otimes \curlywedge^{s} \otimes \mathsf{I}_{s}) \cdot (\mathsf{I}_{s^2} \otimes (\mathsf{I}_{s} \otimes {}^{s}\mathsf{X}^{s}) \cdot (mult \otimes \mathsf{I}_{s}) \otimes \mathsf{I}_{s^2}) \\
a_2 & = & \mathsf{I}_{s^4} \otimes pred \otimes \mathsf{I}_{s} \\
t_2 & = & (\mathsf{I}_{s} \otimes \curlywedge^{s} \otimes \mathsf{I}_{s^3} \otimes \curlywedge^{s}) \cdot (\mathsf{I}_{s^2} \otimes {}^{s}\mathsf{X}^{s^5} \cdot (\mathsf{I}_{s^4} \otimes leq \cdot {}_{1}^{1}\mathsf{X})) \cdot \delta_{s^6;1,1} \\
b_1 & = & (\mathsf{I}_{s^5} \otimes \curlywedge^{s}) \cdot (\mathsf{I}_{s^3} \otimes (\mathsf{I}_{s} \otimes {}^{s}\mathsf{X}^{s}) \cdot (mult \otimes \mathsf{I}_{s}) \otimes \mathsf{I}_{s}) \\
b_2 & = & \mathsf{I}_{s^5} \otimes suc \\
e & = & \mathsf{I}_{s^2} \otimes (\mathsf{I}_{s} \otimes \curlywedge^{s}) \cdot (div \otimes \mathsf{I}_{s}) \otimes \mathsf{I}_{s^2}
\end{array}$$

♡ The initial global-state view is still present here. In spite of the fact that many statements use small parts of the memory, the definitions describe the effect of each statement on the whole memory. ◇

A hierarchy of subclasses of the class of MixNA flownomial expressions over X may be defined by MixNA$_\sigma$, indexed by alternating strings σ over $\{\Sigma, \Pi\}$ as follows:

- MixNA$_\Sigma$ denotes the subclass of AddNA expressions over X. MixNA$_\Pi$ denotes the subclass of MultNA expressions over X.
- MixNA$_{\Sigma\pi}$ denotes the subclass of AddNA expressions over MixNA$_\pi$; MixNA$_{\Pi\sigma}$ denotes the subclass of MultNA expressions over MixNA$_\sigma$

The degree of nesting control and parallel operators in a MixNA expression E, called *copar degree*, is given by the minimal length of the alternating strings σ over $\{\Sigma, \Pi\}$ such that E belongs to MixNA$_\sigma$.

Notice that the copar degree of the starting $BIN1$ specification is given by $\Sigma\Pi$, hence it is 2.

The problem of extracting parallelism from a sequential program may now be formulated as the problem of finding an equivalent representation of the program which exploits the detailed representation in order to maximize the parallelism. In mathematical terms, the problem is to transform a MixNA expression of copar degree $\Sigma\Pi$ into an equivalent expression with an increasing, eventually maximal copar degree.

Efficient code generation. Regarding the feedback operators we shall use the following valid identities:

Mix1. $\mathsf{I}_a \otimes (f \uparrow_{\oplus}^{d}) = (\rho_{a;b,d} \cdot (\mathsf{I}_a \otimes f) \cdot \delta_{a;c,d}) \uparrow_{\oplus}^{a\otimes d}$, for $f : b \oplus d \to c \oplus d$

Mix2. Let E be $({}_2{>\!\!\bullet}_{a\otimes b} \cdot (\mathcal{R}^a \otimes \mathsf{I}_b) \cdot (\mathsf{I}_a \otimes f) \cdot \delta_{a;c,d} \cdot (\mathsf{I}_{a\otimes c} \oplus \mathsf{I}_a \otimes g)) \uparrow_{\oplus}^{a\otimes b}$, where $f : a \otimes b \to c \oplus d$ and $g : d \to b$. Then $E = (\mathcal{R}^a \otimes \mathsf{I}_b) \cdot (\mathsf{I}_a \otimes E \cdot (\flat^a \otimes \mathsf{I}_c))$.

Axiom meaning. The former axiom shows that additive feedback commutes with parallel composition with an identity. The latter one capture the intuitive observation that a "read-only" used variable of a cycle preserves its initial value at the output of the cycle. ◇

It is intuitively clear that the initialization statements i_1, i_2, i_3, i_4 are independent. This may be checked in the formal model as well, i.e.,

$$\begin{aligned} i_1 \cdot i_2 \cdot i_3 \cdot i_4 &= ((\mathcal{R}^s \otimes \mathsf{I}_s \otimes (\flat^s \otimes one) \otimes (\flat^s \otimes one) \otimes \flat^s \otimes (\flat^s \otimes one)) \\ &\quad \cdot (\mathsf{I}_s \otimes {}^s\mathsf{X}^{s^3} \otimes \mathsf{I}_s) \end{aligned}$$

It is a bit more difficult to show that the loops may be done in parallel. Let us start with

$$E_2 = [{}_2{>\!\!\bullet}_{s^6} \cdot t_2 \cdot (\mathsf{I}_{s^6} \oplus b_1 \cdot b_2)] \uparrow_{\oplus}^{s^6}$$

If we consider the previous name-based interpretation, it is clear that E_2 does not depend on the variables n, z_1, y_1. To simplify the computation we use permutations to replace the sequence $(n, k, z_1, z_2, y_1, y_2)$ by $(n, z_1, y_1, k, z_2, y_2)$. Let $\sigma = (\mathsf{I}_{s^2} \otimes {}^s\mathsf{X}^s \otimes \mathsf{I}_{s^2}) \cdot (\mathsf{I}_s \otimes {}^s\mathsf{X}^s \otimes {}^s\mathsf{X}^s \otimes \mathsf{I}_s)$. Then

$$\begin{aligned} \sigma E_2 \sigma^{-1} &= [(\sigma \oplus \mathsf{I}_{s^6}) \cdot {}_2{>\!\!\bullet}_{s^6} \cdot t_2(\mathsf{I}_{s^6} \oplus b_1 \cdot b_2) \cdot (\sigma^{-1} \oplus \mathsf{I}_{s^6})] \uparrow_{\oplus}^{s^6} \\ &= [(\sigma \oplus \sigma) \cdot {}_2{>\!\!\bullet}_{s^6} \cdot t_2(\mathsf{I}_{s^6} \oplus b_1 \cdot b_2) \cdot (\sigma^{-1} \oplus \sigma^{-1})] \uparrow_{\oplus}^{s^6} \\ &= [{}_2{>\!\!\bullet}_{s^6} \cdot \sigma t_2(\sigma^{-1} \oplus \sigma^{-1})(\mathsf{I}_{s^6} \oplus (\sigma b_1 \sigma^{-1}) \cdot (\sigma b_2 \sigma^{-1}))] \uparrow_{\oplus}^{s^6} \end{aligned}$$

Notice that

$$\begin{aligned} &\sigma b_1 \sigma^{-1} = \mathsf{I}_{s^3} \otimes \bar{b}_1 \text{ where } \bar{b}_1 := \mathsf{I}_s \otimes (\mathsf{I}_s \otimes \mathcal{R}^s)(mult \otimes \mathsf{I}_s) \\ &\sigma b_2 \sigma^{-1} = \mathsf{I}_{s^3} \otimes \bar{b}_2 \text{ where } \bar{b}_2 := \mathsf{I}_{s^2} \otimes suc \\ &\sigma t_2(\sigma^{-1} \oplus \sigma^{-1}) = (\mathsf{I}_{s^3} \otimes \bar{t}_2)\delta_{s^3;s^3,s^3} \text{ where} \\ &\bar{t}_2 := (\mathcal{R}^s \otimes \mathsf{I}_s \otimes \mathcal{R}^s) \cdot {}^s\mathsf{X}^{s^4} \cdot (\mathsf{I}_{s^3} \otimes leq \cdot {}^1_1\mathsf{X}) \cdot \delta_{s^3;1,1} \end{aligned}$$

(In the proof of the last equality the distributivity identity D10 is used.) As ${}_2{>\!\!\bullet}_{s^6} = \rho_{s^3;s^3,s^3} \cdot (\mathsf{I}_{s^3} \otimes {}_2{>\!\!\bullet}_{s^3})$, we get

$$\begin{aligned} \sigma E_2 \sigma^{-1} &= (\rho_{s^3;s^3,s^3} \cdot (\mathsf{I}_{s^3} \otimes ({}_2{>\!\!\bullet}_{s^3} \cdot \bar{t}_2 \cdot (\mathsf{I}_{s^3} \oplus \bar{b}_1 \cdot \bar{b}_2))) \cdot \delta_{s^3;s^3,s^3}) \uparrow_{\oplus}^{s^6} \\ &= \mathsf{I}_{s^3} \otimes E_2' \end{aligned}$$

where $E_2' := ({}_2{>\!\!\bullet}_{s^3} \cdot \bar{t}_2 \cdot (\mathsf{I}_{s^3} \oplus \bar{b}_1 \cdot \bar{b}_2)) \uparrow_{\oplus}^{s^3}$ (the last equality follows by Mix1). Finally, with Mix2 one gets

$$E_2' = (\wedge^s \otimes \mathsf{I}_{s^2}) \cdot (\mathsf{I}_s \otimes E_2' \cdot (\vee^s \otimes \mathsf{I}_{s^2}))$$

In a similar way one may transform the first loop to emphasize its independence of the variables z_2, y_2. Denote

$$E_1 = [{}_2\!\succ\!\bullet_{s^6} \cdot t_1 \cdot (\mathsf{I}_{s^6} \oplus a_1 \cdot a_2)] \uparrow_{\oplus}^{s^6} \quad \text{and} \quad \tau = \mathsf{I}_{s^3} \otimes {}^s\mathsf{X}^s \otimes \mathsf{I}_s$$

Then $\tau E_1 \tau^{-1} = E_1' \otimes \mathsf{I}_{s^2}$, where

$$\begin{array}{lcl} E_1' & := & [{}_2\!\succ\!\bullet_{s^4} \cdot \bar{t}_1 \cdot (\mathsf{I}_{s^4} \oplus \bar{a}_1 \cdot \bar{a}_2)] \uparrow_{\oplus}^{s^4} \\ \bar{t}_1 & := & (\wedge^{s^2} \otimes \mathsf{I}_s \otimes \wedge^s) \cdot ({}^{s^5}\mathsf{X}^s \otimes \mathsf{I}_s) \cdot ((\mathsf{I}_s \otimes dif) \cdot leq \otimes \mathsf{I}_{s^4}) \cdot \delta_{1,1;s^4} \\ \bar{a}_1 & := & \mathsf{I}_{s^2} \otimes (\mathsf{I}_s \otimes \wedge^s) \cdot (mult \otimes \mathsf{I}_s) \\ \bar{a}_2 & := & \mathsf{I}_{s^3} \otimes pred \end{array}$$

Both variables n, k are read-only used in the loop modeled by E_1. Using Mix2 one gets

$$E_1' = (\wedge^{s^2} \otimes \mathsf{I}_{s^2}) \cdot (\mathsf{I}_{s^2} \otimes E_1' \cdot (\vee^{s^2} \otimes \mathsf{I}_{s^2}))$$

All these computations and the multiplicative version of axiom B5 (i.e., $(f \otimes \mathsf{I}_c) \cdot (\mathsf{I}_b \otimes g) = f \otimes g$, for $f : a \to b,\ g : c \to d$) show that the loops may be done in parallel. Collecting the results and keeping only the inputs and the outputs which actually interest us (i.e., the inputs n, k and the outputs z_1) we get $(\mathsf{I}_{s^2} \otimes \top_{s^4}) \cdot BIN1 \cdot (\vee^{s^2} \otimes \mathsf{I}_s \otimes \vee^{s^3}) = BIN2$, where the derived program $BIN2$ is given by the expression

$$\begin{array}{rcl} BIN2 & = & \wedge^{s^2} \cdot ((\mathsf{I}_{s^2} \otimes one \otimes \mathsf{I}_s) \cdot E_1' \cdot (\vee^{s^2} \otimes \mathsf{I}_s \otimes \vee^s) \\ & & \qquad \otimes (\mathsf{I}_s \otimes one \otimes one) \cdot E_2' \cdot (\vee^s \otimes \mathsf{I}_s \otimes \vee^s)) \\ & & \cdot\, div \end{array}$$

In this final version the program contains two independent loops which may be done in parallel followed by a final div operation, which requires the exits (synchronization) of both loops. The copar degree of $BIN2$ is given by $\Pi\Sigma\Pi$, hence it is 3.

♡ Notice that the multiplicative network algebra operators have not been heavily used here. For instance, $\uparrow_{\otimes}$ has not been used. ◇

Exercises and Problems (sec. 12.4)

1. Apply the described technique to formalize various existing methods used to optimize the code. (**)
2. Study the hierarchy introduced by the alternating additive-multiplicative layers (the copar degree) on mixed flownomials. Is it strict? (**)

12.5 Duality: III. Space–time

We describe here a third possibility of dualizing network expressions; it corresponds to the interchange of space and time.

Formal presentation. Let us notice that, while there are natural timed models for $\mathsf{Mix_{BS}NA}$, the time aspects are not really covered by the $\mathsf{Mix_{BS}NA}$ operators. Indeed, since there exists a untimed relational model $\mathsf{MixRel}(D)$ for $\mathsf{Mix_{BS}NA}$, briefly described in the previous subsections, it follows that time is independent of the behavioural and architectural aspects of the systems. The timed extension of the $\mathsf{Mix_{BS}NA}$, denoted by $\mathsf{Mix_{BST}NA}$, describes the full range of behavioural, spatial, and temporal aspects of a system.

Space-time duality operators. We shall use the space-time duality operators $^{\Omega}$: $\mathsf{Mult_SNA} \to \mathsf{Mult_TNA}$ and $^{\Theta}$: $\mathsf{Mult_TNA} \to \mathsf{Mult_SNA}$ given by the following transformations. On interfaces, the duality operators act as idempotent monoid isomorphisms

$$
\begin{array}{rclcrcl}
(a \otimes b)^{\Omega} & = & a^{\Omega} \frown b^{\Omega} & \qquad & (a^{\Omega})^{\Theta} & = & a \\
(a \frown b)^{\Theta} & = & a^{\Theta} \otimes b^{\Theta} & & (a^{\Theta})^{\Omega} & = & a
\end{array}
$$

On network algebra operators the duality operators act as follows

$$
\begin{array}{rclcrcl}
(f \otimes g)^{\Omega} & = & f^{\Omega} \frown g^{\Omega} & \qquad & (f \frown g)^{\Theta} & = & f^{\Theta} \otimes g^{\Theta} \\
(f \cdot g)^{\Omega} & = & f^{\Omega} \cdot g^{\Omega} & & (f \cdot g)^{\Theta} & = & f^{\Theta} \cdot g^{\Theta} \\
(f \uparrow^{c}_{\otimes})^{\Omega} & = & (f^{\Omega}) \uparrow^{c^{\Omega}}_{\frown} & & (f \uparrow^{c}_{\frown})^{\Theta} & = & (f^{\Theta}) \uparrow^{c^{\Theta}}_{\otimes} \\
(\mathsf{I}_a)^{\Omega} & = & \mathsf{I}_{a^{\Omega}} & & (\mathsf{I}_a)^{\Theta} & = & \mathsf{I}_{a^{\Theta}} \\
({}^{a}\mathsf{X}^{b})^{\Omega} & = & {}^{a^{\Omega}}_{b^{\Omega}}\mathsf{X} & & ({}^{a}_{b}\mathsf{X})^{\Theta} & = & {}^{a^{\Theta}}\mathsf{X}^{b^{\Theta}} \\
(\curlywedge^{a}_{k})^{\Omega} & = & {}_{a^{\Omega}}{\prec}_{k} & & ({}_{a}{\prec}_{k})^{\Theta} & = & \curlywedge^{a^{\Theta}}_{k} \\
(\curlyvee^{k}_{a})^{\Omega} & = & {}_{k}{\succ}_{a^{\Omega}} & & ({}_{k}{\succ}_{a})^{\Theta} & = & \curlyvee^{k}_{a^{\Theta}}
\end{array}
$$

It is clear that the duality operators are mutually converse, i.e. $(f^{\Omega})^{\Theta} = f$ and $(f^{\Theta})^{\Omega} = f$.

The duality operators are extended to the mixed setting $^{\Omega}$: $\mathsf{Mix_{BS}NA} \to \mathsf{Mix_{BT}NA}$ and $^{\Theta}$: $\mathsf{Mix_{BT}NA} \to \mathsf{Mix_{BS}NA}$ by adding the behavioural dimension. They leave the additive operators unchanged. In addition,

$$
(\delta_{a;b,c})^{\Omega} = \mathsf{d}_{a^{\Omega};b^{\Omega},c^{\Omega}} \qquad\qquad (\mathsf{d}_{a;b,c})^{\Theta} = \delta_{a^{\Theta};b^{\Theta},c^{\Theta}}
$$

Finally, it is naturally extended as an internal transformation in the fully mixed $\mathsf{Mix_{BST}NA}$ setting.

Space-time duality thesis. An interesting space-time duality thesis is described here.

ST-Dual thesis. As we said, in this section we briefly explore the following

> *Space-time duality thesis*
> The world of concurrent, real-time, object-oriented programming may be modeled by $\mathsf{Mix}_{\mathsf{BST}}\mathsf{NA}$ and it is invariant with respect to space-time duality.

The term "concurrent" is used in a generic form here, covering both parallel and piped processes.

The restricted version relating $\mathsf{Mix}_{\mathsf{BS}}\mathsf{NA}$ and $\mathsf{Mix}_{\mathsf{BT}}\mathsf{NA}$ connects the classical imperative programming with pure OO programming. It claims that by interchange of space and time a meaningful concept in classical imperative programming would lead to a corresponding meaningful concept in the pure object-oriented programming world and conversely.

Its status. At the moment this is just an interesting *working hypothesis*. ST-Dual thesis is to be studied in relation with the general computing principles, as well as with respect to the current achievements in concurrent, real-time, object-oriented programming. While many situations from the latter world may easily invalidate the thesis, along the way one may find interesting and nontrivial extensions to solve the mismatch, still keeping the ST-Dual thesis as a valid option. In other words, ST-Dual thesis is considered as a general guiding law. It is not restricted to a current, fixed concurrent object-oriented programming setting, e.g., to Java programming.

Space-time duality picture. Most of the comments on space-time duality will use the picture in Fig. 12.2 as a basic starting point. The picture uses a rectangle where *space-time duality is figured as the symmetry with respect to the top-left/bottom-right diagonal.* The space extent is along the horizontal lines, while the time extent is along the vertical dimension.

The top-right half of the picture is the world of *imperative programs*, IMP-world. This is a clear and well-established world, hence it will have a rather strong and fixed position in all reasonings related to ST-Dual thesis. This part of the figure also contains a sketch of the programs of this world, both at the low level of Turing machines, and at the higher level used in current programming languages. The development of ST-Dual machinery is accompanied with a presentation of certain pieces of this IMP-world. After a careful reconsideration, these notions will be carried over the diagonal to the OO-world. To keep the discussion simpler, no special features of imperative programs will be considered (e.g., modules, recursive procedures, etc.).

The symmetric bottom-left half of the picture is the world of *pure object-oriented programs*, POO-world. This "pure object-oriented" programming style is one where all computations are made by message invocation only. While SMALLTALK may be a possible candidate for this, actually this world is explicitly kept open in order to accommodate the ST-Dual thesis. Since our top-right half of the

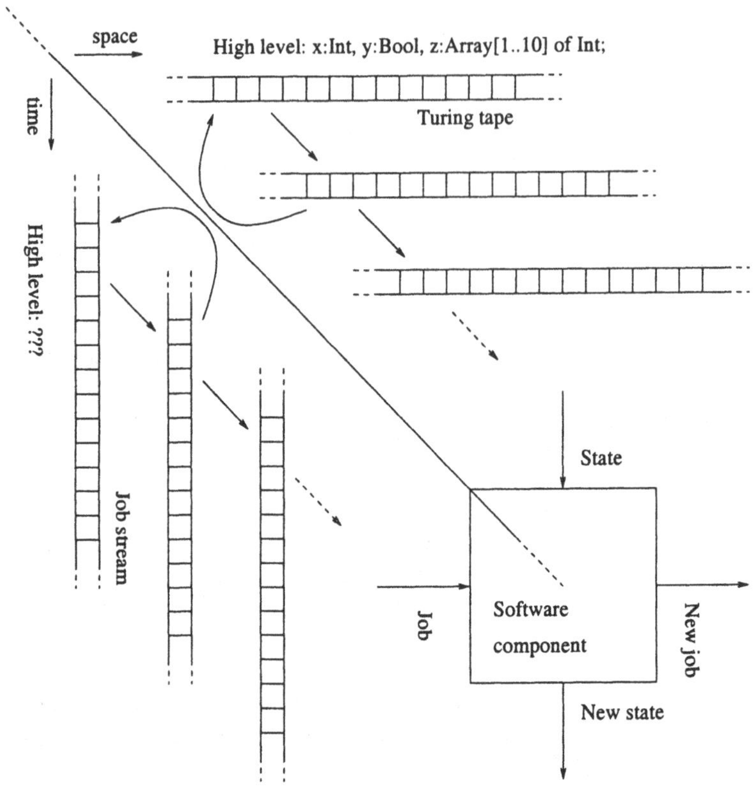

Fig. 12.2. Space-time duality picture

picture is rather strong and fixed we have to consider the bottom-left half as weak and flexible in order to have room for appropriate changes. Step-by-step this POO-world will become clearer and clearer by importing simple facts from the IMP-world via ST-Dual machinery.

Finally, the bottom-right part of the picture tries to find the place of a software component in the full concurrent, real-time, OO-programming world. It is illustrated by a rectangle which intersects both worlds. A component has both a state and a job interface. It does the specific task; after that, both the state and the job class may change. The main novelty may be that not only the state is changed due to a job requirement, but also the job is changed after this process.

Facts supporting ST-Dual thesis. In this subsection we present a few facts supporting the ST-Dual thesis. On the way, we will describe the ST-Dual machinery; it uses simple facts about imperative programs and carries them over the diagonal; interesting facts on interacting systems modeling OO-programs are obtained so far.

Memory tape and job stream. A basic fact to start with is Turing *memory tape*. Turing tape, as is drawn in the picture, is a spatial memory model. It is a linear, discrete, and infinite collection of 0/1 bits.

The dual notion corresponding to memory tape is called *job stream*. By ST-Dual machinery, the above properties show a job stream is to be considered as an infinite, linear collection of P/A (Present/Absent) bits. Job streams are used for the interaction interfaces of the interacting systems.

The memory state of an imperative program consists of various memory cells that are fully accessible to the program at a given time. However, a statement use a finite part of them only; the remaining cells are left unchanged.

A job stream consists of a sequence of actions the object has to process. The object duty is to process some jobs and to report at its output interface the result. A finite part of the job stream has to be used by an agent. The remaining jobs in the stream are left unchanged.

State-transformer vs. job-transformer devices. We make here one more step to unravel the SD-Dual machinery. Let us consider the simple acyclic straight-line programs. They are *simple sequences of statements seen as state-transformers* (no tests, no loops). The dual of a program will be called (interacting) *system*; the dual of a statement will be called *schedule.*

In our simple case here, a program is just a finite sequence of statements which have to be executed at different time slots, one after the other. Each statement has a finite time extent.

By the ST-Dual machinery an interacting system has to be split in pieces corresponding to program statements; we just introduced a name for them: "schedules." Then, an interacting system, at this simple setting, is a (finite) sequence of schedules which are to be executed at different space positions (say by different agents), with a directed linear causal order; that is, the output of one schedule is the input of the next one. Moreover, to be in accord with the IMP-world hypothesis, each schedule has a finite space extent. (In other words, the schedule is supposed to take an input job stream request and to transform it into an output job stream, according to the schedule rule and using a fixed amount of spatial resources.)

This structural way of considering interacting systems seems to be new. It will play a fundamental role in structuring the (P)OO-world.

We want to have a simple, natural semantics for the whole model. Relational semantics is maybe the simplest approach. In the view of this semantics, a program is just a state-transforming relation. Given a current memory state, a statement produces a new memory state. Usually the programs are deterministic, hence at most one new state is produced.

The basic program construction is *sequential composition.* This is so fundamental that a special symbol denoting this operation, nowadays ";", has occurred quite late in programming languages. Reading the program from top to bottom just means to go along with this operation. (Remember that we have no tests, no loops here.)

An important additional fact is that the statements are *generic*, i.e., they act in a uniform, parametric way for a whole class of similar inputs. Then, a statement is seen as a relation between the input and the output states.

What does ST-Dual machinery get out from this IMP-world situation? Interacting systems, in general, and schedules, in particular, are to be seen as *job-transformer relations.* Each interacting system has an input job interface and an output job interface. According to the situation within IMP-world, the *job streams propagate in the system in a directed, causal way,* from one schedule to another. The atom of interaction has to be split, as Abramsky inspiringly pointed out. ST-Dual machinery says that interacting systems should be created in a structural way from schedules in the same way programs are created from statements.

One important fact may be stated now: it is *not* the *inter*action of the agents of primary interest. This will come later. Their sequential composite as a model from cause to effect is the fundamental operation we have to focus on.

Finally, let us now try to gather together job streams in the same way memory instances were collected in program states. The dual notion corresponding to memory state is job *class*. At the current setting a class is just a collection of similar job streams. Then, a schedule acts in a *generic, parametric way* on the job streams of a class and transform them in job streams corresponding to the output class.

Clock vs. balance. The imperative programs are based on global states and global clock. A statement may consist of various small actions that are to be done in order to transform the state into a new one. The global clock assumption means that all the small statements have to be finished in order the statement be finished. If the pieces of the statement are done in parallel, then it is useful to have all the sub-statements of appropriate time length in order to shorten the clock step, consequently, to shorten the overall time running.

The dual concept of the global clock is called *overall balance.* We suggest to use balance as dual to clock, the balance being a device to indicate a certain size of the objects, in the same way the clock is used to indicate the time. The agent does its work on the given input job stream and reports the resulting job stream at its output interface. An action is implemented if all the sub-jobs that have to be done along with this transformation are implemented. If these actions are considered separate, then in order to minimize the resources allocated to the agent one has to use actions of appropriate space requirements. Overall small balance is a good goal for the system schedules, in the same way a global small clock is good for program statements.

Assignments. An assignment statement reads a finite number of memory cells, computes something, and writes the result at the required memory cell(s).

In the dual POO-world one has to use a finite part of the input job stream, so that the object acts when a (finite) collection of job requirements is accumulated. (This clearly has no meaning in a *sequential* OO-world; indeed, the waiting for a new job request in a sequential OO-case will result in a system failure; hence more execution threads are necessary to model this situation.) It does the specific job and reports the result. The time on job streams is local, hence there is no relationship between the time information of the input and the output jobs; hence the current process may wait for as many jobs as he wants before spooling the 1st output job.

One more comment on agent's invocation. The suggested natural extension of the OO view is to consider a method be activated by a collection of method invocations, not by a unique call, as usual. If the agent has enough memory space (which is not a priori possible!), then it may simulate this as follows: it may store the invocation calls in its memory till all the relevant information has been obtained; thereafter it may proceed to satisfy the input request. However, an "atomic read" of the relevant part of the job class is better, in the same way an "atomic read" of the memory state is better than reading memory bits; the former avoids a complicated analysis of piped job required processes, in the same way the latter avoids a complicate analysis of parallel memory access processes.

Memory and job: high level structure. In the previous analysis we have used a very low level model for both the memory space and the job stream. Current imperative programs have more complicated higher spatial structure on memory states. The dual result will be that the homogeneous and atomic view of time will be somehow relaxed, each process having his own type of timed structure on the job streams.

A high level structure on data may be used for imperative programs, e.g., four bit numbers such as 1011 may be used. Similar temporal Arabic-like structures may be used on jobs streams. The corresponding streams are four-time-steps actions PAPP saying that during this time interval the request for the specific action was present/absent/present/present.

Tests. The semantics of tests is well-known: depending on the current memory values of certain variables, the computation run may continue in one or other point of the program; GOTO statements are used here. To this end, program states are labeled.

The dual version is natural, too. A schedule may test for (a part of) its job stream; depending on the current value of this input job request it may decide to forward the job stream to be further processed in one or another point. To this end, GOTO statements and dual labels pointing to job classes are used. This job test appears as a sort of classification: depending on certain information, an input job stream is classified as requiring one sort of further processing, or another.

Programming vs. planning. Programming is the basic activity in the IMP-world. The dual activity may be fairly called *planning.*

An imperative program may be seen as a finite representation of a certain collection of operating sequences people do to get the intended output. These sequences may vary from an input to another. A label in the program, denoting a program state, gathers together memory instances that (1) are reached by the program *at different times,* (2) are similar, and (3) require similar transformations coded by the current program statement. The program control gives a *cover* of these sequences in the following sense: (1) each real computation sequence is syntactically possible, but (2) there may be syntactically possible sequences that are run-time impossible.

The planning activity has the dual properties. Systems of interacting objects are seen as job transformers. The elementary job transformers are the system schedules. A sequence of job scheduling between different objects may lead to the desired output job result. For various input job streams the scheduling sequences may be different. The collection of all these interaction patterns may be infinite, hence they are not too useful for our finite machines in these forms. For this reason we have to collect together job streams which (1) are located *at various places,* (2) are similar, and (3) their processing are similar and may be coded by the current schedule as a collection of methods that do the specific job transformation. In a few words, *the dual concept corresponding to imperative programming is planning (scheduling).* In the vein of this slogan, the design of a schedule structure for the job delegation that obey the input–output job constraint is the main task in this world.

Program vs. schedule correctness. Programs are used as repeated tasks for input data fulfilling certain constraints. The functionality of the program is given by its input–output state relation. A program is correct if it terminates and respects the desired input–output specification.

Systems of interacting objects may be tested for various scenarios of job streams fulfilling certain constrains. It seems natural to look at the amount of work the system has done from this job stream by considering the reported output job stream. The effect on states of the test scenario may be of collateral interest here. The termination property here means that a finite amount of objects/spatial resources are to be used in order to do the input-ooutput job transformation.

Assertions vs. fairness. This ST-Dual thesis throws a new light on the fairness problem, as well. The current approaches either ignore the problem, or propose rather ad-hoc solutions. The ST-Dual machinery says that the problem is important, but the solution is already in our hands: it is the classical program verification via inductive assertions, but presented for the dual world. Fairness is reduced to the use of inductive assertions on job classes: provide assertions (invariants) for job classes at various points of the system and verify that the system schedules satisfy them.

More expressiveness in describing interaction interfaces. A higher timed structure on job classes allows us to express easily certain properties. Usually, as in the ROOM (Real-Time Object-Oriented Method) model [GŞB98], an input, say f, is kept fixed up to the end of a computation step (the signal is enabled during this period). In our current notation this means $f = P \overset{m}{.\,.\,.} P$. If you want to express the fact that 2 steps f is active, then 3 steps g, then 1 step f, this will be easily described as $f =$ PPAAAP and $g =$ AAPPPA. The agent has to check both fields f and g each time before reacting. The situation here is dual to the multiplication of two numbers, say, where one has to check simultaneously the fields of both arguments to get the result.

Parallel vs. piped processes. By the ST-Dual machinery parallel processes are transformed into piped processes and vice-versa. This allows us to use the techniques for parallelizing programs in order to get techniques for piping the schedules. This way, one may decrease the overall required space.

State vs. class labels. State labels are commonly used to describe control flow using GOTO statements. They provide the possibility to make a (rather free) jump from one statement to another. In the same vein, using the ST-Dual machinery one may introduce and use class labels. Then, after processing its input job stream a schedule may pass the resulting job stream to a (fairly arbitrary) new schedule according to the interaction system structure. It may be the case that mobility, the important notion people in Computer Science are now focusing on, is mainly reduced to this feature of interaction systems.

We hope the way to apply the ST-Dual machinery has become clear by now. More such analogies may be easily obtained.

12.6 Object-oriented programs/software components

The full $\mathsf{Mix_{BST}NA}$ setting is a good candidate for a model for software systems based on real-time, concurrent, object-oriented systems. It requires information about the interfaces corresponding to all the involved properties: behaviour, space, and time.

The notation $(a|b|c)$ may be used to describe the situation where a specifies a type for the required job stream, b specifies a type for the control properties, and c a type for the spatial interface of a system component. When the a-part is missing no special job stream is required. The system works from its own resources; it is closed, not open to the environment. Dually, when the c-part is absent, the global state view is understood and no special spatial resources are necessary for a component to be executed. The b-part describe the control features, indicating where a component is active or not.

Vertical composition along the component states is like in imperative programs. Horizontal composition along the process interaction interfaces (job streams) describes the interaction of the components as in an OO interacting system.

Vertical feedback models the repetition of component invocation in time generating classes of similar memory instances – these are the usual "states" (one arrives at the same point of a program, but at a different time). Horizontal feedback models the repetition of components invocation in space generating classes of similar job requirements, – these are usual OO programming "classes" (one arrives at the same point of a program, but at a different space, hence a new object of the class is considered).

For spatial parallelism, a "true parallel" data-flow-like model comes along with the horizontal juxtaposition operation. For temporal parallelism, a "true piped" job-flow-like model comes along with the vertical juxtaposition operation.

Multiplicative operators distribute over the additive ones, while space and time are dual features. These may be used to keep the algebraic laws at a reasonable size.

Comments, problems, bibliographic remarks

Notes. The mixed network algebra model was introduced in 1996 in [Şte97a, Şte97b, Şte98b].

Bicartesian categories are studied in [LS89]. The usefulness of distributive categories for computer science was singled out by Walters [Wal92a, Wal92b, SVW94]; for mathematical applications, see [Coc93]. The interesting old book [AN60] contains the roots of Boolean logic and Galois connection. Relational systems are presented in [Coh81]. A survey on algebraic specification techniques may be found in [Wir90]

Various results on mixing algebraic and coalgebraic methods do exist, especially aiming to describe object-oriented programming, e.g., [Hag87], [GD94], [Rei95], [Jac96], [GM97] and [JR97]; see also [Mes92]. For an introduction to object-oriented programing one may consult [Mey97].

The axiomatization for mixed relations is from [Şte97a, GLŞ00]. Preliminary versions of the category of plans were described in [ŞK97, KŞ98]. In [Şte97b], a small kernel programming language FEST based on MixNA is proposed. An application of the MixNA formalism to ROOM (Real Time Object Oriented Programming) model is described in [GŞB98].

The running example in Section 12.4 is from [Şte98b]. The program was taken from [MP92]; in this book many examples of parallel programs are presented, together with temporal logic based verifications. Another interesting temporal logic based formalism is the duration calculus; see, e.g., [HZ97]. The classical book by Hecht [Hec77] describes flow analysis based algorithms for code optimization. A recent method for parallelizing programs is described in [GP95].

Space-time duality thesis has occurred in the "Dynamic Data-flow Networks" project based on work of M. Broy, R. Grosu, and Gh. Ştefănescu; it was formally stated in February 1999, see [BGŞ99]. (Preliminary attempts were described in [Şte98a, GŞ98].)

Related calculi, closing remarks

> The pure and simple truth is rarely pure and never simple.
>
> Oscar Wilde

In this last chapter we briefly present some more references and comments aiming to compare the NA approach to other algebraic studies which recently occurred in the computer science literature.

Iteration theories. Iteration theories may be seen as a lifting of regular algebra from a matrix to an algebraic theory setting. The starting point of the research of Bloom, Ésik, etal. on iteration theories, as documented in [BÉ93b], was the same as ours: the aim is to get a clean algebraic theory for flowcharts. As a consequence, some parts of Chapters 5–8 of the present book overlap with similar results presented in [BÉ93b].

However, the approaches are somehow orthogonal. In [BÉ93b] the authors present a deep study of iteration, but the acyclic algebraic structure is supposed to be at least an algebraic theory. This hypothesis prohibits the possibility to obtain an algebraic structure for flowchart/network themselves. Our approach is based on a monoidal category structure on acyclic flowcharts/networks, hence a broader range of applications is obtained. Another key difference is the emphasis we are putting on abstract networks, with the pleasant result of having a uniform calculus for various types of networks.

As an auxiliary side-effect, some basic algebraic structures used in [BÉ93b], but introduced there just for technical reasons, have a simple and clear meaning in the current enlarged network algebra setting; e.g., the key Conway theory structure of [BÉ93b] is nothing other than a BNA over an algebraic theory. Last, but not least, the monoidal setting for acyclic networks makes more feasible the study of the mixed models than starting directly with mixed algebraic theories.

Functional programming, rewriting, shared graphs. Functional programming is a well-established field in computer science. It is based on λ-calculus; term rewriting is used as a basic tool for interpreting functional programs. Recent studies have extended the rewriting techniques from term trees to shared graphs; see, e.g., [AK96]. In network algebra terms, this requires the passing from algebraic theories to enriched symocats. Extensions to cyclic shared graphs were presented by M. Hasegawa [Has97a, Has97b], A. Jeffrey [Jef97], H. Miyoshi [Miy98], or A. Corradini and F. Gadducci [CG99]; the resulting algebraic setting is that of $b\delta$-symocats with feedback – usually, these algebras are presented using "trace monoidal categories" name.

Circuits. Various algebraic calculi are used to handle circuits, e.g., Ruby, CIRCAL, etc.; see [Möl98b] for a recent approach, partially using the network algebra machinery. In [Mol88] a sort of 2-dimensional extension of the symocat setting is described to cope with "planar" circuits used for VLSI design.

Relation algebra, allegories. Many studies dedicated to algebraic presentation of relations do exist. They use either the classical algebraic approach that started in the nineteenth century with Pierce and Schröder (see e.g., [Rig48], [TG87], [SS93]; see also [BKS97]), or the modern categorical approach (see, e.g., [Kaw73, Kaw90], [FS90], [BÉŞ95]).

A particularly interesting study is [SV97]. It is based on Q-operations (introduced by Jónsson, 1991) which are defined as follows: for a matrix of relations $r = (r_{i,j})_{i,j\in[n]}$ and $k,l \in [n]$ define $Q_n^{kl}(r)$ by: $(s,t) \in Q_n^{kl}(r)$ iff there are $u_1,\ldots,u_n$ such that $s = u_k, t = u_l$, and $(u_i,u_j) \in r_{i,j}$ for all $i,j \in [n]$. (These Q-operations are nothing other than slight variations of the multiplicative feedback.) Certain multiplicative networks naturally appear in this setting and an axiomatization of representable Q-algebras is presented. The precise relationship between NA and Q-algebras deserves to be clarified.

Action calculi, π-calculus. The adding of parallelism to functional programs was challenging for a long time. An interesting proposal is the π-calculus of Milner, Parrow, and Walker [MPW92]. Milner also proposed action calculi as a more abstract mechanism for interaction; see [Mil94] and [Mif96] for some results using (enriched) BNA structures.

Tiles. In the last few years Ugo Montanari and his students have proposed a very interesting "tile model" as a unifying tool for concurrent programming, see, e.g., [GM99] and [Bru99]. This may be the closest model to the mixed network algebra model presented in Chapter 12. Tiles come with both a vertical composition modeling the usual sequential composition of programs and a horizontal composition for handling program interaction. As a further example, see [BM97], where zero-safe nets are studied; these nets are extensions of Petri nets to a setting where a synchronous decomposition of transitions in small steps is possible.

Chu spaces. In some recent papers [Pra92, Pra94], V. Pratt describes a time-information duality, mainly connected with the Chu space model of distributed computation; see [Gup94] for a presentation of Chu spaces. Pratt describes a duality between (acyclic) automata and certain "schedulers." This duality is somehow a combination of the NA space-time duality and the duality between additive and multiplicative NA models (flowchart and circuits), but the precise relationship between them remains to be clarified.

Other models. Asynchronous automata ([DR95], [Kri00], [Mor00]) are close to the model of parallel programs described in Section 12.4. A Kleene like theorem for asynchronous automata was presented by Zielonka [Zie87]; it may be interesting to get a algebraic version of the Zielonka theorem, similar to the algebraic version of the classical Kleene theorem we have presented in Chapter 8.

Appendices

It is impossible to enjoy idling thoroughly unless one has plenty of work to do.

Jerome K. Jerome

This appendix contains some more technical proofs.

Appendix A: Equivalent BNA presentation (LR-flow)

In this appendix we spell out the definition of LR-flows and prove that they are equivalent to BNAs.

In a BNA composition may be written in terms of juxtaposition, feedback, and right composition with (abstract) bijections of type X as follows:

$$f \cdot g = [(f \star g) \cdot {}^n\mathsf{X}^p] \uparrow^n \qquad \text{for } f : m \to n \text{ and } g : n \to p$$

Consequently, it may be useful to have a presentation of the BNA structure using these operations only. However, the resulting axiom system is rather complicated. Due to this reason, we present here an intermediary characterization of the BNA structure with one more operation: left composition with bijections.

Definition V.1. (LR-flow over $\mathbb{B}\mathrm{i}$) Let $(M, \star, \epsilon)$ be a monoid. We say a structure $T = \{T(m,n)\}_{m,n \in M}$ endowed with:

- Identities $\mathsf{I}_m \in T(m,m)$;
- Juxtaposition $\star : T(m,n) \times T(p,q) \to T(m \star p, n \star q)$;
- Feedback $\uparrow^p : T(m \star p, n \star p) \to T(m,n)$;
- Right-Hand-Side Composition with Morphisms in $\mathbb{B}\mathrm{i}$:
  for $f \in T(m,n)$, $n = n_1 \star \ldots \star n_k$, and $\phi \in \mathbb{B}\mathrm{i}(k,k)$

$$f \triangleright_{n_1,\ldots,n_k} \phi \in T(m,\ n_{\phi^{-1}(1)} \star \ldots \star n_{\phi^{-1}(k)})$$

- Left-Hand-Side Composition with Morphisms in $\mathbb{B}\mathrm{i}$:
  for $f \in T(m,n)$, $m = m_1 \star \ldots \star m_k$, and $\phi \in \mathbb{B}\mathrm{i}(k,k)$

$$\phi \triangleleft_{m_1,\ldots,m_k} f \in T(m_{\phi(1)} \star \ldots \star m_{\phi(k)},\ n)$$

is a LR-*flow over* $\mathbb{B}\mathrm{i}$ (left-right flow over $\mathbb{B}\mathrm{i}$) if it fulfils the axioms in Table V.1.

♠ We also use the following equivalent form of LR9:
LB9′ $f \star g = [{}^1\mathsf{X}^1 \triangleleft_{n,m} (g \star f)] \triangleright_{q,p} {}^1\mathsf{X}^1$, for $f : m \to p,\ g : n \to q$ ♣

Table V.1. Axioms for LR-flow over $\mathbb{B}\mathrm{i}$ (f, g, h denote morphisms in T; ϕ, ψ morphisms in $\mathbb{B}\mathrm{i}$; and P, Q denote interface decompositions and $\phi(P), \psi^{-1}(P)$ their images via the underlined relations).

LB1	$f \star (g \star h) = (f \star g) \star h$
LB2	$\mathsf{I}_\epsilon \star f = f = f \star \mathsf{I}_\epsilon$
LB3a	$(f \rhd_P \phi) \rhd_{\phi(P)} \psi = f \rhd_P (\phi \cdot \psi)$
LB3b	$\phi \lhd_{\psi^{-1}(P)} (\psi \lhd_P f) = (\phi \cdot \psi) \lhd_P f$
LB3c	$\phi \lhd_P (f \rhd_Q \psi) = (\phi \lhd_P f) \rhd_Q \psi$
LB4a	$\mathsf{I}_k \lhd_{m_1,\ldots,m_k} f = f = f \rhd_{n_1,\ldots,n_k} \mathsf{I}_k \qquad (k \geq 0)$
LB4b	$\mathsf{I}_{n_1\star\ldots\star n_k} \rhd_{n_1,\ldots,n_k} \phi = \phi \lhd_{n_{\phi^{-1}(1)},\ldots,n_{\phi^{-1}(k)}} \mathsf{I}_{n_{\phi^{-1}(1)}\star\ldots\star n_{\phi^{-1}(k)}}$
LB5a	$(f \star g) \rhd_{P,Q} (\phi \star \psi) = (f \rhd_P \phi) \star (g \rhd_Q \psi)$
LB5b	$(\phi \star \psi) \lhd_{P,Q} (f \star g) = (\phi \lhd_P f) \star (\psi \lhd_Q g)$
LB6	$\mathsf{I}_{a\star b} = \mathsf{I}_a \star \mathsf{I}_b$
LB9	$(f \star g) \rhd_{m',n'} {}^1\mathsf{X}^1 = {}^1\mathsf{X}^1 \lhd_{n,m} (g \star f)$ for $f : m \to m',\ g : n \to n'$
LR1a	$\phi \lhd_P f \uparrow^p = [(\phi \star \mathsf{I}_1) \lhd_{P,p} f] \uparrow^p$
LR1b	$f \uparrow^p \rhd_P \phi = [f \rhd_{P,p} (\phi \star \mathsf{I}_1)] \uparrow^p$
LR1c	$f \rhd_P (\phi \uparrow^1) = [(f \star \mathsf{I}_p) \rhd_{P,p} \phi] \uparrow^p$ when $P = (n_1, \ldots, n_k)$ and $p = n_{\phi^{-1}(k+1)}$ when $\phi^{-1}(k+1) \leq k$
LR1d	$(\phi \uparrow^1) \lhd_P f = [\phi \lhd_{P,p} (f \star \mathsf{I}_p)] \uparrow^p$ when $P = (n_1, \ldots, n_k)$ and $p = n_{\phi(k+1)}$ if $\phi(k+1) \leq k$
LR2	$f \star g \uparrow^p = (f \star g) \uparrow^q$
LR3	$[f \rhd_{n,p,q} (\mathsf{I}_1 \star {}^1\mathsf{X}^1)] \uparrow^{q\star p} = [(\mathsf{I}_1 \star {}^1\mathsf{X}^1) \lhd_{m,q,p} f] \uparrow^{p\star q}$ for $f : m \star q \star p \to n \star p \star q$
LR4	$f \uparrow^\epsilon = f$
LR5	$f \uparrow^b \uparrow^a = f \uparrow^{a\star b}$
R-refine	$\mathsf{I}_{..} \rhd_{n_1,\ldots,n_i^1\star\ldots\star n_i^r,\ldots,n_k} \phi = \mathsf{I}_{..} \rhd_{n_1,\ldots,n_i^1,\ldots,n_i^r,\ldots,n_k} (\phi(1)_1 \ldots \phi(i)_r \ldots \phi(k)_1)$
L-refine	Similar

A binding convention: The new composition operations $\rhd$ and $\lhd$ have the same binding power as composition, i.e., stronger then $\star$, but weaker than $\uparrow$.

Some comments on the axioms may help for a quicker parsing of Table V.1. The axioms are similar to those used in the BNA definition. They are sometimes multiplied since each composition $f \cdot g$ in the standard BNA setting now has two instances: either f is in T and g bijection, or conversely. This happens with B3, B4, B5, and R1. The side condition in LR1c has the following role: $\phi^{-1}(k+1) \leq k$ excludes the case when $\phi : k+1 \to k+1$ has $\phi(k+1) = k+1$; if this happens, p is arbitrary; if not so, then the sort of the piece of the interface of f to which $k+1$ refers is $n_{\phi^{-1}(k+1)}$. LR1d is similar. The last two axioms are the refinement rules. They show how the bijection is changed when a refinement of the interface decomposition is applied.

Now we can state the main theorem here.

Theorem V.2. *(equivalence of two presentations for BNAs)*
LR-flows and aα-flows coincide; more precisely, the axioms in Table V.1 are equivalent to the BNA axioms.

The proof is reduced to the following two lemmas.

Lemma V.3. *If* $(\{T(m,n)\}_{m,n\in M}, \star, \cdot, \uparrow, \mathsf{I}_m, {}^m\mathsf{X}^n)$ *is a BNA, then* $(\{T(m,n)\}_{m,n\in M}, \star, \triangleleft, \triangleright, \uparrow, \mathsf{I}_m)$ *is a LR-flow, where*

$$f \triangleright_{n_1,\ldots,n_k} \phi = f \cdot (\phi(1)_{n_1} \cdots \phi(k)_{n_k}) \text{ and}$$
$$\psi \triangleleft_{m_1,\ldots,m_k} f = (\psi(1)_{m_{\psi(1)}} \cdots \psi(k)_{m_{\psi(k)}}) \cdot f$$

PROOF: The new compositions $\triangleleft$ and $\triangleright$ are particular instances of the usual BNA composition. Therefore, the LR-flow axioms in Table V.1 of type LB* or LR* are particular instances of the corresponding B* or R* axioms in Table 1.1.

There are two more axioms: R-refine and L-refine. The validity of R-refine follows from the "unique extension property" in Theorem 2.3 as follows. (1) For the left-hand-side term in R-refine, take k different sorts $s_1, \ldots, s_i, \ldots, s_k$. Then, the meaning $(\phi(1)_{n_1} \cdots \phi(i)_{n_i^1\star\ldots\star n_i^r} \cdots \phi(k)_{n_k})$ of ϕ for the left-hand-side term is obtained using Theorem 2.3 with the assignment $n_1, \ldots, n_i^1 \star \ldots \star n_i^r, \ldots, n_k$ for $s_1, \ldots, s_i, \ldots, s_k$. (2) For the right-hand-side term, let $\phi' = (\phi(1)_{s_1} \cdots \phi(i)_{t_i^1\star\ldots\star t_i^r} \cdots \phi(k)_{s_k})$ be the interpretation of ϕ in $\mathbb{B}\mathrm{i}_{\{s_1,\ldots,s_k,t_i^1,\ldots,t_i^l\}}$, according to the valuation $s_1, \ldots, t_i^1 \star \ldots \star t_i^r, \ldots, s_k$ for $s_1, \ldots, s_i, \ldots, s_k$. Then, evaluate ϕ' in T, according to the assignment $n_1, \ldots, n_i^1, \ldots, n_i^r, \ldots, n_k$ for $s_1, \ldots, t_1, \ldots, t_r, \ldots, s_k$. (3) As both evaluations (1) and (2) agree on the generators $s_1, \ldots, s_k$, by Theorem 2.3 the same abstract bijection in T is obtained.

The proof for L-refine is similar. □

Lemma V.4. *If* $(\{T(m,n)\}_{m,n\in M}, \star, \triangleleft, \triangleright, \uparrow, \mathsf{I}_m)$ *is a LR-flow, then* $(\{T(m,n)\}_{m,n\in M}, \star, \cdot, \uparrow, \mathsf{I}_m, {}^m\mathsf{X}^n)$ *is a BNA, with* $\cdot$ *and* X *defined by:*

$$\text{(def)}\quad f \cdot g = [(f \star g) \triangleright_{n,p} {}^1\mathsf{X}^1] \uparrow^n \quad \text{for } m \xrightarrow{f} n \xrightarrow{g} p$$
$${}^p\mathsf{X}^q = \mathsf{I}_{p\star q} \triangleright_{p,q} {}^1\mathsf{X}^1$$

PROOF: We start the proof with a few simple, but useful observations.

(1) A "dual" definition for composition:

$$\text{(def')}\quad f \cdot g = [(21) \triangleleft_{n,m} (g \star f)] \uparrow^n \quad \text{for } f : m \to n \text{ and } g : n \to p$$

holds in T. (21) denotes the transposition $(1,2) \mapsto (2,1)$. (Similar shorthand notations are used below. Since we use bijections on no more than 9 elements, no confussion will arise.) A verification for def' is given below, which uses in turn: LB9' in $=^1$, LB3, LB4a in $=^2$, and def in $=^3$

$$\begin{aligned}
&[(21) \triangleleft_{n,m} (g \star f)] \uparrow^n \\
&\quad =^1 \quad [(21) \triangleleft_{n,m} ([(21) \triangleleft_{m,n} (f \star g)] \triangleright_{n,p} (21))] \uparrow^n \\
&\quad =^2 \quad [(f \star g) \triangleright_{n,p} (21)] \uparrow^n \\
&\quad =^3 \quad f \cdot g
\end{aligned}$$

(2) Extended LR1: LR1z' denotes the rule obtained by a repetitive application of rule LR1z, with $z \in \{a,b,c,d\}$; for example LR1a' is

$$\phi \lhd_P f \uparrow^{p_1\star\ldots\star p_r} = [(\phi \star \mathsf{I}_r) \lhd_{P,p_1,\ldots,p_r} f] \uparrow^{p_1\star\ldots\star p_r}$$

(3) Another R2: the identity

$$R2' \quad f \uparrow^p \star g = [(132) \lhd_{m,p,m'} (f \star g) \rhd_{n,p,n'} (132)] \uparrow^p \quad \text{for } f: m\star p \to n\star p, \quad g: m' \to n'$$

holds in T. A verification is described below using in turn LB9$'$ in $=^1$, LR2 in $=^2$, LR1a, LR1b for $=^3$, LB9$'$ in $=^4$, L,R-refine and LB3 in $=^5$

$$\begin{aligned} & f \uparrow^p \star g \\ &=^1 \quad (21) \lhd_{m',m} (g \star f \uparrow^p) \rhd_{n',n} (21) \\ &=^2 \quad (21) \lhd_{m',m} (g \star f) \uparrow^p \rhd_{n',n} (21) \\ &=^3 \quad [(213) \lhd_{m',m,p} (g \star f) \rhd_{n',n,p} (213)] \uparrow^p \\ &=^4 \quad [(213) \lhd_{m',m,p} [(21) \lhd_{m\star p,m'} (f \star g) \rhd_{n\star p,\ n'} (21)] \rhd_{n',n,p} (213)] \uparrow^p \\ &=^5 \quad [(213) \lhd_{m',m,p} [(312) \lhd_{m,p,m'} (f \star g) \rhd_{n,p,n'} (231)] \rhd_{n',n,p} (213)] \uparrow^p \\ &=^6 \quad [(132) \lhd_{m,p,m'} (f \star g) \rhd_{n,p,n'} (132)] \uparrow^p \end{aligned}$$

(4) Shifting bijections on feedback: let $P = (p_1, \ldots, p_k)$, $p = p_1 \star \ldots \star p_k$, $Q = (q_1 \star \ldots \star q_r)$, $q = q_1 \star \ldots \star q_r$ and $f : m\star p\star s\star t\star q \to n\star p\star s\star t\star q$. Identity LR3 implies

$$[(f \uparrow^q) \rhd_{n,P,t,s} (\mathsf{I}_1 \star \mathsf{I}_k \star {}^1\mathsf{X}^1)] \uparrow^{s\star t}\uparrow^p = [(\mathsf{I}_1 \star \mathsf{I}_k \star {}^1\mathsf{X}^1) \lhd_{m,P,s,t} (f \uparrow^q)] \uparrow^{t\star s}\uparrow^p$$

With LR1a and LR1b this equality gives

$$[f \rhd_{n,P,t,s,Q} (\mathsf{I}_1 \star \mathsf{I}_k \star {}^1\mathsf{X}^1 \star \mathsf{I}_l)] \uparrow^{p\star s\star t\star q} = [(\mathsf{I}_1 \star \mathsf{I}_k \star {}^1\mathsf{X}^1 \star \mathsf{I}_l) \lhd_{m,P,s,t,Q} f] \uparrow^{p\star t\star s\star q}$$

Since every morphism in $\mathbb{B}\mathrm{i}$ is a composite of morphisms of the type $\mathsf{I}_p \star {}^1\mathsf{X}^1 \star \mathsf{I}_q$ it follows that

LR3$'$ $[f \rhd_{n,\phi^{-1}(P)} (\mathsf{I}_1 \star \phi)] \uparrow^p = [(\mathsf{I}_1 \star \phi) \lhd_{m,P} f] \uparrow^q$ for $f \in T(m\star p, n\star q)$, $\phi \in \mathbb{B}\mathrm{i}(k,k)$ and $P = (p_1, \ldots, p_k)$, $p = p_1 \star \ldots \star p_k$, $q = p_{\phi(1)} \star \ldots \star p_{\phi(k)}$

holds in a LR-flow over $\mathbb{B}\mathrm{i}$.

(5) "$\cdot$" extends $\lhd$ and $\rhd$: a bijection $\phi \in \mathbb{B}\mathrm{i}(k,k)$ can be thought of as an element in $T(m,n)$, relative to a decomposition $m = p_1 \star \ldots \star p_k$ or $n = p_{\phi^{-1}(1)} \star \ldots \star p_{\phi^{-1}(k)}$. Namely, it is either $\mathsf{I}_m \rhd_{p_1,\ldots,p_k} \phi$ or $\phi \lhd_{p_{\phi^{-1}(1)},\ldots,p_{\phi^{-1}(k)}} \mathsf{I}_n$. (By LB4a they are equal.) This observation leads to the following (Right-Scalar-Composition, Left-Scalar-Composition) identities:

RSC $\quad f \rhd_{n_1,\ldots,n_k} \phi = f \cdot (\mathsf{I}_n \rhd_{n_1,\ldots,n_k} \phi)$, for $f \in T(m,n)$, $n = n_1 \star \ldots \star n_k$ and $\phi \in \mathbb{B}\mathrm{i}(k,k)$

LSC $\quad \phi \lhd_{n_1,\ldots,n_k} f = (\phi \lhd_{n_1,\ldots,n_k} \mathsf{I}_n) \cdot f$, for $\phi \in \mathbb{B}\mathrm{i}(k,k)$, $n = n_1 \star \ldots \star n_k$ and $f \in T(n,p)$

which are valid in T. We prove RSC. If $P = (n_1, \ldots, n_k)$, $n = n_1 \star \ldots \star n_k$, $p = n_{\phi^{-1}(1)} \star \ldots \star n_{\phi^{-1}(k)}$, and $\phi \in \mathbb{B}\mathrm{i}(k,k)$, then using in turn def in $=^1$, LB5a, refine for $=^2$, LB3 in $=^3$, refine for $=^4$, repeated LR1c for $=^5$, identity in $\mathbb{B}\mathrm{i}$ in $=^6$ one gets

$$\begin{aligned}
&f \cdot (\mathsf{I}_n \rhd_P \phi)\\
&=^1 \; [(f \star (\mathsf{I}_n \rhd_P \phi)) \rhd_{n,p} (21)] \uparrow^n\\
&=^2 \; ([(f \star \mathsf{I}_n) \rhd_{n,P} (\mathsf{I}_1 \star \phi)] \rhd_{n,\phi(P)} (2_1 1_k)) \uparrow^n\\
&=^3 \; ((f \star \mathsf{I}_n) \rhd_{n,P} [(\mathsf{I}_1 \star \phi) \cdot {}^1\mathsf{X}^k]) \uparrow^n\\
&=^4 \; ((f \star \mathsf{I}_n) \rhd_{P,P} [(\mathsf{I}_k \star \phi) \cdot {}^k\mathsf{X}^k]) \uparrow^{n_1 \star \ldots \star n_k}\\
&=^5 \; f \rhd_P [(\mathsf{I}_k \star \phi) \cdot {}^k\mathsf{X}^k] \uparrow^k\\
&=^6 \; f \rhd_P \phi
\end{aligned}$$

LSC may be proved in a similar way using (def′).

Now we check the validity of the axioms from the BNA definition. First, B1 coincides with LB1 and B2 with LB2. For B3, suppose $m \xrightarrow{f} n \xrightarrow{g} p \xrightarrow{h} q$. Then using in turn def in $=^1$, LR2 in $=^2$, LB4a, LR1b in $=^3$, one gets

$$\begin{aligned}
&f \cdot (g \cdot h)\\
&=^1 \; [(f \star [(g \star h) \rhd_{p,q} (21)] \uparrow^p) \rhd_{n,q} (21)] \uparrow^n\\
&=^2 \; [(f \star (g \star h) \rhd_{p,q} (21)) \uparrow^p \rhd_{n,q} (21)] \uparrow^n\\
&=^3 \; [(f \rhd_n (1) \star (g \star h) \rhd_{p,q} (21)) \rhd_{n,q,p} (213)] \uparrow^p \uparrow^n\\
&=^4 \; [(f \star g \star h) \rhd_{n,p,q} (231)] \uparrow^{n \star p}
\end{aligned}$$

and using def in $=^1$, R2′ in $=^2$, LB5, LR1b, LB3 in $=^3$, and LR3, LB3 in $=^4$, one gets

$$\begin{aligned}
&(f \cdot g) \cdot h\\
&=^1 \; [([(f \star g) \rhd_{n,p} (21)] \uparrow^n \star h) \rhd_{p,q} (21)] \uparrow^p\\
&=^2 \; [[(132) \lhd_{m,n,p} [(f \star g) \rhd_{n,p} (21) \star h] \rhd_{p,n,q} (132)] \uparrow^n \rhd_{p,q} (21)] \uparrow^p\\
&=^3 \; [[(132) \lhd_{m,n,p} (f \star g \star h) \rhd_{n,p,q} [(213) \cdot (132) \cdot (213)]] \uparrow^n \uparrow^p\\
&=^4 \; [(f \star g \star h) \rhd_{n,p,q} [(321) \cdot (132)] \uparrow^p \uparrow^n\\
&=^5 \; [(f \star g \star h) \rhd_{n,p,q} (231)] \uparrow^{n \star p}
\end{aligned}$$

hence axiom B3 holds in T.

For B4, suppose $f : m \to n$. Then, with LSC and LB4a one gets

$$\mathsf{I}_m \cdot f = ((1) \lhd_m \mathsf{I}_m) \cdot f = (1) \lhd_m f = f$$

and similarly one may prove that $f \cdot \mathsf{I}_n = f$. Hence, B4 holds in T.

For B5, suppose $m \xrightarrow{f} n \xrightarrow{g} p$ and $m' \xrightarrow{f'} n' \xrightarrow{g'} p'$. Then, using in turn def in $=^1$, LR2, then R2′ in $=^2$, LB5, refine in $=^3$, LB5, LB3 in $=^4$, LB9′, LB5, LB3 in $=^5$, LR3′, LB3 in $=^6$, refine in $=^7$, and def in $=^8$ one gets

$$\begin{aligned}
&(f \cdot g) \star (f' \cdot g')\\
&=^1 \; [(f \star g) \rhd_{n,p} (21)] \uparrow^n \star [(f' \star g') \rhd_{n',p'} (21)] \uparrow^{n'}\\
&=^2 \; [(132) \lhd_{m,n,m' \star n'} [(f \star g) \rhd_{n,p} (21) \star (f' \star g') \rhd_{n',p'} (21)] \rhd_{p,n,p' \star n'} (132)] \uparrow^n \uparrow^{n'}\\
&=^3 \; [[(1342) \lhd_{m,n,m',n'} (f \star g \star f' \star g') \rhd_{n,p,n',p'} (2143)] \rhd_{p,n,p',n'} (1423)] \uparrow^{n' \star n}\\
&=^4 \; [(1342) \lhd_{m,n,m',n'} (f \star g \star f' \star g') \rhd_{n,p,n',p'} (4132)] \uparrow^{n' \star n}\\
&=^5 \; [(1243) \lhd_{m,m',n,n'} (f \star f' \star g \star g') \rhd_{n,n',p,p'} (4312)] \uparrow^{n' \star n}\\
&=^6 \; [(f \star f' \star g \star g') \rhd_{n,n',p,p'} (3412)] \uparrow^{n \star n'}\\
&=^7 \; [(f \star f' \star g \star g') \rhd_{n \star n', p \star p'} (21)] \uparrow^{n \star n'}\\
&=^8 \; (f \star f') \cdot (g \star g')
\end{aligned}$$

Using LSC and RSC, one easily proves that LB9′ implies B9′. For R1, suppose $f : m' \to m$, $g : m \star p \to n \star p$ and $h : n \to n'$. Then using: def in $=^1$; R2′ in $=^2$; LR1, LB3 in $=^3$; LR3, LB3 in $=^4$; identity in 𝔹i in $=^5$; LR1c in $=^6$; refine in $=^7$; and def in $=^8$ one gets

$$\begin{aligned}
& g \uparrow^p \cdot h \\
&=^1 \; [(g \uparrow^p \star h) \triangleright_{n,n'} (21)] \uparrow^n \\
&=^2 \; [[(132) \triangleleft_{m,p,n} (g \star h) \triangleright_{n,p,n'} (132)] \uparrow^p \triangleright_{n,n'} (21)] \uparrow^n \\
&=^3 \; [(132) \triangleleft_{m,p,n} (g \star h) \triangleright_{n,p,n'} (231)] \uparrow^{n \star p} \\
&=^4 \; [(g \star h) \triangleright_{n,p,n'} (321)] \uparrow^{p \star n} \\
&=^5 \; [(g \star h) \triangleright_{n,p,n'} ((3412) \uparrow^1)] \uparrow^{p \star n} \\
&=^6 \; [(g \star h \star \mathsf{I}_p) \triangleright_{n,p,n',p} (3412)] \uparrow^{p \star n \star p} \\
&=^7 \; [(g \star h \star \mathsf{I}_p) \triangleright_{n \star p, n' \star p} (21)] \uparrow^{n \star p} \uparrow^p \\
&=^8 \; [g \cdot (h \star \mathsf{I}_p)] \uparrow^p
\end{aligned}$$

and dually, using (def′) one may prove the other half of the identity R1, i.e.,

$$f \cdot g \uparrow^p = [(f \star \mathsf{I}_p) \cdot g] \uparrow^p$$

Identity R2 coincides to LR2. For R3, suppose $f : m \star p \to n \star q$ and $g : q \to p$. Then using: def in $=^1$; refine in $=^2$; LB9′, LB5 in $=^3$; LB3, refine in $=^4$; LR3′, LB3 in $=^5$; and LR1a in $=^6$ one gets

$$\begin{aligned}
& [f \cdot (\mathsf{I}_n \star g)] \uparrow^p \\
&=^1 \; [(f \star \mathsf{I}_n \star g) \triangleright_{n \star q, n \star p} (21)] \uparrow^{n \star q} \uparrow^p \\
&=^2 \; [(f \star \mathsf{I}_n \star g) \triangleright_{n,q,n,p} (3412)] \uparrow^{n \star q} \uparrow^p \\
&=^3 \; [[(132) \triangleleft_{m \star p, q, n} (f \star g \star \mathsf{I}_n) \triangleright_{n \star q, p, n} (132)] \triangleright_{n,q,n,p} (3412)] \uparrow^{p \star n \star q} \\
&=^4 \; [(132) \triangleleft_{m \star p, q, n} (f \star g \star \mathsf{I}_n) \triangleright_{n,q,p,n} (3421)] \uparrow^{p \star n \star q} \\
&=^5 \; [(f \star g \star \mathsf{I}_n) \triangleright_{n,q,p,n} (4321)] \uparrow^{p \star q \star n} \\
&=^6 \; [(f \star g) \triangleright_{n,q,p} ((4321) \uparrow^1)] \uparrow^{p \star q} \\
&=^7 \; [(f \star g) \triangleright_{n,q,p} (132)] \uparrow^{p \star q}
\end{aligned}$$

and dual, using (def′) one may prove that

$$[(\mathsf{I}_m \star g) \cdot f] \uparrow^q = [(132) \triangleleft_{m,p,q} (f \star g)] \uparrow^{q \star p}$$

From these two identities and LR3 it follows that R3 holds in T.

Finally, we check the axioms for constants. B6 is already included in the definition of a LR-flow over 𝔹i; B7 follows from LB9′; B8 follows using: def in $=^1$; RSC, LB3 in $=^3$; refine in $=^5$ in

$$\begin{aligned}
& ({}^a\mathsf{X}^b \star \mathsf{I}_c) \cdot (\mathsf{I}_b \star {}^a\mathsf{X}^c) \\
&=^1 \; (\mathsf{I}_{a \star b} \triangleright_{a,b} {}^1\mathsf{X}^1 \star \mathsf{I}_c) \cdot (\mathsf{I}_b \star \mathsf{I}_{a \star c} \triangleright_{a,c} {}^1\mathsf{X}^1) \\
&=^2 \; [\mathsf{I}_{a \star b \star c} \triangleright_{a,b,c} ({}^1\mathsf{X}^1 \star \mathsf{I}_1)] \cdot [\mathsf{I}_{b \star a \star c} \triangleright_{b,a,c} (\mathsf{I}_1 \star {}^1\mathsf{X}^1)] \\
&=^3 \; [\mathsf{I}_{a \star b \star c} \triangleright_{a,b,c} [({}^1\mathsf{X}^1 \star \mathsf{I}_1) \cdot (\mathsf{I}_1 \star {}^1\mathsf{X}^1)]] \\
&=^4 \; \mathsf{I}_{a \star b \star c} \triangleright_{a,b,c} {}^1\mathsf{X}^2 \\
&=^5 \; \mathsf{I}_{a \star b \star c} \triangleright_{a,b \star c} {}^1\mathsf{X}^1 \\
&=^6 \; {}^a\mathsf{X}^{b \star c}
\end{aligned}$$

R7 follows by

$$^{a}\mathsf{X}^{a} \uparrow^{a} = [\mathsf{I}_{a\star a} \rhd_{a,a} {}^{1}\mathsf{X}^{1}] \uparrow^{a} = \mathsf{I}_{a} \rhd_{a} ({}^{1}\mathsf{X}^{1} \uparrow^{1}) = \mathsf{I}_{a} \rhd_{a} \mathsf{I}_{1} = \mathsf{I}_{a}$$

With this last check the proof is complete. □

Fact V.5. *(Left Feedback) In the definition of a LR-flow one may use a Left Feedback "$\uparrow$ _" instead of the Right Feedback "_ $\uparrow$" that we have used. In such a case the axioms LR1a–d, LR2 and LR3 are replaced with the following ones:*

$$\begin{array}{rcl}
\phi \lhd_{P} (\uparrow^{p} f) & = & \uparrow^{p} [(\mathsf{I}_{1} \star \phi) \lhd_{p,P} f] \\
(\uparrow^{p} f) \rhd_{P} \phi & = & \uparrow^{p} [f \rhd_{p,P} (\mathsf{I}_{1} \star \phi)] \\
f \rhd_{P} (\uparrow^{1} \phi) & = & \uparrow^{p} [(\mathsf{I}_{p} \star f) \rhd_{p,P} \phi] \\
(\uparrow^{1} \phi) \lhd_{P} f & = & \uparrow^{p} [\phi \lhd_{p,P} (\mathsf{I}_{p} \star f)] \\
(\uparrow^{p} f) \star g & = & \uparrow^{p} (f \star g) \\
\uparrow^{t\star s} (f \rhd_{s,t,n} ({}^{1}\mathsf{X}^{1} \star \mathsf{I}_{1})] & = & \uparrow^{s\star t} [({}^{1}\mathsf{X}^{1} \star \mathsf{I}_{1}) \lhd_{t,s,m} f]
\end{array}$$

with compatibility relationships between p and P in the third and fourth lines similar to those for the right feedback. The connection between these two feedbacks is given by the following relations

$$\uparrow^{p} f = [{}^{1}\mathsf{X}^{1} \lhd_{p,m} f \rhd_{p,n} {}^{1}\mathsf{X}^{1}] \uparrow^{p} \quad \textit{for } f : p \star m \to p \star n$$

$$f \uparrow^{p} = \uparrow^{p} [{}^{1}\mathsf{X}^{1} \lhd_{m,p} f \rhd_{n,p} {}^{1}\mathsf{X}^{1}] \quad \textit{for } f : m \star p \to n \star p$$

□

Appendix B: Lifting BNA from connections to networks

We show here that the BNA/LR-flow structure is preserved when one passes from connecting arrows to abstract networks.

This appendix contains the core of the BNA soundness proof. We show that the LR-flow structure is preserved when one passes from connecting arrows to abstract networks. Coupled with the equivalence result between BNA and LR-flow structures of the previous chapter this solves the BNA soundness problem.

Lemma V.6. *(1) If T is an LR-flow, then the LR-flow axioms in Table V.1, except for LB9′, hold in $[X, T]$.*
(2) LB9′ holds in $[X, T]_{aa}$.

Proof: (1) For LB1, suppose three pairs in $[X,T]$ are given, namely $(\underline{x}, f) : m \to n$ with $\underline{x} : r \to s$, $(\underline{x}', f') : m' \to n'$ with $\underline{x}' : r' \to s'$, and $(\underline{x}'', f'') : m'' \to n''$ with $\underline{x}'' : r'' \to s''$. Then

$$[(\underline{x}, f) \star (\underline{x}', f')] \star (\underline{x}'', f'') = (\underline{x} \star \underline{x}' \star \underline{x}'', \ g)$$

where

$$\begin{array}{rcl}
g & = & (1324) \lhd_{m\star m', s\star s', m'', s''})[(1324) \lhd_{m,s,m',s'} (f \star f') \\
 & & \rhd_{n,r,n',r'} (1324) \star f''] \rhd_{n\star n', r\star r', n'', r''} (1324) \\
 & = & (125346) \lhd_{m,m',s,s',m'',s''} [(132456) \lhd_{m,s,m',s',m'',s''} (f \star f' \star f'') \\
 & & \rhd_{n,r,n',r',n'',r''} (132456)] \rhd_{n,n',r,r',n'',s''} (124536) \\
 & = & (135246) \lhd_{m,s,m',s',m'',s''} (f \star f' \star f'') \rhd_{n,r,n',r',n'',r''} (142536)
\end{array}$$

In a similar way one gets

$$(\underline{x}, f) \star [(\underline{x}', f') \star (\underline{x}'', f'')] = (\underline{x} \star \underline{x}' \star \underline{x}'',\ g)$$

hence axiom LB1 holds in $[X, T]$.

LB2 holds in $[X, T]$ since for $(\underline{x}, f) : m \to n$ with $\underline{x} : r \to s$

$$\begin{aligned} (\underline{x}, f) \star (\lambda, \mathsf{I}_\epsilon) &= (\underline{x},\ (1324) \triangleleft_{m,s,\epsilon,\epsilon} (f \star \mathsf{I}_\epsilon) \triangleright_{n,r,\epsilon,\epsilon} (1324)) \\ &= (\underline{x},\ f \star \mathsf{I}_\epsilon) \qquad \text{by refine} \\ &= (\underline{x}, f) \end{aligned}$$

and similarly for the other equality.

For LB3, suppose we are given a pair $(\underline{x}, f) : m \to n$ with $\underline{x} : r \to s$. Then LB3a holds by

$$\begin{aligned} ((\underline{x}, f) \triangleright_P \phi) \triangleright_{\phi(P)} \psi &= (\underline{x},\ [f \triangleright_{P,r} (\phi \star \mathsf{I}_1)] \triangleright_{\phi(P),r} (\psi \star \mathsf{I}_1)) \\ &= (\underline{x},\ f \triangleright_{P,r} (\phi \cdot \psi \star \mathsf{I}_1)) \\ &= (\underline{x}, f) \triangleright_P (\phi \cdot \psi) \end{aligned}$$

LB3b is dual to LB3a. LB3c follows by

$$\begin{aligned} [(\phi \triangleleft_Q (\underline{x}, f)] \triangleright_P \psi &= (\underline{x},\ [(\phi \star \mathsf{I}_1) \triangleleft_{Q,s} f] \triangleright_{P,s} (\psi \star \mathsf{I}_1)) \\ &= (\underline{x},\ (\phi \star \mathsf{I}_1) \triangleleft_{Q,s} [f \triangleright_{P,s} (\psi \star \mathsf{I}_1)]) \\ &= \phi \triangleleft_Q [(\underline{x}, f) \triangleright_P \psi] \end{aligned}$$

For LB4a, if $(\underline{x}, f) : m \to n$ with $\underline{x} : r \to s$ and $m = m_1 \star \ldots \star m_k$, then

$$\begin{aligned} \mathsf{I}_k \triangleright_{m_1,\ldots,m_k} (\underline{x}, f) &= (\underline{x},\ \mathsf{I}_{k+1} \triangleright_{m_1,\ldots,m_k,s} f) \\ &= (\underline{x}, f) \end{aligned}$$

and similarly for the other equality. For LB4b, if $\phi \in \mathbb{B}\mathrm{i}(k,k)$, then

$$\begin{aligned} (\lambda, \mathsf{I}_{n_1\star\ldots\star n_k}) \triangleright_{n_1,\ldots,n_k} \phi &= (\lambda,\ \mathsf{I}_{n_1\star\ldots\star n_k} \triangleright_{n_1,\ldots,n_k,\epsilon} (\phi \star \mathsf{I}_1)) \\ &= (\lambda,\ \mathsf{I}_{n_1\star\ldots\star n_k} \triangleright_{n_1,\ldots,n_k} \phi) \\ &= (\lambda,\ \phi \triangleleft_{n_{\phi^{-1}(1)},\ldots,n_{\phi^{-1}(k)}} \mathsf{I}_{n_{\phi^{-1}(1)}\star\ldots\star n_{\phi^{-1}(k)}}) \\ &= \phi \triangleleft_{n_{\phi^{-1}(1)},\ldots,n_{\phi^{-1}(k)}} (\lambda, \mathsf{I}_{n_{\phi^{-1}(1)}\star\ldots\star n_{\phi^{-1}(k)}}) \end{aligned}$$

For LB5 suppose we are given the pairs $(\underline{x}, f) : m \to n$ with $\underline{x} : r \to s$, and $(\underline{x}', f') : m' \to n'$ with $\underline{x}' : r' \to s'$. If moreover: $P = (n_1, \ldots, n_k)$ is a decomposition of n, $\phi \in \mathbb{B}\mathrm{i}(k,k)$, and $p = n_{\phi^{-1}(1)} \star \ldots \star n_{\phi^{-1}(k)}$; and $P' = (n'_1, \ldots, n'_{k'})$ is a decomposition of n', $\phi' \in \mathbb{B}\mathrm{i}(k', k')$, and $p' = n'_{\phi'^{-1}(1)} \star \ldots \star n'_{\phi'^{-1}(k')}$, then LB5a follows by

$$\begin{aligned} &[(\underline{x}, f) \star (\underline{x}', f')] \triangleright_{P,P'} (\phi \star \psi) \\ &= (\underline{x} \star \underline{x}',\ [(1324) \triangleleft_{m,s,m',s'} (f \star f') \triangleright_{n,r,n',r'} (1324)] \triangleright_{P,P',r,r'} (\phi \star \phi' \star \mathsf{I}_1 \star \mathsf{I}_1)]) \\ &= (\underline{x} \star \underline{x}',\ (1324) \triangleleft_{m,s,m',s'} (f \star f') \triangleright_{P,r,P',r'} [(1_k 3_1 2_{k'} 4_1) \cdot (\phi \star \phi' \star \mathsf{I}_1 \star \mathsf{I}_1)]) \\ &= (\underline{x} \star \underline{x}',\ (1324) \triangleleft_{m,s,m',s'} (f \star f') \triangleright_{P,r,P',r'} [(\phi \star \mathsf{I}_1 \star \phi' \star \mathsf{I}_1) \cdot (1_k 3_1 2_{k'} 4_1)]) \\ &= (\underline{x} \star \underline{x}',\ (1324) \triangleleft_{m,s,m',s'} [(f \star f') \triangleright_{P,r,P',r'} (\phi \star \mathsf{I}_1 \star \phi' \star \mathsf{I}_1)] \triangleright_{p,r,p',r'} (1324)]) \\ &= (\underline{x},\ f \triangleright_{P,r} (\phi \star \mathsf{I}_1)) \star (\underline{x}',\ f' \triangleright_{P',r'} (\phi' \star \mathsf{I}_1)) \\ &= (\underline{x}, f) \triangleright_P \phi\ \star\ (\underline{x}', f') \triangleright_{P'} \phi' \end{aligned}$$

The other identity LB5b may be proved in a similar way.

Next, the axioms for feedback are checked. We prefer to prove them for the Left Feedback and to use Fact V.5.

For LR1a let $(\underline{x}, f) : p\star m \to p\star n$ with $\underline{x} : r \to s$, $n = n_1 \star \ldots \star n_k$ and $\phi \in \mathbb{B}\mathrm{i}(k,k)$. Then

$$\begin{aligned} [\uparrow^p (\underline{x}, f)] \triangleright_P \phi &= (\underline{x},\ (\uparrow^p f) \triangleright_{P,r} (\phi \star \mathsf{l}_1)) \\ &= (\underline{x},\ \uparrow^p [f \triangleright_{p,P,r} (\mathsf{l}_1 \star \phi \star \mathsf{l}_1)]) \\ &= \uparrow^p [(\underline{x}, f) \triangleright_{p,P} (\mathsf{l}_1 \star \phi)] \end{aligned}$$

LR1b is similar to LR1a. For LR1c suppose $(\underline{x}, f) : m \to n$ with $\underline{x} : r \to s$, $n = n_1 \star \ldots \star n_k$ and $\phi \in \mathbb{B}\mathrm{i}(1\star k, 1\star k)$. If $p = n_{\phi^{-1}(1)-1}$ when $\phi^{-1}(1) \geq 1$, then

$$\begin{aligned} (\underline{x}, f) \triangleright_P (\uparrow^1 \phi) &= (\underline{x},\ f \triangleright_{P,r} ((\uparrow^1 \phi) \star \mathsf{l}_1)) \\ &= (\underline{x},\ f \triangleright_{P,r} \uparrow^1 (\phi \star \mathsf{l}_1)) \\ &= (\underline{x},\ \uparrow^p [(\mathsf{l}_p \star f) \triangleright_{p,P,r} (\phi \star \mathsf{l}_1)]) \\ &= \uparrow^p [(\underline{x},\ \mathsf{l}_p \star f) \triangleright_{p,P} \phi] \\ &= \uparrow^p [((\lambda, \mathsf{l}_p) \star (\underline{x}, f)) \triangleright_{p,P} \phi] \end{aligned}$$

LR1d is similar to LR1c.

For LR2 suppose we are given the pairs $(\underline{x}, f) : p\star m \to p\star n$ with $\underline{x} : r \to s$ and $(\underline{x}', f') : m' \to n'$ with $\underline{x}' : r' \to s'$. Then

$$\begin{aligned} &\uparrow^p (\underline{x}, f) \star (\underline{x}', f') \\ &= (\underline{x} \star \underline{x}',\ (1324) \triangleleft_{m,s,m',s'} (\uparrow^p f \star f') \triangleright_{n,r,n',r'} (1324)) \\ &= (\underline{x} \star \underline{x}',\ \uparrow^p [(\mathsf{l}_1 \star (1324)) \triangleleft_{p,m,s,m',s'} (f \star f') \triangleright_{p,n,r,n',r'} (\mathsf{l}_1 \star (1324))]) \\ &= (\underline{x} \star \underline{x}',\ \uparrow^p [(1324) \triangleleft_{p\star m,s,m',s'} (f \star f') \triangleright_{p\star n,r,n',r'} (1324)]) \\ &= \uparrow^p [(\underline{x}, f) \star (\underline{x}', f')] \end{aligned}$$

For LR3, suppose the pair $(\underline{x}, f) : p\star q\star m \to q\star p\star n$ with $\underline{x} : r \to s$ is given. Then

$$\begin{aligned} \uparrow^{p\star q} [(\underline{x}, f) \triangleright_{q,p,n} (213)] &= (\underline{x}, \uparrow^{p\star q} [f \triangleright_{q,p,n,r} (2134)]) && \text{definition } \uparrow, \cdot_R \\ &= (\underline{x}, \uparrow^{p\star q} [f \triangleright_{q,p,n\star r} (213)]) && \text{refine} \\ &= (\underline{x}, \uparrow^{q\star p} [(213) \triangleleft_{q,p,m\star s} f]) && \text{LR3 in } T \\ &= \uparrow^{q\star p} [(213) \triangleleft_{p,q,m} (\underline{x}, f)] \end{aligned}$$

LB6 clearly holds. For R-refine we take $(\underline{x},\ f) : m \to n$ with $\underline{x} : r \to s$ and $n = n_1 \star \ldots \star (n_i^1 \star \ldots \star n_i^r) \star \ldots \star_k$. Then

$$\begin{aligned} &(\underline{x}, f) \triangleright_{n_1,\ldots,n_i^1\star\ldots\star n_i^r,\ldots,n_k} \phi \\ &= (\underline{x},\ f \triangleright_{n_1,\ldots,n_i^1\star\ldots\star n_i^r,\ldots,n_k,s} (\phi \star \mathsf{l}_1)) \\ &= (\underline{x},\ f \triangleright_{n_1,\ldots,n_i^1,\ldots,n_i^r,\ldots,n_k,s} ((\phi(1)_1 \ldots \phi(i)_r \ldots \phi(k)_1) \star \mathsf{l}_1)) \\ &= (\underline{x},\ f) \triangleright_{n_1,\ldots,n_i^1,\ldots,n_i^r,\ldots,n_k} (\phi(1)_1 \ldots \phi(i)_r \ldots \phi(k)_1) \end{aligned}$$

Finally, L-refine is similar. This ends the proof of the first part of this lemma.

(2) For the second part we have to prove LB9′ in $[X, T]_{a\alpha}$. If $(\underline{x}, f) : m \to n$ with $\underline{x} = x_1 \star \ldots \star x_k : r \to s$ and $(\underline{x}', f') : m' \to n'$ with $\underline{x}' = x_1' \star \ldots \star x_{k'}' : r' \to s'$, then the right-hand-side term of LB9′ is

$$
\begin{aligned}
&(21) \lhd_{m',m} [(\underline{x}', f') \star (\underline{x}, f)] \rhd_{n',n} (21) \\
&= \quad (\underline{x}' \star \underline{x},\ (3124) \lhd_{m',s',m,s} (f' \star f) \rhd_{n',r',n,r} (2314)) \\
&=: \quad (\underline{x}' \star \underline{x},\ g)
\end{aligned}
$$

This last nf-pair is similar via a block transposition with the nf-pair obtained from the evaluation of the left-hand-side term in LB9$'$, namely

$$
\begin{aligned}
&(\underline{x}, f) \star (\underline{x}', f') \\
&= \quad (\underline{x} \star \underline{x}',\ (1324) \lhd_{m,s,m',s'} (f \star f') \rhd_{n,r,n',r'} (1324)) \\
&=: \quad (\underline{x} \star \underline{x}',\ h)
\end{aligned}
$$

More precisely, we show that

$$(\underline{x}' \star \underline{x},\ g) \rightarrow_{(2_{k'}1_k)} (\underline{x} \star \underline{x}',\ h)$$

Condition (i) in the definition of simulation related to the sort preservation property of bijection $(2_{k'}1_k)$ clearly holds. The validity of the second condition (ii) related to the compatibility of connecting relations follows by:

$$
\begin{aligned}
&(1243) \lhd_{m,m',s',s}\, g \rhd_{n,n',r',r} (1243) \\
&= \quad (3142) \lhd_{m',s',m,s} (f' \star f) \rhd_{n',r',n,r} (2413) \\
&= \quad (1324) \lhd_{m,s,m',s'} (f \star f') \rhd_{n,r,n',r'} (1324) \\
&= \quad h
\end{aligned}
$$

□

From the two lemmas above we get the following theorem.

Theorem V.7. *If T is a LR-flow, then $[X,T]_{a\alpha}$ is a LR-flow.* □

With Theorem V.2 we can lift this result to the BNA level. We get the following corollary

Corollary V.8. *If T is a BNA, then $[X,T]_{a\alpha}$ is a BNA.* □

Appendix C: Demonic relation operators

We present here the main changes one has to do on the calculus for relations presented in Chapter 3 to cope with relations under demonic operators.

Relations with demonic operators are used in studies related to predicate transformer semantics of nondeterministic programs, to model the connection wires in synchronous data-flow networks, or in process algebra setting to model the communication between agents and their environments.

The aim of this appendix is to provide equational axiomatizations for various classes of finite relations with demonic (including looping) operators. We present axiomatizations for three types of demonic calculi: i.e., calculi for relations with forward demonic –, backward demonic – or two-way demonic operators.

One point of debate is the definition of (sequential) composition. Most of the work has been done using the "angelic" view, i.e., the situation when a pair (x,z) is in the composite relation $R \cdot S$ if *there exists* an intermediary element y such that $(x, y) \in R$ and $(y, z) \in S$.

Axiomatizations as enriched symocats for various classes of finite relations under angelic operators were presented in Chapter 3. The aim of the present appendix is to provide equational axiomatizations for classes of finite relations using demonic operators. Roughly speaking, the demonic view consist in the slogan:

> *Demonic rule*
>
> If in a branching point an input (resp. output) disconnected path exists, then it destroys the other input (resp. output) connected paths of that point.

By making a distinction between incoming paths, outgoing paths, or both we get three types of demonic calculi, i.e. forward demonic -, backward demonic -, and two-way demonic calculus, respectively.

Relations with backward demonic operators have been used to study predicate transformer semantics for nondeterministic programs, see e.g., [BKS97]. In this theory one is interested in getting the weakest liberal precondition $\mathtt{wlp}(P, \phi)$ which guarantees the termination of the program P in a state fulfilling property ϕ. Now, if one looks, e.g., at the sequential composite of two programs $P \cdot Q$, then a state s guarantees the termination of $P \cdot Q$ iff s guarantees the termination of P *and* in all states which are reached by P starting from s the termination of Q is guaranteed. Using the terms of the above slogan, one may say that a path in P starting from a state s which fails to be forwarded by a terminating path in Q destroys the other paths of P from s.

A case where relations with forward demonic operators arise is the model of connection wires in synchronous data-flow networks (SDFNs, for short), see Chapter 10. This model consists of networks of cells that compute and communicate data in a synchronous way. Such a network has a (local) crash property. That is, if one component fails to deliver any of its output data, an undefined datum results and it destroys the other data it will be engaged in computation with. So using forward demonic relation operators one can determine whether or not a SDFN is correct in the sense that it transforms a vector of input streams into a vector of output streams. In this way forward demonic operators occur in modeling synchronous data-flow networks.

In the axiomatic point of view of the present paper this means the operations we are using are juxtaposition and forward demonic sequential composition and forward demonic feedback. It turned out that a slight change of the axioms in Chapter 3 is needed to provide axiomatizations for relations with forward demonic operators, i.e., the original axioms A1d and R10 have been replaced by R1d^o and R10^o, respectively.

Other work using demonic models exists. We mention [Jos92, dNS95]. In [Jos92] the author reworks Hoare's CSP with a particular demonic interpretation of the communication. In [dNS95] a (CCS-like) process algebra model DIOA for demonic input/output automata is presented and axiomatizations for recursion-free processes are given.

Acyclic forward demonic case. We denote by $\mathbb{R}\mathrm{el}^{fd}$ the algebraic structure of finite relations $\mathbb{R}\mathrm{el}^{fd}(m, n) = \{r : r \subseteq [m] \times [n]\}$. Juxtaposition is the same as in $\mathbb{R}\mathrm{el}$. Composition is changed to

– FORWARD DEMONIC SEQUENTIAL COMPOSITION $\cdot_{\mathrm{fd}}$:
  if $f : m \to n$ and $g : n \to p$, then $f \cdot_{\mathrm{fd}} g : m \to p$ is the relation defined by

$$\begin{array}{rcll} f \cdot_{\mathrm{fd}} g & = & \{\, (i,j): & \exists k \in [n].\ [(i,k) \in f \ \&\ (k,j) \in g] \ \text{ and} \\ & & & \forall k' \in [n].\ [(k',j) \in g \Rightarrow \exists i' \in [m].(i',k') \in f] \,\} \end{array}$$

A connection from in_i to out_j in the angelic composite $f \cdot g$ is a connection in $f \cdot_{\text{fd}} g$, as well, precisely in the case all the maximal paths in $G(f \cdot g)$ ending in out_j starts from an input in_k, for a $k \in [m]$.

Theorem V.9. *The axioms I and III in Table 6.3, but using $A1d^o$ instead of $A1d$ give a sound and complete axiomatization for relations with juxtaposition, forward demonic sequential composition, and $\mathsf{I}, \mathsf{X}, \perp, \wedge, \top, \vee$-constants. (No feedback is used here.)*

PROOF: We follow the proof of the analogous result in the angelic case. Here we adapt the proof for the forward demonic case, but with a less formal presentation. An illustration of the formal computation proofs is given at the end of the appendix.

I. First of all we note that the normal form representation of angelic relations may be used in this demonic case, as well. Hence, each relation $f : m \to n$ may be represented as $f = (\sum_{i\in[m]} \wedge^1_{m_i}) \cdot_{\text{fd}} b \cdot_{\text{fd}} (\sum_{j\in[n]} \vee^{n_i}_1)$, where $b : \Sigma_{i\in[m]} m_i \to \Sigma_{j\in[n]} n_j$ is a bijection and $m_i, n_j \geq 0$.

II. It is clear that each constant may be written in such a normal form. The normal form representation of such an f is not unique due to two reasons: (i) certain permutations may be done in an identification ($\vee^{n_i}_1$) or ramification ($\wedge^1_{m_i}$) point and (ii) more than one path may connect an input–output pair. The proof that two normal form expressions that represent the same relation are equivalent in the specified axiomatic subsystem is the same as in the angelic case.

III. It is obvious that the juxtaposition of two normal forms is a normal form, too. Hence we still have to show that the forward demonic sequential composite of two normal forms may be reduced to a normal form and, moreover, to show that this reduction may be done inside the axiomatic system. We apply the same technique as in the angelic case.

Following the indicated proofs, first of all we single out certain properties A0–A4 which follows from the given axioms. They are as in the angelic case, except for $A1(b)^o$ presented in Table V.2.

Table V.2. Specific axioms for demonic cases

$A1(a)^o$	$\wedge^1_n \cdot (\Sigma_{i\in[n]} \wedge^1_{n_i}) =$	$\wedge^1_{\Sigma_{i\in[n]} n_i}$	if $\forall i \in [n].n_i > 0,$
		$\perp^1 \cdot \top_{\Sigma_{i\in[n]} n_i}$	if $\exists i \in [n].n_i = 0$
$A1(b)^o$	$(\Sigma_{i\in[n]} \vee^{n_i}_1) \cdot \vee^n_1 =$	$\vee^{\Sigma_{i\in[n]} n_i}_1$	if $\forall i \in [n].n_i > 0,$
		$\perp^{\Sigma_{i\in[n]} n_i} \cdot \top_1$	if $\exists i \in [n].n_i = 0$
$A1b^o$	$\wedge^a \cdot (\perp^a \star \mathsf{I}_a) = \perp^a \cdot \top_a$		
$A1d^o$	$(\top_a \star \mathsf{I}_a) \cdot \vee_a = \perp^a \cdot \top_a$		
$R10^o$	$[(\mathsf{I}_a \star \wedge^a) \cdot ({}^a\mathsf{X}^a \star \mathsf{I}_a) \cdot (\mathsf{I}_a \star \vee_a)] \uparrow^a = \perp^a \cdot \top_a$		

As one can see, only property $A1(b)^o$ is different from the angelic case. This difference produces some complications in the proof, as we shall see.

The next step is to apply the above properties in order to normalize the sequential composite of two normal forms, say

$$[(\star_{i\in[m]} \wedge^1_{m_i}) \cdot_{\text{fd}} b \cdot_{\text{fd}} (\star_{j\in[n]} \vee^{n_i}_1)] \cdot_{\text{fd}} [(\star_{i'\in[n]} \wedge^1_{m'_{i'}}) \cdot_{\text{fd}} b' \cdot_{\text{fd}} (\star_{j'\in[p]} \vee^{n'_{j'}}_1)]$$

We start applying A3 from left to right in order to commute the middle factors, i.e., the ramifications with indices $m'_{i'}$ pass over the identifications with indices n_i. Then,

(*) using general results in a symmetric strict monoidal category, these new ramifications produced by applying A3 pass over the bijection b, as well, and finally with A1 are eaten by the ramifications with indices m_i.

A similar way takes the identifications with indices n_i which arrive in front of the identifications with indices $n'_{j'}$. Till now it was similar to the angelic case. But in this point, when we apply A1(b)^o the second case may occur. So, in general we get an expression of the following type

$$(\star_{i\in[m]}\wedge^{1}_{..}) \cdot_{\text{fd}} b'' \cdot_{\text{fd}} (\mathsf{I}_{..} \star \perp^{k}) \cdot_{\text{fd}} b''' \cdot_{\text{fd}} (\star_{j'\in[p]}\vee^{..}_{1})$$

(In order to simplify the writing we have collected all the resulting $\perp$'s in a unique summand by using appropriate bijections; at the same time, certain indices are not given in an explicit way, because their values are not relevant.) Hence certain additional 0-ramifications $\perp$'s occur, but fortunately they are eliminated applying the step (*) above and we get a normal form. □

Cyclic forward demonic case. We complete our basic set of forward demonic operators with

- FORWARD DEMONIC FEEDBACK $\uparrow_{\text{fd}}$: if $f : m+1 \to n+1$, then $f \uparrow^1_{\text{fd}} : m \to n$ is defined by: (1) if $(m+1, n+1) \notin f$, then

$$f \uparrow^1_{\text{fd}} = \{ (i,j) : \begin{array}{l} (i,j) \in f \text{ and } [(m+1,j) \in f \Rightarrow \exists i' \in [m].(i', n+1) \in f] \\ \text{or} \\ (i, n+1) \in f \text{ and } (m+1, j) \in f \ \} \end{array}$$

(2) if $(m+1, n+1) \in f$, then

$$f \uparrow^1_{\text{fd}} = \{ (i,j) : (i,j) \in f \text{ and } (m+1, j) \notin f \ \}$$

It may be easily seen that the demonic feedback $f \uparrow^1_{\text{fd}}$ is included in the usual one $f \uparrow^1$. A more intuitive definition for $f \uparrow^1_{\text{fd}}$ may be given using the graph $G(f \uparrow^1_{\text{fd}})$ as follows: $(i,j) \in f \uparrow^1_{\text{fd}}$ iff (1) $(i,j) \in f \uparrow^1$ and (2) every maximal path in $G(f \uparrow^1)$ ending in out_j is finite and it starts from an in_k, with $k \in [m]$.

The second condition states that one connection $(i,j) \in f \uparrow^1$ is *not* a connection in $f \uparrow^1_{\text{fd}}$ in two cases:

- $(m+1, j) \in f$ and $(m+1, n+1) \in f$; notice that in such a case there exists an infinite path ending in out_j;
- $(m+1, j) \in f$ and $\nexists(k, n+1) \in f$; in such a case $out_{n+1} \to in_{m+1} \to out_j$ is a maximal path which fails to start from an input vertex in_k, with $k \in [m]$.

Hence the direct definition and the one in terms of graphs for the forward demonic feedback are equivalent.

Theorem V.10. *The axioms in I–IV in Table 6.3, but using the new A1d^o, R10^o versions instead of A1d, R10 axioms, give a sound and complete axiomatization for relations with forward demonic operators.*

PROOF: The proof is based on the similar result in the acyclic case (Theorem V.9). The remaining proof may be obtained from the proof of the corresponding result in the angelic case (Theorem 3.15) using axiom R10^o instead of R10 in the specific case 322. □

Two-ways demonic case. In the previous sections we have studied relations with forward demonic operators, namely the case when (a) a maximal i-disconnected incoming path to an output vertex v destroys the access of other incoming paths to v. One may consider the *backward demonic* case, as well, i.e., the case given by the following dual condition: (b) a maximal o-disconnected outgoing path from an input vertex s destroys the creation of other outgoing paths from s. Using duality, one gets an axiomatization for the backward demonic calculus of relations. Finally, one may consider the *two-way demonic* case in which both conditions (a) and (b) are required. The following theorem holds.

Theorem V.11. *An axiomatization for relations with two-way demonic operators is given by the axioms I–IV in Table 6.3, where the $A1b^o/A1d^o/R10^o$ versions for the $A1b/A1d/R10$ axioms are used.*

PROOF: The proof of the fact that a two-way demonic sequential composite $f \cdot_{\mathrm{fbd}} g$ of two normal forms may be reduced to a normal form using properties A0, A1(a)^o, A1(b)^o, A2–A4 is now a bit more complicated. (A1(a)^o is the dual of A1(b)^o.)

1. We start with the same procedure as in the forward demonic case moving the middle ramifications and identifications in $E_0 := f \cdot_{\mathrm{fbd}} g$ near the similar ones which occur in the extremal factors, but now applying A1(a)^o and A1(b)^o. Hence additional terms $\top$'s and $\perp$'s may occur. Let E_1 be the resulting term.
2. Repeat 1, starting with E_1 and obtaining E_2. And so on.

One may see that, at each application of 1, if additional terms occur, then at least one path/edge is deleted from the corresponding graph. Hence, the algorithm stops in a finite number of steps and we get a normal form for $f \cdot_{\mathrm{fbd}} g$. The rest of the proof is similar to the corresponding proof in the forward demonic case. □

Simple algebraic computations. Here we present some simple computations which justifies in an algebraic way the results presented in Example 3.2. We start with the relational term presented before

$$f = (\wedge^1 \star \wedge^1 \star \mathsf{I}_1 \star \perp^1) \cdot (\mathsf{I}_1 \star \vee_1 \cdot (\top_1 \star \mathsf{I}_1) \cdot \vee_1 \star \vee_1 \cdot [(\wedge^2 \cdot (\mathsf{I}_2 \star \vee_1)) \uparrow^1])$$

and compute the normal form in various relational calculi.

Angelic case: First we note that

$$\begin{aligned}
&(\wedge^2 \cdot (\mathsf{I}_2 \star \vee_1)) \uparrow^1_{\oplus}\\
&= [(\wedge^1 \star \wedge^1) \cdot (\mathsf{I}_1 \star {}^1\mathsf{X}^1 \star \mathsf{I}_1) \cdot (\mathsf{I}_2 \star \vee_1)] \uparrow^1_{\oplus}\\
&= \wedge^1 \cdot ([(\mathsf{I}_2 \star \wedge^1) \cdot (\mathsf{I}_1 \star {}^1\mathsf{X}^1 \star \mathsf{I}_1) \cdot (\mathsf{I}_2 \star \vee_1)] \uparrow^1_{\oplus})\\
&= \wedge^1 \cdot (\mathsf{I}_1 \star [(\mathsf{I}_1 \star \wedge^1) \cdot ({}^1\mathsf{X}^1 \star \mathsf{I}_1) \cdot (\mathsf{I}_1 \star \vee_1)] \uparrow^1_{\oplus})\\
&= \wedge^1 \cdot (\mathsf{I}_1 \star \mathsf{I}_1)\\
&= \wedge^1
\end{aligned}$$

Now we have a kind of sequential composite of two normal forms. This is normalized in the usual way:

$$\begin{aligned}
f &= (\wedge^1 \star \wedge^1 \star \mathsf{I}_1 \star \perp^1) \cdot (\mathsf{I}_1 \star \vee_1 \cdot (\top_1 \star \mathsf{I}_1) \cdot \vee_1 \star \vee_1) \cdot \wedge^1)\\
&= (\wedge^1 \star \wedge^1 \star \mathsf{I}_1 \star \perp^1) \cdot (\mathsf{I}_1 \star \vee_1 \cdot \mathsf{I}_1 \star \vee_1 \cdot \wedge^1)\\
&= (\wedge^1 \star \wedge^1 \star \mathsf{I}_1 \star \perp^1) \cdot (\mathsf{I}_1 \star \vee_1 \star (\wedge^1 \star \wedge^1) \cdot (\mathsf{I}_1 \star {}^1\mathsf{X}^1 \star \mathsf{I}_1) \cdot (\vee_1 \star \vee_1))\\
&= (\wedge^1 \star \wedge^1 \star \mathsf{I}_1 \star \perp^1) \cdot (\mathsf{I}_1 \star \mathsf{I}_2 \star \wedge^1 \star \wedge^1) \cdot (\mathsf{I}_4 \star {}^1\mathsf{X}^1 \star \mathsf{I}_1) \cdot (\mathsf{I}_1 \star \vee_1 \star \vee_1 \star \vee_1)\\
&= (\wedge^1 \star \wedge^1 \cdot (\mathsf{I}_1 \star \wedge^1) \star \wedge^1 \star \perp^1) \cdot (\mathsf{I}_4 \star {}^1\mathsf{X}^1 \star \mathsf{I}_1) \cdot (\mathsf{I}_1 \star \vee_1 \star \vee_1 \star \vee_1)\\
&= (\wedge^1 \star \wedge^1_3 \star \wedge^1 \star \perp^1) \cdot (\mathsf{I}_4 \star {}^1\mathsf{X}^1 \star \mathsf{I}_1) \cdot (\mathsf{I}_1 \star \vee_1 \star \vee_1 \star \vee_1)
\end{aligned}$$

What we have got is just the normal form of the resulting relation computed in the angelic case.

Forward demonic case: As in the previous case we have

$$(\wedge^2 \cdot_{fd} (I_2 \star V_1)) \uparrow^1_{fd}$$
$$\dots$$
$$= \wedge^1 \cdot_{fd} (I_1 \star [(I_1 \star \wedge^1) \cdot_{fd} ({}^1X^1 \star I_1) \cdot_{fd} (I_1 \star V_1)] \uparrow^1_{fd})$$
$$= \wedge^1 \cdot_{fd} (I_1 \star \bot^1 \cdot_{fd} \top_1)$$

Then, using the new axioms we get

$$\begin{array}{ll} & f \\ = & (\wedge^1 \star \wedge^1 \star I_1 \star \bot^1) \cdot_{fd} (I_1 \star V_1 \cdot_{fd} (\top_1 \star I_1) \cdot_{fd} V_1 \star V_1 \cdot_{fd} [\wedge^1 \cdot_{fd} (I_1 \star \bot^1 \cdot_{fd} \top_1)]) \\ = & (\wedge^1 \star \wedge^1 \star I_1 \star \bot^1) \cdot_{fd} (I_1 \star V_1 \cdot_{fd} \bot^1 \cdot_{fd} \top_1 \star V_1 \cdot_{fd} \wedge^1 \cdot_{fd} (I_1 \star \bot^1) \star \top_1) \\ = & (\wedge^1 \star \wedge^1 \star I_1 \star \bot^1) \cdot_{fd} (I_1 \star \bot^2 \star I_2) \cdot_{fd} (I_1 \star \top_1 \star V_1 \cdot_{fd} I_1 \star \top_1) \\ = & (\wedge^1 \cdot_{fd} (I_1 \star \top_1) \star \wedge^1 \cdot_{fd} (\bot^1 \star I_1) \star I_1 \star \bot^1) \cdot_{fd} (I_1 \star \top_1 \star V_1 \star \top_1) \\ = & (I_1 \star I_1 \star I_1 \star \bot^1) \cdot_{fd} (I_1 \star \top_1 \star V_1 \star \top_1) \end{array}$$

which is the normal form of the resulting relation computed in the forward demonic case.

Backward demonic case: As in the previous forward demonic case we have

$$(\wedge^2 \cdot_{bd} (I_2 \star V_1)) \uparrow^1_{bd} = \wedge^1 \cdot_{bd} (I_1 \star \bot^1 \cdot_{bd} \top_1)$$

Then,

$$\begin{array}{ll} & f \\ = & (\wedge^1 \star \wedge^1 \star I_1 \star \bot^1) \cdot_{bd} (I_1 \star V_1 \cdot_{bd} (\top_1 \star V_1) \cdot_{bd} V_1 \star V_1 \cdot_{bd} [\wedge^1 \cdot_{bd} (I_1 \star \bot^1 \cdot_{bd} \top_1)]) \\ = & (\wedge^1 \star \wedge^1 \star I_1 \star \bot^1) \cdot_{bd} (I_1 \star V_1 \cdot_{bd} I_1 \star V_1 \cdot_{bd} \wedge^1 \cdot_{bd} (I_1 \star \bot^1) \star \top_1) \\ = & (\wedge^1 \star \wedge^1 \star I_1 \star \bot^1) \cdot_{bd} (I_1 \star V_1 \star V_1 \cdot_{bd} \bot^1 \cdot_{bd} \top_1 \star \top_1) \\ = & (\wedge^1 \star \wedge^1 \star I_1 \star \bot^1) \cdot_{bd} (I_1 \star V_1 \star \bot^2 \star \top_1 \star \top_1) \\ = & (\wedge^1 \star \wedge^1 \cdot_{bd} (I_1 \star \bot^1) \star \bot^1 \star \bot^1) \cdot_{bd} (I_1 \star V_1 \star \top_2) \\ = & (\wedge^1 \star \bot^1 \cdot_{bd} \top_1 \star \bot^2) \cdot_{bd} (I_1 \star V_1 \star \top_2) \\ = & (\wedge^1 \star \bot^1 \star \bot^2) \cdot_{bd} (I_1 \star (I_1 \star \top_1) \cdot_{bd} V_1 \star \top_2) \\ = & (\wedge^1 \star \bot^3) \cdot_{bd} (I_1 \star I_1 \star \top_2) \end{array}$$

which is the normal form of the expected result computed in the backward demonic case.

Two-way demonic case: The feedback part gives the same result as in previous demonic cases. We may compute

$$\begin{array}{l} f \\ = (\wedge^1 \star \wedge^1 \star I_1 \star \bot^1) \cdot_{fbd} (I_1 \star V_1 \cdot_{fbd} (\top_1 \star I_1) \cdot_{fbd} V_1 \star V_1 \cdot_{fbd} [\wedge^1 \cdot_{fbd} (I_1 \star \bot^1 \cdot_{fbd} \top_1)]) \\ = (\wedge^1 \star \wedge^1 \star I_1 \star \bot^1) \cdot_{fbd} (I_1 \star V_1 \cdot_{fbd} \bot^1 \cdot_{fbd} \top_1 \star V_1 \cdot_{fbd} [\wedge^1 \cdot_{fbd} (I_1 \star \bot^1)] \star \top_1) \\ = (\wedge^1 \star \wedge^1 \star I_1 \star \bot^1) \cdot_{fbd} (I_1 \star \bot^2 \star \top_1 \star V_1 \cdot_{fbd} \bot^1 \cdot_{fbd} \top_1 \star \top_1) \\ = (\wedge^1 \star \wedge^1 \star I_1 \star \bot^1) \cdot_{fbd} (I_1 \star \bot^2 \star \top_1 \star \bot^2 \star \top_1 \star \top_1) \\ = (\wedge^1 \cdot_{fbd} (I_1 \star \bot^1) \star \wedge^1 \cdot_{fbd} (\bot^1 \star \bot^1) \star \bot^1 \star \bot^1) \cdot_{fbd} (I_1 \star \top_1 \star \top_2) \\ = (\bot^1 \cdot_{fbd} \top_1 \star \bot^1 \star \bot^2) \cdot_{fbd} (I_1 \star \top_3) \\ = (\bot^1 \star \bot^3) \cdot_{fbd} (\top_1 \star \top_3) \\ = \bot^4 \cdot_{fbd} \top_4 \end{array}$$

which is the normal form of the expected result computed in the two-way demonic case.

Appendix D. Generating congruences

In this appendix a technical construction for congruences generated by sets of conditional equations is described at the general level of universal algebra. (However, it has been used only in the particular case of the enzymatic rule.)

The enzymatic rule is given in terms of conditional equations. This makes our technical life a lot harder. As we have seen in Chapter 4, in order to capture natural equivalence relations on networks we have to consider the congruence relations generated by commuting relations C_{xy} in the class of congruence relations satisfying the enzymatic rule Enz_{xy}. How may such a congruence be generated? Below we give a general construction for universal algebras with conditional equations.

Generating congruences. Let A be a universal algebra of sort S and with operations in Σ. For a relation $R \subseteq A \times A$, various closures may be defined, e.g.,

$Ref(R)$	reflexive closure of R;
$Sym(R)$	symmetric closure of R;
$Op(R)$	the closure of R to operations;
$Trans(R)$	transitive closure of R;
$C(R)$	the congruence relation generated by R;
$C_J(R)$	the congruence satisfying the set J of implications generated by R.

Before stating the algorithm we recall a few standard notations and definitions on universal algebras.
(1) A *conditional equation* is a tuple $\langle X;\ t_1 = t_1' \,\&\ldots\&\, t_r = t_r' \Rightarrow t = t'\rangle$, where $t_1, t_1' \in T_\Sigma(X)_{s_1}, \ldots, t_r, t_r' \in T_\Sigma(X)_{s_r}, t, t' \in T_\Sigma(X)_s$.
(2) A congruence relation $\sim$ on A *satisfies* this equation if $[f^\#(t_1) \sim f^\#(t_1')$ and $\ldots$ and $f^\#(t_r) \sim f^\#(t_r')$ implies $f^\#(t) \sim f^\#(t')]$ for an arbitrary function $f : X \to A$, where $f^\#$ denotes its (unique) extension to terms.
(3) Finally, *A satisfies the conditional equation* if equality on A satisfies it.

The construction of the congruence relation $C(R)$ generated by a relation R is well-known:

$$C(R) = Trans(Op(Sym(Ref(R))))$$

This result is extended to algebras with conditional equations as follows. Define inductively the congruences:

- $C_J^0(R) = C(R)$;
- $C_J^{k+1} = C(R \cup E_k)$, for $k \geq 0$, where
$$E_k = \{(f^\#(t), f^\#(t')) : \exists \langle X;\ t_1 = t_1' \,\&\ \ldots\ \&\, t_r = t_r' \Rightarrow t = t'\rangle \in J \text{ and } f : X \to A \text{ such that } (f^\#(t_j), f^\#(t_j')) \in C_J^k(R), \forall j \in [r]\}$$

Proposition V.12. *The sequence of congruences $C_J^k(R)$ is increasing and $C_J(R) = \bigcup_{k\geq 0} C_J^k(R)$.*

PROOF: By induction we show that $R \subseteq C_J^0(R) \subseteq C_J^1(R) \subseteq \ldots \subseteq C_J^k(R) \subseteq \ldots$

The first two inclusions obviously hold. For the inductive step, let us notice that $C_J^k(R) \subseteq C_J^{k+1}(R) \Rightarrow E_k \subseteq E_{k+1} \Rightarrow C_J^{k+1}(R) \subseteq C_J^{k+2}(R)$. Since $(C_J^k(R))_{k\in\mathbb{N}}$ is an increasing sequence of congruences it follows that $C_J(R) = \bigcup_{k\geq 0} C_J^k(R)$ is a congruence, too.

Finally, notice that $C_J(R)$ satisfies the conditional equations in J. Indeed, let $\langle X;\ t_1 = t'_1 \ \&\ \ldots\ \&\ t_r = t'_r \Rightarrow t = t'\rangle \in J$ be a conditional equation and $f : X \to A$ a valuation. Suppose moreover that $(f^{\#}(t_j), f^{\#}(t'_j)) \in C_J(R), \forall j \in [r]$. As $C^k_J(R)$ is increasing, there is an index k_0 such that all these pairs are in $C^{k_0}_J(R)$. Then $(f^{\#}(t), f^{\#}(t')) \in E_{k_0} \subseteq C^{k_0+1}_J(R) \subseteq C_J(R)$. □

Appendix E: Automata, complements

Unique solution for guarded equations. The equation $X = X$ has an infinite number of solutions. The same is true also for equations $X = AX + B$, with A, B languages with $\lambda \in A$. An equation $X = AX + B$ is *guarded* (against this nondeterminism) if $\lambda \notin A$; then it has a unique solution

$$X = A^*B$$

This way one arrives at the following inference rule: let α, β, γ be regular expressions such that "β does not contain the empty word." (This condition has a formal, syntactic form, but we omit it here.) Then the inference rule, called "unique solution for a system of guarded equations," is defined by:

$$\text{(Unic)} \quad \alpha = \beta\alpha + \gamma \Rightarrow \alpha = \beta^*\gamma$$

The invariance rule (Inv) may be replaced by (Unic) rule and one still has a complete system, i.e., the axioms for idempotent Conway theories plus (Unic) are sound and complete for the algebra of regular events. This follows from the fundamental theorem by the extension of "the unique solution for systems of guarded equations" to "the least solution," in the case of nondeterminate systems. However, Kozen has shown that this inference rule has no good algebraic properties (is not invariant at the substitution).

Minimization of nondeterministic automata. It is easy to see that there exists minimal nondeterministic automata. For instance, the automata naturally associated with the following regular acyclic expressions $ab + b(a + b)$ and $(a + b)b + ba$, or to the cyclic ones aa^+ and a^+a, are equivalent, minimal, but not isomorphic.

"ad-hoc" techniques of minimizing the state number, illustrated in an example. Let us start with the language $L = b^2a^* + a^2 + aa^*b$.

(1) The associated minimal dfa has 8 states (it is not nfa-minimal); it may be specified using the following system of equations (the initial state is 1):

$$\begin{array}{ll}
1 = a2 + b6 & L_1 = a^2 + aa^*b + b^2a^* \\
2 = a3 + b5 & L_2 = a + a^*b \\
3 = a4 + b5 + \lambda & L_3 = 1 + a^*b \\
4 = a4 + b5 & L_4 = a^*b \\
5 = a8 + b8 + \lambda & L_5 = 1 \\
6 = a8 + b7 & L_6 = ba^* \\
7 = a7 + b8 + \lambda & L_7 = a^* \\
8 = a8 + b8 & L_8 = 0
\end{array}$$

One can delete the "collapse" state 8; the result is a partial, deterministic automaton. Next, one can see that $L_3 = L_4 + L_5$, hence the state 3 may be deleted, together with the outgoing edges, while the incoming edges of 3 are redirected to go to both states 4 and 5. Finally, one gets an automaton with 6 states (and 9 edges).

(2) Another possibility is to start with the dual language $L^{\vee} = a^*b^2 + a^2 + ba^*a$, get the minimal partial deterministic automaton, and finally, take the dual of the result. That way one gets a nfa with 3 initial states 1,3, and 6; the languages $\check{L}$ *towards inputs* associated with these states are those corresponding to the following system of equations:

$$
\begin{array}{ll}
1 = a1 + a2 & \check{L}_1 = a^* \\
2 = b8 & \check{L}_2 = aa^* + b \\
3 = b2 + b4 & \check{L}_3 = 1 \\
4 = b5 + b6 + b7 & \check{L}_4 = b \\
5 = a5 + a6 & \check{L}_5 = a^*b^2 \\
6 = a7 + \lambda & \check{L}_6 = 1 + a^*b^2 \\
7 = a8 & \check{L}_7 = a + a^*b^2 \\
8 = \lambda & \check{L}_8 = baa^* + a^2 + a^*b^2
\end{array}
$$

One can see that $\check{L}_6 = \check{L}_3 + \check{L}_5$; therefore, the state 6 may be deleted, together with its incoming edges, while the outgoing edges of 6 are redirected to start from both 5 and 3. One gets:

$$
\begin{array}{ll}
1 = a1 + a2 & \check{L}'_1 = a^* \\
2 = b8 & \check{L}'_2 = aa^* + b \\
3 = b2 + b4 + a7 & \check{L}'_3 = 1 \\
4 = b5 + b7 & \check{L}'_4 = b \\
5 = a5 + a7 & \check{L}'_5 = a^*b^2 \\
7 = a8 & \check{L}'_7 = a + a^*b^2 \\
8 = \lambda & \check{L}'_8 = baa^* + a^2 + a^*b^2
\end{array}
$$

Notice that $\check{L}'_3 \subseteq \check{L}'_1$. One can add the edges $3 \xrightarrow{a} 1$ and $3 \xrightarrow{a} 2$ such that the new language $\check{L}''_3$ fully covers the language $\check{L}'_1$. That way, state 1 may be eliminated from the set of initial states. Finally, one gets the following system with a unique initial state 3:

$$
\begin{array}{ll}
1 = a1 + a2 & \check{L}''_1 = a^*a \\
2 = b8 & \check{L}''_2 = aa^* + b \\
3 = a1 + a2 + b2 + b4 + a7 & \check{L}''_3 = 1 \\
4 = b5 + b7 & \check{L}''_4 = b \\
5 = a5 + a7 & \check{L}''_5 = a^*b^2 \\
7 = a8 & \check{L}''_7 = a + a^*b^2 \\
8 = \lambda & \check{L}''_8 = baa^* + a^2 + a^*b^2
\end{array}
$$

In this system $\check{L}''_2 = \check{L}''_1 + \check{L}''_3$. This shows that state 2 can be eliminated, together with its incoming edges, while the outgoing edges are redirected to start from both states 1 and 3. One gets the following equivalent system:

$$
\begin{array}{ll}
1 = a1 + b8 & \check{L}''_1 = a^*a \\
3 = a1 + b8 + b4 + a7 & \check{L}''_3 = 1 \\
4 = b5 + b7 & \check{L}''_4 = b \\
5 = a5 + a7 & \check{L}''_5 = a^*b^2 \\
7 = a8 & \check{L}''_7 = a + a^*b^2 \\
8 = \lambda & \check{L}''_8 = baa^* + a^2 + a^*b^2
\end{array}
$$

This is a nfa with 6 states (and 11 edges).

(3) A final possibility is to start directly from the regular expression and to build an equivalent nfa corresponding to the following system with initial state 1:

$1 = a4 + a6 + b1$
$2 = b3$
$3 = a3 + \lambda$
$4 = b5$
$5 = \lambda$
$6 = a6 + b5$

This is a nfa with 6 states (and 8 edges).

Appendix F: Data-flow networks; checking NA axioms

The BNA structure of $\mathsf{SPF}(M)$. We prove here that

Theorem V.13. $(\mathsf{SPF}(M), \otimes, \cdot, \uparrow, \mathsf{I}_a, {}^a\mathsf{X}^b, \wedge, \vee, \top)$ *is a dβ-symocat with feedback.*

PROOF: It is easy to see that all the axioms without the feedback operator hold. Also the axioms involving the feedback operation are easy to verify, except perhaps for R5, for which we give a detailed proof here. It states that a simultaneous multiple feedback is equivalent to repeated unary feedbacks, i.e.,

$$(f \uparrow^{c\otimes d}) = (f \uparrow^{d}) \uparrow^{c}$$

for $f \in \mathsf{SPF}(M)(a \otimes c \otimes d,\ b \otimes c \otimes d)$.

Let $f \in \mathsf{SPF}(M)(a \otimes c \otimes d,\ b \otimes c \otimes d)$. Then: $f \uparrow^{c\otimes d} \in \mathsf{SPF}(M)(a, b)$ is defined by

$$(f \uparrow^{c\otimes d})(x) = y$$

where $y = \sqcup_{k\geq 1} y_k$ and y_k, z_k, w_k are inductively defined by

$$(y_k, z_k, w_k) = f(x, z_{k-1}, w_{k-1}) \quad \text{for } k \geq 1$$

where $z_0 = \langle\,\rangle, w_0 = \langle\,\rangle$. Denote $z := \sqcup_{k\geq 1} z_k$ and $w := \sqcup_{k\geq 1} w_k$.

Similarly, $(f \uparrow^{d}) \uparrow^{c} \in \mathsf{SPF}(M)(a, b)$ is defined as follows:

$$((f \uparrow^{d}) \uparrow^{c})(x) = \overline{y}$$

where $\overline{y} = \sqcup_{i\geq 1} \overline{y}_i$ and $\overline{y}_i, \overline{z}_i$ are inductively defined by

$$(\overline{y}_i, \overline{z}_i) = (f \uparrow^{d})(x, \overline{z}_{i-1}) \text{ for } i \geq 1$$

where $\overline{z}_0 = \langle\,\rangle$, hence by the definition of $f \uparrow^{d}$ there are elements $\tilde{y}_{i,j}, \tilde{z}_{i,j}, \tilde{w}_{i,j}$ for $i, j \geq 1$ such that for all $i \geq 1$:

$$\overline{y}_i = \sqcup_{j\geq 1} \tilde{y}_{i,j}, \quad \overline{z}_i = \sqcup_{j\geq 1} \tilde{z}_{i,j} \quad \text{and}$$

$$(\tilde{y}_{i,j}, \tilde{z}_{i,j}, \tilde{w}_{i,j}) = f(x, \overline{z}_{i-1}, \tilde{w}_{i,j-1}) \quad \text{for } j \geq 1$$

where $\tilde{w}_{i,0} = \langle\,\rangle$. It is obvious that each sequence $(\tilde{y}_{i,j}), (\tilde{z}_{i,j})$, and $(\tilde{w}_{i,j})$ is increasing on both indices i, j, hence the following notation makes sense: $\overline{z} := \sqcup_{i\geq 1} \overline{z}_i$ and $\overline{w} := \sqcup_{i\geq 1} \overline{w}_i$, where for $i \geq 1:\ \ \overline{w}_i := \sqcup_{j\geq 1} \tilde{w}_{i,j}$.

(1) $f\uparrow^{c\otimes d}\sqsubseteq (f\uparrow^d)\uparrow^c$. First note that

$$(y_k, z_k, w_k) \sqsubseteq (\tilde{y}_{k,k}, \tilde{z}_{k,k}, \tilde{w}_{k,k}), \quad \forall k \geq 1$$

Indeed, for $k = 1$ it follows by $(y_1, z_1, w_1) = f(x, z_0, w_0) = f(x, \langle\ \rangle, \langle\ \rangle) = f(x, \overline{z}_0, \tilde{w}_{1,0}) = (\tilde{y}_{1,1}, \tilde{z}_{1,1}, \tilde{w}_{1,1})$. If it holds for k, then it holds for $k+1$ by: $(y_{k+1}, z_{k+1}, w_{k+1}) = f(x, z_k, w_k) \sqsubseteq f(x, \tilde{z}_{k,k}, \tilde{w}_{k,k}) \sqsubseteq f(x, \overline{z}_k, \tilde{w}_{k,k}) \sqsubseteq f(x, \overline{z}_k, \tilde{w}_{k+1,k}) = (\tilde{y}_{k+1,k+1}, \tilde{z}_{k+1,k+1}, \tilde{w}_{k+1,k+1})$. We get $(f\uparrow^{c\otimes d})(x) = y = \sqcup_{k\geq 1} y_k \sqsubseteq \sqcup_{k\geq 1}\tilde{y}_{k,k} = \sqcup_{i\geq 1}\sqcup_{j\geq 1}\tilde{y}_{i,j} = \sqcup_{i\geq 1}\overline{y}_i = \overline{y} = ((f\uparrow^d)\uparrow^c)(x)$.

(2) $(f\uparrow^d)\uparrow^c\sqsubseteq f\uparrow^{c\otimes d}$. We prove by double induction that

$$(\tilde{y}_{i,j}, \tilde{z}_{i,j}, \tilde{w}_{i,j}) \sqsubseteq (y, z, w), \quad \forall i, j \geq 1$$

First note that $(y, z, w) = f(x, z, w)$. Indeed, $f(x, z, w) = f(x, \sqcup_{k\geq 1} z_k, \sqcup_{k'\geq 1} w_{k'}) = \sqcup_{k\geq 1} f(x, z_k, w_k) = \sqcup_{k\geq 1}(y_{k+1}, z_{k+1}, w_{k+1}) = (y, z, w)$.
Now we check the inductive implication. If $i = 1$ and $j = 1$, then $(\tilde{y}_{1,1}, \tilde{z}_{1,1}, \tilde{w}_{1,1}) = f(x, \overline{z}_0, \tilde{w}_{1,0}) = f(x, \langle\rangle, \langle\rangle) = (y_1, z_1, w_1) \sqsubseteq (y, z, w)$. The passing from j to $j+1$ follows by $(\tilde{y}_{1,j+1}, \tilde{z}_{1,j+1}, \tilde{w}_{1,j+1}) = f(x, \overline{z}_0, \tilde{w}_{1,j}) = f(x, \langle\rangle, \tilde{w}_{1,j}) \sqsubseteq f(x, z, w) = (y, z, w)$. For the inductive step from i to $i+1$, if moreover $j = 1$, then $(\tilde{y}_{i+1,1}, \tilde{z}_{i+1,1}, \tilde{w}_{i+1,1}) = f(x, \overline{z}_i, \tilde{w}_{i+1,0}) = f(x, \sqcup_{j'\geq 1}\tilde{z}_{i,j'}, \langle\rangle) \sqsubseteq f(x, z, w) = (y, z, w)$ and the passing from j to $j+1$ is similar as in the previous case $i = 1$. □

The structure of $\mathcal{P}$SPF(M). In the main text it was proved that BNA axioms holds in $\mathcal{P}$SPF(M). For the branching constants we have only partial results. Here we give a proof for Theorem 10.12 regarding the soundness of the standard NA axioms in the $\mathcal{P}$SPF(M) model; for the reader's convenience we repeat the statement here.

Theorem V.14. *($\mathcal{P}$SPF, $\otimes, \cdot, \uparrow, \mathsf{I}, \mathsf{X}, \curlywedge, \flat, \curlyvee, \dagger$) obeys the additional axioms A1–A2, A4–A6, A8–A9, A12–A19 and F3–F4 in Table 6.3, where $\curlywedge, \flat, \curlyvee, \dagger$ replace $\wedge, \bot, \vee, \top$, respectively. For the remaining axioms, only one inclusion holds, i.e., "$\subset$" for A3, A7, A11 and F5 and "$\supset$" for A10.*

PROOF: First of all, we explain the interplay between the branching constants. The meaning of $\wedge$ and $\vee$ as split and merge constants, respectively, is taken for granted. In order to have a theory which is closed under the feedback operation, we have to see which is the result of the application of the feedback to such constants.

It is easy to see that $\underset{\phi}{\curlywedge^s}\uparrow^s = \dagger_s$ for all oracles ϕ, hence $\curlywedge^s\uparrow^s = \dagger_s$. This equality reflects the fact that our feedback is the least fixed point solution. For the other constant one may see that $\overset{\phi}{\curlyvee}_s\uparrow^s = \flat^s$ for all oracles ϕ. (For each oracle ϕ, the merge function $\overset{\phi}{\curlyvee}_s$ is continuous, hence $\overset{\phi}{\curlyvee}_s\uparrow^s$ is a well-defined function and has to be equal to the unique function $\flat^s : s \to 0$.) All these amount to saying that every set of branching constants including the split and merge constants and closed to the network algebra operations contains $\{\curlywedge, \curlyvee, \flat, \dagger\}$.

We use extended oracles $\phi : \omega \to \{1, \ldots, k\}$ for $k \geq 1$. For instance, the meaning of such an oracle in the case of the split constant $\underset{\phi}{\curlywedge^s_k}$ is to show the number of the output channel where the current token is sent to. Similarly for the merge constant.

Axioms A14–A15 and A18–A19 hold by definition. On the other hand, it is easy to see that A12–A13 and A16–A17 hold. Hence we may restrict ourself to the analysis of the remaining axioms in the case of single channels, i.e. $a = s \in S$.

For axiom A1 it is enough to see that both terms are equal to $\vee_s^3$. Clearly, $(\overset{\phi'}{\vee}_s \otimes \mathsf{I}_s) \cdot \overset{\phi''}{\vee}_s = \overset{\phi}{\vee}{}_s^3$ where ϕ is the 3-oracle obtained from ϕ' and ϕ'' according to the left-hand-side formula. Similarly for the right-hand-side term. The proof is finished showing that a 3-oracle may be simulated by a 2-oracle in both ways corresponding to the left-hand-side and right-hand-side term of the identity, respectively. A5 may be proved in a similar way. For A2 and A6 it is enough to replace an oracle $\phi : \omega \to \{1,2\}$ by the oracle $\overline{\phi}$ obtained interchanging numbers 1 and 2. Axioms A4 holds since for all oracles ϕ one has $\overset{\phi}{\vee}_s \cdot \downarrow^s = \downarrow^s \otimes \downarrow^s$. Axiom A8 also holds. (The splitting of an empty stream is a couple of empty streams.) Clearly, $\top_s \cdot \downarrow^s = \mathsf{I}_0$, hence A9 is valid. Finally, axioms F3 and F4 are valid, as we have already seen in the beginning part of the proof.

In the remaining part of the proof we show that the other axioms do not hold.

For A3, one may see that $[(\top_s \otimes \mathsf{I}_s)\ \overset{\phi}{\vee}_s\ ](x)$ is the prefix of x up to the maximal token k such that $\phi(1) = \ldots = \phi(k) = 2$. Hence A3 is not valid, but the "$\supseteq$" inclusion holds. On the other hand, it is interesting to note that varying ϕ and keeping fixed x we get the *prefix closure* of x.

For the dual axiom A7, one may see that $[\ \underset{\phi}{\wedge}{}^s\ (\perp^s \otimes \mathsf{I}_s)](x)$ is the sub-stream of x given by those positions k for which $\phi(k) = 2$. Hence A7 fails, but the inclusion "$\supseteq$" holds. In this case, varying ϕ and keeping fixed x we get the *sub-stream closure* of x.

For A10 one may see that $E(\phi', \phi'', \psi', \psi'') = (\ \underset{\phi'}{\wedge}{}^s \otimes \underset{\phi''}{\wedge}{}^s\ )(\mathsf{I}_s \otimes {}^s\mathsf{X}^s \otimes \mathsf{I}_s)(\ \overset{\psi'}{\vee}_s \otimes \overset{\psi''}{\vee}_s\ )$ generate a larger class of stream processing functions than $F(\sigma,\tau) = \overset{\sigma}{\vee}_s \cdot \underset{\tau}{\wedge}{}^s$. Indeed, (1) $E(\phi', \phi'', \psi', \psi'')(1^\frown 2^\frown \ldots, a^\frown b^\frown \ldots) = (2^\frown \ldots, b^\frown \ldots)$ for $\phi' = 1^\frown 2 \ldots;\ \ \phi'' = 1^\frown 2 \ldots;\ \ \psi' = 2 \ldots;\ \ \psi'' = 1^\frown 1^\frown \ldots$. On the other hand, this output is not possible for $F(\sigma,\tau)(1^\frown 2^\frown \ldots, a^\frown b^\frown \ldots)$ since the first output on at least one channel here is in the set $\{1, a\}$. (2) Conversely, it may be seen that $F(\sigma,\tau)$ may be simulated by $E(\phi', \phi'', \psi', \psi'')$ if one takes ϕ' and ϕ'' as certain restrictions of τ and ψ' and ψ'' as certain restrictions of σ. More precisely, for an oracle α and a subset of natural numbers $A \subseteq \omega$ let us denote by $\alpha|_A$ the oracle obtained by restricting α to A, i.e. if A consists of the elements $a_1 < a_2 < \ldots$ then $\alpha|_A(i) = \alpha(a_i)$ for $i = 1, 2, \ldots$. Now $\phi' = \tau|_{\sigma^{-1}(1)}$, $\phi'' = \tau|_{\sigma^{-1}(2)}$, $\psi' = \sigma|_{\tau^{-1}(1)}$, $\psi'' = \sigma|_{\tau^{-1}(2)}$ (In case certain oracles as above are finite, we may extend them to infinite oracles in an arbitrary way and the result holds.)

With respect to A11, one may easily see that $E(\phi, \psi) = \underset{\phi}{\wedge}{}^s \cdot \overset{\psi}{\vee}_s$ generates a set of functions which properly includes I_s.

Finally, the left-hand side of F5 specifies a bag, hence the corresponding set of functions properly include I_s. □

Appendix G: Axiomatizing mixed relations

Theorem V.15. *Let M be an $x_1x_2x_3x_4$-enriched symmetric semiringal category and $h : S^{\oplus,\otimes} \to Ob(M)$ a homomorphism of $(\oplus, \otimes, 0, 1)$-algebras ($Ob(M)$ denotes the ob-*

jects of M). Then there is a unique homomorphism of $x_1x_2x_3x_4$-enriched symmetric semiringal categories $H : x_1x_2x_3x_4\text{-MixRels}_S \to M$ which extends h.

PROOF: (Sketch) We say a relation $f : a \to b$ is written in the *additive mixed normal form* (briefly, additive `mixnf`) if it has the following shape ($red\_$ is the inverse of $dis\_$)

$$f = dis_a \cdot g \cdot (r^1 \oplus \ldots \oplus r^k) \cdot h \cdot red_b$$

where

- g^{-1}, h are additive normal form terms which represent functions — that is, g is additive *ad*-term and h is additive *da*-term, and
- $r^1, \ldots, r^k$ are multiplicative normal form terms.

That means, g is a *daaa*-relation, h is an *adaa*-relation and $r^1, \ldots, r^k$ are *aadd*-relations. Moreover, in the particular case when g and h are identities and $k = 1$ the resulting relational term is a sum of multiplicative normal forms, while in the case $r^1, \ldots, r^k$ are identities, the result is an additive normal form.

We separate the axioms in

M1 $(M, \oplus, \cdot, \mathsf{I}, \mathsf{X}, {\bullet\!\!<}_k, {}_k{>\!\!\bullet})$ fulfils the angelic additive axioms of type A1–A4 (see p.192);

M2 $(M, \otimes, \cdot, \mathsf{I}, \mathsf{X}, \mathsf{R}_k, \mathsf{V}^k)$ fulfils the corresponding forward-demonic multiplicative of type A1–A4 (see Appendix C);

M3 the scalar–vectorial axioms for the additive and multiplicative branching constants.

From the enriched semiringal category axioms it follows that:

M4 the additive branching constants ${\bullet\!\!<}_k, {}_k{>\!\!\bullet}$ commute with arbitrary multiplicative terms (i.e., the additive strong axioms hold whenever f is a term over $(M, \otimes, \cdot, \mathsf{I}, \mathsf{X}, \mathsf{R}_k, \mathsf{V}^k)$);

M5 $({}_a{\bullet\!\!<}_k \otimes {}_a{\bullet\!\!<}_l) \cdot \delta_{a,\overset{k}{\ldots},a;a,\overset{k}{\ldots},a} = {}_{a\otimes a}{\bullet\!\!<}_{kl}$
${}_k{>\!\!\bullet}_a \otimes {}_l{>\!\!\bullet}_a = \delta_{a,\overset{k}{\ldots},a;a,\overset{k}{\ldots},a} \cdot {}_{kl}{>\!\!\bullet}_{a\otimes a}$

The proof of the completeness part consists of two steps:

(a) each term may be brought to an additive normal form `mixnf` using the axioms; and

(b) two additive `mixnf` forms which represent the same relation may be transformed one into the other using the axioms.

For (a), we prove that sum, product and composition of additive mixed normal forms may be brought to a normal form via the axioms.

1. $f \oplus f'$: If $f = dis_a \cdot g \cdot r \cdot h \cdot red_b$ and $f' = dis_{a'} \cdot g' \cdot r' \cdot h' \cdot red_{b'}$ are two additive `mixnfs`, then using the distributivity of $\oplus$ over $\cdot$ one gets $f \oplus f' = (dis_a \oplus dis_{a'}) \cdot (g \oplus g') \cdot (r \oplus r') \cdot (h \oplus h') \cdot (red_b \oplus red_{b'})$. Since $dis_a \oplus dis_{a'} = dis_{a \oplus a'}$ and $red_b \oplus red_{b'} = red_{b \oplus b'}$ the resulting term gives an additive `mixnf` for sum.

2. $f \otimes f'$: Suppose $f = dis_a \cdot g \cdot r \cdot h \cdot red_b$ and $f' = dis_{a'} \cdot g' \cdot r' \cdot h' \cdot red_{b'}$ are two additive `mixnfs`. Using the distributivity of $\otimes$ over $\cdot$ one gets $f \otimes f' = (dis_a \otimes dis_{a'}) \cdot (g \otimes g') \cdot (r \otimes r') \cdot (h \otimes h') \cdot (red_b \otimes red_{b'})$. Next, by the distributivity of $\otimes$ over $\oplus$, each

term $g \otimes g'$, $r \otimes r'$, or $h \otimes h'$ may be written as a sum of (tensor) products. For the middle term $r \otimes r'$ it is clear that the tensor product of two multiplicative normal forms $r_i \otimes r'_j$ may be brought to a multiplicative nf by the distributivity of $\otimes$ over $\cdot$. The first (resp. last) term is a sum of additive branching constants. By the distributivity of $\otimes$ over $\oplus$, one gets sums of terms $c \otimes c'$, where c and c' are $\bullet\!\!<_k$ (resp. ${}_k\!>\!\!\bullet$) constants; then by axioms M5 one finally gets an additive `mixnf`. (This method applies to the bijections in g or h, as well. E.g., they may be written as a composite of sums of identities I and transpositions X.) The *dis* and *red* constants in between the factors are annihilated, while the ones from the top (resp. the bottom) contribute to the final *dis* (resp. *red*) factor.

3. $f \cdot f'$: Suppose $f = dis_a \cdot g \cdot r \cdot h \cdot red_b$ and $f' = dis_b \cdot g' \cdot r' \cdot h' \cdot red_c$ are two additive `mixnfs`. We apply the standard procedure of [CŞ91] to normalize the composite of two additive normal forms. By M1, one may commute h and g' to get $f \cdot f' = dis_a \cdot g \cdot r \cdot a'_1(g') \cdot \phi \cdot c_1(h) \cdot r' \cdot h' \cdot red_c$ where $a'_1(g')$ (resp. $c_1(h)$) is an appropriate sum of the same type as g' (resp. h) and ϕ is an additive bijective term. Next, by the strong axioms M4, $a'_1(g')$ commutes with r (resp. $c_1(h)$ computes with r'), hence one gets $f \cdot f' = dis_a \cdot g \cdot a'_2(g') \cdot b_1(r) \cdot \phi \cdot b'_1(r') \cdot c_2(h) \cdot h' \cdot red_c$ where $a'_2(g')$ (resp. $b_1(r), b'_1(r')$, or $c_2(h)$) is an appropriate sum of the same type as g' (resp. r, r', or h). By axioms M1 (for X) ϕ may be commuted with $b_1(r)$, say, and thereafter it may be incorporated into $a'_2(g')$. One gets a new term $f \cdot f' = dis_a \cdot [g \cdot a'_3(g')] \cdot [b_2(r) \cdot b'_1(r')] \cdot [c_2(h) \cdot h'] \cdot red_c$ where $a'_3(g')$ (resp. $b_2(r)$) is an appropriate sum of the same type as g' (resp. r). By M1 axioms the first and the last $[\ldots]$ factors may be brought to an appropriate additive nf. The middle $[\ldots]$ factor is a composite of sums of multiplicative nfs, and by the distributivity of $\oplus$ over $\cdot$ it may be written as a sum of composites of multiplicative nfs. By M2 axioms each term of the sum may be brought to a multiplicative nf, hence an additive `mixnf` for $f \cdot f'$ is finally obtained.

For (b), one has to notice that two additive normal form mixed terms which represent the same mixed relation may be transformed one into the other using the standard procedure in Chapter 3. (The reduction is based on the fact that the set of multiplicative relations r_i correspond to the multiplicative worlds, hence their set is the same in both representations.) □

Appendix H: Discats as sysecats

Proposition V.16. *A bicartesian category is a semiringal category (and it obeys the strong axioms for* ${}_k\!>\!\!\bullet$ *and* R_k*) whenever the following are isomorphisms:*

$$\rho_{a;b,c} := [\mathsf{I}_a \otimes (\mathsf{I}_b \oplus {}_0\!>\!\!\bullet_c) \oplus \mathsf{I}_a \otimes ({}_0\!>\!\!\bullet_b \oplus \mathsf{I}_c)] \cdot {}_2\!>\!\!\bullet_{a\otimes(b\oplus c)}$$
$$\rho_{a;} := {}_0\!>\!\!\bullet_{a\otimes 0}$$

PROOF: We insert a few proofs here, namely for D4 and D7. The other axioms may be proved in a similar way.

D4 is ${}^{a\otimes c}_{a\otimes b}\mathsf{X} \cdot \rho_{a;c,b} = \rho_{a;b,c} \cdot (\mathsf{I}_a \otimes {}^{c}_{b}\mathsf{X})$

Note that

$$
\begin{aligned}
&{}^{a\otimes c}_{a\otimes b}\mathsf{X}\cdot\rho_{a;c,b}\\
&\quad={}^{a\otimes c}_{a\otimes b}\mathsf{X}\cdot[\mathsf{I}_a\otimes(\mathsf{I}_c\oplus{}_0{\succ\!\bullet}_b)\oplus\mathsf{I}_a\otimes({}_0{\succ\!\bullet}_c\oplus\mathsf{I}_b)]\cdot{}_2{\succ\!\bullet}_{a\otimes(c\oplus b)}\\
&\quad=[\mathsf{I}_a\otimes({}_0{\succ\!\bullet}_c\oplus\mathsf{I}_b)\oplus\mathsf{I}_a\otimes(\mathsf{I}_c\oplus{}_0{\succ\!\bullet}_b)]\cdot{}^{a\otimes(c\oplus b)}_{a\otimes(c\oplus b)}\mathsf{X}\cdot{}_2{\succ\!\bullet}_{a\otimes(c\oplus b)}\\
&\quad=[\mathsf{I}_a\otimes({}_0{\succ\!\bullet}_c\oplus\mathsf{I}_b)\oplus\mathsf{I}_a\otimes(\mathsf{I}_c\oplus{}_0{\succ\!\bullet}_b)]\cdot{}_2{\succ\!\bullet}_{a\otimes(c\oplus b)}
\end{aligned}
$$

On the other hand,

$$
\begin{aligned}
&\rho_{a;b,c}\cdot(\mathsf{I}_a\otimes{}^{c}_{b}\mathsf{X})\\
&\quad=[\mathsf{I}_a\otimes(\mathsf{I}_b\oplus{}_0{\succ\!\bullet}_c)\oplus\mathsf{I}_a\otimes({}_0{\succ\!\bullet}_b\oplus\mathsf{I}_c)]\cdot{}_2{\succ\!\bullet}_{a\otimes(b\oplus c)}\cdot(\mathsf{I}_a\otimes{}^{c}_{b}\mathsf{X})\\
&\quad=[(\mathsf{I}_a\otimes(\mathsf{I}_b\oplus{}_0{\succ\!\bullet}_c))\cdot(\mathsf{I}_a\otimes{}^{c}_{b}\mathsf{X})\oplus(\mathsf{I}_a\otimes({}_0{\succ\!\bullet}_b\oplus\mathsf{I}_c))\cdot(\mathsf{I}_a\otimes{}^{c}_{b}\mathsf{X})]\cdot{}_2{\succ\!\bullet}_{a\otimes(c\oplus b)}\\
&\qquad\qquad\text{(by }{}_2{\succ\!\bullet}\text{-strong)}\\
&\quad=[\mathsf{I}_a\otimes({}_0{\succ\!\bullet}_c\oplus\mathsf{I}_b)\oplus\mathsf{I}_a\otimes(\mathsf{I}_c\oplus{}_0{\succ\!\bullet}_b)]\cdot{}_2{\succ\!\bullet}_{a\otimes(c\oplus b)}
\end{aligned}
$$

D7 is

$$
\begin{aligned}
&[({}^{a'}\mathsf{X}^{b}\oplus{}^{a''}\mathsf{X}^{b})\cdot\rho_{b;a',a''}\cdot{}^{b}\mathsf{X}^{a'\oplus a''}\oplus({}^{a'}\mathsf{X}^{c}\oplus{}^{a''}\mathsf{X}^{c})\cdot\rho_{c;a',a''}\cdot{}^{c}\mathsf{X}^{a'\oplus a''}]\cdot\rho_{a'\oplus a'';b,c}\\
&=(\mathsf{I}_{a'\otimes b}\oplus{}^{a'\otimes c}_{a''\otimes b}\mathsf{X}\oplus\mathsf{I}_{a''\otimes c})\cdot(\rho_{a';b,c}\oplus\rho_{a'';b,c})\cdot({}^{a'}\mathsf{X}^{b\oplus c}\oplus{}^{a''}\mathsf{X}^{b\oplus c})\cdot\rho_{b\oplus c;a',a''}\cdot{}^{b\oplus c}\mathsf{X}^{a'\oplus a''}
\end{aligned}
$$

LHS

$$
\begin{aligned}
&=[({}^{a'}\mathsf{X}^{b}\oplus{}^{a''}\mathsf{X}^{b})\cdot[\mathsf{I}_b\otimes(\mathsf{I}_{a'}\oplus{}_0{\succ\!\bullet}_{a''})\oplus\mathsf{I}_b\otimes({}_0{\succ\!\bullet}_{a'}\oplus\mathsf{I}_{a''})]\cdot{}_2{\succ\!\bullet}_{b\otimes(a'\oplus a'')}\cdot\underline{{}^{b}\mathsf{X}^{a'\oplus a''}}\\
&\quad\oplus({}^{a'}\mathsf{X}^{c}\oplus{}^{a''}\mathsf{X}^{c})\cdot[\mathsf{I}_c\otimes(\mathsf{I}_{a'}\oplus{}_0{\succ\!\bullet}_{a''})\oplus\mathsf{I}_c\otimes({}_0{\succ\!\bullet}_{a'}\oplus\mathsf{I}_{a''})]\cdot{}_2{\succ\!\bullet}_{c\otimes(a'\oplus a'')}\cdot\underline{{}^{c}\mathsf{X}^{a'\oplus a''}}]\\
&\quad\cdot[\mathsf{I}_{a'\oplus a''}\otimes(\mathsf{I}_b\oplus{}_0{\succ\!\bullet}_c)\oplus\mathsf{I}_{a'\oplus a''}\otimes({}_0{\succ\!\bullet}_b\oplus\mathsf{I}_c)]\cdot{}_2{\succ\!\bullet}_{(a'\oplus a'')\otimes(b\oplus c)}\\
&\qquad\qquad\text{(the underlined terms are shifted to the front and annihilated)}\\
&=[((\mathsf{I}_{a'}\oplus{}_0{\succ\!\bullet}_{a''})\otimes\mathsf{I}_b\oplus({}_0{\succ\!\bullet}_{a'}\oplus\mathsf{I}_{a''})\otimes\mathsf{I}_b)\cdot{}_2{\succ\!\bullet}_{(a'\oplus a'')\otimes b}\\
&\quad\oplus((\mathsf{I}_{a'}\oplus{}_0{\succ\!\bullet}_{a''})\otimes\mathsf{I}_c\oplus({}_0{\succ\!\bullet}_{a'}\oplus\mathsf{I}_{a''})\otimes\mathsf{I}_c)\cdot{}_2{\succ\!\bullet}_{(a'\oplus a'')\otimes c}\\
&\quad\cdot[\mathsf{I}_{a'\oplus a''}\otimes(\mathsf{I}_b\oplus{}_0{\succ\!\bullet}_c)\oplus\mathsf{I}_{a'\oplus a''}\otimes({}_0{\succ\!\bullet}_b\oplus\mathsf{I}_c)]\cdot{}_2{\succ\!\bullet}_{(a'\oplus a'')\otimes(b\oplus c)}\\
&\qquad\qquad\text{(by }{}_2{\succ\!\bullet}\text{-strong)}\\
&=[(\mathsf{I}_{a'}\oplus{}_0{\succ\!\bullet}_{a''})\otimes(\mathsf{I}_b\oplus{}_0{\succ\!\bullet}_c)\oplus({}_0{\succ\!\bullet}_{a'}\oplus\mathsf{I}_{a''})\otimes(\mathsf{I}_b\oplus{}_0{\succ\!\bullet}_c)\\
&\quad\oplus(\mathsf{I}_{a'}\oplus{}_0{\succ\!\bullet}_{a''})\otimes({}_0{\succ\!\bullet}_b\oplus\mathsf{I}_c)\oplus({}_0{\succ\!\bullet}_{a'}\oplus\mathsf{I}_{a''})\otimes({}_0{\succ\!\bullet}_b\oplus\mathsf{I}_c)]\cdot{}_4{\succ\!\bullet}_{(a'\oplus a'')\otimes(b\oplus c)}
\end{aligned}
$$

RHS

$$
\begin{aligned}
&=(\mathsf{I}_{a'\otimes b}\oplus{}^{a'\otimes c}_{a''\otimes b}\mathsf{X}\oplus\mathsf{I}_{a''\otimes c})\cdot([\mathsf{I}_{a'}\otimes(\mathsf{I}_b\oplus{}_0{\succ\!\bullet}_c)\oplus\mathsf{I}_{a'}\otimes({}_0{\succ\!\bullet}_b\oplus\mathsf{I}_c)]\cdot{}_2{\succ\!\bullet}_{a'\otimes(b\oplus c)}\\
&\quad\oplus[\mathsf{I}_{a''}\otimes(\mathsf{I}_b\oplus{}_0{\succ\!\bullet}_c)\oplus\mathsf{I}_{a''}\otimes({}_0{\succ\!\bullet}_b\oplus\mathsf{I}_c)]\cdot{}_2{\succ\!\bullet}_{a''\otimes(b\oplus c)})\\
&\quad\cdot(\underline{{}^{a'}\mathsf{X}^{b\oplus c}}\oplus\underline{{}^{a''}\mathsf{X}^{b\oplus c}})\\
&\quad\cdot[\mathsf{I}_{b\oplus c}\otimes(\mathsf{I}_{a'}\oplus{}_0{\succ\!\bullet}_{a''})\oplus\mathsf{I}_{b\oplus c}\otimes({}_0{\succ\!\bullet}_{a'}\oplus\mathsf{I}_{a''})]\cdot{}_2{\succ\!\bullet}_{(b\oplus c)\otimes(a'\oplus a'')}\cdot{}^{b\oplus c}\mathsf{X}^{a'\oplus a''}\\
&\qquad\qquad\text{(the underlined terms are shifted to the right and annihilated)}\\
&=(\mathsf{I}_{a'\otimes b}\oplus{}^{a'\otimes c}_{a''\otimes b}\mathsf{X}\oplus\mathsf{I}_{a''\otimes c})\cdot([\mathsf{I}_{a'}\otimes(\mathsf{I}_b\oplus{}_0{\succ\!\bullet}_c)\oplus\mathsf{I}_{a'}\otimes({}_0{\succ\!\bullet}_b\oplus\mathsf{I}_c)\\
&\quad\oplus\mathsf{I}_{a''}\otimes(\mathsf{I}_b\oplus{}_0{\succ\!\bullet}_c)\oplus\mathsf{I}_{a''}\otimes({}_0{\succ\!\bullet}_b\oplus\mathsf{I}_c)])\cdot({}_2{\succ\!\bullet}_{a'\otimes(b\oplus c)}\oplus{}_2{\succ\!\bullet}_{a''\otimes(b\oplus c)})\\
&\quad\cdot[(\mathsf{I}_{a'}\oplus{}_0{\succ\!\bullet}_{a''})\otimes\mathsf{I}_{b\oplus c}\oplus({}_0{\succ\!\bullet}_{a'}\oplus\mathsf{I}_{a''})\otimes\mathsf{I}_{b\oplus c}]\cdot{}_2{\succ\!\bullet}_{(a'\oplus a'')\otimes(b\oplus c)}\\
&=(\mathsf{I}_{a'\otimes b}\oplus{}^{a'\otimes c}_{a''\otimes b}\mathsf{X}\oplus\mathsf{I}_{a''\otimes c})[(\mathsf{I}_{a'}\oplus{}_0{\succ\!\bullet}_{a''})\otimes(\mathsf{I}_b\oplus{}_0{\succ\!\bullet}_c)\oplus(\mathsf{I}_{a'}\oplus{}_0{\succ\!\bullet}_{a''})\otimes({}_0{\succ\!\bullet}_b\oplus\mathsf{I}_c)\\
&\quad\oplus({}_0{\succ\!\bullet}_{a'}\oplus\mathsf{I}_{a''})\otimes(\mathsf{I}_b\oplus{}_0{\succ\!\bullet}_c)\oplus({}_0{\succ\!\bullet}_{a'}\oplus\mathsf{I}_{a''})\otimes({}_0{\succ\!\bullet}_b\oplus\mathsf{I}_c)]\cdot{}_4{\succ\!\bullet}_{(a'\oplus a'')\otimes(b\oplus c)}\\
&=LHS
\end{aligned}
$$

$\square$

Appendix I: Decomposing morphisms in discats

Mixalgebra vs. relational structure. Here we compare mixalgebra signatures and usual relational structure signatures, e.g. of [Coh81]. Obviously, a relational structure is a mixalgebra. We describe certain conditions which assure they are equivalent.

In the one-sorted case, given a mixalgebra operation $f : A^{i(f)} \to o(f).A$ one may define

- a generalized $o(f)$-valued $i(f)$-ary test $f^c = f \cdot (o(f).\mathfrak{R}_0^A) : A^{i(f)} \to o(f)$ and

- a usual $i(f)$-ary operation $f^s = f \cdot {}_{o(f)}{\succ\!\!\bullet}_A : A^{i(f)} \to A$.

Lemma V.17.

$$\mathfrak{R}_2^{k.A}((k.\mathfrak{R}_0^A) \otimes {}_k{\succ\!\!\bullet}_A)\delta_{1,\dots,1;A} = \mathsf{I}_A$$

PROOF:

$$\begin{aligned}
&\mathfrak{R}_2^{k.A}((k.\mathfrak{R}_0^A) \otimes {}_k{\succ\!\!\bullet}_A)\delta_{1,\dots,1;A}\\
= &\{{\succ\!\!\bullet}_A \text{ in terms of } {\succ\!\!\bullet}_1\}\\
&\mathfrak{R}_2^{k.A}((k.\mathfrak{R}_0^A) \otimes \rho_{1,\dots,1;A}({}_k{\succ\!\!\bullet}_1 \otimes \mathsf{I}_A))\delta_{1,\dots,1;A}\\
= &\{\text{distributivity}\}\\
&\mathfrak{R}_2^{k.A}[\rho_{1,\dots,1;A}((k.\mathsf{I}_1) \otimes \mathfrak{R}_0^A)\delta_{1,\dots,1;1} \otimes \rho_{1,\dots,1;A}({}_k{\succ\!\!\bullet}_1 \otimes \mathsf{I}_A)]\delta_{1,\dots,1;A}\\
= &\{\text{commutation for } \mathfrak{R}\}\\
&\rho_{1,\dots,1;A}\mathfrak{R}_2^{k\otimes A}[(\mathsf{I}_k \otimes \mathfrak{R}_0^A) \otimes ({}_k{\succ\!\!\bullet}_1 \otimes \mathsf{I}_A)]\delta_{1,\dots,1;A}\\
= &\{\text{scalar-vectorial axioms, including } {}_k{\succ\!\!\bullet}_1 = \mathfrak{R}_0^k\}\\
&\rho_{1,\dots,1;A}\mathfrak{R}_2^{k\otimes A}[(\mathsf{I}_k \otimes \mathfrak{R}_0^A) \otimes (\mathfrak{R}_0^k \otimes \mathsf{I}_A)]\delta_{1,\dots,1;A}\\
= &\{\text{(multiplicative) algebraic theory identity}\}\\
&\rho_{1,\dots,1;A}\mathsf{I}_{k\otimes A}\delta_{1,\dots,1;A}\\
= &\{\ \}\\
&\mathsf{I}_A
\end{aligned}$$

□

Proposition V.18. *In a distributive category setting, given f^c, f^s as above, the original mixalgebra operation f may be recovered by the formula*

(*) $\quad f = \mathfrak{R}_2^{A^{i(f)}}(f^c \otimes \mathsf{I}_{A^{i(f)}})\delta_{1,\dots,1;A^{i(f)}}(o(f).f^s).$

PROOF:

$$\begin{aligned}
&\mathfrak{R}_2^{A^{i(f)}}(f^c \otimes \mathsf{I}_{A^{i(f)}})\delta_{1,\dots,1;A^{i(f)}}(o(f).f^s)\\
= &\{\text{insert an identity}\}\\
&\mathfrak{R}_2^{A^{i(f)}}(f^c \otimes \mathsf{I}_{A^{i(f)}})\delta_{1,\dots,1;A^{i(f)}}(o(f).(\mathsf{I}_1 \otimes f^s))\\
= &\{\text{distributivity rule}\}\\
&\mathfrak{R}_2^{A^{i(f)}}(f^c \otimes \mathsf{I}_{A^{i(f)}})((o(f).\mathsf{I}_1) \otimes f^s)\delta_{1,\dots,1;A}\\
= &\{\text{monoidal rules in the multiplicative setting}\}\\
&\mathfrak{R}_2^{A^{i(f)}}(f^c \otimes f^s)\delta_{1,\dots,1;A}\\
= &\{\text{definitions } f^c, f^s\}\\
&\mathfrak{R}_2^{A^{i(f)}}(f \cdot (o(f).\mathfrak{R}_0^A) \otimes f \cdot {}_{o(f)}{\succ\!\!\bullet}_A)\delta_{1,\dots,1;A}\\
= &\{\mathfrak{R}_2 \text{ strong commutation rule}\}\\
&f \cdot \mathfrak{R}_2^{o(f).A}((o(f).\mathfrak{R}_0^A) \otimes {}_{o(f)}{\succ\!\!\bullet}_A)\delta_{1,\dots,1;A}
\end{aligned}$$

= {Lemma V.17}

f □

Roughly speaking, the given mixalgebra operation f may be simulated by the following procedure: (1) copy the input; (2) compare input elements via f^c; (3) distribute the input copy to each possible test output; and (4) apply the state transforming function f^s.

It is worthwhile to emphasize the rules we have used so far. The proof heavily uses the distributivity axiom and the strong commutation axiom for the multiplicative ramification constant. Since the distributivity rule already implies the strong axiom for the additive identification constant, it follows that the setting where the proof takes place is mainly that of a distributive category.

This decomposition of a mixalgebra operation into a test and a usual algebra operation is useful since it clearly show the limit of the relational structure formalism: the reduction of mixalgebra operations to relational structures cannot be done in a setting where one cannot freely copy a state (e.g., in the case of a resource sensitive logic, as it is for instance Girard linear logic; in a quantum computation setting; or in our general semiringal categories), or when the distributivity rule is not valid.

The mixalgebra signature may also be useful in the many-sorted case. In that case the state transformation part f^s of a mixalgebra operation consists of a tuple of operations f_j^s with different ranges. By itself, each operation f_j^s is partial, but their tuple is total.

Path decomposition. Another useful decomposition is in terms of paths. To a mixalgebra operation $f : A_{s_1} \times \ldots \times A_{s_m} \to A_{t_1} \star \ldots \star A_{t_n}$, with $s_i \in S$ and $t_i \in S \cup \{1\}$ one may associate some path-operations

$$f_j : A_{s_1} \times \ldots \times A_{s_m} \to A_{t_j} \star 1$$

removing the values from the output summands different from j, and identifying all the resulting summands equal to 1. One may interpret such an operation f_j as the corestriction of f to the j-th component, if this is defined, else skip.

In the case the mixalgebra signature is such that a morphism $: 1 \to 0$ exists in the corresponding distributive category,

Notice that the free distributive catagory MixFn has no such morphisms.

then the original mixalgebra operation may be described using the associated path operations. In the lemma below we give the proof in the case $n = 2$; the proof in the general case is similar.

Lemma V.19. *(1) If a distributive category has a morphism $k : 1 \to 0$, then a morphism $f : A \to B \star C$ may be described using the associated path operations f_j as follows*

$$f = \mathcal{R}_2^A (f_1 \otimes f_2) \delta_{B,1;1,C} (\mathsf{I}_B \star \mathcal{R}_0^{B \otimes C} \cdot k \star k \star \mathsf{I}_C)$$

(2) A similar decomposition exists when the mixalgebra signature is such that there is a ground term $c_a : 1 \to a$, for each sort a, e.g.,

$$f = \mathcal{R}_2^A (f_1 \otimes f_2) \delta_{B,1;1,C} (\mathsf{I}_B \star (\mathcal{R}_0^{B \otimes C} \cdot c_C \star c_C \star \mathsf{I}_C) \cdot {}_3\!\succ\!\!\bullet_C)$$

PROOF:

$\mathcal{R}_2^A (f_1 \otimes f_2) \delta_{B,1;1,C} (\mathsf{I}_B \star \mathcal{R}_0^{B \otimes C} \cdot k \star k \star \mathsf{I}_C)$

$= \mathcal{R}_2^A (f(\mathsf{I}_B \star \mathcal{R}_0^C) \otimes f(\mathcal{R}_0^B \star \mathsf{I}_C)) \delta_{B,1;1,C} (\mathsf{I}_B \star \mathcal{R}_0^{B \otimes C} \cdot k \star k \star \mathsf{I}_C)$

$$
\begin{aligned}
&= f\mathcal{R}_2^{B\star C}((\mathsf{I}_B \star \mathcal{R}_0^C) \otimes (\mathcal{R}_0^B \star \mathsf{I}_C))\delta_{B,1;1,C}(\mathsf{I}_B \star \mathcal{R}_0^{B\otimes C} \cdot k \star k \star \mathsf{I}_C) \\
&= f(\mathcal{R}_2^B \star {}_0\!\succ\!\!\bullet_{B\otimes C} \star {}_0\!\succ\!\!\bullet_{C\otimes B} \star \mathcal{R}_2^C) \\
&\quad \cdot \rho_{B,C;B,C}((\mathsf{I}_B \star \mathcal{R}_0^C) \otimes (\mathcal{R}_0^B \star \mathsf{I}_C))\delta_{B,1;1,C}(\mathsf{I}_B \star \mathcal{R}_0^{B\otimes C} \cdot k \star k \star \mathsf{I}_C) \\
&= f(\mathcal{R}_2^B \star {}_0\!\succ\!\!\bullet_{B\otimes C} \star {}_0\!\succ\!\!\bullet_{C\otimes B} \star \mathcal{R}_2^C) \\
&\quad \cdot (\mathsf{I}_B \otimes \mathcal{R}_0^B \star \mathsf{I}_B \otimes \mathsf{I}_C \star \mathcal{R}_0^C \otimes \mathcal{R}_0^B \star \mathcal{R}_0^C \otimes \mathsf{I}_C)(\mathsf{I}_B \star \mathcal{R}_0^{B\otimes C} \cdot k \star k \star \mathsf{I}_C) \\
&= f(\mathsf{I}_B \star \mathsf{I}_0 \star \mathsf{I}_0 \star \mathsf{I}_C) \\
&= f
\end{aligned}
$$
□

Appendix J: Plans as free discats

Theorem V.20. Plan(Σ) *is the free distributive category generated by* Σ.

PROOF: We start with the freeness property, then we show that Plan is a discat, indeed.

Interpretation. First we show how an interpretation on plans may be uniquely inferred from the interpretation of the (property) variables. Note that the obvars play no role here.

It is enough to define the interpretation in the case of a unique input obvar. Then it may be naturally and uniquely extended to general plans.

The interpretation on plans may be inductively defined using the maximal length of a term occurring in a plan, either in a leaf term or in a guard. Technically, it is useful to define the length of a guard $f^c(t_1,\ldots,t_m) = i$ as $lg(f_i^s(t_1,\ldots,t_m)) - 0.5$.

Let P be a plan. Let $s_i := f_i^s(t_1,\ldots,t_m)$ or $c_i := [f^c(t_1,\ldots,t_m) = i]$ be a term of a maximal length occurring in P. Since the plan is *complete*, the terms $s_j := f_j^s(t_1,\ldots,t_m)$ and/or $c_j := [f^c(t_1,\ldots,t_m) = j]$ occur in P, for each $j \in [n]$, in the same guard context. (By the completeness condition, this is a part of a larger guard of an elementary plan; but our term was maximal, hence no proper extension may exist.) But our starting terms were of maximal length, hence the new ones cannot be prefixes of longer terms. Hence one may decompose the plan as $rest \cdot \phi$, where ϕ contains identities and a term

$$f(x_1,\ldots,x_m)\cdot(\alpha_1 \oplus \ldots \oplus \alpha_m)$$

with $\alpha_j \in \{\mathsf{I}_1, \mathcal{R}_0^1\}$ being I_1 for s_i leaf term and $\mathcal{R}_0^1$ for c_j without a corresponding leaf term s_j.

Let us note that the terms α_j are well defined. Indeed, each term s_j and/or c_j occur in a unique elementary plan, for the given guard context. A basic property used here is the *determinism* of the plans: for a given (maximal) guard, there is a a unique elementary plan containing it. (If two elementary plans have different obvars and contain the same c_j in the given context, then there is a guard condition which separate them and this is impossible.)

The remaining plan may be obtained as follows:

- in an elementary plan containing $f_j^s(t_1,\ldots,t_m)$, this term is replaced by $x_1 := t_1,\ldots,x_m := t_m$, the corresponding condition $f^c(t_1,\ldots,t_m) = j$ is dropped from the guard, and (as we said) α_j is set to I_1;
- in the case the guard $f^c(t_1,\ldots,t_m) = j$ occurs without a corresponding leaf term $f_j^s(t_1,\ldots,t_m)$, then we insert $x_1 := t_1,\ldots,x_m := t_m$ in the leaf term, drop the

condition $f^c(t_1, \ldots, t_m) = j$ from the guard and set α_j to $\mathfrak{R}_0^1$. (The place where they are inserted does not matter too much: they are later destroyed. To have a unique definition one has to make a choice, e.g. to put them on the rightmost part of the tuple. If more such guards are in an elementary plan they are ordered from left to right according to the order of the corresponding conditions in the guard.)

After this transformation the equal elementary plans are identified.

One may iterate this transformation for all terms of the maximal length. Then one gets a decomposition

$$P = P' \cdot [\ldots f^1(x_1^1, \ldots, x_{m_1}^1) \cdot (\alpha_1^1 \oplus \ldots \oplus \alpha_{m_1}^1) \ldots f^r(x_1^r, \ldots, x_{m_r}^r) \cdot (\alpha_1^r \oplus \ldots \oplus \alpha_{m_r}^r) \ldots]$$

The decomposition is unique and the remained plan P' has a shorter length.

Iterating the procedure above we get a decomposition on layers, and finally obtained a plan of depth 0, hence a mixfunction. The placement of the operation symbols in layers is unique, and with some standard conventions as above, the placement of the operation symbols on layers is unique, too.

This decomposition is then use to define the interpretation of a plan in an arbitrary mixalgebra, provided the interpretation of the propvars is fixed.

Morphism. Next, one has to show that the interpretation preserves the distributive category operation. This is very tedious, and we do not expect any new insight from doing it fully formal. If a nicer definition of the plan will be found, an easier check of this properties may be done.

Discat structure. To show that the plans satisfy the discat structure we fully need the conditions on the elementary plans we have already required; actually they were introduced just in order to assure such a structure may be put on top of Plan.

The symmetric semiringal category axioms are easy to check. Note that nothing special is required to prove the distributivity rule.

Finally, we show that Plan is a $\otimes$-algebraic theory and a $\oplus$-algebraic cotheory, hence it has products and coproducts.

- the rule $f \cdot \mathfrak{R}_0^{C \oplus D} = \mathfrak{R}_0^{A \otimes B}$ holds true since the plans use only *relevant* guard conditions; more precisely, when one composes $f \cdot \mathfrak{R}_0^{C \oplus D}$, all the leaf terms are removed, hence by the additional transformation related to the composition, all the guards are removed, too and the plan corresponding to $\mathfrak{R}_0^{A \otimes B}$ is obtained;

- the rule $f \cdot \mathfrak{R}_2^{C \oplus D} = \mathfrak{R}_2^{A \otimes B} \cdot (f \otimes f)$ holds true since the plans use *noncontradictory* conditions in the guards; indeed, in the plan corresponding to the right-hand-side term, the branching corresponding to contradictory conditions in the parallel composition of the f-s plans are deleted, and the corresponding leaf terms are considered undefined, and the same result is obtained for the left-hand-side term;

- the rule ${}_0{\bullet\!\!<}_{A \otimes B} 0 \cdot f = {}_0{\bullet\!\!<}_{C \oplus D}$ is valid since the plans have a *demonic* multiplicative behaviour; indeed, the plan corresponding to leaf terms in ${}_0{\bullet\!\!<}_{A \otimes B}$ are all equal to *dd* and this darkness is propagated, hence the equality is valid;

- finally, ${}_2{>\!\!\bullet}_{A \otimes B} \cdot f = (f \oplus f) \cdot {}_2{>\!\!\bullet}_{C \oplus D}$ hold true just by applying the definition of plan composition. □

Bibliography

[Abr96] S. Abramsky. Retracing some paths in process algebra. In *Proceedings of CONCUR'96*, volume 1119 of *Lecture Notes in Computer Science*. Springer-Verlag, Berlin, 1996.

[AFI98] L Aceto, W. Fokkink, and A. Ingolfsdotter. On a question of A. Salomaa. The equational theory of regular expressions over a singleton is not finitely based. *Theoretical Computer Science*, 209:163–178, 1998.

[AK96] Z.M. Ariola and J.W. Klop. Equational term graph rewriting. *Fundamenta Informaticae*, 26:207–240, 1996.

[AM80] M.A. Arbib and E.G. Manes. Partially additive categories and flow diagram semantics. *Journal of Algebra*, 62:203–227, 1980.

[AN60] Arnauld and Nicole. *La Logique ou l'art de penser.* Guillaume Defprez, Paris, 1660. La cinquieme edition, 1683; Reprinted, Flammarion, Paris, 1970.

[Arn78] V.I. Arnold. *Ordinary differential equations.* Editura Ştiinţifică şi Enciclopedică, Bucharest, 1978. (Romanian translation of the Russion edition, Nauka, Moscow 1971).

[Arn94] A. Arnold. *Finite transition systems.* Prentice-Hall, 1994. First published in French by Masson, Paris, 1992.

[BA81] J.D. Brock and W.B. Ackermann. Scenarios: A model of non-determinate computation. In J. Diaz and I. Ramos, editors, *Proceedings of Formalization of Programming Concepts*, volume 107 of *Lecture Notes in Computer Science*, pages 252–259. Springer-Verlag, 1981.

[Bai76] E.S. Bainbridge. Feedback and generalized logic. *Information and Control*, 31:75–96, 1976.

[Bar87] M. Bartha. A finite axiomatization of flowchart schemes. *Acta Cybernetica*, 8:203–217, 1987.

[BB92] J.C.M. Baeten and J.A. Bergstra. Discrete time process algebra. In W.R. Cleaveland, editor, *Proceedings of CONCUR'92*, volume 630 of *Lecture Notes in Computer Science*, pages 401–420. Springer-Verlag, 1992.

[BBK93] J.C.M. Baeten, J.A. Bergstra, and J.W. Klop. Decidability of bisimulation equivalence for processes generating context-free languages. *Journal of the ACM*, 40:653–682, 1993.

[BBP94] J.A. Bergstra, I. Bethke, and A. Ponse. Process algebra with iteration. *The Computer Journal*, 60:109–137, 1994.

[BBŞ95] J.C.M. Baeten, J.A. Bergstra, and Gh. Ştefănescu. Process algebra with feedback. In A. Ponse, M. de Rijke, and Y. Venema, editors, *Modal Logic and Process Algebra*, volume 53 of *CSLI Lecture Notes*, pages 13–37. Stanford, 1995.

[BÉ85] S.L. Bloom and Z. Ésik. Axiomatizing schemes and their behavior. *Journal of Computer and System Sciences*, 31:375–393, 1985.

[BÉ93a] S.L. Bloom and Z. Ésik. Equational axioms for regular sets. *Mathematical Structures in Computer Science*, 3:1–24, 1993.

[BÉ93b] S.L. Bloom and Z. Ésik. *Iteration Theories: The Equational Logic of Iterative Processes.* EATCS Monographs in Theoretical Computer Science. Springer-

Verlag, Berlin, 1993.

[BÉ97] S.L. Bloom and Z. Ésik. The equational logic of fixed points. *Theoretical Computer Science*, 1997.

[BÉŞ95] S.L. Bloom, Z. Ésik, and Gh. Ştefănescu. Notes on the equational theories of relations. *Algebra Universalis*, 33:98–126, 1995.

[BÉT93] S.L. Bloom, Z. Ésik, and D. Taubner. Iteration theory of synchronization trees. *Information and Computation*, 102:1–55, 1993.

[BEW80a] S.L. Bloom, C.C. Elgot, and J.B. Wright. Solutions of the iteration equation and extensions of the scalar iteration operation. *SIAM Journal of Computing*, 9:26–45, 1980.

[BEW80b] S.L. Bloom, C.C. Elgot, and J.B. Wright. Vector iteration in pointed algebraic theories. *SIAM Journal of Computing*, 9:525–540, 1980.

[BGM99] R. Bruni, F. Gadducci, and U. Montanari. Normal forms for partitions and relations. In J.L. Fiadeiro, editor, *Proceedings of WADT'98, 13th Workshop on Recent Trends in Algebraic Specification Techniques*, *Lecture Notes in Computer Science*. Springer-Verlag, 1999. To appear.

[BGŞ99] M. Broy, R. Grosu, and Gh. Ştefănescu. On space-time duality in concurrent object-oriented programming. Draft, August 1999.

[BJ66] C. Böhm and G. Jacopini. Flow diagrams, Turing machines and languages with only two formation rules. *Communications of the ACM*, 9(5):366–371, 1966.

[BK84] J.A. Bergstra and J.W. Klop. Process algebra for synchronous communication. *Information and Control*, 60:109–137, 1984.

[BK85] J.A. Bergstra and J.W. Klop. Algebra of communicating processes with abstraction. *Theoretical Computer Science*, 37:77–121, 1985.

[BK94] J.A. Bergstra and J.W. Klop. The algebra of recursively defined processes and the algebra of regular processes. In A. Ponse, C. Verhoef, and S.F.M. van Vlijmen, editors, *Proceedings of ACP'94*, volume 630 of *Workshops in Computing*, pages 1–25. Springer-Verlag, London, 1994. Also: Report IW 235/83, Mathematical Centre, Amsterdam, 1983.

[BKS97] C. Brinks, W. Kahl, and G. Schmidt, editors. *Relational Methods in Computer Science*. Springer Wien, New York, 1997.

[BM97] R. Bruni and U. Montanari. Zero-safe nets, or transition synchronization made simple. *Electronic Notes in Theoretical Computer Science*, 7:20 pages, 1997.

[BMŞ97] J.A. Bergstra, C.A. Middelburg, and Gh. Ştefănescu. Network algebra for asynchronous dataflow. *International Journal of Computer Mathematics*, 65:57–88, 1997.

[Böh84] A.P.W. Böhm. *Dataflow Computation*, volume 6 of *CWI Tracts*. Center for Mathematics and Computer Science, Amsterdam, 1984.

[Boo47] G. Boole. *The mathematical analysis of logic*. Macmillan, Barclay, & Macmillan, Cambridge, 1847. Reprinted, Oxford, Basil Blackwell, 1948.

[BP96] M.A. Bezem and A. Ponse. Two finite specifications of a queue. *Theoretical Computer Science*, 177:487–507, 1996.

[Bra84] W. Brauer. *Automaten-theorie*. B.G. Teubner, Stuttgart, 1984. Russian translation, 1987.

[Bro83] M. Broy. Fixed point theory for communication and concurrency. In D. Bjørner, editor, *Formal Description of Programming Concepts II*, pages 125–147. North-Holland, 1983.

[Bro88] M. Broy. Nondeterministic dataflow programs: How to avoid the merge anomaly. *Science of Computer Programming*, 10:65–85, 1988.

[Bro93] M. Broy. Functional specification of time sensitive communicating systems. *ACM Transactions on Software Engineering and Methodology*, 2:1–46, 1993.

[Bru99] R. Bruni. *Tile Logic for Synchronized Rewriting of Concurrent Systems*. PhD thesis, Dipartimento di Informatica, Universita di Pisa, 1999. Report TD-1/99.

[Brz64] J.A. Brzozowski. Derivatives of regular expressions. *Journal of ACM*, 11:481–

494, 1964.

[BŞ94a] J.A. Bergstra and Gh. Ştefănescu. Bisimulation is two-way simulation. *Information Processing Letters*, 52:285–287, 1994.

[BŞ94b] J.A. Bergstra and Gh. Ştefănescu. Network algebra for synchronous and asynchronous dataflow. Preprint LGPS-122, Department of Philosophy, Utrecht University, 1994.

[BŞ96] M. Broy and Gh. Ştefănescu. The algebra of stream processing functions. Technical Report TUM-I9620 and SFB-Bericht Nr. 342/11/96 A, Institute of Informatics, Technical University Munich, 1996. To appear in *Theoretical Computer Science*, volume 251.

[BŞ98] J.A. Bergstra and Gh. Ştefănescu. Network algebra with demonic relation operators. *Revue Roumaine de Mathematiques Pures et Applique*, 43(5-6):503–520, 1998.

[BvW98] R.-J. Back and J. von Wright. *Refinement calculus.* Graduate Texts in Computer Science. Springer-Verlag, New York, 1998.

[BW90a] J.C.M. Baeten and W.P. Weijland. *Process Algebra.* Cambridge Tracts in Theoretical Computer Science. Cambridge University Press, 1990.

[BW90b] M. Barr and C. Wells. *Category theory for Computing Science.* Prentice-Hall, New York, 1990.

[BWM94] H. Barendregt, H. Wupper, and H. Mulder. Computable processes. Technical Report CSI-R9405, Computing Science Institute, Catholic University of Nijmegen, 1994.

[CEW58] I.M. Copy, C.C. Elgot, and J.B. Wright. Realization of events by logical nets. *Journal of the ACM*, 5:181–196, 1958.

[CG99] A. Corradini and F. Gadducci. Rewriting on cyclic structures: equivalence between the operational and the categorical description. *Theoretical Informatics and Applications*, 1999. To appear.

[CM72] H.M.S. Coxeter and W.O. Moser. *Generators and relations for discrete groups.* Springer-Verlag, Berlin, 1972. 3rd edition.

[Coc93] J.R.B. Cockett. Introduction to distributive categories. *Mathematical Structures in Computer Science*, 3:277–307, 1993.

[Coh81] P.M. Cohn. *Universal Algebra.* D. Reidel Publishing Company, Dordrecht, Holland, 1981. Revised edition.

[Con71] J.H. Conway. *Regular Algebra and Finite Machines.* Chapman and Hall, 1971.

[Cou83] B. Courcelle. Fundamental properties of infinite trees. *Theoretical Computer Science*, 25:95–169, 1983.

[CŞ88a] V.E. Căzănescu and Gh. Ştefănescu. A formal representation of flowchart schemes I. *Analele Universitătii Bucuresti, Matematică - Informatică*, 37:33–51, 1988.

[CŞ88b] V.E. Căzănescu and Gh. Ştefănescu. Bi-flow-calculus. Unpublished manuscript, 1988.

[CŞ89] V.E. Căzănescu and Gh. Ştefănescu. A formal representation of flowchart schemes II. *Studii si Cercetări Metematice (Mathematical Reports)*, 41:151–167, 1989.

[CŞ90a] V.E. Căzănescu and Gh. Ştefănescu. Towards a new algebraic foundation of flowchart scheme theory. *Fundamenta Informaticae*, 13:171–210, 1990.

[CŞ90b] V.E. Căzănescu and Gh. Ştefănescu. A note on axiomatizing flowchart schemes. *Acta Cybernetica*, 9:349–359, 1990.

[CŞ91] V.E. Căzănescu and Gh. Ştefănescu. Classes of finite relations as initial abstract data types I. *Discrete Mathematics*, 90:233–265, 1991.

[CŞ92] V.E. Căzănescu and Gh. Ştefănescu. A general result of abstract flowchart schemes with applications to the study of accessibility, reduction and minimization. *Theoretical Computer Science*, 99:1–63, 1992. Fundamental Study.

[CŞ94] V.E. Căzănescu and Gh. Ştefănescu. Classes of finite relations as initial abstract data types II. *Discrete Mathematics*, 126:47–65, 1994.

[CŞ95] V.E. Căzănescu and Gh. Ştefănescu. Feedback, iteration and repetition. In Gh. Păun, editor, *Mathematical aspects of natural and formal languages*, pages 43–62. World Scientific, Singapore, 1995. Also in: *Preprint Series in Mathematics*, No.42, Department of Mathematics, The National Institute for Scientific and Technical Creation, Bucharest, 1988.

[CU82] V.E. Căzănescu and C. Ungureanu. Again on advice on structuring compilers and proving them correct. Preprint Series in Mathematics 75, Department of Mathematics, The National Institute for Scientific and Technical Creation, Bucharest, 1982.

[DBS+95] J. Desharnais, N. Belkhiter, S.B.M. Sghaier, F. Tchier, A. Jaoua, A. Mili, and N. Yaguia. Embedding a demonic semilattice in a relation algebra. *Theoretical Computer Science*, 149:333–360, 1995.

[DE95] J. Desel and J. Esparza. *Free choice Petri nets*, volume 40 of *Cambridge Tracts in Theoretical Computer Science*. Cambridge University Press, 1995.

[DF98] R. Diaconescu and K. Futatsugi. *CafeOBJ Report: The Language, Proof Techniques, and Methodologies for Object-Oriented Algebraic Specification*, volume 6 of *AMAST Series in Computing*. World Scientific, 1998.

[DMN97] J. Desharneis, A. Mili, and T.T. Nguyen. Refinement and demonic semantics. In Brinks et al. [BKS97].

[dNS95] R. de Nicola and R. Segala. A process algebraic view of input/output automata. *Theoretical Computer Science*, 138:391–424, 1995.

[DR95] V. Diekert and G. Rozenberg, editors. *The book of traces*. World Scientific, Singapore, 1995.

[DW81] M.D. Davis and E.J. Weyuker. *Computability, complexity, and languages*. Academic Press, 1981.

[EBT78] C.C. Elgot, S.L. Bloom, and R. Tindell. On the algebraic structure of rooted trees. *Journal of Computer and System Sciences*, 16:362–399, 1978.

[Eil74] S. Eilenberg. *Automata, languages, and machines, Vol. A*. Academic Press, New York, 1974.

[Elg75] C.C. Elgot. Monadic computation and iterative algebraic theories. In H.E. Rose and J.C. Sheperdson, editors, *Proceedings of Logic Colloquium '73*, volume 80 of *Studies in Logic and the Foundations of Mathematics*, pages 175–230. North-Holland, 1975.

[Elg76a] C.C. Elgot. Matricial theories. *Journal of Algebra*, 42:391–421, 1976.

[Elg76b] C.C. Elgot. Structured programming with and without GO TO statements. *IEEE Transactions on Software Engineering*, SE-2:41–53, 1976.

[Elg77] C.C. Elgot. Some geometrical categories associated with flowchart schemes. In *Proceedings of FCT'77*, LNCS, pages 256–259. Springer-Verlag, 1977.

[ES82] C.C. Elgot and J.C. Shepherdson. An equational axiomatization of the algebra of reducible flowchart schemes. In S.L. Bloom, editor, *Selected papers, C.C. Elgot*. Springer-Verlag, 1982.

[Ési80] Z. Ésik. Identities in iterative and rational algebraic theories. *Computational Linguistic and Computational Languages*, 7:183–207, 1980.

[FS90] P.J. Freyd and A. Scedrov. *Categories, allegories*. Volume 39 of *North-Holland Mathematical Library*. North-Holland, 1990.

[GD94] J.A. Goguen and R. Diaconescu. Towards an algebraic semantics for the object paradigm. In *Recent Trends in Data Type Specification*, volume 785 of *Lecture Notes in Computer Science*, pages 1–34. Springer-Verlag, 1994.

[Gib95] J. Gibbons. An initial-algebra approach to directed acyclic graphs. In *Proc. MPC'95*, volume 947 of *Lecture Notes in Computer Science*, pages 282–303. Springer-Verlag, 1995.

[Gin68] A. Ginzburg. *Algebraic theory of automata*. ACM Monograph Series. Academic Press, New York/London, 1968.

[Gin79] S. Ginali. Regular trees and free iterative theories. *Journal of Computer and System Sciences*, 18:228–242, 1979.

[Gir87] J.-Y. Girard. Linear logic. *Theoretical Computer Science*, 50, 1987.

[GLŞ00] R. Grosu, D. Lucanu, and Gh. Ştefănescu. Mixed relations as enriched semiringal categories. *Journal of Universal Computer Science*, 6(1):112–129, 2000.

[GM97] J.A. Goguen and G. Malcolm. A hidden agenda. Technical Report CS97-538, University of California at San Diego, 1997.

[GM99] F. Gadducci and U. Montanari. The tile model. In *Papers dedicated to R. Milner festschrift*. The MIT Press, Cambridge, to appear, 1999. Also: Technical Report TR-96-27, Department of Computer Science, University of Pisa, 1996.

[Gog74] J.A. Goguen. On homomorphism, correctness, termination, unfoldments and equivalence of flow diagram programs. *Journal of Computer and System Sciences*, 8:333–365, 1974.

[GP95] M. Girkar and C.D. Polychronopoulos. Extracting task-level parallelism. *ACM Transactions on Programming Languages and Systems*, 17:600–634, 1995.

[GR92] V. Garg and M.T. Ragunath. Concurrent regular expressions and their relationship to Petri nets. *Theoretical Computer Science*, 96:285–304, 1992.

[Gre75] S. Greibach. *Theory of program structures: schemes, semantics, verification. Lecture Notes in Computer Science*. Springer-Verlag, Berlin, 1975.

[GŞ98] R. Grosu and Gh. Ştefănescu. On space-time duality in computing. Draft, November 1998.

[GŞB98] R. Grosu, Gh. Ştefănescu, and M. Broy. Visual formalism revised. In *Proceeding of the CSD'98 (International Conference on Application of Concurrency to System Design, March 23-26, 1998, Fukushima, Japan)*, pages 41–51. IEEE Computer Society Press, 1998.

[GTWW77] J.A. Goguen, J.W. Thatcher, E.G. Wagner, and J.B. Wright. Initial algebra semantics and continuous algebras. *Journal of ACM*, 24:68–95, 1977.

[Gup94] V. Gupta. *Chu spaces: A model for concurrency*. PhD thesis, Department of Computer Science, Stanford University, 1994.

[Hag87] T. Hagino. *A Categorical Programming Language*. PhD thesis, Department of Computer Science, University of Edinburgh, 1987. Report CST-47-87; ECS-LFCS-87-38.

[Has97a] M. Hasegawa. *Models of Sharing Graphs: A Categorical Semantics of let and letrec*. PhD thesis, Department of Computer Science, University of Edinburgh, 1997. Also available at Springer-Verlag, Berlin, in the *Distinguished PhD Dissertation Series*, 1999.

[Has97b] M. Hasegawa. Recursion from cyclic sharing: Traced monoidal categories and models of cyclic lambda calculi. In *Proc. 3rd International Conference on Typed Lambda Calculi and Applications*. Springer-Verlag, Berlin, 1997.

[Hec77] M. Hecht. *Flow analysis of computer programs*. The Computer Science Library, Programming Language Series, Vol. 5. North-Holland, New York, 1977.

[Hen88] M. Hennessy. *Algebraic theory of processes*. Foundations of Computing. The MIT Press, Cambridge, 1988.

[Hot65] H. Hotz. Eine algebraiserung des syntheseproblems fur schaltkreise. *EIK*, 1:185–205; 209–231, 1965.

[HZ97] M.R. Hansen and C. Zhou. Duration calculus: Logical foundations. *Formal Aspects of Computing*, 9:283–330, 1997.

[Jac96] B. Jacobs. Objects and classes, co-algebraically. In B. Freitag, C.B. Jones, C. Lengauer, and H.-J. Schek, editors, *Object-Orientation with Parallelism and Persistence*, pages 83–103. Kluwer Acad. Publ., 1996.

[Jef97] A. Jeffrey. Premonoidal categories and a graphical view of programs. Preprint; Author's home-page, University DePaul, 1997.

[Jon94] B. Jonsson. A fully abstract trace model for dataflow and asynchronous networks. *Distributed Computing*, 7:197–212, 1994.

[Jos92] M. Josephs. Receptive process theory. *Acta Informatica*, 29:17–31, 1992.

[JR97] B. Jacobs and J. Rutten. A tutorial on (co)algebras and (co)induction. *EATCS Bulletin*, 62:222–259, 1997.

[JSV96] A. Joyal, R. Street, and D. Verity. Traced monoidal categories. *Proceedings of the Cambridge Philosophical Society*, 119:447–468, 1996.

[Kah74] G. Kahn. The semantics of a simple language for parallel processing. In J.L. Rosenfeld, editor, *Proceedings of Information Processing '74*, pages 471–475, 1974.

[Kan95] M. Kanovich. Petri nets, Horn programs, linear logic and vector games. *Annals of Pure and Applied Logic*, 75:107–135, 1995.

[Kas95] C. Kassel. *Quantum Groups.* Graduate Texts in Mathematics, Vol. 155. Springer Verlag, New York, 1995.

[Kau95] L.H. Kauffman, editor. *Knots and applications.* Series on Knots and Everything, Vol. 6. World Scientific, Singapore, 1995.

[Kaw73] Y. Kawahara. Relations in categories with pullback. *Mem. Fac. Sci. Kyushu Univ. Ser. A*, 27:149–173, 1973.

[Kaw90] Y. Kawahara. Pushout-complements and the basic concepts of grammars in toposes. *Theoretical Computer Science*, 77:267–289, 1990.

[KB99] M. Koutny and E. Best. Operational and denotational semantics for the box algebra. *Theoretical Computer Science*, 211:1–83, 1999.

[Kel78] R.M. Keller. Denotational models for parallel programs with nondeterminate operators. In E. Neuhold, editor, *Formal Description of Programming Concepts*, pages 337–366. North-Holland, 1978.

[Kle56] S.C. Kleene. Representation of events in nerve nets and finite automata. In C.E. Shannon and J. McCarthy, editors, *Automata Studies*, volume 34 of *Annals of Mathematical Studies*, pages 3–41. Princeton University Press, 1956.

[Klo87] C.D. Kloos. *Semantics of Digital Circuits. Lecture Notes in Computer Science*, Vol. 285. Springer Verlag, Berlin, 1987.

[Knu73] D. Knuth. *The art of computer programming, Vol. 3 Sorting and Searching.* Addison-Wesley, Reading, MA, 1973.

[Kok87] J. Kok. A fully abstract semantics for data flow nets. In J.W. de Bakker, A.J. Nijman, and P.C. Treleaven, editors, *Proceedings of PARLE '87*, volume 259 of *Lecture Notes in Computer Science*, pages 351–368, 1987.

[Kot78] V.E. Kotov. *Introduction to the theory of program schemes.* Nauka, Novosibirsk, 1978. (Russian).

[Kot84] V.E. Kotov. *Petri nets.* Nauka, Moscow, 1984. (Russian).

[Koz91] D. Kozen. A completeness theorem for Kleene algebras and the algebra of regular events. In *Proc. 6th LICS Symposium*, pages 214–225. IEEE Computer Society Press, 1991. Journal version appeared in: *Information and Computation*, 110:366–390, 1994.

[KP86] R. Keller and P. Panangaden. Semantics of networks with nondeterminate operators. *Distributed Computing*, 1:235–245, 1986.

[Kri00] P. Krishnan. Distributed timed automata. *Electronic Notes in Theoretical Computer Science*, 28, 2000. Proc. of the FCT'99 Workshop on Distributed Systems.

[Kro91] D. Krob. Complete systems of B-rational identites. *Theoretical Computer Science*, 89:207–343, 1991.

[KS85] W. Kuich and A. Salomaa. *Semirings, Automata, Languages.* Springer-Verlag, Berlin, 1985.

[KŞ98] Y. Kawahara and Gh. Ştefănescu. Plans, mixalgebras and distributive categories. Draft; also handouts for the GKLI lectures, Munich, 1998, January 1998.

[Laf92] Y. Lafont. Penrose diagrams and 2-dimensional rewriting. In *Proceedings of the LMS Symposium on Application of Categories in Computer Science*, pages 191–201. Cambridge University Press, 1992.

[Lan71] S. Mac Lane. *Categories for the working mathematician.* Springer-Verlag,

Berlin, 1971.

[Law63] F.W. Lawvere. Functorial semantics of algebraic theories. *Proceedings of the National Academy of Sciences, U.S.A.*, 50:869–872, 1963.

[LB67] S. Mac Lane and G. Birkhoff. *Algebra*. AMS Chelsea Publishing, 1967. 3rd Edition, 1999.

[LS89] J. Lambek and P.J. Scott. *Introduction to higher order categorical logic*. Cambridge University Press, Cambridge, 1989.

[MA86] E.G. Manes and M.A. Arbib. *Algebraic approaches to program semantics*. Springer-Verlag, Berlin, 1986.

[Mad96] R.D. Maddux. Relation-algebraic semantics. *Theoretical Computer Science*, 160:1–85, 1996.

[Man74] Z. Manna. *Mathematical theory of computation*. McGraw-Hill, New York, 1974.

[Man92] E.G. Manes. *Predicate transformer semantics*. Cambridge University Press, Cambridge, 1992.

[Mes92] J. Meseguer. Conditional rewriting logic as a unified model of concurrency. *Theoretical Computer Science*, 96:73–155, 1992.

[Mey97] B. Meyer. *Object-Oriented Software Construction*. Prentice-Hall, 1997. 2nd edition.

[Mif96] A. Mifsud. *Control Structures*. PhD thesis, Department of Computer Science, University of Edinburgh, 1996.

[Mil79] R. Milner. Flowgraphs and flow algebra. *Journal of the ACM*, 26:794–818, 1979.

[Mil80] R. Milner. *A Calculus of Communicating Systems*, volume 92 of *Lecture Notes in Computer Science*. Springer-Verlag, Berlin, 1980.

[Mil89] R. Milner. *Communication and concurrency*. Prentice-Hall International, 1989.

[Mil94] R. Milner. Action calculi V : Reflexive molecular forms. Author's home page, Department of Computer Science, University of Edinburgh, 1994.

[Miy98] H. Miyoshi. Rewriting logic for cyclic sharing structures. Author's home-page, Shukutoku University, 1998.

[MM90] J. Meseguer and U. Montanari. Petri nets are monoids. *Information and Computation*, 88:105–155, 1990.

[Mol88] P. Molitor. Free net algebras in VLSI-Theory. *Fundamenta Informaticae*, 11:117–142, 1988.

[Möl94] B. Möller. Ideal streams. In E.-R. Olderog, editor, *Programming Concepts, Methods and Calculi. IFIP Transactions A-56*, pages 39–58. North-Holland, 1994.

[Möl98a] B. Möller. Modal and temporal operators on partial orders. Draft paper; presented to the 4th RelMiCS seminar, 1998.

[Möl98b] B. Möller. Towards deductive hardware design. In B. Möller and J.V. Tucker, editors, *Prospects for Hardware Design, Lecture Notes in Computer Science*, Vol. 1546, pages 421–468. Springer-Verlag, 1998.

[Mor00] R. Morin. Hierarchy of asynchronous automata. *Electronic Notes in Theoretical Computer Science*, 28, 2000. Proc. of the FCT'99 Workshop on Distributed Systems.

[MP92] Z. Manna and A. Pnueli. *The temporal logic of reactive and concurrent systems*. Springer-Verlag, Berlin, 1992.

[MPW92] R. Milner, J. Parrow, and D. Walker. A calculus of mobile processes. *Information and Computation*, 100:1–40; 41–77, 1992.

[Old91] E.-R. Olderog. *Nets, terms and formulas*. Cambridge Tracts in Theoretical Computer Science. Cambridge University Press, 1991.

[Par81] D. Park. Concurrency and automata on infinite sequences. In *Proceedings of the 5th GI Conference, Lecture Notes in Computer Science*, pages 167–183. Springer-Verlag, 1981.

[Par83] D. Park. The fairness problem and nondeterministic computing networks. In

Foundations of Computer Science IV: Part 2, Semantics and Logic, volume 159 of *Mathematical Center Tracts*, pages 133–161. Mathematical Center, Amsterdam, 1983.

[Par85] J. Parrow. *Fairness Properties in Process Algebra with Applications in Communication Protocol Verification*. PhD thesis, Department of Computer Science, Uppsala University, 1985.

[Par87] J. Parrow. Synchronization flow algebra. Technical Report LFCS-87-35, Laboratory for the Foundation of Computer Science, University of Edinburgh, 1987.

[Par93] J. Parrow. Structural and behavioral equivalence of networks. *Information and Computation*, 107:58–90, 1993.

[Pet62a] C.A. Petri. Fundamentals of a theory of asynchronous information flow. In *Proceedings IFIP Congress 62 (Munich, 1962)*, pages 386–390. North-Holland, Amsterdam, 1962.

[Pet62b] C.A. Petri. *Kommunikation mit Automaten*. PhD thesis, Institute für Instrumentelle Mathematik, Bonn, Germany, 1962.

[PP79] L. Popescu and N. Popescu. *Theory of categories*. Editura Academiei, Bucureşti and Sijthoff & Noordhoff International Publishers, 1979.

[Pra92] V.R. Pratt. The duality of time and information. In *Proceedings CONCURR'92*, volume 630 of *Lecture Notes in Computer Science*, pages 237–253. Springer-Verlag, Berlin, 1992.

[Pra94] V.R. Pratt. Time and information in sequential and concurrent computation. In *Proceedings TPPP'94 (Theory and Practice of Parallel Programming), Sendai, Japan, Lecture Notes in Computer Science*. Springer-Verlag, Berlin, 1994.

[Pre99] V. Preoteasa. A relation between unambiguous regular expressions and abstract data types. *Fundamenta Informaticae*, 1999. To appear.

[Red64] V.N. Redko. On defining relations for the algebra of regular events. *Ukrainian Mathematical Journal*, 16:120–126, 1964. (Russian).

[Rei85] W. Reisig. *Petri Nets. An Introduction*. EATCS Monographs in Theoretical Computer Science. Springer-Verlag, Berlin, 1985.

[Rei95] H. Reichel. An approach to object semantics based on terminal coalgebras. *Mathematical Structures in Computer Science*, 5:129–152, 1995.

[Rig48] J. Riguet. Relations binaires, fermetures, correspondences de Galois. *Bull. Soc. Math. France*, 76:114–155, 1948.

[RT94] J. Rutten and D. Turi. Initial and final coalgebra semantics for concurrency. Technical report, CWI, Amsterdam, 1994.

[Rud74] S. Rudeanu. *Boolean functions and equations*. North-Holland, Amsterdam/London, 1974.

[Rus89] J. Russell. Full abstraction for nondeterministic dataflow networks. In *Proceedings of FoCS '89*. IEEE Computer Society Press, 1989.

[Sal66] A. Salomaa. Two complete axiom systems for the algebra of regular events. *Journal of ACM*, 13:158–169, 1966.

[Sel98] P. Selinger. A note on Bainbridge's powerset construction. Author's home-page, University of Michigan, 1998.

[ŞK97] Gh. Ştefănescu and Y. Kawahara. Distributive categories and mixed network algebras. In *Proc. Joint Conference on Discrete Mathematics and Applied Mathematics, Seta, Japan, 18-20 December, 1997*, pages 29–34. 1997.

[SN85] J. Staples and V. Nguyen. A fixed point semantics for nondeterministic data flow. *Journal of the ACM*, 32:411–444, 1985.

[Şor97] R. Şoricuţ. Verifying ring buffer protocols in process algeba. In *Proc. Current Trends in Cybernetics and Philosophy of Science, Oradea, 1996*, pages 317–326. Europa Nova, Bucharest, 1997.

[SS93] G. Schmidt and T. Ströhleim. *Relations and graphs*. EATCS Monographs in Theoretical Computer Science. Springer-Verlag, Berlin, 1993.

[Sta87] E.W. Stark. Concurrent transition system semantics of process networks. In

Fourteenth ACM Symposium on Principles of Programming Languages, pages 199–210. IEEE Computer Society Press, 1987.

[Sta92] E.W. Stark. A calculus of dataflow networks. In *Proceedings of LICS '92*, pages 125–136. IEEE Computer Society Press, 1992.

[Şte86a] Gh. Ştefănescu. An algebraic theory of flowchart schemes. In P. Franchi-Zannettacci, editor, *Proceedings 11-th Colloquium on Trees in Algebra and Programming, CAAP'86*, volume 214 of *Lecture Notes in Computer Science*, pages 60–73. Springer–Verlag, 1986.

[Şte86b] Gh. Ştefănescu. Feedback theories (a calculus for isomorphism classes of flowchart schemes). Technical Report 24, Department of Mathematics, The National Institute for Scientific and Technical Creation, April 1986. Published in: *Revue Roumaine de Mathematiques Pures et Applique*, 35:73–79, 1990.

[Şte87a] Gh. Ştefănescu. On flowchart theories: Part I. The deterministic case. *Journal of Computer and System Sciences*, 35(2):163–191, 1987.

[Şte87b] Gh. Ştefănescu. On flowchart theories: Part II. The nondeterministic case. *Theoretical Computer Science*, 52(3):307–340, 1987.

[Şte90] Gh. Ştefănescu. Feedback theories (a calculus for isomorphism classes of flowchart schemes). *Revue Roumaine de Mathematiques Pures et Applique*, 35:73–79, 1990. Previously distributed as INCREST Preprint No.24/1986.

[Şte91] Gh. Ştefănescu. *Determinism and nondeterminism in program scheme theory: algebraic aspects.* PhD thesis, Faculty of Mathematics, University of Bucharest, 1991. (Romanian).

[Şte94] Gh. Ştefănescu. Algebra of flownomials: Part I. Binary flownomials; basic theory. Technical Report TUM-I9437 and SFB-Bericht Nr. 342/16/94 A, Institute of Informatics, Technical University Munich, 1994.

[Şte97a] Gh. Ştefănescu. Axiomatizing mixed relations. In *Proceedings of 3rd RelMiCS Seminar, Hammamet, Tunisia*, pages 177–186. University of Science, Technology and Medicine of Tunis, 1997.

[Şte97b] Gh. Ştefănescu. A short tour on FEST. In *Proc. Current Trends in Cybernetics and Philosophy of Science, Oradea, 1996*, pages 327–332. Europa Nova, Bucharest, 1997. Also: Preprint 38/1996, Institute of Mathematics, Romanian Academy.

[Şte98a] Gh. Ştefănescu. On space-time duality in computing: Imperative programming versus wave computation. In *Abstracts, 4th Relational Methods in Computer Science Seminar*, pages 197–202. Stefan Banach Mathematical Centre, Warsaw, 1998.

[Şte98b] Gh. Ştefănescu. Reaction and control I. Mixing additive and multiplicative network algebras. *Logic Journal of the IGPL*, 6(2):349–368, 1998.

[SV97] V. Stebletsova and Y. Venema. Axioms for Jónsson's Q-algebras. In *Proceedings of 3rd RelMiCS Seminar, Hammamet, Tunisia*, pages 215–224. University of Science, Technology and Medicine of Tunis, 1997.

[SVW94] N. Sabadini, S. Vigna, and R.F.C. Walters. A note on recursive functions. 3rd author's home page, School of Mathematics and Statistics, University of Sydney, 1994.

[Sza80] M.E. Szabo. *Algebra of proofs.* North-Holland, Amsterdam, 1980.

[Tan96] A.S. Tanenbaum. *Computer networks.* Prentice-Hall, 1996.

[Tar81] R.E. Tarjan. A unified approach to path problems. *Journal of the ACM*, 28(3):577–593, 1981.

[TG87] A. Tarski and S. Givant. *A formalization of set theory without variables.* American Mathematical Society, 1987.

[Tom72] I. Tomescu. A matrix method for determining all pairs of compatible states of a sequential machine. *IEEE Transactions on Computers*, C-11:502–503, 1972.

[TWW79] J.W. Thatcher, E.G. Wagner, and J.B. Wright. Notes on algebraic fundamentals for theoretical computer science. In *Foundation of Computer Science III:*

Part 2. Language, logic, semantics, pages 83–164. Mathematical Center, Amsterdam, 1979.

[TZ92] J.V. Tucker and J.I. Zucker. Theory of computation over stream algebras and its applications. In *Proceedings of MFCS '92*, volume 629 of *Lecture Notes in Computer Science*, pages 62–80. Springer-Verlag, 1992.

[vG90] R.J. van Glabbeek. *Comparative Concurrency Semantics and Refinement of Actions*. PhD thesis, Department of Mathematics and Computer Science, University of Amsterdam, 1990.

[vGV93] R.J. van Glabbeek and F.W. Vaandrager. Modular specification of process algebras. *Theoretical Computer Science*, 113:293–348, 1993.

[Wal92a] R.F.C. Walters. *Categories and Computer Science*. Cambridge University Press, Cambridge, 1992. Also: Carslaw Publications, 1991.

[Wal92b] R.F.C. Walters. An imperative language based on distributive categories. *Mathematical Structures in Computer Science*, 2:249–256, 1992.

[Win87] G. Winskel. Petri nets, algebras, morphisms and composability. *Information and Computation*, 72:197–238, 1987.

[Wir90] M. Wirsing. Algebraic specifications. In *Handbook of Theoretical Computer Science, Vol. 2*. North-Holland, Amsterdam, 1990.

[WN94] G. Winskel and M. Nielsen. Models for concurrency. In S. Abramsky, D. Gabbay, and T. Maibaum, editors, *Handbook of Logic in Computer Science, Vol. 3 Semantic Structures*. Clarendon Press, Oxford, 1994.

[Yan58] Yu.I. Yanov. On logical schemata of algorithms. In V.E. Liapunoff, editor, *Problems of Cybernetics, Vol. 1*. Fitmatgiz, M, 1958. (Russian).

[Zie87] W. Zielonka. Notes on finite asynchronous automata. *Theoretical Informatics and Applications*, 21:99–135, 1987.

List of tables

List of figures

Index

Other titles in the DMTCS series:

Combinatorics, Complexity, Logic: Proceedings of DMTCS '96
D. S. Bridges, C. S. Calude, J. Gibbons, S. Reeves, I. Witten (Eds)
981-3083-14-X

Formal Methods Pacific '97: Proceedings of FMP '97
L. Groves and S. Reeves (Eds)
981-3083-31-X

The Limits of Mathematics: A Course on Information Theory and the Limits of Formal Reasoning
Gregory J. Chaitin
981-3083-59-X

Unconventional Models of Computation
C. S. Calude, J. Casti and M. J. Dinneen (Eds)
981-3083-69-7

Quantum Logic
K. Svozil
981-4021-07-5

International Refinement Workshop and Formal Methods Pacific '98
J. Grundy, M. Schwenke and T. Vickers (Eds)
981-4021-16-4

Computing with Biomolecules: Theory and Experiments
Gheorghe Paun (Ed)
981-4021-05-9

People and Ideas in Theoretical Computer Science
C. S. Calude (Ed)
981-4021-13-X

Combinatorics, Computation and Logic: Proceedings of DMTCS'99 and CATS'99
C. S. Calude and M. J. Dinneen (Eds)
981-4021-56-3

Polynomials: An Algorithmic Approach
M. Mignotte and D. Stefanescu
981-4021-51-2

The Unknowable
Gregory J. Chaitin
981-4021-72-5

Sequences and Their Applications: Proceedings of SETA '98
C. Ding, T. Helleseth and H. Niederreiter (Eds)
1-85233-196-8

Finite versus Infinite: Contributions to an Eternal Dilemma
Cristian S. Calude and Gheorghe Paun (Eds)
1-85233-251-4